OXFORD LOGIC GUIDES: 48

Series Editors

D.M. GABBAY
A.J. MACINTYRE
D. SCOTT

OXFORD LOGIC GUIDES

10. Michael Hallett: *Cantorian set theory and limitation of size*
17. Stewart Shapiro: *Foundations without foundationalism*
18. John P. Cleave: *A study of logics*
21. C. McLarty: *Elementary categories, elementary toposes*
22. R.M. Smullyan: *Recursion theory for metamathematics*
23. Peter Clote and Jan Krajícek: *Arithmetic, proof theory, and computational complexity*
24. A. Tarski: *Introduction to logic and to the methodology of deductive sciences*
25. G. Malinowski: *Many valued logics*
26. Alexandre Borovik and Ali Nesin: *Groups of finite Morley rank*
27. R.M. Smullyan: *Diagonalization and self-reference*
28. Dov M. Gabbay, Ian Hodkinson, and Mark Reynolds: *Temporal logic: Mathematical foundations and computational aspects, volume 1*
29. Saharon Shelah: *Cardinal arithmetic*
30. Erik Sandewall: *Features and fluents: Volume I: A systematic approach to the representation of knowledge about dynamical systems*
31. T.E. Forster: *Set theory with a universal set: Exploring an untyped universe, second edition*
32. Anand Pillay: *Geometric stability theory*
33. Dov M. Gabbay: *Labelled deductive systems*
35. Alexander Chagrov and Michael Zakharyaschev: *Modal Logic*
36. G. Sambin and J. Smith: *Twenty-five years of Martin-Lőf constructive type theory*
37. María Manzano: *Model theory*
38. Dov M. Gabbay: *Fibring Logics*
39. Michael Dummett: *Elements of Intuitionism, second edition*
40. D.M. Gabbay, M.A. Reynolds and M. Finger: *Temporal Logic: Mathematical foundations and computational aspects, volume 2*
41. J.M. Dunn and G. Hardegree: *Algebraic Methods in Philosophical Logic*
42. H. Rott: *Change, Choice and Inference: A study of belief revision and nonmonotoic reasoning*
43. Johnstone: *Sketches of an Elephant: A topos theory compendium, volume 1*
44. Johnstone: *Sketches of an Elephant: A topos theory compendium, volume 2*
45. David J. Pym and Eike Ritter: *Reductive Logic and Proof Search: Proof theory, semantics and control*
46. D.M. Gabbay and L. Maksimova: *Interpolation and Definability: Modal and Intuitionistic Logics*
47. John L. Bell: *Set Theory: Boolean-valued models and independence proofs, third edition*
48. Laura Crosilla and Peter Schuster: *From Sets and Types to Topology and Analysis: Towards practicable foundations for constructive mathematics*

From Sets and Types to Topology and Analysis

Towards Practicable Foundations for Constructive Mathematics

Edited by

LAURA CROSILLA
Dipartimento di Filosofia, Università di Firenze

and

PETER SCHUSTER
Mathematisches Institut, Universität München

CLARENDON PRESS • OXFORD
2005

OXFORD
UNIVERSITY PRESS

Great Clarendon Street, Oxford OX2 6DP

Oxford University Press is a department of the University of Oxford.
It furthers the University's objective of excellence in research, scholarship,
and education by publishing worldwide in

Oxford New York

Auckland Cape Town Dar es Salaam Hong Kong Karachi
Kuala Lumpur Madrid Melbourne Mexico City Nairobi
New Delhi Shanghai Taipei Toronto

With offices in

Argentina Austria Brazil Chile Czech Republic France Greece
Guatemala Hungary Italy Japan Poland Portugal Singapore
South Korea Switzerland Thailand Turkey Ukraine Vietnam

Oxford is a registered trade mark of Oxford University Press
in the UK and in certain other countries

Published in the United States
by Oxford University Press Inc., New York

First published 2005

British Library Cataloguing in Publication Data
Data available

Library of Congress Cataloguing in Publication Data
Data available

Typeset by SPI Publisher Services, Pondicherry, India
Printed in Great Britain
on acid-free paper by
Biddles Ltd., King's Lynn. Norfolk

ISBN 0-19-856651-4 978-0-19-856651-9

1 3 5 7 9 10 8 6 4 2

PREFACE

This volume arose from the workshop, with the same title, held from 12 to 16 May 2003 at Venice International University. The idea of organizing that conference came to mind when, in a sunny spring lunchtime in the park around the Neue Pinakothek in München, we discussed the apparent distance that has emerged between the practice of constructive mathematics and the formal systems devised as a foundation for it.

The workshop proved that in constructive mathematics there is a particular need for a closer link between practice and foundations, but also that the first steps had already been made in this direction. It furthermore showed that on each side there are quite distinct styles and approaches which deserve further opportunities for comparison and incentives for collaboration. We hope that the meeting has contributed to a process of reconciliation which will prove fruitful.

The purpose of this volume is to further promote awareness and discussion of the issue of bridging the foundations and practice of constructive mathematics. To this end, we have brought together articles representing the state of the art in type-theoretic and set-theoretic foundations, as well as in the practice of constructive mathematics *à la* Bishop and of formal topology. Some further progress that has been achieved since the time of the workshop, and which was partly prompted by it, could be included.

We have decided to divide the contributions into two main groups, foundations and practice, each of which has two parts: type theory and set theory for the first, and analysis and topology for the second. Due to the very nature of the topic, this can only be a rough separation, which must be improper in many respects. Most papers are fairly self-contained and include a useful list of references, so that they can also be read as a high-level introduction to specific topics in constructive mathematics and formal foundations. Some articles are quite concise, due to space constraints, but may provoke further reading.

For the reader's convenience, we provide an introductory chapter in which, apart from a succinct description of each field together with the basic references, we review the respective contributions from a general perspective. The volume opens with a short recollection of the first encounter by Douglas Bridges with Errett Bishop, a lively document which shows the human side of two mathematicians who contributed considerably to the subject of this volume.

Florence L.C.
Munich P.S.

February 2005

ACKNOWLEDGEMENTS

The workshop this volume emerged from was particularly sponsored by the *Volkswagen–Stiftung*, which generously contributed to the travel and accommodation expenses of many participants. Additional support originated in the *Istituto Italiano per gli Studi Filosofici* and the *Deutsche Forschungsgemeinschaft* through the latter's *Graduiertenkolleg Logik in der Informatik* at Munich, whose speaker, Helmut Schwichtenberg, showed a lot of interest from the beginning. Local logistic assistance was provided by the *Venice International University* staff, and a helping hand often came from the participants of the workshop.

Angus Macintyre established our contacts with the publisher, and Alison Jones from *Oxford University Press* sometimes needed to be patient with us. We are indebted to Douglas Bridges, Andrea Cantini, Michael Rathjen, and Giuseppe Rosolini for their respective advices on editorial matters. Editing this book would have been impossible without the technical assistance by Júlia Zappe. Last but not least, we are most grateful to the authors and anonymous referees, without whose respective contributions this volume would not have emerged.

ERRETT BISHOP

It is never easy to capture the essence of a man so richly endowed as was Errett Bishop; it is almost impossible when, as in my case, one's direct contacts with him were limited to a sporadic, topically limited correspondence over 10 years, a brief meeting in a cafe in San Diego, and a five-week stay in his home while writing parts of a book together.

My first contact with Errett was an indirect one when I was a graduate student. As so often with a major event in life (not that I realized how major it would be at the time), I recall the exact date and place of our first meeting: 11 April 1968, in James Thin's bookstore, Edinburgh. There, on an impulse, I bought *Foundations of Constructive Analysis* (FCM) on the only occasion that I ever saw it displayed for sale. Errett's uncompromising introduction and 'constructivist manifesto' soon fascinated me, and the more he spoke through the book, the clearer became the reasons for my unease, shading into a vague dissatisfaction, with the mathematics of my then PhD project on von Neumann algebras, a subject full of proofs of the 'suppose the contrary, and use Zorn's lemma to construct [*sic*] a maximal family of projections with such-and-such a property' form.

Four or five years later, when working on a DPhil on constructive mathematics, I hesitatingly submitted to Errett a few typed pages of my work, dealing with a number of potentially useful properties equivalent to a result that (as I subsequently learned) was recursively false. His kind response encouraged me to further correspondence about my work over the next few years, a period in which, according to his letters and as a result of the hostile reception that he had frequently met when presenting his constructive mathematics to colleagues across the States, he had become almost completely withdrawn from mathematics. But it was not until, 'by a set of curious chances', Springer Verlag had put Errett and me in touch about the possible revision of FCM that we finally met in the flesh, in a cafe when my wife and I were passing through San Diego around New Year in 1980. At that meeting he reacted again with kindness to the idea of a very junior mathematician becoming Boswell to his Johnson; so began the collaboration which led, in 1985, to the appearance of our book *Constructive Analysis*.

Errett did not want FCM to be republished in its original form, and was unwilling to undertake himself the work that he deemed necessary to justify the reissue of the book. The revisions and additions to FCM proved so substantial that we eventually decided to regard our book as a new one, with a new title. From the outset, although the book was heavily based on FCM, Errett insisted that I be billed as co-author since I was going to do almost all the writing. This seemed characteristic of his encouragement of younger mathematicians.

Our *modus scribendi* was for me to put together a draft of a section and then submit it to Errett for his (often radical) alterations. The drafting was no easy task, since FCM was written in a highly condensed style and there were several subsequent papers, almost all technically hard and in one case containing major errors that were not easy to correct, that had to be incorporated; so Errett was often called upon to clarify something or, in the case of the defective paper, to produce proofs that were not the absolute nonsense that had appeared in print. It always impressed me how clearly and promptly he returned the correct or completed arguments to me (this was, remember, in the days of airmail, when the time between sending a letter from England to California and receiving a reply was a minimum of about ten days). If there was a delay in his response, it was typically caused by the pains he took to lecture and assess a large calculus class (he assessed each of the many students by an individual interview).

It was only when, in November–December 1982, I spent five weeks in the Bishops' home in La Jolla, that I realized fully how fast a mathematician Errett was. He would write out long, new proofs in pencil on a pad on his knee, without even scribbled notes as a guide. (One proof that I recall being presented to me thus was the highly complicated one of the fundamental lemma of the chapter on Banach algebras.) After Errett had died, Paul Halmos (who had been his PhD supervisor at Chicago) told me that he had known one other mathematician who was as fast a thinker as Errett: namely, John von Neumann, for whom Paul had been a research assistant at one stage. Need I say more?

A burning question for me, as for many who have read his work, was 'How did Errett come to constructive mathematics in the first place?' He told me of two events that helped to precipitate him in that direction. The first was when he had been lecturing on truth tables in one of those low-level courses that he described unforgettably as 'mathematics for gracious living'. A couple of bright students had come to him after the lecture, to argue against his definition of (material) implication and in favour of the notion that the truth of the antecedent in some sense causes that of the consequent; this got Errett thinking about implication, a topic that bothered him right up to his death. The second event occurred in his own research in several complex variables, a subject in which he had a huge reputation before entering the constructive domain: in trying to visualize some hypersurfaces, he had come to the conclusion (perhaps through some kind of Brouwerian example?) that those surfaces could not be constructed in any real sense; from which he was led to ask what it meant to assert their existence.[1]

Both Errett and Jane Bishop were extremely kind and generous to me when I was, for rather a long time, a guest in their home. We had many interesting conversations, all revealing the depth and breadth of Errett's interests, some showing amusing misconceptions about British society (for example, the not uncommon American belief that the Queen has real political power). I learned that Errett was an avid collector of rocks—there was a wonderful collection of geodes in the house—and (this I learned the hard way) a good and very competitive tennis player.

During my stay in La Jolla, Errett complained of stomach pains, which he attributed to an old hernia problem. Alas, it was much more: shortly after I returned to England, he wrote me a wonderful, poignant letter explaining that he had been given four months to live. Clearly, his involvement with the book was over, but he wrote the most wonderful

[1] Here I should state that this is a rough memory of a brief conversation more than 20 years ago; like all memories, its detail may not exactly reflect the original.

words about what I had done for him and encouraging me to finish the work. That letter, which, like all my correspondence with him, I still have, was the last I heard from Errett

It took another two years for me to finish the book. At the final stage, I hunted through my *Oxford Dictionary of Quotations* to see if there was something that, as an epigraph, would serve as a fitting tribute to Errett. The one I eventually fixed upon (and which delighted Jane Bishop) was this:

> *vivida vis animi pervicit, et extra processit longe*
> *Flammantia moenia mundi atque omne immensum*
> *Peragravit, mente animaque*

> the vital strength of this spirit won through, and he made his way
> far outside the flaming walls of the world and ranged over the
> measureless whole, in both mind and spirit

As an epitaph for a mathematician who single-handedly showed that deep mathematics could be developed constructively, and thereby pulled the subject back from the edge of the grave, I can think of none better.

Douglas Bridges

CONTENTS

LIST OF CONTRIBUTORS

Peter Aczel Departments of Computer Science and Mathematics, University of Manchester

Ulrich Berger University of Wales Swansea

Vasco Brattka Department of Mathematics & Applied Mathematics, University of Cape Town

Douglas Bridges Department of Mathematics & Statistics, University of Canterbury

A. Bucalo DSI, Università di Milano, Italy

Thierry Coquand Chalmers, University of Göteborg, Sweden

Laura Crosilla Dipartimento di Filosofia, Università di Firenze

Christopher Fox Department of Mathematics, University of Manchester

Nicola Gambino University of Cambridge

Peter Hancock 7 Cluny Avenue, Edinburgh EH10 4RN

Robin Havea Department of Mathematics & Computing Science, University of the South Pacific

Hajime Ishihara School of Information Science, Japan Advanced Institute of Science and Technology

Henri Lombardi Equipe de Mathématiques, CNRS UMR 6623, UFR des Sciences et Techniques, Université de Franche-Comté

Maria Emilia Maietti Dipartimento di Matematica Pura ed Applicata, Università di Padova

Sara Negri University of Helsinki

Erik Palmgren Department of Mathematics, Uppsala University

Jan von Plato University of Helsinki

Michael Rathjen Department of Mathematics, Ohio State University

G. Rosolini DISI, Università di Genova, Italy

Marie-Françoise Roy IRMAR (UMR CNRS 6625), Université de Rennes 1

Giovanni Sambin Dipartimento di Matematica Pura ed Applicata, Università di Padova

Peter Schuster Mathematisches Institut, Universität München

Monika Seisenberger University of Wales Swansea

Anton Setzer Department of Computer Science, University of Wales Swansea

Alex Simpson LFCS, School of Informatics, University of Edinburgh, UK

Bas Spitters Department of Computer Science, University of Nijmegen

Thomas Streicher Fachbereich 4 Mathematik, TU Darmstadt

Hiroki Takamura Japan Advanced Institute of Science and Technology

Steven Vickers School of Computer Science, University of Birmingham

Luminiţa Vîţă Department of Mathematics & Statistics, University of Canterbury

1

INTRODUCTION

LAURA CROSILLA AND PETER SCHUSTER

> Perhaps the times are not conducive to introspection. Mathematics flourishes as never before. Its scope is immense, its quality high. Mathematicians flourish as never before. Their profession is respectable, their salaries good. Mathematical methods are more fashionable than ever before. Witness the surge of interest in mathematical logic, mathematical biology, mathematical economics, mathematical psychology—in mathematical investigations of every sort. The extent to which many of these investigations are premature or unrealistic indicates the deep attraction mathematical exactitude holds for the contemporary mind.
>
> And yet there is dissatisfaction in the mathematical community.
>
> ERRETT BISHOP [1, p. vii]

1.1 Foundations for Constructive Mathematics

The present volume addresses the relationship between the foundations and practice of constructive mathematics, by presenting on the one hand some very recent contributions to constructive analysis and formal topology, and on the other hand studies which underline the capabilities and expressiveness of various formal systems which have been introduced as foundations for constructive mathematics. Deliberately, a wide range of results, foundational styles, and techniques has been collected to give a general overview of the state of the art of constructive mathematics.

In our opinion the problem of providing foundations for constructive mathematics is part of a twofold more general problem. First of all it is a philosophical question. A certain variety of mathematics, say classical, intuitionistic, recursive or constructive, is developed as a consequence of specific philosophical views. These may, for example, relate to the ontological status of mathematical objects, our knowledge of them, or the characteristics of our mathematical activity. A philosophical analysis is then needed to clarify and explore the underlying ideas which are often assumed only implicitly; to justify the validity of the general conception; and to determine the consequences of specific assumptions. We may call this the philosophical problem.

Secondly, it is a mathematical question. One aims at identifying a collection of basic mathematical principles on which a specific kind of mathematics rests. In other words, the quest for foundations is here not the search for a coherent and consistent collection of ideas which are linked with the mathematical practice in question, either as motivation, justification or consequences; it is, rather, the pursuit of a fundamental collection of axioms or principles, which may be expressed in a formal or informal way, and will allow one to work in a specific mathematical framework.

Clearly the two tasks are strongly interrelated and, we believe, should rely on each other. Their mutual interaction has driven much of the investigation at the very origin of constructive thinking, especially the activity of Brouwer. His criticism of the now classical approach to mathematics is essentially the result of a long-term reflection on issues such as the meaning of mathematical activity and the status of mathematical truth.

It appears that as far as mathematicians in practice are concerned, the impetus for discussing foundational issues in this sense has not seen a revival, apart from rare exceptions, since the early 20th century. The philosophical motivation for doing constructive mathematics has rather been considered a source of confusion and obscurity at the expense of the mathematical content. In our perspective a renewed debate of the philosophical issues is instead highly desirable, especially in the case of constructive mathematics. Constructive mathematics is the result of a fresh perspective which becomes natural when one begins to explain what is usually taken for granted; the philosophical view therefore seems essential for its acceptance.[1]

Investigations into foundations have, over time, assumed a rather technical aspect, with emphasis on the fundamental mathematical assumptions which allow for the development of certain mathematical styles. The question is often even more specific as the fundamental principles take the form of axioms in a formal system. While for classical mathematics set theory, usually in the form of the Zermelo–Fraenkel axiomatization, has widely been accepted as a foundation, in the case of intuitionistic and constructive mathematics the more definite issue of formal foundational systems has emerged relatively late. Many proposals in this direction can in fact be traced back to the late 1960s and early 1970s, and seem to have been prompted by the publication of Bishop's *Foundations of Constructive Analysis* [1] in 1967.

Formal systems for constructive mathematics might have emerged relatively late also as a consequence of a tendency already present in the early days of intuitionism. Brouwer actually was rather hostile towards logic and the formalization of mathematics, probably because of his conception of mathematical activity as a mental pre-linguistic construction. It was only through the work of Heyting that intuitionistic logic was brought into a Hilbert-style formal system. It is remarkable that in *Intuitionism. An Introduction* [3, p. 4–9], the same Heyting states that 'intuitionism proceeds independently of the formalization, which can but follow after the mathematical construction'. In fact, 'as the meaning of a word can never be fixed precisely enough to exclude every possibility of misunderstanding, we can never be mathematically sure that the formal system expresses correctly our mathematical thoughts'. Heyting's position can be better understood if we take into account his view that 'the characteristic of mathematical thought is that it does not convey truth about the external world, but is only concerned with mental constructions'.

From its very origin the relationship between constructive mathematical practice and any kind of formalism was thus deemed to be quite complex. Also Bishop's monograph

[1] In the Preface to his *Elements of Intuitionism*, Dummett more emphatically writes: '...intuitionism will never succeed in its struggle against rival, and more widely accepted, forms of mathematics unless it can win the philosophical battle.' [2, p. ix].

contains negative remarks both on the formalist programme and on the role of logic in the actual development of constructive mathematics [1, p. 4–6]:

The successful formalization of mathematics helped keep mathematics on a wrong course. The fact that space has been arithmetized loses much of its significance if space, number, and everything else are fitted into a matrix of idealism where even the positive integers have an ambiguous computational existence. Mathematics becomes the game of sets, which is a fine game as far as it goes, with rules that are admirably precise. The game becomes its own justification, and the fact that it represents a highly idealized version of mathematical existence is universally ignored.

... it is the fault of the logicians that many mathematicians who think they know something of the constructive point of view have in mind a dinky formal system or, just as bad, confuse constructivism with recursive function theory.

He then adds:

A bugaboo of both Brouwer and the logicians has been compulsive speculation about the nature of the continuum. In the case of the logicians this leads to contortions in which various formal systems, all detached from reality, are interpreted within one another in the hope that the nature of the continuum will somehow emerge.

Bishop nonetheless seems to assume a different position in the following years, as in [4, p. 60] he proposes a formal system for constructive mathematics:

Another important foundational problem is to find a formal system that will efficiently express existing predictive mathematics.[2] I think we should keep the formalism as primitive as possible, starting with a minimal system and enlarging it only if the enlargement serves a genuine mathematical need. In this way the formalism and the mathematics will hopefully interact to the advantage of both.

A new era of foundational systems for constructive mathematics began in the early 1970s, and was inspired by Bishop's work. The challenge arose to find coherent settings in which to express the mathematical principles underlying Bishop's constructive analysis [1]. In a relatively short period this challenge was taken up by, among others, Feferman (explicit mathematics), Martin-Löf (type theory), Friedman, Myhill, and Aczel (intuitionistic and constructive set theory).[3]

In this volume we have gathered some of the most recent contributions to some of the foundational styles proposed in the last three decades and have placed them next to very recent results in constructive analysis and formal topology. We hope that this will constitute a reference which, presenting the state of the art of both mathematical

[2]The term 'predictive mathematics' is explained by Bishop on page 53 of the same article: 'Thus by "constructive" I shall mean a mathematics that describes or predicts the results of certain finitely performable, albeit hypothetical computations within the set of integers. If a word is needed to delimit this special variety of constructivism, I propose the term "predictive".'

[3]While Martin-Löf's type theory was intended as 'a full scale system for formalizing intuitionistic mathematics as developed, for example, in the book by Bishop' [5, 6], Martin-Löf's earlier monograph [7] was written independently of Bishop's, which 'the author [Martin-Löf] did not get access to until after this manuscript [Martin-Löf's] was finished' [7, p. 13].

practice and formal foundation, will foster both their interaction and the discussion (including the philosophical discussion) of foundational issues.

1.2 Bridging Foundations and Practice

It might appear far too ambitious a project to bring the foundations and the practice of constructive mathematics closer to each other. On the one hand, no one could expect a working mathematician to accept having his creativity and personal style bound by any form of axiomatic or formal constraint. On the other hand, foundational systems have naturally become objects of study in themselves, and the need to show their adequacy for a certain kind of mathematics has consequently become less compelling for their practitioners. A very similar but more noticeable evolution has taken place in the case of classical mathematics and Zermelo–Fraenkel set theory, the latter having produced extremely sophisticated results far beyond the immediate issue of serving as a foundation.

In the case of constructive mathematics, we feel that there is a stronger need for a tight accord between practice and foundation. This is a consequence of the peculiar nature of the subject, which requires a very precise formulation of all the principles it assumes. It is in fact characteristic of all systems based on a logic weaker than the classical logic to have a higher sensitivity to the formulation of their statements. Hence for constructive settings the role of formal foundations appears to be more relevant, a formal system being a tool for a deeper insight into the mathematical notions and for a clearer formulation of the assumptions made.

Not surprisingly a certain amount of constructive mathematics in recent years has been dedicated to the relationship between constructively distinct concepts which are classically equivalent, a good example being the various ways to define compactness for metric spaces [8–11]. Often, in the classical setting, a certain mathematical concept gives rise to a series of equivalent characterizations, which only splits into a plurality of different concepts when viewed from the constructive perspective. Far from being a defect, this is a benefit which allows us to distinguish notions whose differences are opaque from the classical point of view.

Ishihara's contribution to this volume goes further in this direction. He develops a weak base system over which he continues, now in a formal setting, the so-called constructive reverse mathematics that he has started in recent years. The programme is similar in spirit to (classical) reverse mathematics as performed by Friedman and Simpson [12]. There are two more contributions which particularly witness the convergence of foundations and practice. First, Palmgren analyses the interplay between the notions of (uniform) continuity in Bishop's analysis and in formal topology, making use of distinctively type-theoretic features of the latter. Secondly, Aczel and Fox study separation properties in general topology as developed within constructive set theory rather than using the more common type-theoretic setting of formal topology.

Evidence for the workability of bridging foundations and practice has thus been provided along several lines. This evidence has been reinforced by developments that took place after the contributions to this volume were completed. One example is the recent construction, carried out by Palmgren, Ishihara, and Aczel, of quotients in formal topology and in topology as based on constructive set theory.

1.3 Different but Related Foundations

When in the 1970s various formal systems for developing constructive mathematics emerged, they clearly were more than just formal systems. Each of them was a specific incarnation of a certain conceptual framework for developing constructive mathematics, while every framework of this kind centred around one or several main systems. Examples of this phenomenon are the constructive set theory *à la* Aczel with the principal system CZF, and Martin-Löf type theory as the theory of sets with one reflecting universe. The characteristic of being open to extensions and restrictions was particularly clear in the case of type theory, for which a number of systems was soon put forward: intensional, extensional, with or without universes, and with or without inductively defined types. Variants of CZF have also been proposed in the meantime, this multiplicity responding to the need for extra flexibility that a single system can barely handle.

Each of the formal frameworks that have been introduced as a foundation for constructive mathematics represents a different perspective on the role of foundations. Constructive type theory, for instance, with its characteristic predominance of the propositions-as-types paradigm, stresses the computational content of mathematical statements. It also underlines the pure constructive essence of mathematical statements via Martin-Löf's so-called meaning explanations. Constructive set theory, on the other hand, was put forward with the intention to provide a user-friendly system for the development of Bishop-style constructive mathematics; whence the emphasis is on the simplicity of the language and the similarities with classical ZF.

The distinctions between the various frameworks for constructive mathematics are nonetheless not sharp, as one can perceive separate collections of systems as only different views of the same conceptual structure. For instance, Martin-Löf type theory and constructive set theory have been brought to a common ground through the seminal work by Aczel on the type-theoretic interpretation of constructive set theory [13–15] and subsequent work by, among others, Aczel and Gambino [16] and Rathjen and Tupailo [17]. What can certainly be observed is that Martin-Löf type theory and constructive set theory share a similar attitude with regard to which notions should be avoided, and which ones on the contrary allow us to remain in the realm of constructivity. These systems, in all their formulations, not only adhere to intuitionistic logic but also comply to a form of (generalized) predicativism.[4] In this respect they essentially

[4]Predicative theories have been characterized in proof-theoretic terms by Schütte and Feferman as those theories to which we can associate the proof-theoretic ordinal Γ_0. The essential idea is that the proof-theoretic techniques used for the ordinal analysis up to Γ_0 are of a predicative nature, building up from below. A more liberal approach to predicativity allows for theories whose strength matches that of so-called (iterated) inductive definitions. The issue of a predicative approach to mathematics has been in recent years especially put forward by Feferman. In [18] Rathjen proposes a study of the upper bounds of Martin-Löf type theory, there considered as an emblematic constructive and predicative theory. The investigations are carried out in relation to a so-called revised Hilbert programme. It should also be mentioned that in recent years Jäger has

differ from the approach of so-called intuitionistic set theories (in particular the system IZF) and the calculus of constructions, in which full impredicativity is endorsed.

The task of distinguishing and separating different frameworks of ideas, as well as assessing their adequacy in capturing the essence of constructivity, is certainly hard and prone to disputes. The reader may compare the different perspectives from which Maietti and Sambin, Rathjen, and Simpson have contributed to this volume.

In addition to the different conceptions of what constitutes a constructive theory, there are distinct methods for studying them. For example the methods of proof theory, like syntactic interpretations, could be used to capture the strength of a specific system and compare it with other constructive or classical theories. Alternatively one could use category theory to classify theories on the basis of their models, and contrast them with each other. The contribution by Rathjen belongs to the first approach, while those by Simpson and Gambino follow the second.

The personal view of the editors is that different foundational approaches and methods are welcome. Each of them has its distinguishing objectives, and contributes in its peculiar form to the common aim of unravelling constructive foundations. A confrontation between the respective achievements can but be beneficial.

1.4 Practicable and Weak Foundations

In an attempt to foster the discussion on foundations, we now propose some possible criteria which could drive the development of practicable foundations for constructive mathematics. First, the formal tool one selects should be as general as possible, to allow us to formalize different varieties of constructive mathematics and, ideally, to represent different conceptions of foundations. At this point the observation already quoted from Bishop [4] applies, that the formalism should be kept as primitive as possible, starting with a minimal system and enlarging it only when this is prompted by genuine mathematical need. Secondly, the development of mathematics should not be made unnecessarily complicated by syntactic details. In Feferman's words [20], one should have your cake and eat it too.

With regard to the first point, we observe that just as a weaker logic allows for a deeper insight into mathematical notions, also weaker set or type theories allow for clearer distinctions. An obvious benefit of moving to a weaker system is the higher level of generality thus obtained, a stronger theory still being accessible from a weaker one by adding specific principles. In a weaker system one can elucidate the essential features that are shared by different foundational traditions, and can access all of them from a common ground. In this context it is noteworthy that in the last three decades evidence has been given, especially through the work of Feferman and the (classical) reverse mathematics programme of Friedman and Simpson, that most of the constructions carried out in the main body of mathematics do not require us to assume essentially impredicative notions. An ideal formal system therefore ought to be sufficiently strong for a large part of mathematics but at the same time weak enough to allow overall control

introduced the notion of a meta-predicative theory. This characterizes systems which, though impredicative, are still amenable to a proof-theoretical analysis that does not use any impredicative method [19].

of the ingredients of a mathematical proof. An additional virtue of weaker frameworks is that they prompt a finer concept analysis which forces more perspicuous proofs.

The other criterion, the simplicity of formulation, has been particularly emphasized by Myhill in [21]. Constructive set theory was meant to isolate the principles underlying Bishop's conception of mathematics and at the same time enable one to express constructive mathematics in a workable way, making the formalization as trivial as in the classical case.[5] It appears that in recent years the simple formulation of constructive set theory, in particular of Aczel's CZF and its variants, has brought it to play a prominent role in the foundations for constructive mathematics. We remark, though, that constructive set theory fails to exhibit a genuinely constructive meaning on its own, for which it has to rely on its interpretation in type theory. We thus believe that the most can be achieved from an open-minded approach to foundations in which the two systems are taken as two sides of a single framework of ideas.

1.5 Martin-Löf Type Theory

Constructive type theory was founded by Per Martin-Löf 'with the philosophical motive of clarifying the syntax and semantics of intuitionistic mathematics' [23]. It is a rich and expressive theory whose fundamental concept is that of (dependent) type. The theory gains its full strength by admitting inductive data types and (a hierarchy of) type universes.

One of the benefits of type theory is that it directly stems from a profound analysis of constructivity, so that its adequacy as a foundation for constructive mathematics can hardly be called into question. Its retaining from the very beginning a clear philosophical inclination constitutes an element of distinction from other foundational systems, like constructive set theory. The theory in fact appears to have arisen from reflecting first on the semantics of constructive notions, syntactic details being laid down later.

Martin-Löf's constructive type theory has several sources. There are Russell's fundamental idea of a type as a domain of quantification, Church's typed lambda calculus, and Gödel's *Dialectica* interpretation. Martin-Löf [5] also recognizes the importance of later work by Scott, Kreisel, and Goodman (theories of constructions). Clear affinities are also discernible with Gentzen's natural deduction in the form adopted by Prawitz. The relevance that the proposition-as-types paradigm has for type theory correlates it with the AUTOMATH project led by de Bruijn [24].

It is an important consequence of the proposition-as-types paradigm that in type theory the analysis of logical notions is not isolated but on the contrary carried out as part of the more general task of clarifying the notion of set. This seems in accordance with the priority Brouwer gave to mathematics over logic. The logical rules are in fact

[5] Bishop [4] and Goodman and Myhill [22] propose using appropriate versions of Gödel's theory of functionals with quantifiers and choice as formal systems for [1]. Referring to those proposals, Myhill writes [21]: 'We refuse to believe that things have to be this complicated—the argumentation of [1] looks very smooth and seems to follow directly from a certain conception of what sets, functions, etc. are, and we wish to discover a formalism which isolates the principles underlying this conception in the same way that Zermelo–Fraenkel set theory isolates the principles underlying classical (nonconstructive) mathematics. We want these principles to be such as to make the process of formalization completely trivial, as it is in the classical case.'

elegantly obtained from the more general rules for sets, by means of the Curry–Howard correspondence. They stand out for particular clarity if compared with the informal justification of the operators in intuitionistic logic which is usually referred to as the Brouwer–Heyting–Kolmogorov interpretation. The correspondence has also played an important role in undertaking to use type theory as a programming language, the latter bringing to concrete reality Bishop's claim of the computational relevance of constructive mathematics.

The literature on type theory is quite extensive and varied in style, some parts emphasizing the mathematical, others the computational, aspects. Fundamental references are Martin-Löf's [5, 6, 23, 25], whereas [26] is an introduction, with a comprehensive bibliography, to type theory as a programming language. We now review the contributions to this volume which are related to type theory.

An impredicative setting is chosen by Streicher in his analysis of the notion of universe. Universes figure prominently in type theory as well as in category theory, and correspond to inaccessibility conditions in set theory. Martin-Löf [5] sets up a hierarchy of reflecting universes to increase the expressive power of type theory. Streicher now defines a notion of universe, and of a hierarchy of universes, in the setting of toposes. He suggests a topos with natural number object, and a hierarchy of universes containing this object, as a reasonable model for impredicative constructive mathematics. He then investigates the question whether universes do exist in toposes, and shows that hierarchies of universes can be found in all sheaf and realizability toposes. Within the paper there arise connections with the respective contributions by Rathjen, Simpson, and Gambino.

A completely different approach is chosen by Maietti and Sambin. Their contribution presents work in progress on the formulation of a minimal predicative type theory. Two conditions are imposed on a system to be considered as a minimal foundation for constructive mathematics: it has to satisfy the proofs-as-programs paradigm, as the authors define it, and needs to be compatible with the most relevant foundational theories, including Martin-Löf type theory, the calculus of constructions, and Zermelo–Fraenkel set theory. The authors explain their motivation for introducing a new foundational system and also discuss the relation a minimal theory should have with principles like the type-theoretic axiom of choice (AC) and Church's thesis (CT). A variant of constructive type theory is then proposed which satisfies the conditions of a minimal theory, and in which propositions are introduced independently from the notion of set.

The authors conjecture that their minimal theory is also compatible with AC and CT, though these principles should not be directly derivable within the theory, the non-derivability of AC being a consequence of the separation imposed between propositions and types. The combination of compatibility and non-derivability of these principles is essential to guarantee the satisfaction of the compatibility requirement for a minimal theory.

Type theory as a programming language is the main topic of the contribution by Hancock and Setzer. They present an extension of Martin-Löf type theory by coinductive definitions and use it for the development of interactive programs. This has been the focus of their work in recent years and has given rise to a series of papers, the present one being particularly expository and comprehensive. Inductive and coinductive definitions have been at the centre of extensive study in recent years, especially in theoretical computer science.

Hancock and Setzer further provide helpful information on the literature, clarifying its relation with their own proposal. They take inspiration from Haskell's input/output monad and represent within type theory various kinds of interactions between the user and the real world (with or without history) and the programs (client or server oriented). Weakly final coalgebras are introduced for a notion of polynomial functor, and the relationship with guarded induction is discussed.

Computational practice is also of central importance for the contribution by Berger and Seisenberger. The context is that of program extraction from proofs, which—as presented in the paper—is based on two fundamental ideas. The first one is that a constructive proof of a $\forall\exists$ statement may be used to write a program for which the statement itself is a specification. According to the Curry–Howard correspondence, types correspond to propositions and elements to proofs, this being formalized in a precise way by means of realizability. The second idea is to combine realizability with a refined version of Gödel's double negation interpretation, to build a mechanical tool for extracting constructive proofs from classical theorems. Ideally the technique provides a means to constructivize possibly large parts of classical mathematics, and to extract information (or programs) from them in an effective way. The aim is therefore very much in the spirit of Bishop's conception of constructive mathematics.

Berger and Seisenberger also give a sketch of the techniques involved in program extraction, from either constructive or classical proofs, by a case study on Higman's lemma. They also report on the implementation of this case study in a concrete proof-development system. Of particular interest is their use, in the determination of the realizers, of a refined analysis of dependent choice principles in the classical proof.

The relation between classical and constructive proofs is a fundamental topic of the contribution by Negri and von Plato. They recall their method for converting axioms into rules in a sequent calculus style. The rules may be obtained in two distinct ways, obeying a dual pattern. In the present paper they apply this technique to find a correspondence between what they call classical and constructive proofs. To this purpose they consider dual mathematical notions, examples of which are equality and apartness or, in a geometric context, parallelism and convergence. They then show how the method may be used in various cases to produce an automatic transformation from proofs with one notion to proofs with the dual one: that is, from classical to constructive proofs and *vice versa*.

A detailed analysis is made of which conditions need to be met for the transformation to apply. As a consequence, meta-theoretical results may also be imported from one theory to its dual. Negri and von Plato also study the permutability of mathematical rules, obtaining an extension of Herbrand's theorem which holds for a class of theories.

1.6 Constructive Set Theory

In [21] Myhill proposed a theory called constructive set theory (CST), which mirrors Bishop's informal set theory and bridges it with the tradition of Zermelo–Fraenkel set theory. His intention was somewhat different from Martin-Löf's in developing type theory, as constructive set theory was intended as a foundational system for carrying out constructive mathematics in a user-friendly way, having in mind a user familiar with classical set theory. CST was preceded by other attempts to formalize constructive mathematics in a Zermelo–Fraenkel style but based on intuitionistic logic. In particular, Friedman and Myhill had proposed various systems [27–29], the best known of which is Friedman's so-called intuitionistic Zermelo–Fraenkel set theory (IZF). This is characterized by allowing for impredicative principles of full separation and power set, with the result that, in proof-theoretic terms, its strength equals that of classical ZF. In [21] Myhill takes a different route by making the crucial observation that to develop constructive analysis in Bishop's way one does not need to appeal to impredicative set constructors. He therefore proposes a system which avoids full separation and power set.

Myhill's system was subsequently modified by Aczel [13] so as to obtain a subsystem of classical ZF which is called constructive Zermelo–Fraenkel set theory (CZF). Apart from the language and the choice principles, CZF subsumes most of the essential characteristics of CST; in particular, it is predicative according to the more liberal conception of predicativity for which generalized inductive definitions are permitted. One of the benefits of CZF is that it makes use of the extremely simple language of ZF. As a consequence, the set-theoretic course of doing mathematics can also be followed in the development of constructive mathematics. One only needs to ensure that the limit of intuitionistic logic is not trespassed and, more importantly, that one makes exclusive use of the set-theoretic axioms which do belong to the constructive realm.

A lot is now known about CZF and some natural extensions and subsystems, as well as their proof-theoretic strength. Starting from Aczel's interpretation, the relationship between type theory and set theory has been clarified. Unfortunately, no introductory presentation of CZF is available in print at the moment, the topic being confined to single research articles at a fairly high level. We refer instead to the draft [30] of a forthcoming book by Aczel and Rathjen, which we hope will soon go to press. Beeson's monograph [31], though not up to date, is still useful for constructive and intuitionistic set theory in general. For recent work on the category-theoretic aspects of constructive set theory see the references provided in the contributions to this volume by Simpson and Gambino.

Before we review their articles, let us concentrate on Rathjen's. He explores, from a proof-theoretic point of view, extensions of constructive set theory by a principle GID which asserts the existence, as sets, of greatest and least fixed points of generalized inductive definitions. Inductive and coinductive definitions play a prominent role in theoretical computer science, and have also been investigated in the context of Martin-Löf type theory (see the contribution by Hancock and Setzer). Recently, coinductive definitions have made their appearance also in the context of formal topology, for instance when it comes to generate a positivity predicate [32].

Rathjen now shows that the addition of the principle of GID is far from being innocent. It boosts the proof-theoretic strength of CZF so as to equal a rather strong fragment of second order arithmetic. It is nonetheless more innocuous than adding to the theory the highly impredicative principles of full separation and power set, from which GID can be deduced. Apart from this result, the paper hints at proof-theoretic techniques which are ubiquitous in the analysis of constructive set theory, and avails itself of recent work conducted in proof theory and related to the μ-calculus.

In his contribution on recently discovered aspects of constructive set theory, Simpson considers various systems, ranging from a basic constructive set theory (BCST), which is of the same proof-theoretic strength as Peano arithmetic, to fully impredicative theories obtained by adding unrestricted separation or power set to BCST. He uses category-theoretic tools, in the style of the algebraic set theory put forward by Joyal and Moerdijk [33], to establish relationships between these theories. Every system is shown to possess a sound and complete class of models which enjoy a simple axiomatization; some systems also have models which occur naturally in mathematics. The author claims that the presence of models of this kind justifies scientific interest for the theories he singles out.

The issue of weak systems is also discussed from the same perspective, and simultaneously the suggestion is made of relaxing our view on what constitutes constructivity, by allowing for full impredicativity as well. Simpson's paper provides a non-technical introduction to a flourishing area in category-theoretic foundations, and bears connections with the results contributed by Gambino and by Streicher.

Gambino studies models of a variant, called CST, of the system CZF, with exponentiation replacing subset collection (see Rathjen's contribution for a list of the most relevant axioms of CZF). The focus of the article is on the development of new models for CST based on categories of presheaves. Gambino's approach is related to Simpson's, as it concentrates at the beginning on categories of classes rather than sets, following the lines of algebraic set theory. He defines a general notion of what is a categorical model for CST, and shows that categories of presheaves provide examples of such models. To do so, he considers presheaves as functors with values in a category of classes. Gambino also hints at an application to independence results.

The article provides an example of the advantages and difficulties which one faces when bringing to a predicative context ideas already explored in an impredicative environment. It thus bears similarities with some of the articles on formal topology contributed to this volume. In all these cases a clearer and, especially, a more general analysis is gained as compared with the impredicative context.

1.7 Bishop-Style Constructive Mathematics

In his *Foundations of Constructive Analysis* [1] from 1967, Bishop left aside most of the philosophical issues of constructivism. Motivated by fairly pragmatic reasons, he presented large portions of mathematical analysis within a constructive framework, with the aim to be

> a piece of constructivist propaganda, designed to show that there does exist a satisfactory alternative [to classical mathematics].

The movement of intuitionism failed, according to Bishop,

> to convince the mathematical public that abandonment of the idealistic viewpoint would not sterilize or cripple the development of mathematics. Brouwer and other constructivists were much

more successful in their criticisms of classical mathematics than in their efforts to replace it with something better. Many mathematicians familiar with Brouwer's objections to classical mathematics concede their validity but remain unconvinced that there is any satisfactory alternative.

He also writes:

Our program is simple: to give numerical meaning to as much as possible of classical abstract analysis. Our motivation is the well-known scandal, exposed by Brouwer (and others) in great detail, that classical mathematics is deficient in numerical meaning.

A relevant aspect of Bishop's framework is that no use is made of principles which are incompatible with classical mathematics. The typical practitioner of Bishop-style mathematics restricts his attention neither to recursively defined real numbers nor to Brouwerian choice sequences. His mathematics looks like classical mathematics, but at the same time shows a faithful commitment to the constant use of intuitionistic logic and constructive reasoning. The similarities with the classical approach have made Bishop-style mathematics accessible to the wider community of mathematicians, thus crossing the boundary of the limited circle of adherents to intuitionism.

Bishop was originally concerned with real and complex analysis, including considerable portions of functional analysis. In his later years he revised his book together with Bridges [34]. Bridges and Richman [35] made explicit that Bishop-style constructive mathematics simultaneously generalizes ZFC-based classical mathematics, Markov-style recursive mathematics, and intuitionistic mathematics *à la* Brouwer. Since the early 1970s a lot of work has been done also on constructive algebra in the spirit of Bishop [36]. Formal aspects of constructive mathematics were considered in the monographs by Beeson [31] and Troelstra and van Dalen [37].

Since 1980 several conferences either have been explicitly dedicated to constructive mathematics in Bishop's sense or else have included it as a principal topic [38, 39, 41]. More recent developments of Bishop-style constructive mathematics include the choice-free approach started by Richman in [42] and the theory of apartness spaces begun by Bridges and Vîţă with [43]; neither of these is represented in this volume. The new direction in constructive algebra taken by Coquand and Lombardi is represented by their contribution with Roy; as it cannot be thought of without its background in formal topology, we will discuss it in that context. Before doing so, we review the contributions to constructive analysis.

To accomplish a finer analysis of mathematical concepts, Ishihara develops a formal system based on intuitionistic logic, and applies it to detect the differences between various classically indistinguishable compactness properties of metric spaces. His aim is to single out the logical and set-theoretic principles which can be proved equivalent to a given mathematical assertion on the basis of the underlying system. The general idea of classifying the theorems of mathematics in this way can be traced back to the concept of Brouwerian counterexamples, which serve for distinguishing the non-constructive theorems by deducing constructively untenable principles from them. The use of equivalences to arrive at a more detailed map of the non-constructive part of mathematics seems to go back to Mandelkern [8].

The new line of research followed by Ishihara has taken the name of constructive reverse mathematics due to similarities with the classical programme of reverse mathematics [12], which is based on subsystems of classical second order arithmetic. Constructive reverse mathematics needs to rely on a clear-cut foundational system which has to be weak enough, as otherwise one could not detect all relevant assumptions or could demonstrate equivalences which do not hold in weaker systems. For instance, the picture of compactness properties drawn by Veldman [44] and Loeb [45] in intuitionistic reverse mathematics differs slightly from that in constructive reverse mathematics; some differences are highlighted by Berger [46].

The interplay between intuitionistic mathematics and Bishop-style constructive mathematics is crucial for Spitters's contribution. Brouwer's continuity principle (CP) is inconsistent with the decidability of Σ_1^0-statements, which Bishop called the limited principle of omniscience (LPO). In fact, as a reformulation of the trichotomy of real numbers LPO allows for defining discontinuous functions, whose presence conflicts with CP. While CP is needed to prove many theorems of intuitionistic analysis, LPO suffices for a considerable portion of classical analysis. A more advanced instance of this phenomenon is Ulam's classical theorem that every measurable subset of a complete separable metric space can be approximated by compact subsets.

Spitters first gives an intuitionistic proof of Ulam's theorem, and then shows how one can avoid invoking CP using a method which is now known as Ishihara's trick, and whose application yields a proof that entirely lies within Bishop-style constructive mathematics. In the case of Ulam's theorem, Ishihara's trick guarantees that at a certain point the proof splits into two branches. Either one has LPO at one's disposal, and thus is in a sufficiently classical context, or one is granted the variant of CP that is needed to proceed roughly as in the intuitionistic setting. It is noteworthy that since LPO conflicts with CP, the first of the two proof branches made possible by Ishihara's trick is invisible from the perspective of intuitionistic mathematics.

Takamura's contribution is also of interest in view of the Bridges–Hellman debate on whether constructive mathematics is able to cope with quantum mechanics, which in our view was decided, positively, by Bridges [47]. Since quantum mechanics is bound up with the theory of operator algebras on Hilbert spaces, the latter was a challenge for constructive mathematics from the very beginning. Bishop gave a constructive proof of the Gelfand representation theorem for a special kind of commutative C*-algebra, and Spitters has recently brought the general commutative case to a certain perfection. Takamura now proves that an arbitrary C*-algebra together with a state possesses a cyclic representation. A constructive point-set result of this kind is furthermore interesting because Banaschewski and Mulvey [48] have studied at length and in depth the point-free approach.

While the Hahn–Banach theorem in its customary exact form is essentially nonconstructive [9], there is a constructively valid approximate version due to Bishop. An appropriate translation can be proved in formal topology [49], and in constructive reverse mathematics this theorem is related to the fan theorem.[6] Bridges and Havea now

[6]This is forthcoming work by Josef Berger and Hajime Ishihara.

show that Bishop's variant suffices to constructivize the proof ingredients of Sinclair's theorem on the spectral range of a Hermitian element of a Banach algebra. Their work indicates once more that exact solutions whose existence can merely be guaranteed by classical logic are often unnecessary to prove a statement of a concrete character even when they occur in most classical proofs.

Although essentially non-constructive theorems of real analysis are used in the customary proofs of the uniqueness of the topology on the real Euclidean space of a fixed dimension, Bridges and Vîţă manage to avoid invoking any fragment of the law of excluded middle at least when the given topology is locally convex. Their proof uses barely more than elementary geometric reasoning combined with approximation techniques typical of constructive analysis, but contains an algorithm for computing the desired homeomorphism. The authors acknowledge that using a foundational system, from which they still refrain, would ensure that no unwanted principles are invoked during the proof. Although large parts of constructive mathematics may be formalized in principle, we have some doubts that this is a routine task in every case.

According to Bridges [50], constructive analysis may serve as a framework for any algorithmic approach to analysis. Although they are not yet completely clarified, strong links must exist between Bishop-style and computable analysis, which is the framework for Brattka's contribution. In fact, when it comes to defining a function in computable analysis, one seems to end up with a discontinuity—in conflict with the Kreisel–Lacombe–Shoenfield–Tsejtin theorem—as soon as one invokes a non-constructive distinction-by-cases, which is licit *a priori* in computable analysis as it is based on classical logic.

While computable normed spaces are necessarily separable, non-separable spaces occur naturally in classical functional analysis, for example as the dual spaces of common separable spaces. To include some of them, Brattka proposes the wider notion of a general computable normed space, and proves Landau's characterization of the duals of sequence spaces as a test case. This is possible by moving from sequential computability structures, which are bound up with separable spaces, to represented spaces as used by Weihrauch and others: that is, spaces whose points are indexed by infinite strings over some alphabet.

Brattka's work is in parallel with a recent tendency in constructive analysis. While for Bishop all metric spaces were separable, and he even considered inseparable metric spaces as an instance of 'pseudo-generality', nowadays one sometimes tries to dispense with separability, especially when one wants to dispense with countable choice. To keep to separable spaces, on the other hand, helps to minimize the use of arbitrary subsets, and thus to remain in the predicative realm, which advantage Bishop might have had in mind when choosing his preference.

1.8 Formal Topology

The framework of formal topology was put forward in the mid-1980s by Martin-Löf and Sambin [51] in order to make available to Martin-Löf type theory the concepts of point-free topology. Formal topology constitutes a constructive and predicative generalization of the theory of frames and locales, which have been studied since the 1950s, though usually in a classical or impredicative way. According to Sambin [52, p. 359], '... a predicative notion of frame is ... nothing but the notion of formal topology'.

Formal topology was founded for foundational concerns. At the origin of formal topology we see two motivations for working without points: the philosophical one that in many cases a point is a concept of an infinitary and somewhat ideal character, and the related methodological one that in general the points of a topological space are objects of a higher type, and thus do not always form a completed concrete entity. Formal topology is concerned with the actual data (neighbourhoods, observables, etc.) rather than with ideal exact objects (points, states, etc.); the former are thought of as only approximations to the latter.

Standard references for locale theory are the monographs by Johnstone [53] and Vickers [54], with the former's survey paper [55] motivating point-free topology in general. While Sambin's original article [51] is still a good source for the early developments in formal topology, for a recent account of the state of the art in this field see [52]. The proceedings of the second workshop on formal topology [56] include forerunners of some of the articles on formal topology contributed to the present volume, all of which we now review.

Palmgren's contribution helps us to understand better, by taking into account formal topology, what a continuous function ought to be in constructive mathematics. As Waaldijk [57] has pointed out, the concept of continuous function used by Bishop is rather unsatisfactory unless one adopts the fan theorem. Palmgren first shows that every continuous function on the real line in Bishop's sense can be represented by a continuous mapping on the formal topology of the reals. He next proves that a function on a compact interval of formal reals is uniformly continuous whenever it is induced by a continuous function on the formal topology of this compact interval. His achievements, for which he makes use of the formal version of the Heine–Borel theorem proved by Cederquist and Negri, have been generalized by Curi [58] to mappings on formal metric spaces.

Aczel and Fox study separation properties in topology as done on the basis of the formal system CZF. Aczel began his constructive theory of topological spaces in [59] where he provided an analogue of the adjunction between topological spaces and locales, showed that every formal topology gives rise to a topological space in his sense, and studied whether the formal points form a set. An advantage of Aczel's approach is that he has at his disposal a language for classes, and thus can speak of the whole of (formal) points even if this is not a set.

The main motivation Aczel and Fox give for studying separation properties is that they constitute the first step in learning topology after metric and uniform spaces. A further motivation is that considering separation properties helps us to understand whether the formal points of a formal topology form a set [40, 59–63]. For instance, Palmgren has shown that the formal points form a set whenever they are all maximal [62], which means nothing but T_1.

Aczel's concept of a topological space is closely related to the notion of basic pair, which in turn is the principal constituent of the so-called basic picture started by Gebellato and Sambin in [64]. Roughly speaking, this is a systematic method of studying the interplay between the aspects of topological spaces (and continuous mappings) that involve points, and their point-free counterparts. Considering the basic picture led to a more general version of formal topology [52], emancipating it from its origins in locale theory.

Bucalo and Rosolini [65] put the concept of basic pair under scrutiny. Among other things, they observed that the category of basic pairs is the Freyd completion of the category of (sets with) relations, which confirms that relations are crucial for understanding topology. They also endow the category of basic pairs with a symmetric monoidal structure, whose tensor is given by the (relational) product of relations. Moreover, the functor of opens from topological spaces to basic pairs induces a functor to the commutative comonoids of the latter, which takes products to tensors.

As for locales, this functor has a right adjoint, and the adjunction restricts to an equivalence whose domain consists of the sober topological spaces. A precise description of the range of this functor is the principal achievement of Bucalo's and Rosolini's contribution to the present volume. They prove that a basic pair is given, up to isomorphism, by a sober space precisely when it creates a topology on its first component. While this means, impredicatively, that a basis of a topology can be defined on this candidate for a space, the authors characterize the corresponding property of basic pairs in entirely point-free terms.

Exponentiation is a particular challenge for point-free topology: one cannot simply put a topology on the collection of continuous mappings, which task might be nontrivial either. Hyland first classified the exponentiable locales as the locally compact ones. Maietti now proves, in her contribution to this volume, the counterpart of Hyland's result for inductively generated formal topologies. Her work demonstrates how formal topology profits from category theory, whose use is of course indispensable when it comes to exponentiation. As the contribution by Bucalo and Rosolini, it is a good example of how impredicative reasoning can be unwound predicatively, with great patience but with the benefit of a quite explicit proof.

Maietti's method of showing that every locally compact formal topology is exponentiable splits into two steps: she first does a proof for 'formal covers' (that is, formal topologies without positivity structure), and then transfers her result to 'formal topologies' (that is, formal topologies with a unary positivity predicate). The second part is an application of the coreflection of 'formal covers' into 'formal topologies' constructed by the author together with Valentini, while the first part is based on a predicative definition of the 'way below' relation between opens, due to Curi, that makes it possible to adopt Hyland's ideas. The same splitting technique works when Maietti verifies the somewhat easier converse, that every exponentiable formal topology is locally compact.

The major part of Vickers's article is dedicated to giving a general localic proof of Tychonoff's theorem without the axiom of choice. In contrast to the formal-topological treatment by Negri and Valentini, who followed Coquand's first paper on this subject, the index set of the product under consideration need not be decidable in Vicker's proof. While a more recent approach by Coquand is based on the assumption that each locale under consideration is presented by a distributive lattice of generators, Vickers does not assume the presence of any such presentation. In passing, he highlights the differences and connections between the point-set and the point-free approach to topology, and between the major varieties of the latter, locale theory and formal topology. In particular, Vickers gives some details of Sambin's aforementioned picture that formal topology is a predicative way to put locale theory.

The recent activities by Coquand, Lombardi, and others aim at a partial realization of Hilbert's programme in commutative algebra. In simpler terms, they want to prove theorems constructively and at the same low type level at which they can be formulated. To this end they need to reduce the complexity of some concepts, an important example being the Krull dimension of a commutative ring. Although the traditional definition of this discrete invariant seems quite intuitive, it needs to quantify over all prime ideals of the given ring, which are subsets of a rather arbitrary kind.

To circumvent this problem, Coquand and Lombardi created first an equational and then, in their contribution with Roy to this volume, an inductive characterization of Krull dimension, without primes, which carries over to this setting the inductive concept of dimension going back to Brouwer, Menger, and Urysohn. Further origins of these achievements are Joyal's point-free presentation of the Zariski spectrum as a distributive lattice and the notion of Krull dimension for lattices due to Español. The new predicative definition of Krull dimension for commutative rings has led to elementary constructive proofs of famous theorems on Noetherian rings and even to generalizations to the non-Noetherian case [66].

References

1. Bishop, E. (1967). *Foundations of Constructive Analysis*. McGraw–Hill, New York.
2. Dummett, M. (2000). *Elements of Intuitionism*. 2nd ed., Oxford University Press.
3. Heyting, A. (1956). *Intuitionism. An Introduction.* North-Holland, Amsterdam.
4. Bishop, E. (1970). Mathematics as a numerical language. In: *Intuitionism and Proof Theory* (eds A. Kino, J. Myhill, and R. E. Vesley), pp. 53–71. North-Holland, Amsterdam.
5. Martin-Löf, P. (1975). An intuitionistic theory of types: predicative part. In: *Logic Colloquium '73* (eds H. E. Rose, and J. C. Sheperdson), pp. 73–118. North-Holland, Amsterdam.
6. Martin-Löf, P. An intuitionistic theory of types. [67, pp. 127–172]
7. Martin-Löf, P. (1970). *Notes on Constructive Mathematics.* Almqvist & Wiksell, Stockholm.
8. Mandelkern, M. (1988). Limited omniscience and the Bolzano–Weierstraß principle. *Bull. London Math. Soc.*, **20**, 319–320.
9. Ishihara, H. (1990). An omniscience principle, the König lemma and the Hahn–Banach theorem. *Z. Math. Logik Grundlag. Math.*, **36**, 237–240.
10. Bridges, D., Ishihara, H., Schuster, P. (2002). Compactness and continuity, constructively revisited. In: *Computer Science Logic.* 16th International Workshop, CSL. 11th Annual Conf. of the EACSL. Edinburgh, Scotland (ed J. Bradfield), pp. 89–102, Proceedings. Lecture Notes Comput. Sci. **2471**. Springer, Berlin and Heidelberg.

11. Ishihara, H., Schuster, P. (2004). Compactness under constructive scrutiny. *Math. Logic Quart.*, **50**, 540–550.
12. Simpson, S. G. (1999). *Subsystems of Second Order Arithmetic.* Springer, Berlin and Heidelberg.
13. Aczel, P. (1978). The type theoretic interpretation of constructive set theory. In: *Logic Colloquium '77* (eds A. MacIntyre, L. Pacholski, and J. Paris), pp. 55–66. North-Holland, Amsterdam.
14. Aczel, P. The type theoretic interpretation of constructive set theory: choice principles. [39, 1–40]
15. Aczel, P. (1986). The type theoretic interpretation of constructive set theory: inductive definitions. In: *Logic, Methodology, and Philosophy of Science VII* (eds R. Barcan Marcus, G. J. W. Dorn, and P. Weingartner), pp. 17–49. North-Holland, Amsterdam.
16. Aczel, P., Gambino, N. (2002). Collection principles in dependent type theory. In: *Types for Proofs and Programs.* Proc. Internat. Workshop TYPES 2000. Durham, United Kingdom (eds P. Callaghan, Z. Luo, J. McKinna, and R. Pollack), pp. 1–23. Lecture Notes Computer Science **2277**. Springer, Berlin and Heidelberg.
17. Rathjen, M., Tupailo, S. Characterizing the interpretation of set theory in Martin-Löf type theory. *Ann. Pure Appl. Logic*, to appear.
18. Rathjen, M. The constructive Hilbert program and the limits of Martin-Löf type theory. *Synthese*, to appear.
19. Jäger, G. (2005) Metapredicative and explicit Mahlo: a proof–theoretic perspective. In: *Logic Colloquium 2000* (ed. R. Cori), pp. 272–293. A. K. Peters, Wellesley, MA.
20. Feferman, S. (1979). Constructive theories of functions and classes. In: *Logic Colloquium '78* (eds M. Boffa, D. van Dalen, and K. McAloon), pp. 159–224. North-Holland, Amsterdam.
21. Myhill, J. (1975). Constructive set theory. *J. Symbolic Logic*, **40**, 347–382.
22. Goodman, N., Myhill, J. (1972). The formalization of Bishop's constructive mathematics. In: *Toposes, Algebraic Geometry, and Logic* (ed W. Lawvere), pp. 83–96. Halifax, Nova Scotia, 1971. Proceedings. Lecture Notes in Math. **274**. Springer, Berlin.
23. Martin-Löf, P. (1982). Constructive mathematics and computer programming. In: *Logic, Methodology, and Philosophy of Science VI* (eds J. J. Cohen, J. Loś, H. Pfeiffer, and K.-P. Podewksi), pp. 153–175. North-Holland, Amsterdam.
24. de Bruijn, N. G. (1980). A survey of the project AUTOMATH. In: *To H.B. Curry: Essays on Combinatory Logic, Lambda Calculus, and Formalism* (eds J. P. Seldin, and J. R. Hindley), pp. 589–606. Academic Press, New York.
25. Martin-Löf, P. (1984). *Intuitionistic Type Theory.* Notes by G. Sambin of a series of lectures given in Padua, June 1980. Bibliopolis, Napoli.
26. Nordström, B., Petersson, K., Smith, J. (1990). *Programming in Martin-Löf's Type Theory. An introduction.* Oxford University Press.
27. Friedman, H. (1973). The consistency of classical set theory relative to a set theory with intuitionistic logic. *J. Symbolic Logic*, **38**, 315–319.

28. Myhill, J. (1973). Some properties of intuitionistic Zermelo–Fraenkel set theory. In: *Cambridge Summer School in Mathematical Logic (1971)*. pp. 206–231. Lecture Notes in Math. **337**. Springer, Berlin.
29. Friedman, H. (1977). Set theoretic foundations for constructive analysis. *Ann. Math.*, **105**, 1–28.
30. Aczel, P., Rathjen, M. *Notes on Constructive Set Theory.* Preprint no. 40, Institut Mittag–Leffler, 2000/2001
31. Beeson, M. J. (1985). *Foundations of Constructive Mathematics.* Springer, Berlin and Heidelberg.
32. Martin-Löf, P., Sambin, G. (2003). Generating positivity by coinduction. Preprint no. 8, University of Padua.
33. Joyal, A., Moerdijk, I. (1995). *Algebraic Set Theory*. Cambridge University Press.
34. Bishop, E., Bridges, D. (1985). *Constructive Analysis.* Springer, Berlin and Heidelberg.
35. Bridges, D., Richman, F. (1987). *Varieties of Constructive Mathematics.* Cambridge University Press.
36. Mines, R., Ruitenburg, W., Richman, F. (1987). *A Course in Constructive Algebra.* Springer, New York.
37. Troelstra, A. S., van Dalen, D. (1988). *Constructivism in Mathematics. An Introduction.* Two volumes. North-Holland, Amsterdam.
38. Richman, F. (1981), ed., *Constructive Mathematics.* Lecture Notes in Math. **873**. Springer, Berlin and Heidelberg.
39. Troelstra, A. S., van Dalen, D. (1982). eds., *The L.E.J. Brouwer Centenary Symposium.* North-Holland, Amsterdam.
40. Curi, G. The points of (locally) compact regular formal topologies. [41, 39–54]
41. Schuster, P., Berger, U., Osswald, H. (2001). eds., *Reuniting the Antipodes. Constructive and Nonstandard Views of the Continuum.* Proceedings 1999 Venice Symposion. Synthese Library **306**. Kluwer, Dordrecht.
42. Richman, F. (2000). The fundamental theorem of algebra: a constructive development without choice. *Pacific J. Math.*, **196**, 213–230.
43. Bridges, D., Vîţă, L. (2003). Apartness spaces as a framework for constructive topology. *Ann. Pure Appl. Logic*, **119**, 61–83.
44. Veldman, W. (2005) The fan theorem as an axiom and as a contrast to Kleene's alternative. Report No. 0509, Department of Mathematics, Radboud University, Nijmegen.
45. Loeb, I. (2005). Equivalents of the (weak) fan theorem. *Ann. Pure Appl. Logic*, **132**, 51–66.
46. Berger, J. (2005). The fan theorem and uniform continuity. In: *Computability in Europe 2005. New Computational Paradigms* (eds. S. B. Cooper, B. Loewe, and L. Torenvliet), pp. 18–22. Proceedings, Amsterdam. Lecture Notes Computer Science **3526**. Springer, Berlin.
47. Bridges, D. (1995). Constructive mathematics and unbounded operators—a reply to Hellman, *J. Philosophical Logic*, **24**, 549–561.

48. Banaschewski, B., Mulvey, C. (2000). The spectral theory of commutative C*-algebras: the constructive Gelfand–Mazur theorem. *Quaest. Math.*, **23**, 465–488.
49. Cederquist, J. Coquand, T. Negri, S. The Hahn–Banach theorem in type theory. [67, 57–72]
50. Bridges, D. (1999). Constructive mathematics: a foundation for computable analysis. *Theoret. Comput. Sci.*, **219**, 95–109.
51. Sambin, G. (1987). Intuitionistic formal spaces—a first communication. In: *Mathematical Logic and its Applications.* (ed D. Skordev), pp. 187–204. Plenum, New York and London.
52. Sambin, G. (2003). Some points in formal topology. *Theoret. Comput. Sci.*, **305**, 347–408.
53. Johnstone, P. (1982). *Stone Spaces*. Cambridge University Press.
54. Vickers, S. (1989). *Topology via Logic.* Cambridge University Press.
55. Johnstone, P. (1983). The point of pointless topology. *Bull. Amer. Math. Soc. (N.S.)*, **8**, 41–53.
56. Banaschewski, B., Coquand, T., Sambin, G., eds., Second Workshop on Formal Topology. Venice, 2002. Proceedings. *Ann. Pure Appl. Logic* (special issue), **137** (2006), no. 1–3.
57. Waaldijk, F. On the foundations of constructive mathematics. Forthcoming.
58. Curi, G. (2003). *Geometry of Observations.* Doctorate Dissertation, University of Siena.
59. Aczel, P. Aspects of general topology in constructive set theory. [56]
60. Curi, G. (2003). Constructive metrisability in point–free topology. *Theoret. Comput. Sci.*, **305**, 85–109.
61. Curi, G. On the collection of points of a formal space. [56]
62. Palmgren, E. Maximal and partial points in formal spaces. [56]
63. Palmgren, E. Predicativity problems in point–free topology. In: *Proceedings of the Annual European Summer Meeting of the Association for Symbolic Logic*, Helsinki, Finland, 2003 (eds V. Stoltenberg–Hansen and J. Jäänänen), pp. 14–20, Lecture Notes in Logic 24, ASL (to appear)
64. Sambin, G., Gebellato, S. (1999). A preview of the basic picture: a new perspective on formal topology. In: *Types for Proofs and Programs.* Proc. Internat. Workshop TYPES '98, Kloster Irsee, Germany (eds Th. Altenkirch, W. Naraschewski, and B. Reus), pp. 194–207. Lecture Notes Computer Science **1657**. Springer, Berlin.
65. Bucalo, A., Rosolini, G. Completions, comonoids, and topological spaces. [56]
66. Coquand, T., Lombardi, H., Quitté, C. (2004). Generating non–Noetherian modules constructively. *Manuscripta Math.*, **115**, 513–520.
67. Sambin, G., Smith, J., eds., (1998). Twenty–five Years of Constructive Type Theory. Venice, 1995. Proceedings. Oxford University Press.
68. Richman, F. (1990). Intuitionism as generalization. *Philos. Math.*, **5**, 124–128.

PART I

Foundations

2

GENERALIZED INDUCTIVE DEFINITIONS IN CONSTRUCTIVE SET THEORY

MICHAEL RATHJEN

Abstract

The intent of this paper is to study generalized inductive definitions on the basis of Constructive Zermelo–Fraenkel set theory, **CZF**. In theories such as classical Zermelo–Fraenkel set theory, it can be shown that every inductive definition over a set gives rise to a least and a greatest fixed point, which are sets. This principle, notated **GID**, can also be deduced from **CZF** plus the full impredicative separation axiom or **CZF** augmented by the power set axiom. Full separation and a fortiori the power set axiom, however, are entirely unacceptable from a constructive point of view. It will be shown that while **CZF** + **GID** is stronger than **CZF**, the principle **GID** does not embody the strength of any of these axioms. **CZF** + **GID** can be interpreted in Feferman's Explicit Mathematics with a least fixed point principle. The proof-theoretic strength of the latter theory is expressible by means of a fragment of second order arithmetic.

2.1 Introduction

In set theory, a *monotone inductive definition* over a given set A is derived from a mapping

$$\Psi : \mathcal{P}(A) \to \mathcal{P}(A)$$

that is monotone, i.e., $\Psi(X) \subseteq \Psi(Y)$ whenever $X \subseteq Y \subseteq A$. Here $\mathcal{P}(A)$ denotes the class of all subsets of A. The set inductively defined by Ψ, Ψ^∞ is the smallest set Z such that $\Psi(Z) \subseteq Z$. Due to the monotonicity of Ψ such a set exists (on the basis of the axioms of **ZF** say).

But even if the operator is non-monotone it gives rise to a non-monotone inductive definition. The classical view is that the inductively defined set is obtained in stages by iteratively applying the corresponding operator to what has been generated at previous stages along the ordinals until no new objects are generated in this way. More precisely, if $\Upsilon : \mathcal{P}(A) \to \mathcal{P}(A)$ is an arbitrary mapping then the the set-theoretic definition of the set inductively defined by Υ is given by

$$\Upsilon^\infty := \bigcup_\alpha \Upsilon^\alpha,$$

$$\Upsilon^\alpha := \Upsilon\left(\bigcup_{\beta<\alpha} \Upsilon^\beta\right) \cup \bigcup_{\beta<\alpha} \Upsilon^\beta,$$

where α ranges over the ordinals.

Inductive definitions feature prominently in set theory, proof theory, constructivism, and computer science. The question of constructive justification of Spector's consistency proof for analysis prompted the study of formal theories featuring inductive definitions (cf. [1]). In the 1970s, proof-theoretic investigations (cf. [2]) focused on theories of iterated positive and accessibility inductive definitions with the result that their strength is the same regardless of whether intuitionistic or classical logic is being assumed.

The concept of an *inductive type* is also central to Martin-Löf's constructivism [3, 4]. Inductive types such as the types of natural numbers and lists, W-types and type universes are central to the expressiveness and mathematical strength of Martin-Löf type theory.

The objective of this paper is to study generalized inductive definitions on the basis of Constructive Zermelo–Fraenkel set theory, **CZF**, a framework closely related to Martin-Löf type theory. In theories such as classical Zermelo–Fraenkel set theory (**ZF**), it can be shown that every inductive definition over a set gives rise to a least and a greatest fixed point, which are sets. This principle, denoted **GID**, can also be deduced from **CZF** plus the full impredicative separation axiom or **CZF** augmented by the power set axiom. However, full separation and a fortiori the power set axiom are entirely unacceptable from a constructive point of view. It will be shown that while **CZF** + **GID** is stronger than **CZF**, the principle **GID** does not embody the strength of any of these axioms. A rough lower bound for the strength of **CZF** + **GID** is established by translating an intuitionistic μ-calculus into **CZF** + **GID**. An upper bound for the strength of this theory is obtained through an interpretation in Feferman's *Explicit Mathematics* with a least fixed point principle. The proof-theoretic strength of the latter theory is expressible by means of a fragment of second order arithmetic based on $\mathbf{\Pi}^1_2$ comprehension.

The paper is organized as follows. Section 2.2 shows that **CZF** provides a flexible framework for inductively defined classes and and reviews the basic results. Moreover, the general inductive definition principle is introduced therein. Section 2.3 is concerned with lower bounds while Section 2.4 is devoted to finding an upper bound.

2.2 Inductive Definitions in CZF

CZF provides an excellent framework for reasoning about inductive definitions. The next subsection will briefly review the language and axioms for **CZF**.

2.2.1 *The system* ***CZF***

The language of **CZF** is the same first order language as that of classical Zermelo–Fraenkel set theory, **ZF**, whose only non-logical symbol is $\in$. The logic of **CZF** is intuitionistic first order logic with equality. Among its non-logical axioms are *Extensionality*, *Pairing* and *Union* in their usual forms. **CZF** has additionally axiom schemata which we will now proceed to summarize.

Infinity: $\exists x \forall u \big[u \in x \leftrightarrow \big(\emptyset = u \vee \exists v \in x\ u = v+1\big)\big]$ where $v + 1 = v \cup \{v\}$.

Set Induction: $\forall x[\forall y \in x \phi(y) \rightarrow \phi(x)] \rightarrow \forall x \phi(x)$

Bounded Separation: $\forall a \exists b \forall x[x \in b \leftrightarrow x \in a \wedge \phi(x)]$

for all *bounded* formulae ϕ. A set-theoretic formula is *bounded* or *restricted* or Δ_0 if it is constructed from prime formulae using $\neg$, $\wedge$, $\vee$, $\rightarrow$, $\forall x \in y$ and $\exists x \in y$ only.

Strong Collection: For all formulae ϕ,

$$\forall a\big[\forall x \in a \exists y \phi(x, y) \ \rightarrow \ \exists b\, [\forall x \in a\, \exists y \in b\, \phi(x, y) \wedge \forall y \in b\, \exists x \in a\, \phi(x, y)]\big]$$

Subset Collection: For all formulae ψ,

$$\begin{aligned} &\forall a \forall b \exists c \forall u\, \big[\forall x \in a\, \exists y \in b\ \psi(x, y, u) \ \rightarrow \\ &\quad \exists d \in c\, [\forall x \in a\, \exists y \in d\, \psi(x, y, u) \wedge \forall y \in d\, \exists x \in a\, \psi(x, y, u)]\big]. \end{aligned}$$

Subset Collection can be expressed in a less obtuse way as a single axiom by using the notion of *fullness*.

Definition 2.1 *As per usual, we use $\langle x, y\rangle$ to denote the ordered pair of x and y. We use* $\mathbf{Fun}(g)$, $\mathbf{dom}(R)$, $\mathbf{ran}(R)$ *to convey that g is a function and to denote the domain and range of any relation R, respectively.*

For sets A, B let $A \times B$ be the cartesian product of A and B, that is the set of ordered pairs $\langle x, y\rangle$ with $x \in A$ and $y \in B$. Let ${}^{A}B$ be the class of all functions with domain A and with range contained in B. Let $\mathbf{mv}({}^{A}B)$ be the class of all sets $R \subseteq A \times B$ satisfying $\forall u \in A\, \exists v \in B\, \langle u, v\rangle \in R$. The expression $\mathbf{mv}({}^{A}B)$ should be read as the collection of multivalued functions from the set A to the set B. A set C is said to be full in $\mathbf{mv}({}^{A}B)$ if $C \subseteq \mathbf{mv}({}^{A}B)$ and

$$\forall R \in \mathbf{mv}({}^{A}B)\, \exists S \in C\ S \subseteq R.$$

Over the axioms of **CZF** with Subset Collection omitted, Subset Collection is equivalent to *Fullness*, that is to say the statement $\forall x \forall y \exists z$ *z is full in* $\mathbf{mv}({}^{x}y)$ (cf. [5]).

2.2.2 *Inductively defined classes in* **CZF**

Here we shall review some facts showing that **CZF** accommodates inductively defined classes. We begin with a general approach due to [6] which reflects most directly the generative feature of inductive definitions by viewing them as a collection of rules for generating mathematical objects.

Definition 2.2 *An inductive definition is a class of ordered pairs. If Φ is an inductive definition and $\langle x, a\rangle \in \Phi$ then we write*

$$\frac{x}{a}\,\Phi$$

and call $\frac{x}{a}\,\Phi$ an (inference) step of Φ, with set x of premisses and conclusion a. For any class Y, let

$$\Gamma_{\Phi}(Y) = \left\{a \mid \exists x \left(x \subseteq Y \ \wedge \ \frac{x}{a}\,\Phi\right)\right\}.$$

Thus $\Gamma_{\Phi}(Y)$ consists of all conclusions that can be deduced from a set of premisses comprised by Y using a single Φ-inference step. A class Y is Φ-closed if $\Gamma_{\Phi}(Y) \subseteq Y$.

Y is Φ-correct if $Y \subseteq \Gamma_\Phi(Y)$. Note that Γ_Φ is monotone; i.e. for classes Y_1, Y_2, whenever $Y_1 \subseteq Y_2$, then $\Gamma_\Phi(Y_1) \subseteq \Gamma_\Phi(Y_2)$.

We define the class inductively defined by Φ to be the smallest Φ-closed class, and denote it by $\mathbf{I}_*(\Phi)$. In other words, $\mathbf{I}_*(\Phi)$ is the class of Φ-theorems. Likewise, we define the class coinductively defined by Φ to be the greatest Φ-closed class, and denote it by $\mathbf{I}^*(\Phi)$. For precise definitions of $\mathbf{I}_*(\Phi)$ and $\mathbf{I}^*(\Phi)$ in the language of set theory we refer to the two main results about inductively and coinductively defined classes given below. They also state that these classes always exist.

An *ordinal* is a transitive set whose elements are transitive also. As per usual, we use variables $\alpha, \beta, \gamma, \ldots$ to range over ordinals.

Theorem 2.3 (**CZF**) *(Class Inductive Definition Theorem) For any inductive definition Φ there is a smallest Φ-closed class $\mathbf{I}_*(\Phi)$.*

Moreover, there is a class $J \subseteq \mathbf{ON} \times V$ such that

$$\mathbf{I}_*(\Phi) = \bigcup_\alpha J^\alpha,$$

and for each α,

$$J^\alpha \;=\; \Gamma_\Phi\left(\bigcup_{\beta\in\alpha} J^\beta\right).$$

J is uniquely determined by the above, and its stages J^α will be denoted by Γ^α_Φ.

Proof. [7], Section 4.2 or [8], Theorem 5.1. □

The next result uses the *Relativized Dependent Choices Axiom*, **RDC**. It asserts that for arbitrary formulae ϕ and ψ, whenever $\forall x\big[\phi(x) \to \exists y\big(\phi(y) \wedge \psi(x, y)\big)\big]$ and $\phi(b_0)$, then there exists a function f with $\mathbf{dom}(f) = \omega$ such that $f(0) = b_0$ and $(\forall n \in \omega)\big[\phi(f(n)) \wedge \psi(f(n), f(n+1))\big]$.

Theorem 2.4 (**CZF+RDC**) *(Class Coinductive Definition Theorem) For any inductive definition Φ there is a greatest Φ-closed class $\mathbf{I}^*(\Phi)$. Moreover, $\mathbf{I}^*(\Phi)$ can be characterized as the class of Φ-correct sets, i.e.,*

$$\mathbf{I}^*(\Phi) = \bigcup\{x \mid x \subseteq \Gamma_\Phi(x)\}.$$

Proof. [9], 6.5 or [10], 5.17. □

2.2.3 *Inductively defined sets in **CZF+REA***

Working in **CZF** alone, it is in general not possible to deduce that an inductively defined class actually constitutes a set. To be able to show that certain inductive definitions give rise to sets, Aczel proposed to add the *Regular Extension Axiom*, **REA**, to **CZF** (cf. [11]). **REA** is an axiom which is validated by the interpretation of set theory in Martin-Löf type theory, too. It is related to the W-type in type theory and can also be viewed as a "large" set axiom. In this subsection we present a body of results about so-called bounded inductive definitions which have sets as least fixed points providing one adopts **REA** or the slightly weaker **wREA**.

Definition 2.5 *A is inhabited if* $\exists x\, x \in A$. *An inhabited set A is regular if A is transitive, and for every* $a \in A$ *and set* $R \subseteq a \times A$ *if* $\forall x \in a\, \exists y\, (\langle x, y\rangle \in R)$, *then there is a set* $b \in A$ *such that* $\forall x \in a\, \exists y \in b\, (\langle x, y\rangle \in R) \ \wedge\ \forall y \in b\, \exists x \in a\, (\langle x, y\rangle \in R)$. *We write* $\mathbf{Reg}(C)$ *to express that C is regular.* **REA** *is the principle*

$$\forall x\, \exists y\, (x \subseteq y \ \wedge\ \mathbf{Reg}(y)).$$

For the purposes of inductive definitions, a weakened notion of regularity suffices. A transitive inhabited set C is weakly regular if for any $u \in C$ *and* $R \in \mathbf{mv}({}^{u}C)$ *there exists a set* $v \in C$ *such that* $\forall x \in u\, \exists y \in v\, \langle x, y\rangle \in R$. *We write* $\mathbf{wReg}(C)$ *to express that C is weakly regular. The Weak Regular Extension Axiom,* **wREA**, *is as follows: Every set is a subset of a weakly regular set.*

Definition 2.6 *We call an inductive definition* Φ *local if* $\Gamma_\Phi(X)$ *is a set for all sets X.*

We define a class B to be a bound for Φ *if whenever* $\frac{x}{a}\,\Phi$ *then x is an image of a set* $b \in B$; *i.e. there is a function from b onto x. We define* Φ *to be (regular, weakly regular) bounded if*

1. $\{y \mid \frac{x}{y}\,\Phi\}$ *is a set for all sets x,*
2. Φ *has a bound that is a (regular, weakly regular) set.*

Proposition 2.7 (**CZF**)

(i) Every bounded inductive definition Φ *is local; i.e.* $\Gamma_\Phi(X)$ *is a set for each set X.*

(ii) If Φ *is a weakly regular bounded inductive definition then* $\mathbf{I}_*(\Phi)$ *is a set.*

Proof. [8], 8.6, 8.7. □

Theorem 2.8 (**CZF** + **wREA**) *If* Φ *is a bounded inductive definition then* $\mathbf{I}_*(\Phi)$ *is a set.*

Proof. [11], 5.2. □

Definition 2.9 **(Examples)** *Let A be a class.*

1. $\mathbf{H}(A)$ *is the smallest class X such that for each set a that is an image of a set in A*

$$a \in \mathcal{P}(X) \Rightarrow a \in X.$$

Note that $\mathbf{H}(A) = \mathbf{I}_*(\Phi)$ *where* Φ *is the class of all pairs* $\langle a, a\rangle$ *such that a is an image of a set in A.*

2. *If R is a subclass of* $A \times A$ *such that* $R_a = \{x \mid xRa\}$ *is a set for each* $a \in A$ *then* $\mathbf{WF}(A, R)$ *is the smallest subclass X of A such that*

$$\forall a \in A\, [R_a \subseteq X \Rightarrow a \in X].$$

Note that $\mathbf{WF}(A, R) = \mathbf{I}_*(\Phi)$ *where* Φ *is the class of all pairs* $\langle R_a, a\rangle$ *such that* $a \in A$.

3. *If B_a is a set for each $a \in A$ then $\mathbf{W}_{a \in A} B_a$ is the smallest class X such that*

$$a \in A \ \wedge \ f : B_a \to X \ \Rightarrow \ \langle a, f \rangle \in X.$$

Note that $\mathbf{W}_{x \in A} B_a = \mathbf{I}_*(\Phi)$ where Φ is the class of all pairs $\langle \mathbf{ran}(f), \langle a, f \rangle \rangle$ such that $a \in A$ and $f : B_a \to V$.

Corollary 2.10 (**CZF** + **wREA**). *If A is a set then*

1. $\mathbf{H}(A)$ *is a set;*
2. *if $R \subseteq A \times A$ such that $R_a = \{x \mid xRa\}$ is a set for each $a \in A$ then $\mathbf{WF}(A, R)$ is a set;*
3. *if B_a is a set for each $a \in A$ then $\mathbf{W}_{a \in A} B_a$ is a set.*

Proof. These inductive definitions are bounded and thus give rise to sets by 2.8. □

2.2.4 *General inductive definitions*

Let Φ be an arbitrary inductive definition. What are the minimum requirements that Φ should satisfy if $\mathbf{I}_*(\Phi)$ and $\mathbf{I}^*(\Phi)$ are to be sets? It is surely expected that $\Gamma_\Phi(X)$ be a set for every set X; so Φ ought to be local. But locality is not enough as the following example shows: The power set inductive definition $Pow := \{\langle x, a \rangle \mid a \subseteq x\}$ is provably local in **ZF** but $\mathbf{I}_*(Pow)$ is a proper class (provably in **ZF**), namely the class of all sets V. The second requirement we shall adopt is that Φ be *conclusion bounded*, i.e., there is a set A such whenever $\frac{x}{y}\ \Phi$ then $y \in A$. Such a set will be called a *conclusion bound* for Φ.

Definition 2.11 *Let* **GID** *be the principle (scheme) asserting that if Φ is a local and conclusion bounded inductive definition then $\mathbf{I}_*(\Phi)$ and $\mathbf{I}^*(\Phi)$ are sets.*

Lemma 2.12 *(i)* **CZF** + *Full Separation* $\vdash$ **GID**.

(ii) **CZF** + **Pow** $\vdash$ **GID**, *where* **Pow** *stands for the Powerset Axiom.*

Proof. (i) is obvious by Theorems 2.3 and 2.4.

(ii): Let Φ be a local inductive definition with conclusion bound A. $\mathcal{P}(A)$ is a set by **Pow** and for every $X \subseteq A$, $\Gamma_\Phi(X)$ is a set. Hence, using Strong Collection there exists a function f with domain $\mathcal{P}(A)$ such that $f(X) = \Gamma_\Phi(X)$ for all $X \in \mathcal{P}(A)$. As a result, $\mathbf{I}_*(\Phi)$ and $\mathbf{I}^*(\Phi)$ are sets by Δ_0 Separation as $\mathbf{I}_*(\Phi) = \{u \in A \mid (\forall X \in \mathcal{P}(A))[f(X) \subseteq X \to u \in X]\}$ and $\mathbf{I}^*(\Phi) = \{u \in A \mid (\exists X \in \mathcal{P}(A))[X \subseteq f(X) \ \wedge \ u \in X]\}$. □

CZF + **Pow** is an extremely strong theory. It is stronger than classical nth order arithmetic for all n, since by means of ω many iterations of the power set operation (starting with ω) one can build a model of intuitionistic type theory within **CZF** + **Pow**. The Gödel–Gentzen negative translation can be extended so as to provide an interpretation of classical type theory with extensionality in intuitionistic type theory (cf. [12]). But more than that can be shown. Iterating the power set operation $\omega + \omega$ times one obtains the set $V_{\omega+\omega}$ which can be demonstrated to be a model of intuitionistic Zermelo set theory. The latter theory is of the same strength as classical Zermelo set theory (see [13], 2.3.1). Thus **CZF** + **Pow** is even stronger than classical Zermelo set theory. The situation with **CZF** + *Full Separation* is not as bad. The latter theory is actually of the same strength as full second order arithmetic. On the other hand, **CZF** is of modest proof-theoretic strength, namely of that of Kripke–Platek set theory or the theory of non-iterated inductive definitions. We will prove that **CZF** + **GID** is in strength related to a subsystem of second order arithmetic based on $\mathbf{\Pi}^1_2$ comprehension. Thus **CZF** + **GID** is considerably stronger than **CZF** but also has only a fraction of the strength of **CZF** + *Full Separation* and **CZF** + **Pow**.

The following gives an equivalent rendering of **GID**.

Definition 2.13 *The scheme* **MFP** *is defined as follows: Let* $\varphi(x, y)$ *be a formula of set theory and* A *be a set. If*

$$\forall x \subseteq A\, \exists! y\, [y \subseteq A \ \wedge\ \varphi(x, y)] \ \wedge \tag{2.1}$$

$$\forall x, x', y, y' \subseteq A \left[\varphi(x, y) \ \wedge\ \varphi(x', y') \ \wedge\ x \subseteq x' \ \Rightarrow\ y \subseteq y'\right], \tag{2.2}$$

then there exists sets $I_*, I^* \subseteq A$ *such that*

$$\begin{aligned} &\varphi(I_*, I_*) \ \wedge\ \forall x, y \subseteq A\, [\varphi(x, y) \ \wedge\ y \subseteq x \ \Rightarrow\ I_* \subseteq x] \ \wedge \\ &\varphi(I^*, I^*) \ \wedge\ \forall x, y \subseteq A\, [\varphi(x, y) \ \wedge\ x \subseteq y \ \Rightarrow\ x \subseteq I_*]. \end{aligned} \tag{2.3}$$

Proposition 2.14 (**CZF**) **GID** *and* **MFP** *are equivalent.*

Proof. First assume **GID** and suppose A is a set such that (2.1) and (2.2) hold. We specify an inductive definition Φ by

$$\Phi := \{\langle x, u\rangle \mid x \subseteq A \ \wedge\ \exists y \subseteq A\, [\varphi(x, y) \ \wedge\ u \in y]\}.$$

On account of (2.1) and (2.2), Φ is local. As Φ is also conclusion bounded by A, $\mathbf{I}_*(\Phi)$ and $\mathbf{I}^*(\Phi)$ are sets due to **GID**. Letting $I_* := \mathbf{I}_*(\Phi)$ and $I^* := \mathbf{I}^*(\Phi)$, one easily checks that (2.3) is satisfied.

Conversely, assume **MFP** and let Φ be a local inductive definition with conclusion bound A. Define $\varphi(x, y)$ by $y = \Gamma_\Phi(x)$. Then (2.1) follows from the locality of Φ and (2.2) is obvious by the definition of Γ_Φ. Hence we may apply **MFP** to conclude that there exists sets I_* and I^* such that $\Gamma_\Phi(I_*) = I_*$, $\Gamma_\Phi(I^*) = I^*$, $\forall x \subseteq A\, [\Gamma_\Phi(x) \subseteq x \Rightarrow I_* \subseteq x]$, and $\forall x \subseteq A\, [x \subseteq \Gamma_\Phi(x) \Rightarrow x \subseteq I^*]$. Consequently we have $I_* = \mathbf{I}_*(\Phi)$ and $I^* = \mathbf{I}^*(\Phi)$. □

2.3 Lower Bounds

To calibrate a first lower bound for the strength of **CZF** + **GID** we shall introduce some fairly recent results about an intuitionistic μ-calculus which is shown to be interpretable in **CZF** + **GID**.

2.3.1 *The μ-calculus*

The μ-calculus extends the concept of an inductive definition. It is basically an algebra of monotone functions over the power class of the domain of a first order structure (or over a complete lattice), whose basic constructors are first order definable operators, functional composition and least and greatest fixed point operators. The μ-calculus arose from numerous works of logicians and computer scientists. It originated with Scott and DeBakker [14] and was developed by Hitchcock and Park [15], Park [16], Kozen [17], Pratt [18], and others (see [19]). The μ-calculus is used in verification of computer programs and provides a toolbox for modelling a variety of phenomena, from finite automata to alternating automata on infinite trees and infinite games with finitely presentable winning conditions. Here we will be interested in the μ-calculus over the natural numbers. The μ-definable sets over the natural numbers were first described by Lubarsky [20]. He determined their complexity in the constructible hierarchy and showed that their ordinal ranks in that hierarchy can reach rather large countable ordinals. In the following we denote by $\mathbf{ACA}_0(\mathcal{L}^\mu)$ an axiomatic theory whose language is an extension of that of the classical μ-calculus over $\mathbb{N}$, $\mathcal{L}^\mu$ (see [20]), by set quantifiers and comprehension for first order properties. This version was axiomatized by Möllerfeld [21]. The letters 'ACA' stand for 'arithmetic comprehension axiom' and the subscript 0 indicates that the induction principle on natural numbers holds for sets rather than arbitrary classes. The name '$\mathbf{ACA}_0(\mathcal{L}^\mu)$' for this theory is somewhat misleading as its comprehension axioms allow for the formation of non-arithmetic sets. However, we will stick to this notation for 'historical' reasons.

Definition 2.15 *The language of* $\mathbf{ACA}_0(\mathcal{L}^\mu)$ *builds on the language of Peano arithmetic,* **PA**. *It has variables* $x, y, z, \ldots, X, Y, Z, \ldots$ *ranging over numbers and sets of numbers, respectively. The terms of* **PA** *will be referred to as number terms. Number terms, set terms and formulae of the language* $\mathcal{L}^\mu$ *are defined as follows.*

1. *The terms of* **PA** *are number terms of* $\mathcal{L}^\mu$.
2. *Set variables are set terms.*
3. $\perp$ *is a formula.*
4. *If* s *and* t *are number terms then* $s = t$ *is a formula.*
5. *If* s *is a number term and* S *is a set term then* $s \in S$ *is a formula.*
6. *If* φ_0 *and* φ_1 *are formulas then* $\varphi_0 \wedge \varphi_1$, $\varphi_0 \vee \varphi_1$ *and* $\varphi_0 \to \varphi_1$ *are formulas.*
7. *If* ψ *is a formula then* $\forall x \psi$ *and* $\exists x \psi$ *are formulas.*
8. *If* ψ *is a formula then* $\forall X \psi$ *and* $\exists X \psi$ *are formulas.*
9. *If* φ *is an* X*-positive first order formula then* $\mu x X.\varphi$ *is a set term.*

In the definition above we call a formula *first order* if it does not contain set quantifiers $\exists X, \forall X$. For X a set variable an expression $\mathfrak{E}$ is said to be *X-positive (X-negative)* if every occurrence of X in $\mathfrak{E}$ is positive (negative). In classical logic we can restrict ourselves to the connectives $\neg, \wedge, \vee$ and then X is positive in a formula φ if every occurrence of X in φ is in the scope of an even number of negations. But as we shall also be concerned with the intuitionistic μ-calculus, we define this notion inductively as follows:

(1) X is X*-positive*;

(2) Y is both X*-positive* and X*-negative* if Y is a set variable different from X;

(3) $\bot$ and $s = t$ are also both X*-positive* and X*-negative*;

(4) $s \in S$ is X*-positive (-negative)* iff S is;

(5) polarity does not change with $\wedge$, $\vee$, quantifiers and the μ-symbol;

(6) and, finally, $\varphi_0 \rightarrow \varphi_1$ is X*-positive (-negative)* iff φ_0 is X*-negative (-positive)* and φ_1 is X*-positive (-negative)*.

For set terms S, T, $S \subseteq T$ is the formula $\forall x(x \in S \rightarrow x \in T)$.

Definition 2.16 *The axioms of* $\mathbf{ACA}_0(\mathcal{L}^\mu)$ *are the following:*

1. *The axioms of* **PA**.
2. *(Induction)* $\forall X\,\big(0 \in X \;\wedge\; \forall u(u \in X \rightarrow u + 1 \in X) \;\rightarrow\; \forall u\, u \in X\big)$.
3. *(First order comprehension)* $\exists Z \forall x[x \in Z \leftrightarrow \varphi(x)]$ *for every first order formula* φ *in which the set variable* Z *does not appear free.*
4. *(Least fixed point axiom)*

$$\forall x[x \in P \leftrightarrow \varphi(x, P)] \;\wedge\; \forall Y\big[\forall x\big(\varphi(x, Y) \rightarrow x \in Y\big) \rightarrow P \subseteq Y\big] \qquad (2.4)$$

where P *stands for the set term* $\mu x X.\varphi$.

$\mathbf{ACA}_0(\mathcal{L}^\mu)$ *is based on classical logic. The system with the underlying logic changed to intuitionistic logic will be denoted by* $\mathbf{ACA}_0^i(\mathcal{L}^\mu)$.

The theories with the full induction scheme **IND** *will be denoted by* $\mathbf{ACA}(\mathcal{L}^\mu)$ *and* $\mathbf{ACA}^i(\mathcal{L}^\mu)$*, respectively.* **IND** *is the scheme*

$$\psi(0) \;\wedge\; \forall x[\psi(x) \rightarrow \psi(x+1)] \;\rightarrow\; \forall x \psi(x)$$

for all formulas ψ.

That X is positive (negative) in ψ will be denoted by $\psi(X^+)$ ($\psi(X^-)$). Positivity is a guarantor of monotonicity, while negativity guarantees antimonotonicity.

Lemma 2.17 *For every* X*-positive formulas* $\psi(X^+)$ *and and every* X*-negative formula* $\theta(X^-)$ *of* $\mathbf{ACA}_0(\mathcal{L}^\mu)$ *we have:*

(i) $\mathbf{ACA}_0^i(\mathcal{L}^\mu) \vdash \forall X \forall Y\,[X \subseteq Y \;\wedge\; \psi(X) \;\rightarrow\; \psi(Y)]$.

(ii) $\mathbf{ACA}_0^i(\mathcal{L}^\mu) \vdash \forall X \forall Y\,[X \subseteq Y \;\wedge\; \theta(Y) \;\rightarrow\; \theta(X)]$.

Proof. Use induction on the complexity of the formulas. □

At first blush, the μ-calculus appears to be innocent enough. Though a first order formula $\varphi(X^+, x)$ may contain complicated μ-terms, it might seem that these act solely as parameters and therefore one could obtain $\mu x X.\varphi(X^+, x)$ via an ordinary first order arithmetic inductive definition in these parameters, so that all the μ-definable sets would turn out to be sets recursive in finite iterations of the hyperjump. But this is far from being true. The μ-calculus allows for nestings of least fixed point operators. Better yet, there can be feedback. This provides the major difficulty in understanding the expressive power of $\mathcal{L}^\mu$. To illustrate the complexity of nested set terms in $\mathcal{L}^\mu$, let $\theta(X^+, Y^-, Z^+, W^-)$ be a first order formula of $\mathcal{L}^\mu$. Then the following are set terms: $\mu z Z.\theta$, $\mu y Y. w \notin \mu z Z.\theta$, $\mu x X.\mu y Y. w \notin \mu z Z.\theta$, $\mu w W.\mu x X.\mu y Y. w \notin \mu z Z.\theta$.

In the μ-calculus one can also define the *greatest fixed point* constructor ν: If $\varphi(X^+, x)$ is first order, $\nu x X.\varphi(X^+, x)$ is $\{u \mid u \notin \mu x X.\neg\varphi(\neg X, x)\}$. The appropriate measure for the complexity of μ-terms was determined by Lubarsky [20]. μ and ν can be viewed as higher order quantifiers giving rise to complexity classes $\mathbf{\Sigma}_n^\mu$ and $\mathbf{\Pi}_n^\mu$ of $\mathcal{L}^\mu$ formulas which measure the alternations of μ and ν.

The pivotal proof-theoretic connection between $\mathbf{ACA}_0(\mathcal{L}^\mu)$ and $\mathbf{ACA}_0^i(\mathcal{L}^\mu)$ was established by Tupailo.

Theorem 2.18 (Tupailo) $\mathbf{ACA}_0(\mathcal{L}^\mu)$ *can be interpreted in* $\mathbf{ACA}_0^i(\mathcal{L}^\mu)$ *via a double negation translation.*

Proof. [22]. □

2.3.2 *Fragments of second order arithmetic*

The proof-theoretic strength of theories is commonly calibrated using standard theories and their canonical fragments. In classical set theory this linear line of consistency strengths is couched in terms of large cardinal axioms while for weaker theories the line of reference systems traditionally consist in second order arithmetic and its fragments, owing to Hilbert's and Bernays' [23] observation that large chunks of mathematics can already be formalized in second order arithmetic.

Definition 2.19 *The language $\mathcal{L}_2$ of second order arithmetic contains number variables $x, y, z, u, \ldots$, set variables $X, Y, Z, U, \ldots$ (ranging over subsets of $\mathbb{N}$), the constant 0, function symbols Suc, $+$, $\cdot$, and relation symbols $=, <, \in$. Suc stands for the successor function. Terms are built up as usual. For $n \in \mathbb{N}$, let $\bar{n}$ be the canonical term denoting n. Formulae are built from the prime formulae $s = t$, $s < t$, and $s \in X$ using $\wedge, \vee, \neg, \forall x, \exists x, \forall X$ and $\exists X$ where s, t are terms. Note that equality in $\mathcal{L}_2$ is only a relation on numbers. However, equality of sets will be considered a defined notion, namely $X = Y$ if and only if $\forall x[x \in X \leftrightarrow x \in Y]$. As per usual, number quantifiers are called bounded if they occur in the context $\forall x(x < s \to \ldots)$ or $\exists x(x < s \wedge \ldots)$ for a term s which does not contain x. The $\mathbf{\Sigma}_0^0$-formulae are those formulae in which all quantifiers are bounded number quantifiers. For $k > 0$, $\mathbf{\Sigma}_k^0$-formulae are formulae of the form $\exists x_1 \forall x_2 \ldots Q x_k \phi$, where ϕ is $\mathbf{\Sigma}_0^0$; $\mathbf{\Pi}_k^0$-formulae are those of the form $\forall x_1 \exists x_2 \ldots Q x_k \phi$. The union of all $\mathbf{\Pi}_k^0$- and $\mathbf{\Sigma}_k^0$-formulae for all $k \in \mathbb{N}$ is the class of arithmetical or $\mathbf{\Pi}_\infty^0$-formulae. The $\mathbf{\Sigma}_k^1$-formulae ($\mathbf{\Pi}_k^1$-formulae) are the formulae $\exists X_1 \forall X_2 \ldots Q X_k \phi$ (resp. $\forall X_1 \exists X_2 \ldots Q x_k \phi$) for arithmetical ϕ.*

The basic axioms in all theories of second order arithmetic are the defining axioms of $0, 1, +, \cdot, <$ and the induction axiom

$$\forall X(0 \in X \wedge \forall x(x \in X \to x + 1 \in X) \to \forall x(x \in X)),$$

respectively, the scheme of induction

$$\mathbf{IND} \qquad \phi(0) \wedge \forall x(\phi(x) \rightarrow \phi(x+1)) \rightarrow \forall x \phi(x),$$

where ϕ *is an arbitrary* $\mathcal{L}_2$*-formula. We consider the axiom scheme of* $\mathcal{C}$*-comprehension for formula classes* $\mathcal{C}$ *which is given by*

$$\mathcal{C}\textbf{-CA} \qquad \exists X \forall u(u \in X \leftrightarrow \phi(u))$$

for all formulae $\phi \in \mathcal{C}$ *in which* X *does not occur.*

For each axiom scheme **Ax** *we denote by* (**Ax**) *the theory consisting of the basic arithmetical axioms, the scheme* $\mathbf{\Pi}^0_\infty$-**CA**, *the scheme of induction and the scheme* **Ax**. *If we replace the scheme of induction by the induction axiom, we denote the resulting theory by* $(\mathbf{Ax})_0$. *An example for these notations is the theory* $(\mathbf{\Pi}^1_1\textbf{-CA})$ *which contains the induction scheme, whereas* $(\mathbf{\Pi}^1_1\textbf{-CA})_0$ *only contains the induction axiom in addition to the comprehension scheme for* $\mathbf{\Pi}^1_1$*-formulae.*

In the framework of these theories one can introduce defined symbols for all primitive recursive functions. In particular, let $\langle , \rangle : \mathbb{N} \times \mathbb{N} \longrightarrow \mathbb{N}$ *be a primitive recursive and bijective pairing function. The* x^{th} *section of* U *is defined by* $U_x := \{y : \langle x, y\rangle \in U\}$. *Observe that a set* U *is uniquely determined by its sections on account of* $\langle , \rangle$*'s bijectivity. Any set* R *gives rise to a binary relation* $\prec_R$ *defined by* $y \prec_R x := \langle y, x\rangle \in R$. *Using the foregoing coding, we can formulate the schema of Bar induction*

$$\mathbf{BI} \qquad \forall X\big[\mathbf{WF}(\prec_X) \ \wedge \ \forall u\big(\forall v \prec_X u \phi(v) \rightarrow \phi(u)\big) \ \rightarrow \ \forall u \phi(u)\big]$$

for all formulae ϕ, *where* $\mathbf{WF}(\prec_X)$ *expresses that* $\prec_X$ *is well-founded, i.e.,* $\mathbf{WF}(\prec_X)$ *stands for the formula* $\forall Y \big[\forall u\big[(\forall v \prec_X u \ v \in Y) \rightarrow u \in Y\big] \rightarrow \forall u \ u \in Y\big]$.

The strength of $\mathbf{ACA}_0(\mathcal{L}^\mu)$ can be expressed by means of a fragment of second order arithmetic.

Theorem 2.20 (Möllerfeld) $\mathbf{ACA}_0(\mathcal{L}^\mu)$ *and* $(\mathbf{\Pi}^1_2\textbf{-CA})_0$ *have the same proof-theoretic strength. The theories prove the same* $\mathbf{\Pi}^1_1$*-sentences of second order arithmetic.*

Proof. [21], 10.6. □

2.3.3 *A first lower bound*

Theorem 2.21 *The theory* $\mathbf{ACA}^i(\mathcal{L}^\mu)$ *can be interpreted in* $\mathbf{CZF}+\mathbf{GID}$. *Specifically, if* θ *is a statement of second order arithmetic and* $\mathbf{ACA}^i(\mathcal{L}^\mu) \vdash \theta$ *then* $\mathbf{CZF}+\mathbf{GID} \vdash \theta$.

Proof. We will first embed $\mathbf{ACA}^i(\mathcal{L}^\mu)$ into a conservative extension of $\mathbf{CZF}+\mathbf{GID}$ with class terms. The set-theoretic language with class terms allows one to build a class term $\{u \mid \varphi(u)\}$ whenever φ is a formula of the (extended) language. Moreover, for every class term $\{u \mid \varphi(u)\}$ and variable x, $x \in \{u \mid \varphi(u)\}$ and $x = \{u \mid \varphi(u)\}$ are formulae. For class terms $\{u \mid \varphi(u)\}$ and $\{u \mid \psi(u)\}$, the expressions $\{u \mid \varphi(u)\} \in \{u \mid \psi(u)\}$ and $\{u \mid \varphi(u)\} = \{u \mid \psi(u)\}$ are considered to be abbreviations for

$\exists y[y = \{u \mid \varphi(u)\} \wedge y \in \{u \mid \psi(u)\}]$ and $\exists y[y = \{u \mid \varphi(u)\} \wedge y = \{u \mid \psi(u)\}]$, respectively. The extension of **CZF** + **GID** via class terms has the additional axioms

$$\forall z[z \in \{u \mid \varphi(u)\} \leftrightarrow \varphi(z)], \tag{2.5}$$

whereas the other axioms are just the axioms of **CZF** + **GID** in the original language without class terms. Formulae in the class language are easily translated back into the official language of set theory by using the direction '$\rightarrow$' of (2.5).

The translation * from the language of $\mathbf{ACA}^i(\mathcal{L}^\mu)$ into the language with class terms will be given next. For number terms s, t, $(s = t)^*$ is the usual translation of such formulae of **PA** into the set-theoretic language. For a set variable X let $X^* := X$ and for a μ-term $\mu x X.\varphi(X^+, x)$ let $\big(\mu x X.\varphi(X^+, x)\big)^*$ be the class term $\mathbf{I}_*(\Phi)$ (according to 2.3), where

$$\Phi := \{\langle X, x\rangle \mid \varphi^*(X, x) \wedge X \subseteq \omega \wedge x \in \omega\}.$$

The translation of the remaining set terms and formulae is as follows: $\bot^* := (0 = 1)^*$; $(\psi \Box \theta)^* := \psi^* \Box \theta^*$ if $\Box = \wedge, \vee, \rightarrow$; $(\forall x \psi)^* := (\forall x \in \omega)\psi^*$; $(\forall X \psi)^* := (\forall X \subseteq \omega)\psi^*$.

Next we aim at showing that for all formulae $\theta(\vec{X}, \vec{y})$ of $\mathbf{ACA}^i(\mathcal{L}^\mu)$ with all free variables exhibited (where $\vec{X} = X_1, \ldots, X_n$, $\vec{y} = y_1, \ldots, y_r$) we have:

$$\begin{aligned} &\text{If } \ \mathbf{ACA}_0^i(\mathcal{L}^\mu) \vdash \theta(\vec{X}, \vec{y}) \\ &\text{then } \mathbf{CZF} + \mathbf{GID} \vdash \vec{X} \subseteq \omega \wedge \vec{y} \in \omega \rightarrow \theta^*(\vec{X}, \vec{y}). \end{aligned} \tag{2.6}$$

Closer scrutiny reveals that the translation leaves positive (negative) occurrences positive (negative). Therefore it is easy to show that the *-translation of the fixed point axioms (2.4) are provable in **CZF** + **GID**. The only axioms requiring special considerations are the axioms for arithmetical comprehension. **CZF** has only Δ_0 Separation. But in general, the $*$-translations of a first order formula of $\mathbf{ACA}_0(\mathcal{L}^\mu)$ is not Δ_0. This is where **GID** and also Strong Collection (in the guise of Replacement) will be needed. By induction on the build-up of first order formulas $\theta(x)$ and μ-terms $\mu X x.\varphi(X^+, x)$, we show that $\{x \in \omega \mid \theta^*(x)\}$ and $\big(\mu x X.\varphi(X^+, x)\big)^*$ are sets. More formally, we shall demonstrate that

$$\mathbf{CZF} + \mathbf{GID} \vdash \vec{V} \subseteq \omega \wedge \vec{y} \in \omega \rightarrow \exists z\, z = \{x \mid \theta^*(x)\}, \tag{2.7}$$

$$\mathbf{CZF} + \mathbf{GID} \vdash \vec{V} \subseteq \omega \wedge \vec{y} \in \omega \rightarrow \exists u\, u = \big(\mu x X.\varphi(X^+, x)\big)^*, \tag{2.8}$$

where $\vec{V}, \vec{y}$ include all free variables of $\theta(x)$ and $\varphi(X^+, x)$ other than x, X. We shall now work in **CZF**+**GID**, assuming that $\vec{V} \subseteq \omega \wedge \vec{y} \in \omega$. Note that $\big(\mu x X.\varphi(X^+, x)\big)^*$ is the class term $\mathbf{I}_*(\Phi)$, where

$$\Phi = \{\langle X, x\rangle \mid \varphi^*(X, x) \wedge X \subseteq \omega \wedge x \in \omega\}.$$

By the meta-inductive assumption (regarding the complexity of μ-terms and formulae), we then have that $\{x \in \omega \mid \varphi^*(X, x)\}$ is a set for all sets $X \subseteq \omega$. Owing to the positivity of X in φ we get that

$$\Gamma_\Phi(X) = \{x \in \omega \mid \varphi^*(X, x)\},$$

showing that Φ is local. Since ω is a conclusion bound for Φ, we get that $\mathbf{I}_*(\Phi)$ is a set.

Now let $\theta(y)$ be first order. Then there are μ-terms $P_1, \ldots, P_r$ whose free number variables are among $\vec{y} = y_1, \ldots, y_k$, and a Δ_0 formula $\vartheta(x, u_1, \ldots, u_r)$ of set theory such that $\theta^*(x)$ is of the form $\vartheta(x, P_1^*, \ldots, P_r^*)$. Note that the number variables $\vec{y}$ may get captured by quantifiers in θ and then will also get quantified in ϑ. By the meta-inductive assumptions, $P_i^*(\vec{y}/\vec{n})$ is a set for all $\vec{n} \in \omega^k$. Thus, using Replacement there are functions $f_1, \ldots, f_r$ with domain ω^k such that $f_i(\vec{n}) = P_i^*(\vec{y}/\vec{n})$ for all $\vec{n} \in \omega^k$. If we now replace every subformula $u \in P_i^*$ of $\vartheta(x, P_1^*, \ldots, P_r^*)$ by $u \in f_i(\vec{y})$ we obtain a Δ_0 formula $\eta(x)$ such that $(\forall x \in \omega)[\eta(x) \leftrightarrow \theta^*(x)]$. Thus, as $\{x \in \omega \mid \eta(x)\}$ is a set by Δ_0 Separation, $\{x \in \omega \mid \theta^*(x)\}$ is a set, too.

As the $*$-translations of instances of **IND** are easily deduced in **CZF**, we have shown that all translations of axioms of $\mathbf{ACA}^i(\mathcal{L}^\mu)$ are provable in **CZF** + **GID**, so that (2.6) ensues. □

Since the theory $\mathbf{ACA}^i(\mathcal{L}^\mu)$ is stronger than $\mathbf{ACA}_0^i(\mathcal{L}^\mu)$ and the latter is of the same strength as $(\mathbf{\Pi}_2^1\text{-}\mathbf{CA})_0$ we get the following:

Corollary 2.22 **CZF** + **GID** *is stronger than* $(\mathbf{\Pi}_2^1\text{-}\mathbf{CA})_0$.

2.3.4 *Better lower bounds*

In view of Theorem 2.20 one might conjecture that $\mathbf{ACA}(\mathcal{L}^\mu)$ and $(\mathbf{\Pi}_2^1\text{-}\mathbf{CA})$ share the same strength. This is however not the case. As Lubarsky [20] showed, the nestings of the μ-terms provide the correct measure for the expressive power of formulas of the μ-calculus. $\mathbf{ACA}(\mathcal{L}^\mu)$ still only allows for finite nestings of the μ-operator while in $(\mathbf{\Pi}_2^1\text{-}\mathbf{CA})$ we can interpret transfinite nestings of such terms. It follows from [24] Corollary 3.16 and Lemma 4.9 that a monotone μ-calculus with transfinite nestings for all ordinals less than ε_0 can be embedded into $(\mathbf{\Pi}_2^1\text{-}\mathbf{CA})$. Conversely, using the techniques of [25] and [24] it can also be shown that $(\mathbf{\Pi}_2^1\text{-}\mathbf{CA})$ is proof-theoretically reducible to such a system. Details will be presented in [26]. Moreover, in [26] it will also be demonstrated that the intuitionistic and classical versions of these theories with $< \alpha$ iterated μ-terms, dubbed $\mathbf{ACA}(\mathcal{L}_{<\alpha}^{\mu,mon})$ and $\mathbf{ACA}^i(\mathcal{L}_{<\alpha}^{\mu,mon})$, respectively, are of the same strength. Here one allows ordinals α from an arbitrary primitive recursive ordinal representation system. As the theories $\mathbf{ACA}^i(\mathcal{L}_{<\alpha}^{\mu,mon})$ can be translated into **CZF**+**GID** as long as α is a provable ordinal of the latter theory, we get that **CZF**+**GID** is stronger than $(\mathbf{\Pi}_2^1\text{-}\mathbf{CA})$. We also get the following result.

Theorem 2.23 *If* $\prec$ *is a primitive recursive well-ordering such that* **CZF** + **GID** $\vdash$ $\mathbf{WF}(\prec)$ *then* **CZF** + **GID** *is at least as strong as* $(\mathbf{\Pi}_2^1\text{-}\mathbf{CA}) + \mathbf{TI}(\prec)$, *where* $\mathbf{TI}(\prec)$ *stands for the schema*

$$\forall x[(\forall u \prec x\, \varphi(u)) \rightarrow \varphi(x)] \rightarrow \forall x \varphi(x)$$

for all $\mathcal{L}_2$*-formulae* $\varphi(x)$.

Proof. This will follow from results in [26]. □

Examples of orderings $\prec$ for which **CZF**+**GID** $\vdash$ **WF**($\prec$) holds are well-orderings of length Γ_0, the proof-theoretic ordinal of ($\mathbf{\Pi}^1_1$-**CA**) and so forth.

2.4 An Upper Bound

How can we obtain an upper bound for the strength of **CZF** + **GID**? The usual proof of **GID** utilizes full separation or the outlandishly strong power-set axiom. As detailed before, **CZF** + *Full Separation* is reducible to second order arithmetic. But it turns out that a much more reasonable upper bound can be found.

Theorem 2.24 *The theory* **CZF**+**REA**+**GID** *can be reduced to* ($\mathbf{\Pi}^1_2$-**CA**)+**BI**. *Specifically, every* $\mathbf{\Pi}^0_2$ *statement of arithmetic provable in* **CZF** + **REA** + **GID** *is provable in* ($\mathbf{\Pi}^1_2$-**CA**) + **BI**.

Proof. The reduction is achieved in two steps. The first consists of an interpretation of **CZF** + **REA** + **GID** in Feferman's Explicit Mathematics augmented with a least fixed point operator, dubbed $\mathbf{T_0}$ + **UMID** + $\mathbb{V}$, by emulating the formulae-as-types interpretation of **CZF** in Martin-Löf type theory (cf. [5]). The second step is to reduce the latter theory to ($\mathbf{\Pi}^1_2$-**CA**) + **BI**. This is achieved by way of model constructions for explicit mathematics from [27] together with partial cut-elimination for systems of explicit mathematics combined with asymmetric interpretations controlled by a hierarchy of operators as introduced in [24]. The latter result will appear in [28].

Here we shall focuss on the first step. Due to page limitations for this paper we will have to be concise. We will mainly use the formalization of the system of explicit mathematics, $\mathbf{T}_0$, as presented in [29, 30], but for precise reference we will use the formalization given in [24], except that we call *types* what was called *classifications* in [24]. The language of $\mathbf{T_0}$, $\mathcal{L}_{T_0}$, is two-sorted, with individual variable $a, b, c, \ldots, x, y, z, \ldots$ and *type* variables $A, B, C, \ldots, X, Y, Z, \ldots$. Elementhood of an object a in a type X will be conveyed by $a : X$. $\mathbf{T_0}$ has a constant **j** (for join) for constructing the disjoint sum of a family of types indexed over a type. Also, $\mathbf{T_0}$ has constants that allow for the formation of types by elementary comprehension. In addition to the usual constants of $\mathbf{T_0}$, we will assume that $\mathcal{L}_{T_0}$ has a constant **lfp**. The principle that every monotone operation f on types has a least fixed point **lfp**(f), which is a type, will be denoted by **UMID**. Moreover, we will add a unary predicate $\mathbb{V}$ to the language which serves the purpose of providing a proper class of objects over which to interpret the quantifiers of **CZF**. $\mathbb{V}$ serves the same purpose as the large type of iterative sets in Aczel's interpretation. $\mathbb{V}$ is not allowed to occur in elementary formulae. There are two axiomatic principles associated with $\mathbb{V}$:

$$\forall X \forall f \left[\forall u : X\, \mathbb{V}(fu) \rightarrow \mathbb{V}(\langle X, f\rangle)\right], \tag{2.9}$$

$$\forall X \forall f \left[\mathbb{V}(\langle X, f\rangle) \wedge [\forall u : X\, \varphi(fu)] \rightarrow \varphi(\langle X, f\rangle)\right] \rightarrow \forall y\, [\mathbb{V}(y) \rightarrow \varphi(y)] \tag{2.10}$$

for all formulae φ, where $\langle x, y\rangle$ is $\mathbf{p}xy$ with $\mathbf{p}$ being the constant for the pairing operation of the applicative part of $\mathbf{T_0}$. Note also that $\mathbb{V}$ will not be the extension of a type.

The theory $\mathbf{T_0}+\mathbf{UMID}$ with the predicate $\mathbb{V}$ and the axioms (2.9) and (2.10) will be denoted by $\mathbf{T_0}+\mathbf{UMID}+\mathbb{V}$. It will also be shown in [28] that the latter system can be reduced to $(\mathbf{\Pi}^1_2\text{-}\mathbf{CA})+\mathbf{BI}$. We will use variables $\alpha, \beta, \gamma, \ldots$ to range over $\mathbb{V}$, that is to say over the objects x such that $\mathbb{V}(x)$. The induction principle (2.10) for $\mathbb{V}$ implies that every α is a pair $\langle X, f\rangle$ where X is a type and $\forall u : X\, \mathbb{V}(fx)$. We put $\bar{\alpha} := \mathbf{p_0}\alpha = X$ and $\tilde{\alpha} := \mathbf{p_1}\alpha = f$, where $\mathbf{p_0}, \mathbf{p_1}$ are the projection constants pertaining to $\mathbf{p}$.

Employing join and elementary comprehension, one can construct in $\mathbf{T_0}$ the types of *disjoint sum* $\sum_{x:A} B_x$ and *dependent product* $\prod_{x:A} B_x$ of a family of types $(B_x)_{x:A}$ indexed over a type A. With the aid of the recursion theorem of $\mathbf{T_0}$ one then proceeds to define type-valued operations $\dot{=}$ and $\dot{\in}$ on $\mathbb{V}\times\mathbb{V}$ via the equations

$$\dot{=}(\alpha,\beta) \simeq \prod_{x:\bar{\alpha}}\sum_{y:\bar{\beta}} \dot{=}(\tilde{\alpha}(x),\tilde{\beta}(y)) \times \prod_{y:\bar{\beta}}\sum_{x:\bar{\alpha}} \dot{=}(\tilde{\alpha}(x),\tilde{\beta}(y)),$$

$$\dot{\in}(\alpha,\beta) \simeq \sum_{y:\bar{\beta}} \dot{=}(\alpha,\tilde{\beta}(y)).$$

The totality of these operations on $\mathbb{V}\times\mathbb{V}$ is an immediate consequence of the induction principle (2.10).

We shall write $\alpha \mathbin{\dot{=}} \beta$ and $\alpha \mathbin{\dot{\in}} \beta$ for $\dot{=}(\alpha,\beta)$ and $\dot{\in}(\alpha,\beta)$, respectively. We are now in a position to assign to each formula $\theta(v_1,\ldots,v_n)$ of set theory (with all free variables among those shown) and $\alpha_1,\ldots,\alpha_n$ from $\mathbb{V}$ a class $[\![\theta(\alpha_1,\ldots,\alpha_n)]\!]$ of objects uniformly in $\vec{\alpha} := \alpha_1,\ldots,\alpha_n$. In doing so we shall exploit the use of class notation. Given a formula $\phi(x)$ of $\mathbf{T_0}$ we form the class $\{x \mid \phi(x)\}$. As per usual, given a class A, we use the expression $t \in A$ to convey that t is in the class A. Of course, formally classes do not exist in $\mathbf{T_0}$ and expressions involving them must be thought of as abbreviations for expressions not involving them. The definition of $[\![\theta(\alpha_1,\ldots,\alpha_n)]\!]$ proceeds along the construction of θ as follows:

$$
\begin{aligned}
[\![\alpha=\beta]\!] &:= \{u \mid u : (\alpha \mathbin{\dot{=}} \beta)\}\\
[\![\alpha\in\beta]\!] &:= \{u \mid u : (\alpha \mathbin{\dot{\in}} \beta)\}\\
[\![\bot]\!] &:= \emptyset \qquad \text{(the empty type)}\\
[\![\theta_1(\vec{\alpha}) \vee \theta_2(\vec{\alpha})]\!] &:= \{\langle 0,u\rangle \mid u \in [\![\theta_1(\vec{\alpha})]\!]\} \cup \{\langle 1,v\rangle \mid v \in [\![\theta_2(\vec{\alpha})]\!]\}\\
[\![\theta_1(\vec{\alpha}) \wedge \theta_2(\vec{\alpha})]\!] &:= \{\langle u,v\rangle \mid u \in [\![\theta_1(\vec{\alpha})]\!] \wedge v \in [\![\theta_2(\vec{\alpha})]\!]\}\\
[\![\theta_1(\vec{\alpha}) \to \theta_2(\vec{\alpha})]\!] &:= \{e \mid (\forall u \in [\![\theta_1(\vec{\alpha})]\!])(eu \in [\![\theta_2(\vec{\alpha})]\!])\}\\
[\![(\forall x \in \beta)\theta(\beta,\vec{\alpha})]\!] &:= \{e \mid (\forall i : \bar{\beta})(ei \in [\![\theta(\tilde{\beta}i,\vec{\alpha})]\!])\}\\
[\![(\exists x \in \beta)\theta(\beta,\vec{\alpha})]\!] &:= \{\langle i,u\rangle \mid i : \bar{\beta} \wedge u \in [\![\theta(\tilde{\beta}i,\vec{\alpha})]\!]\}\\
[\![(\forall x\theta(x,\vec{\alpha})]\!] &:= \{e \mid \forall\beta(e\beta \in [\![\theta(\beta,\vec{\alpha})]\!])\}\\
[\![(\exists x\theta(x,\vec{\alpha})]\!] &:= \{\langle\beta,u\rangle \mid u \in [\![\theta(\beta,\vec{\alpha})]\!]\}
\end{aligned}
$$

A pivotal property of the above interpretation is that for every Δ_0 formula $\theta(\vec{x})$ whose free variables are among $\vec{x} = x_1, \ldots, x_r$, there is a type-valued operation $\vec{\alpha} \mapsto A(\vec{\alpha})$ (total on $\mathbb{V}^r$) such that $\forall u[u \in [\![\theta(\vec{\alpha})]\!] \leftrightarrow u : A(\vec{\alpha})]$. Furthermore, using the constructions for embedding **CZF** + **REA** into type theory (cf. [11, 31]), one constructs for every formula $\theta(\vec{x})$ of set theory a closed application term t_θ such that

$$\mathbf{CZF} + \mathbf{REA} \vdash \theta(\vec{x}) \Rightarrow \mathbf{T_0} + \mathbb{V} \vdash \forall\vec{\alpha}\big(t_\theta\vec{\alpha} \in [\![\theta(\vec{\alpha})]\!]\big). \tag{2.11}$$

Upon nearer examination, the details for the proof of (2.11) are almost the same as in [31], Theorem 4.13 and Theorem 4.33.

In what follows, we shall write $e \Vdash \theta(\vec{\alpha})$ rather than $e \in [\![\theta(\vec{\alpha})]\!]$. We want to extend (2.11) to include **GID**. By Proposition 2.14 it suffices to construct for every instance $\theta(\vec{w})$ of **MFP** an application term t_θ such $t_\theta\vec{\alpha} \Vdash \theta(\vec{\alpha})$ holds for all $\vec{\alpha}$. So suppose

$$e \Vdash \forall x \subseteq \beta\, \exists! y\, [y \subseteq \beta \ \wedge \ \varphi(x, y)], \tag{2.12}$$

$$d \Vdash \forall x, x', y, y' \subseteq \beta \left[\varphi(x, y) \ \wedge \ \varphi(x', y') \ \wedge \ x \subseteq x' \ \Rightarrow \ y \subseteq y'\right]. \tag{2.13}$$

We define X to be a *subtype* of Y, denoted $X \mathrel{\overset{\circ}{\subseteq}} Y$, by $\forall u \,:\, X\, u \,:\, Y$. Let $B := \bar{\beta}$ and suppose $X \mathrel{\overset{\circ}{\subseteq}} B$. Then $\beta_X := \langle X, \tilde{\beta}\rangle$ is in $\mathbb{V}$ and there is a closed application term $\mathbf{t_s}$ (independent of X) such that $\mathbf{t_s} \Vdash \beta_X \subseteq \beta$. Hence, by (2.1) we get

$$e\, \beta_X \mathbf{t_s} \Vdash \exists! y\, [y \subseteq \beta \ \wedge \ \varphi(\beta_X, y)]. \tag{2.14}$$

We can further effectively construct closed application terms $\mathbf{t_0}, \mathbf{t_1}, \mathbf{t_2}$ such that from (2.14) we obtain that $\mathbf{t_1}\, e\, \beta_X \Vdash \delta_X \subseteq \beta$ and $\mathbf{t_2}\, e\, \beta_X \Vdash \varphi(\beta_X, \delta_X)$, where $\delta_X := \mathbf{t_0}\, e\, \beta_X$. Unravelling the meaning of $\mathbf{t_1}\, e\, \beta_X \Vdash \delta_X \subseteq \beta$, we find that there are closed application terms $\mathbf{q_0}, \mathbf{q_1}$ such that for all $i \,:\, \overline{\delta_X}$ we have $\mathbf{q_0}\, e\, \beta_X\, i \,:\, \bar{\beta}$ and $\mathbf{q_1}\, e\, \beta_X\, i \Vdash \widetilde{\delta_X}(i) = \tilde{\beta}(\nu(i))$, where $\nu(i) := \mathbf{q_0}\, e\, \beta_X\, i$. Using Elementary Comprehension, there exists a subtype C_X of B such that

$$\forall j\big(j \,:\, C_X \leftrightarrow \big[j \,:\, \bar{\beta} \ \wedge \ (\exists i \,:\, \overline{\delta_X})\exists z\, [z \Vdash \tilde{\beta}(j) = \tilde{\beta}(\nu(i))]\big]\big).$$

Moreover, C_X can be effectively obtained from X, β, e, that is to say, there exists a closed application term $\mathbf{r}$ such that $\mathbf{r}\beta\, e\, X \simeq C_X$. Put $f := \mathbf{r}\beta\, e$. If one now also takes (2.13) into account, one can ferret out that f is a monotone operation on subtypes of B. Whence, using **UMID**, $\mathbf{lfp}(f)$ is a subtypes of B which names the least fixed point of f. Similarly one can effectively obtain the greatest fixed point of f as in the classical μ-calculus. So there is another closed application term **gfp** such that $\mathbf{gfp}(f)$ is a type denoting the greatest fixed point of f. Finally we define $\beta_* := \langle \mathbf{lfp}(f), \tilde{\beta}\rangle$ and $\beta^* := \langle \mathbf{gfp}(f), \tilde{\beta}\rangle$. It remains to verify that we can effectively construct a closed application term ℓ such that $\ell\, e\, d\, \beta \Vdash \theta(\beta_*, \beta^*)$, where $\theta(\beta_*, \beta^*)$ is the formula of (2.3) with $I_* := \beta_*$ and $I^* := \beta^*$. This is tedious but straightforward. As there is no space left we leave that to the reader. □

The upshot of this paper is that **CZF** + **GID** is sandwiched between ($\mathbf{\Pi}^1_2$-**CA**) and ($\mathbf{\Pi}^1_2$-**CA**) + **BI**.

Corollary 2.25 *The proof-theoretic strength of* **CZF** + **GID** *is greater than that of* $(\boldsymbol{\Pi}^1_2\textbf{-CA})$ *but* **CZF** + **REA** + **GID** *is not stronger than* $(\boldsymbol{\Pi}^1_2\textbf{-CA}) + \mathbf{BI}$.

Acknowledgements

This material is based upon work supported by the National Science Foundation under Award No. DMS-0301162.

References

1. Kreisel, G. (1963). *Generalized inductive definitions,* In: Stanford Report on the Foundations of Analysis Section III Mimeographed, Stanford.
2. Buchholz, W., Feferman, S., Pohlers, W., and Sieg, W. (1981). *Iterated Inductive Definitions and Subsystems of Analysis*. Springer, Berlin.
3. Martin-Löf, P. (1975). An intuitionistic theory of types: predicative part. In: *Logic Colloquium '73*, (eds H. E. Rose and J. Sheperdson), pp. 73–118. North-Holland, Amsterdam.
4. Martin-Löf, P. (1984). *Intuitionistic Type Theory*. Bibliopolis, Naples.
5. Aczel, P. (1978). The type theoretic interpretation of constructive set theory. In: *Logic Colloquium '77*, (eds A. MacIntyre, L. Pacholski, and J. Paris), pp. 55–66. North-Holland, Amsterdam.
6. Aczel, P. (1977). An introduction to inductive definitions. In: *Handbook of Mathematical Logic*, (ed. J. Barwise), pp. 739–782. North-Holland, Amsterdam.
7. Aczel, P. (1982). The type theoretic interpretation of constructive set theory: choice principles. In: *The L. E. J. Brouwer Centenary Symposium*, (eds A. S. Troelstra and D. van Dalen), pp. 1–40. North-Holland, Amsterdam.
8. Aczel, M. and Rathjen P. (2001). *Notes on Constructive Set Theory*, Technical Report 40, Institut Mittag-Leffler (The Royal Swedish Academy of Sciences). *http://www.ml.kva.se/preprints/archive2000-2001.php*
9. Aczel, P. (1988). *Non-well-founded sets*. CSLI Lecture Notes 14 (CSLI Publications, Stanford).
10. Rathjen, M. (2003). The anti-foundation axiom in constructive set theories. In: *Games, Logic, and Constructive Sets*, CSLI Publications, Stanford. *Logic, Language and Computation 9th CSLI*, (eds G. Mints and R. Muskens).
11. Aczel, P. (1986). The type theoretic interpretation of constructive set theory: Inductive definitions. In: *Logic, Methodology and Philosophy of Science VII*, (eds R. B. et al. Marcus), pp. 17–49. North-Holland, Amsterdam.
12. Myhill, J. (1974). "Embedding classical type theory in intuitionistic type theory" a correction. Axiomatic set theory. Proceedings of Symposia in Pure Mathematics, Volume XIII, Part II (American Mathematical Society, Providence), pp. 185–188.
13. Gambino, N. (2000). *Types and sets: A study of the jump to full impredicativity*. Department of Computer Science, Manchester University 112 pages.
14. Scott, D. and DeBakker, J. W. (1969). *A theory of programs*. Unpublished manuscript. IBM, Vienna.

15. Hitchcock, P. and Park, D. M. R. (1973). *Induction rules and termination proofs*. Procceedings of the 1st International Colloquium on Automata, Languages, and Programming, pp. 225–251. North-Holland, Amsterdam.
16. Park, D. M. R. (1970). Fixpoint induction and proof of program semantics. In: *Machine Intelligence V*, (eds Meltzer, Mitchie), pp. 59–78. Edinburgh University Press, Edinburgh.
17. Kozen, D. (1983). Results on the propositional μ-calculus. *Theoretical Computer Science*, **27**, 333–354.
18. Pratt, V. (1981). A decidable μ-calculus. In: *Proceedings 22nd IEEE Symposium on Foundations of Computer Science*, pp. 421–427.
19. Arnold, D. and Niwinski, A. (2001). *Rudiments of the μ-calculus*. Elsevier, Amsterdam.
20. Lubarsky, B. (1993). μ-definable sets of integers. *Journal of Symbolic Logic*, **58**, 291–313.
21. Möllerfeld, M. (2002). Generalized inductive definitions. the μ-calculus and $\mathbf{\Pi}^1_2$-comprehension. Doctorial thesis, University of Münster.
22. Tupailo, S. (2004). On the intuitionistic strength of monotone inductive definitions. *Journal of Symbolic Logic*, **69**, 790–798.
23. Hilbert, D. and Bernays, P. (1938). *Grundlagen der Mathematik II*. Springer, Berlin.
24. Rathjen, M. (1998). Explicit mathematics with the monotone fixed point principle. *Journal of Symbolic Logic*, **63**, 509–542.
25. Rathjen, M. (1996). Monotone inductive definitions in explicit mathematics. *Journal of Symbolic Logic*, **61**, 125–146.
26. Rathjen, M. and Tupailo, S. *On the Strength of monotone fixed point principles in systems of intuitionistic explicit mathematics*. In preparation.
27. Rathjen, M. (1999). Explicit mathematics with the monotone fixed point principle. II: Models. *Journal of Symbolic Logic*, **64**, 517–550.
28. Rathjen, M. *Last words on explicit mathematics with the least fixed point principle*. In preparation.
29. Feferman, S. (1975). A language and axioms for explicit mathematics. In: *Algebra and Logic*, (ed J. N. Crossley), pp. 87–139. Lecture Notes in Math. **450**. Springer, Berlin.
30. Feferman, S. (1979). Constructive theories of functions and classes. In: *Logic Colloquium '78*, (eds M. Boffa, D. van Dalen, and K. McAloon), pp. 159–224. North-Holland, Amsterdam.
31. Rathjen, M. (2004). The formulae-as-classes interpretation of constructive set theory. To appear. In: *Proof Technology and Computation*, Proceedings of the International Summer School Marktoberdorf 2003. IOS Press, Amsterdam.
32. Barwise, J. (1975). *Admissible Sets and Structures*. Springer-Verlag, Berlin, Heidelberg, New York.
33. Beeson, M. (1985). *Foundations of Constructive Mathematics*. Springer-Verlag, Berlin, Heidelberg, New York, Tokyo.

3

CONSTRUCTIVE SET THEORIES AND THEIR CATEGORY-THEORETIC MODELS

ALEX SIMPSON

Abstract

We advocate a pragmatic approach to constructive set theory, using axioms based solely on set-theoretic principles that are directly relevant to (constructive) mathematical practice. Following this approach, we present theories ranging in power from weaker predicative theories to stronger impredicative ones. The theories we consider all have sound and complete classes of category-theoretic models, obtained by axiomatizing the structure of an ambient category of classes together with its subcategory of sets. In certain special cases, the categories of sets have independent characterizations in familiar category-theoretic terms, and one thereby obtains a rich source of naturally occurring mathematical models for (both predicative and impredicative) constructive set theories.

3.1 Introduction

3.1.1 *Constructive set theories*

The modern era of constructive mathematics began with the publication of Bishop's seminal book [1], in which he set out to redress 'the well-known scandal... that classical mathematics is deficient in numerical meaning.' In pursuit of this aim, Bishop undertook the ambitious programme of reformulating the core of mathematical analysis in such a way that all statements and proofs would have direct numerical content. To achieve this, Bishop adopted Brouwer's constructive interpretation of the logical primitives, a step which, in turn, demanded that proofs be carried out using the restricted principles of intuitionistic logic. In addition, Bishop banished from his mathematical ontology much of the machinery of abstract set theory. Instead, he insisted on a very concrete (and intensional) notion of set,[1] and he made use of only the simplest and most tangible constructions on sets (e.g. product, function space). Even quotient sets were apparently considered harmful.[2] Bishop's motivation for such parsimony was to make his mathematics as straightforward and realistic as possible, and few would argue against his outstanding success in this enterprise.

Nevertheless, experience in the decades since the publication of Bishop's book has taught us that there is, after all, no a priori conflict between the concepts of abstract

[1] 'A set is defined by describing exactly what must be done in order to construct an element of the set and what must be done in order to show that two elements are equal' [*ibid.* p. 6].

[2] 'The axiom of choice is [in classical mathematics] used to extract elements from equivalence classes where they should never have been put in the first place' [*ibid.* p. 9].

set theory and the numerical meaningfulness demanded by constructive mathematics. During this period, a wide range of set theories, based on comfortably familiar axioms, and variously labelled as *constructive* or *intuitionistic*, have been thoroughly investigated in the literature, see e.g. [2–6]. While not all such theories are accepted as truly constructive by every constructive mathematician, each can, via the definition of an appropriate realizability interpretation, be shown to meet the basic requirement of numerical meaningfulness. Furthermore, Aczel [4] has shown how even the most strident constructivist can be reconciled to those set theories that only contain axioms based on predicative principles, on account of the possibility of translating such set theories into Martin-Löf's type theory [7], a foundation whose constructive credentials are universally accepted.

The *formal* compatibility of set theory and constructive mathematics licenses the introduction of abstract set-theoretic methods into the practice of *informal* constructive mathematics. A first goal of this paper is to survey, in as straightforward a way as possible, the formal principles for reasoning about sets constructively that are actually relevant to the ways in which sets are used in ordinary informal constructive mathematics. This leads to a different emphasis from the usual presentations of constructive set theories. In standard axiomatizations, the contents of the axiomatized universe are rigidly delimited,[3] and theories are often strongly prescriptive about the nature of sets themselves.[4] In contrast, we leave our mathematical universe entirely open (it may contain whatever entities the constructive mathematician ever feels the need to construct), and we axiomatize the situation in which sets are included in the universe alongside everything else that may be there. Also, we leave our notion of set as unconstrained as possible, while remaining consistent with the ways in which sets are actually used in mathematical practice.

In Sections 3.3–3.5, we present axiomatizations of a number of set theories, motivated by the above desire to meet the pragmatic needs of constructive mathematicians who wish to use abstract set-theoretic methods in their mathematics. The theories range in power from weaker predicative theories to stronger theories incorporating full-blown impredicativism. Accordingly, not every constructivist will accept all the theories we discuss, but our aim is to survey the range of options available rather than to restrict attention to any one fixed interpretation of constructivism. From a formal point of view, the theories we consider are all intertranslatable with existing intuitionistic set theories studied in the literature. The novelty in our formulation is due entirely to taking seriously the pragmatic motivation, discussed above, of basing our theories solely on principles that are directly useful for the purpose of incorporating set-theoretic reasoning into the informal practice of constructive mathematics.

3.1.2 *Category-theoretic models*

The usual motivation for the activity of constructive mathematics is philosophical. Constructive mathematicians question the very meaningfulness of classical mathematics, and constructivization is pursued in order to restore the numerical meaning that is

---

[3]In Myhill's CST [2], every element of the universe is required to be a set, a function or a number; in Aczel's CZF [4], as in classical ZF, all elements are sets.

[4]E.g., in CZF (again as in classical ZF) sets are required to be well-founded.

lacking in classical maths. Depending upon the personal beliefs of the mathematician in question, any one of many different brands of constructivism may be adopted. For example, a sceptical constructive mathematician will probably accept only methods that are predicative in nature.

While such philosophical motivations are undoubtedly important (indeed crucial to those who are in possession of constructivist beliefs), one wonders what further rationale there is for the constructivization of mathematics beyond the purely solipsistic one of wishing to achieve personal reassurance over the justifiability of mathematical methods. To address this, we consider two questions in the style of Kreisel.

1. What are the positive mathematical consequences that arise as a result of developing mathematics constructively as opposed to classically?
2. What additional benefits are obtained if one develops constructive mathematics using predicative methods only?

The first question has an obvious answer. Constructive mathematics is a form of mathematics that is computationally meaningful in its entirety. In particular, a constructive proof of a statement about numbers (whether integer, real or complex) provides direct algorithmic evidence for the statement. Thus constructive mathematics can be usefully performed, even by a classical mathematician, as a means for extracting computational information from proofs.

The second question may also be answered along similar lines. Predicative methods are (significantly) weaker in proof-theoretic strength than impredicative ones. Thus the use of predicative methods guarantees, for example, that any number-theoretic function extracted from a proof will grow slowly in comparison with the wildly inflationary functions definable using impredicative methods. To some, the existence of such bounds on growth may provide an adequate non-philosophical reason for considering predicative constructive mathematics as being of interest. However, as the basic systems of predicative constructive mathematics are at least as rich as intuitionistic first-order arithmetic, the bounds they guarantee are hopelessly infeasible from a computational point of view. In the author's view, the existence of such infeasible bounds does not provide a compelling reason for preferring predicative brands of constructivism to impredicative ones.

There is, nonetheless, a second, entirely different, approach to answering the questions above. Various constructive (or intuitionistic) formal systems arise as *internal languages* of natural notions of category with structure. For example, (extensional) Martin-Löf type theory corresponds to the internal language of locally cartesian closed categories [8], and higher-order intuitionistic type theory is the the internal language of elementary toposes [9]. In the special case of Grothendieck toposes, which first arose in algebraic geometry, one can even interpret full intuitionistic Zermelo–Fraenkel set theory, see e.g. [10]. Locally cartesian closed categories, elementary toposes and Grothendieck toposes are ubiquitous in mathematics. Thus the intuitionistic theories mentioned above are given a non-ideological *raison d'être* through their relationship with such natural kinds of category.

The second goal of this paper is to demonstrate that there is analogous non-ideological motivation for the various constructive set theories considered in Sections 3.3–3.5. To address this, we present, in Section 3.6, a general category-theoretic framework for modelling the full range of different constructive set theories. The framework is based on the *algebraic set theory* of [11], as further developed in [12–14]. Within this framework, each of the constructive set theories axiomatized in Sections 3.3–3.5 has a corresponding notion of model defined by a reasonable collection of category-theoretic axioms.

It is a standard view that the use of weaker constructive set theories, such as predicative theories, represents a minimalist approach to mathematics, in the sense that it is based on reducing the assumptions on which mathematics is based to a minimal core. The consideration of category-theoretic models presents the same practice in an alternative 'maximalist' light. It is a triviality that the weaker a set theory the more inclusive its associated class of category-theoretic models. Thus, from a non-ideological viewpoint, the use of predicative constructive methods can be justified on the grounds of maximalizing the class of available models within which mathematical arguments can be interpreted. In the author's view, this provides a quite acceptable answer to the second question above.

We have argued that constructive set theories can be justified as being of non-ideological mathematical interest because they have associated classes of category-theoretic models. The force of this argument is strengthened considerably in cases in which one can show that the category-theoretic models include familiar mathematical structures that occur in ordinary mathematics. In Section 3.7, we state results in this direction, taken from [13, 14]. These results characterize, in familiar category-theoretic terms, the possible categories of sets associated with certain different constructive set theories. In particular, for the theory we call BCST + Pow, an impredicative constructive set theory, the associated categories of sets are exactly the elementary toposes (with natural numbers object). Similarly, for the predicative theory BCST + Exp, the associated categories of sets are exactly the locally cartesian closed (lcc) pretoposes (again with natural numbers object). Lcc pretoposes generalize elementary toposes, and there are naturally occurring mathematical examples of lcc pretoposes that are not toposes. It follows that there exist naturally occurring mathematical models of the predicative set theory BCST + Exp that do not model the impredicative powerset axiom. The possibility of using constructive set-theoretic methods for reasoning within such naturally occurring models provides the desired non-ideological justification for the use of predicative constructive methods. It would be most interesting to see a convincing mathematical application of the soundness of these methods within such models.

3.2 Sets in Constructive Mathematics

As traditionally practised, the process of constructive mathematics involves the construction of mathematical objects and, of course, the proof of properties of these mathematical objects. The mathematical universe should thus be viewed as an open-ended collection of entities, to which one, from time to time, adds new objects, as and when such objects are created.[5]

[5] It should be remarked that this view of an open-ended universe may also be seen as consistent with the practice of classical mathematics.

In addition to working with mathematical objects individually, it is frequently useful to consider collections of such objects, that is subcollections of the universe, each determined by a given defining property. We shall refer to such collections as *classes*. Of course, the membership of a class is not fixed, once and for all by its defining property, but rather has to be understood within the context of an open-ended universe (note, for example, that the entire universe is itself a class). The use of quantification over such open-ended classes is common mathematical practice. For example, one may perfectly well state and prove a property of all groups, without rigidly fixing the membership of the class of all groups.

Sets are a central ingredient in the definition of many kinds of abstract mathematical structure (for example, group). Indeed, the inclusion of sets in the mathematical universe is indispensable for the pursuit of abstract constructive mathematics. Our aim is to isolate pragmatic principles governing the way in which sets can be usefully included in the universe.

Sets can be motivated in many ways; most commonly via the *iterative conception of set* [15], which, if adopted, leads to (a constructive version of) von Neumann's cumulative hierarchy as the universe of sets. This familiar notion of set does not, however, correspond to how sets are used in mathematical practice. According to the iterative conception, sets are given by well-founded membership trees in which each node is itself a set, whereas, in practice, sets are merely distinguished collections of mathematical objects. It need not be the case that every element of a set is itself a set. Furthermore, the presumption that membership trees are necessarily well-founded is never used in ordinary mathematics.[6]

We suggest that the use of sets, as actually occurs in mathematical practice, is governed by just three basic laws.

1. A set is a distinguished collection of mathematical objects, itself rendered as a mathematical object.
2. A set is uniquely determined by its collection of members.
3. The class of sets is closed under a basic range of useful operations on collections.

The first law above is naturally considered alongside a mathematical universe that may already contain preconstructed mathematical objects that are not themselves sets (for example, natural numbers), and to which other newly constructed non-sets[7] may possibly be added in future. Of course, the law is also consistent with having a mathematical universe in which sets are the only type of mathematical object in existence; but the assumption that all objects are sets has no basis in mathematical practice.

The second law is simply the axiom of extensionality for sets. Extensionality is an essential component of the concept of set as it is actually used in mathematics.

[6]Indeed, for some applications, it is convenient to explicitly contravene this assumption and consider non-well-founded sets [16].

[7]We prefer the term 'non-set' to 'atom' or 'urelement', as, according to our view of constructive mathematics, there is nothing to preclude non-sets from being composite (and hence non-atomic) structures.

The third law is obviously wide open to interpretation. In the sequel, we make the third law precise in a number of different ways, leading to several set theories of differing strength. Not all these set theories will be acceptable to every constructive mathematician. Indeed, it is in the choice of interpretation of the third law that the constructive mathematician nails his or her individual beliefs to the door.

3.3 Axioms for Constructive Set Theory

The goal of this section is to present axioms for working with sets in constructive mathematics. These axioms are motivated by the three laws of sets that we identified above. The formal setting for the axiomatization is intuitionistic first-order logic with equality, where variables $x, y, z, \ldots$, and hence quantifiers, are to be understood as ranging over the entire (open-ended) universe.

In order to emphasize the pragmatic side, we shall, as far as possible, present the axioms in the informal style in which they would actually be used when doing constructive mathematics. To give a formal basis to such informal formulations, it is convenient to introduce notation for classes.

We write $\phi[x]$ for a formula ϕ with the free variable x distinguished; and having distinguished x in ϕ in this way, we write $\phi[y]$ for $\phi[y/x]$. Note that ϕ may contain free variables other than x; also x is not required to actually occur in ϕ. A *class* is defined by a formula $\phi[x]$, representing the class of all x satisfying ϕ, for which we use the notation:

$$\{x \mid \phi\}.$$

We use $X, Y, \ldots$ as meta-variables for classes. If X is the class defined by $\phi[x]$, then we write $y \in X$ to mean $\phi[y]$. Given a class X, a formula $\phi[x, y]$ (with two distinguished free variables x, y) can be seen as defining an X-indexed family of classes $\{Y_x\}_{x \in X}$, where Y_x is the class $\{y \mid \phi\}$, for any $x \in X$. The union of such a family of classes is defined as the class:

$$\bigcup_{x \in X} Y_x \;=\; \{y \mid \exists x .\, x \in X \wedge \phi\}.$$

To incorporate sets into the theory, we include two predicates in the language: a unary predicate $\mathcal{S}(x)$, stating that x is a set; and a binary predicate $x \in y$, stating that x is a member of the set y.[8] These linguistic constructs reflect the first law of sets in Section 3.2. The natural reading of membership above, in which y is forced to be a set, is imposed by an axiom:[9]

Membership If $x \in y$ then y is a set.

The second law of sets in Section 3.2 is implemented by the expected extensionality axiom, formulated in such a way as to be appropriate for a universe possibly containing non-sets.

[8] Note that we are using $\in$ as a predicate in $x \in y$ and as meta-notation in $x \in X$. This will cause no problems. The only situation in which there is any ambiguity is when the class X is a set, in which case the two interpretations coincide.

[9] As is usual, all axioms are implicitly universally quantified over their free variables.

Extensionality If A, B are sets and, for all z, $z \in A$ iff $z \in B$ then $A = B$.

Here, and henceforth, we use $A, B, \ldots$ as variables specifically ranging over the class $\{x \mid \mathcal{S}(x)\}$ of all sets.[10]

The third law of sets from Section 3.2 is addressed by the four (in the first instance) axioms below, which state basic closure properties of the class of sets. These closure properties are all intuitively reasonable if one takes the notion of set to be given by the following informal conception: a *set* is a class that is *absolute* in the sense that its membership is fixed once and for all, independently of any future extension that may be made to the universe. We call this idea the *absolute conception of set.*

Emptyset The empty class $\{x \mid \bot\}$ is a set. We write $\emptyset$ for this set.

Pairing The class $\{x \mid x = y \vee x = z\}$ is a set. We write $\{y, z\}$ for this set.

Equality The class $\{x \mid x = y \wedge x = z\}$ is a set. We write δ_{yz} for this set (the Kronecker delta notation seems appropriate here).

Indexed Union If A is a set and $\{B_x\}_{x \in A}$ is an A-indexed family of sets then $\bigcup_{x \in A} B_x$ is a set.

Although the informal readings of these axioms are clear, it is worth spelling out their formal incarnations. To this end, it is useful to define a quantifier $\mathcal{Z}x.\,\phi$ to be read as 'the class $\{x \mid \phi\}$ is a set' or as 'there are set-many x satisfying ϕ'. Formally, $\mathcal{Z}x.\,\phi$, abbreviates:

$$\exists A.\, \mathcal{S}(A) \;\wedge\; \forall x.\, (x \in A \;\leftrightarrow\; \phi),$$

where A does not occur free in ϕ. Note that the variable x is bound in the statement '$\{x \mid \phi\}$ is a set', thus, for example, the implicit universal quantification in the Pairing and Equality axioms is just over y and z. The formal versions of the Emptyset, Pairing and Equality axioms are now obvious. The Indexed Union axiom expands to[11]

$$\mathcal{S}(A) \;\wedge\; (\forall x \in A.\, \mathcal{Z}y.\,\phi) \;\rightarrow\; \mathcal{Z}y.\, \exists x \in A.\, \phi.$$

In the presence of the other axioms, Indexed Union is equivalent to a combination of the two axioms below, which, although standard, are less mathematically appealing.

Union $(\mathcal{S}(A) \wedge \forall x \in A.\, \mathcal{S}(x)) \;\rightarrow\; \mathcal{Z}y.\, \exists x \in A.\, y \in x.$

Replacement $\mathcal{S}(A) \wedge (\forall x \in A.\, \exists! y.\, \phi) \;\rightarrow$
$$\exists B.\, \mathcal{S}(B) \;\wedge\; (\forall x \in A.\, \exists y \in B.\, \phi) \;\wedge\; (\forall y \in B.\, \exists x \in A.\, \phi),$$

where $\exists!$ is the 'there exists a unique' quantifier. The (straightforward) proof of this equivalence can be found in [6], where, in its recognition, the Indexed Union axiom is called Union-Replacement.

We call the set theory axiomatized above BCST$^-$, which stands for Basic Constructive Set Theory without Infinity.[12] This will be the base theory upon which all our

[10] $A, B, \ldots$ are first-order variables, not meta-variables. Nevertheless, we shall also allow set variables $A, B, \ldots$ to be treated as classes in the obvious way.

[11] The relative quantifiers $\forall x \in A$ and $\exists x \in A$ have the obvious meanings.

[12] This name was suggested by Michael Warren.

other theories will be built. In spite of its apparent weakness, a surprising number of set-theoretic constructions can be squeezed out of the axioms of BCST^-, as we now proceed to show.

The most conspicuous omission from BCST^- is any form of the axiom of separation. Given a set A and formula $\phi[x]$, one can form the subclass $\{x \in A \mid \phi\}$ of A. An *instance of separation* is the assertion, for a particular $\phi[x]$ and set A, that the class $\{x \in A \mid \phi\}$ is itself a set. We shall see that a useful range of instances of separation are derivable in BCST^-.

To prepare for what follows, observe that propositions ϕ are in correspondence with subclasses of the singleton set $\{\emptyset\}$. On the one hand, any proposition ϕ determines the subclass $\{w \in \{\emptyset\} \mid \phi\}$, where w is a variable that does not occur free in ϕ. Conversely, any subclass X of $\{\emptyset\}$ determines the proposition $\emptyset \in X$. We say that the proposition ϕ is *restricted* if the class $\{w \in \{\emptyset\} \mid \phi\}$ is a set, and we write $!\phi$ for the statement that ϕ is restricted.[13] The notion of restricted formula entirely determines the valid instances of separation.

Proposition 3.1 *In* BCST^-*, the following are equivalent, for any set A and property* $\phi[x]$*:*

1. *The class* $\{x \in A \mid \phi\}$ *is a set.*
2. *For all* $x \in A$ *it holds that* $!\phi$.

To help familiarize the reader with the kind of reasoning needed to reduce ordinary set-theoretic constructions to the axioms of BCST^-, we give, as the only proof included in the paper, the short proof of this proposition.

Proof. First we observe that sets in BCST^- are closed under binary intersection. This follows from the Indexed Union and Equality axioms because $A \cap B = \bigcup_{x \in A} \bigcup_{y \in B} \delta_{xy}$.

Now assume that $\{x \in A \mid \phi[x]\}$ is a set. Then, for any $x \in A$, the class $\{x\} \cap \{x \in A \mid \phi[x]\}$ is a set, i.e. $\{w \in \{x\} \mid \phi[x]\}$ is a set. By Replacement, the class $\{w \in \{\emptyset\} \mid \phi[x]\}$ is also a set, i.e. $!\phi$.

Conversely, suppose that $!\phi$, for all $x \in A$. Then, by Replacement, for every $x \in A$, it holds that $\{w \in \{x\} \mid \phi\}$ is a set. By Indexed Union, $\bigcup_{x \in A}\{w \in \{x\} \mid \phi[x]\}$ is a set, i.e. $\{x \in A \mid \phi[x]\}$ is a set, as required. □

The proposition below establishes useful closure conditions on restricted properties, and hence, by the result above, yields rules for deriving valid instances of separation in BCST^-.

Proposition 3.2 (cf. [6]) *The following all hold in* BCST^-.

1. $!(x = y)$.
2. *If* $S(A)$ *then* $!(x \in A)$.
3. *If* $!\phi$ *and* $!\psi$ *then* $!(\phi \wedge \psi)$, $!(\phi \vee \psi)$, $!(\phi \rightarrow \psi)$ *and* $!(\neg\phi)$.
4. *If* $S(A)$ *and* $\forall x \in A.\, !\phi$ *then* $!(\forall x \in A.\, \phi)$ *and* $!(\exists x \in A.\, \phi)$.
5. *If* $\phi \vee \neg\phi$ *then* $!\phi$.

[13] Formally, $!\phi$ is thus an abbreviation for $\text{Ǝ}w.\ w = \emptyset \wedge \phi$, where w is not free in ϕ.

Here, statement 5 generalizes the standard fact that, in classical set theory, full separation follows from Replacement. Using intuitionistic logic, full separation is not a consequence of Replacement, but separation does remain derivable for so-called decidable formulas (those formulas ϕ for which $\phi \vee \neg\phi$ holds).

Statements 1–4 above can be used to build up restricted formulas from atomic formulas $x = y$ and $x \in A$ (assuming A is a set), by closing under the propositional connectives and under quantification in which quantifiers are bounded by sets. Note that an atomic formula $\mathcal{S}(x)$ need not be restricted. One could remedy this by adding a further axiom to BCST^- (see [13]), but such an axiom makes a somewhat arbitrary assumption about sets and does not seem useful for the purpose of carrying out constructive mathematics.

What is indisputably useful for doing mathematics is the construction of various standard derived sets and classes. We next consider to what extent BCST^- is able to cope with the most basic of these: cartesian products, exponentials (i.e. function spaces), powersets and quotient sets.

Given classes X, Y, a product class $X \times Y$ can be defined by

$$X \times Y \;=\; \{p \mid \exists x \in X.\, \exists y \in Y.\, p = (x, y)\},$$

where (x, y) can be taken to be the usual Kuratowski pairing $(x, y) = \{\{x\}, \{x, y\}\}$ (see [6] for a proof that this is a sensible intuitionistic notion of pairing).

Proposition 3.3 *In* BCST^-*, if* A, B *are sets then so is* $A \times B$.

Given a set A and a class X, we can define a class X^A of all functions from A to X by

$$X^A \;=\; \{f \mid \mathcal{S}(f) \;\wedge\; (\forall z \in f.\, z \in A \times X) \;\wedge\; (\forall x \in A.\, \exists! y.\, (x, y) \in f)\}.$$

The assumption that A is a set is important because, although the above definition also makes sense if A is a class, by Replacement, X^A can be inhabited only when A is a set. BCST^- is too weak to prove that B^A is a set, for every set B.

Given a class X the 'power*class*' $\mathcal{P}(X)$ of all sub*sets* of X is defined by

$$\mathcal{P}(X) \;=\; \{A \mid \mathcal{S}(A) \;\wedge\; \forall x \in A.\, x \in X\}.$$

In BCST^-, it cannot be shown that $\mathcal{P}(A)$ is a set, for every set A.

Finally, we turn to quotients. Suppose X is a set and $\sim$ is a subclass of $X \times X$ determining an equivalence relation on X. Under the condition that $\sim$ is a set, i.e. that $\sim\, \in \mathcal{P}(X \times X)$, we can form the quotient class

$$X/{\sim} \;=\; \{B \in \mathcal{P}(X) \mid \exists x.\, x \in B \;\wedge\; \forall y \in X.\, x \sim y \;\leftrightarrow\; y \in B\}.$$

Proposition 3.4 *In* BCST^-*, if* A *is a set and* $\sim\, \in \mathcal{P}(A \times A)$ *is an equivalence relation then* $A/{\sim}$ *is a set.*

The development above highlights both strengths and weaknesses of BCST^-. To address the weaknesses, we consider the following list of further axioms (and schema) that will be used to extend BCST^-, in order to obtain stronger theories.

Exponentiation (Exp) If A, B are sets then so is B^A.
Powerset (Pow) If A is a set then so is $\mathcal{P}(A)$.
Separation (Sep) For any property $\phi[x]$ and set A, the class $\{x \in A \mid \phi\}$ is a set.

We briefly examine the credentials of each of these axioms.

Exp The Exponentiation axiom is standardly taken as one of the main pillars of the predicative constructive set theories, as found, e.g. in [2,4,6]. The axiom may be seen to be in accordance with the absolute conception of set (see Section 3.3). However, such an understanding relies on viewing the notion of function between two sets as being fixed, even though one might, in future, add new entities to the universe (for example, very large sets) that would enable one to prove the existence of functions whose existence could not otherwise be established.

Pow It is straightforward to show that the Powerset axiom implies Exponentiation in BCST^-. The converse is provable using classical logic, but not intuitionistic logic. Indeed, in contrast to Exponentiation, Powerset is unequivocally impredicative. Using it, one may define new subsets of a set by quantifying over the collection (which is now a set) of all subsets of the set, a collection that includes the subset being defined. One can, nevertheless, still make a case that this axiom also conforms with the absolute conception of set: it seems a reasonable enough assumption that the collection of subsets (i.e. absolute subclasses) of a given set is fixed once and for all as soon as the latter set is defined.

Sep The Separation axiom is again impredicative. One may use Separation to define a subset using an unbounded quantifier ranging over all sets, including the set being defined. Further, Separation is not compatible with the absolute conception of set. Using Separation, one may define sets using unbounded quantification over the universe, and such quantifications may readily be affected by future extensions to the universe. However, Separation does agree with a rather looser idea of sets corresponding to those classes that are in some (slightly vague) sense 'small'.

Because Powerset implies Exponentiation, BCST^- has (including itself) six different extensions using combinations of the above axioms, all of which can indeed be shown to be distinct. We shall discuss several such extensions in Section 3.5; but first we consider the, thus far omitted, axiom of Infinity.

3.4 Infinity and Induction

The axiom of Infinity is most naturally incorporated in BCST^- by extending the first order language with three new constants $N, 0, s$.[14] In the context of weak set theories such as BCST^-, one has to be careful in the formulation of Infinity, as induction over

[14] Using such an extended language is inessential, because the axiom of Infinity could instead be prefixed by '$\exists N.\,\exists 0.\,\exists s.$', in which case derived consequences of Infinity also need to be prefixed by corresponding existential statements.

the natural numbers does not appear strong enough to derive definition by primitive recursion.[15] Because of this, we follow Lawvere and turn primitive recursion into the defining property of the natural numbers:[16]

Infinity (Inf) N is a set, $0 \in N$, $s \in N^N$ and, for every set A, element $z \in A$, and function $f \in A^A$, there exists a unique function $h \in A^N$ satisfying: $h(0) = z$ and $h(s(x)) = f(h(x))$ for all $x \in N$.

We write BCST for the theory BCST$^-$+ Inf.

It is not hard to show that Infinity implies that $0 \neq s(x)$ for all $x \in N$, and that $s(x) = s(y)$ implies $x = y$ for all $x, y \in N$. Also, it is straightforward to derive an induction principle for restricted formulas:

!-Ind For any $\phi[x]$ such that $\forall x \in N.\ !\phi$, if $\phi[0]$ and $\forall x \in N.\ \phi \rightarrow \phi[s(x)]$ then $\forall x \in N.\ \phi$.

However, the full induction schema below is not derivable, even in the theory BCST + Pow.[17]

Ind For any $\phi[x]$, if $\phi[0]$ and $\forall x \in N.\ \phi \rightarrow \phi[s(x)]$ then $\forall x \in N.\ \phi$.

Indeed, one can show (cf. [13]) that the statement:

$$\exists I.\ \emptyset \in I \ \wedge\ \forall x \in I.\ \mathcal{S}(x)\ \wedge\ x \cup \{x\} \in I,$$

which asserts the existence of an infinite set containing the von Neumann numerals, is derivable in BCST + Ind but not in BCST + Pow.[18]

It is obvious that the full induction schema, Ind, is derivable in the theory BCST + Sep. However, in contrast to Separation, there is nothing impredicative about full induction. Indeed full induction is a standard feature of (predicative) constructive set theories. In Myhill's CST [2], Ind is assumed as a basic axiom. In Aczel's CZF [4,6], full induction is a consequence of a 'set induction' principle, which is a constructively acceptable way of stating the well-foundedness of sets.

The distinction between !-Ind and Ind is part of a wider phenomenon. For induction in general, whether it be ordinary induction over the natural numbers, or any given variety of transfinite induction, one may distinguish two types of induction principle. *Restricted induction principles*, for example !-Ind above, hold for restricted formulas only, or equivalently for sub*sets* of the domain of induction. Such principles can be expressed by single formulas. In contrast, *full induction principles*, for example Ind, hold for arbitrary formulas, or equivalently for arbitrary sub*classes* of the domain. Such principles are expressed by schemata. In general, full induction principles are stronger

---

[15] I am grateful to Peter Aczel for pointing this out to me.

[16] Various simpler formulations of Infinity are possible in the presence of the Exp, Pow or Sep axioms.

[17] Because it is a schema, in order to formulate full induction in a language without the constants $N, 0, s$ it suffices to use one of the standard set-theoretic encodings of the numerals. This avenue is not available for formulating Inf in the absence of Ind, see footnote 18 below.

[18] Thus it would not be equivalent to formulate the Infinity axiom of BCST using a set-theoretic encoding of the natural numbers.

than the corresponding restricted ones. However, in the presence of full separation, the two forms of induction principle coincide.

In this chapter, we shall only be concerned with induction over the natural numbers. We shall assume the Infinity axiom as basic, and we shall consider Ind as a possible additional axiom (schema) to add to theories.

3.5 Constructive Set Theories

We take BCST as our basic constructive set theory, on top of which we shall add combinations of the axioms Exp, Pow, Sep and Ind.

According to the foregoing analysis, the strongest predicative set theory that can be obtained as a combination of the above axioms is:

$$\mathrm{BCST} + \mathrm{Exp} + \mathrm{Ind}.$$

This theory corresponds very closely to Myhill's theory CST [2].[19]

It is a widely held (though not universal) view that a theory has to be predicative in order to be constructive. In the opinion of the author, this view represents an overly narrow demarcation of constructivism. The hallmarks of constructive mathematics are surely that mathematical definitions should be interpretable concretely as constructions of mathematical objects, each with some inherent numerical (or at least computational) meaning, and that mathematical proofs should provide associated computational information, as required by the constructive interpretation of mathematical statements.

To the predicative constructivist, impredicative definitions are, because of their intrinsic cyclicity, not sufficiently definite to unambiguously furnish the requisite mathematical construction together with its associated computational information. Accordingly, such a constructivist doubts the very meaningfulness of impredicative forms of definition. Nevertheless, to the many that do not share such (necessarily subjective) doubts, there are impredicative type theories, such as the Calculus of Constructions and its extensions [17, 18], which provide very plausible computational readings of impredicative methods of definition. In the view of the author, such calculi give perfectly constructive interpretations to (at least some forms of) impredicative definition.

In the light of the above, we should not a priori banish the impredicative axioms Pow and Sep from appearing in so-called constructive set theories. In view of its arguable adherence to the absolute conception of set, the theory

$$\mathrm{BCST} + \mathrm{Pow} + \mathrm{Ind}$$

seems of particular interest as a candidate impredicative constructive set theory. However, the theory

$$\mathrm{BCST} + \mathrm{Pow} + \mathrm{Sep},$$

which is the strongest theory obtainable as a combination of the above axioms, also suggests itself as being mathematically natural. Indeed this latter theory is exactly the

[19] There are technical differences in formulation, but the theories are intertranslatable.

theory IST from [12], and thus equivalent in strength to the well-known intuitionistic set theory IZF (in its variant with Replacement rather than Collection).

We have not yet provided any justification that the above impredicative set theories do merit the appellation 'constructive', where this is to be understood in the sense described above. Such justification might be given by way of a translation into an impredicative type theory (along the lines of [4, 19, 20]), or, more directly, by defining notions of construction and computation designed specifically for the purpose of justifying the set theories. We leave these tasks as a challenge for future work. In any case, the existence of realizability models of BCST + Pow + Sep (any realizability model of IZF is one such) means that, from a classical point of view, BCST + Pow + Sep is compatible with the idea that proofs should have computational meaning. Thus the possibility of obtaining a constructive (in our sense) justification seems plausible enough.

We end this section by discussing two further axioms. Collection is a strengthening of Replacement:[20]

Collection (Coll) $\mathcal{S}(A) \wedge (\forall x \in A.\, \exists y.\, \phi) \rightarrow$
$\exists B.\, \mathcal{S}(B) \wedge (\forall x \in A.\, \exists y \in B.\, \phi) \wedge (\forall y \in B.\, \exists x \in A.\, \phi)$.

Collection has played an important rôle in the meta-mathematics of constructive set theories, see [5], and is a basic axiom of Aczel's CZF [4, 6]. Our reason for postponing Collection to this point is that we are not convinced that it is a genuinely useful axiom for applications in constructive mathematics (as opposed to meta-mathematics). Moreover, it has a different character from the other axioms in asserting the existence of a set that is not uniquely characterized by the properties it is required to satisfy.

Finally, for completeness, we consider the manifestly non-constructive Law of the Excluded Middle:

LEM $\phi \vee \neg\phi$.

The addition of LEM collapses most of the distinctions between the axioms previously discussed. Indeed, we have:

$$\mathrm{BCST} + \mathrm{Exp} + \mathrm{LEM} \;=\; \mathrm{BCST} + \mathrm{Pow} + \mathrm{Sep} + \mathrm{Ind} + \mathrm{LEM}$$

and this theory is intertranslatable with classical Zermelo-Fraenkel set theory. What is remarkable here is that the theory BCST + Exp is equiconsistent with first-order arithmetic, and thus of very low proof-theoretic strength. So, as is well known, from a proof-theoretic point of view, the Law of the Excluded Middle is a far from harmless addition to a constructive set theory.

3.6 Categories of Classes

The crucial idea for obtaining category-theoretic models of set theory is that, rather than axiomatizing the structure of the category of sets on its own, one should instead

[20] This axiom is often called Strong Collection, because the $\forall y.\exists x.$ clause is absent in the traditional formulation of Collection. However, in the presence of Separation, both stronger and weaker versions are equivalent, and, in the absence of Separation, the weaker is a less natural axiom. Thus it seems reasonable to call the stronger one Collection.

axiomatize the structure of the category of sets together with that of its surrounding supercategory of classes. This approach was first followed by Joyal and Moerdijk in their pioneering book on *algebraic set theory* [11], and has since been further developed in [12–14, 21–24]. The goal of this section is to give a uniform and accessible presentation of the approaches taken in [12–14]. Our exposition is aimed at a reader with a basic knowledge of category theory only. Accordingly, we do not give all technical details. For these, the reader must refer to the original sources.

We first axiomatize the structure we require of a category $\mathcal{C}$ to act like a category of classes, and then below we axiomatize the properties of its distinguished full subcategory $\mathcal{S}$ of *small* objects, which will be the sets amongst the classes. For intuition, the reader should simply think of $\mathcal{C}$ as actually being the category of all classes, with class functions as morphisms, with $\mathcal{S}$ as the full subcategory of sets, and with sets and classes satisfying the properties of one of the constructive set theories presented earlier (e.g. BCST$^-$).

The category $\mathcal{C}$ is assumed to be a *positive Heyting category* in the sense of [25, A1.4.4]. For the purposes of this paper, all that the reader needs to know about this structure is the following. The category $\mathcal{C}$ has a terminal object $\mathbf{1}$, binary (hence finite) products $X \times Y$, an equalizer $m: X \rightarrowtail Y$ for every parallel pair $f, g: Y \longrightarrow Z$, an initial object $\mathbf{0}$, and binary (hence finite) coproducts $X + Y$. Every morphism $f: X \longrightarrow Y$ factors as

$$f = X \stackrel{e_f}{\twoheadrightarrow} \mathrm{Img}(f) \stackrel{m_f}{\rightarrowtail} Y,$$

where e_f is a regular epimorphism[21] and m_f is a monomorphism. Intuitively, regular epis are to be thought of as surjective maps, so $\mathrm{Img}(f)$ represents the image of the map f, and m_f exhibits the image as a subobject of Y. In general, subobjects $P \rightarrowtail X$ can be understood as representing predicates over X, and intuitionistic first-order logic with equality can be used to reason about such predicates. For example, given a predicate $P \rightarrowtail X \times Y$, one can form the following two subobjects of X,

$$\{x \in X \mid \exists y \in Y.\ P(x, y)\} \rightarrowtail X \qquad \{x \in X \mid \forall y \in Y.\ P(x, y)\} \rightarrowtail X,$$

and these obey the expected logical laws. This logic of subobjects is called the *internal logic* of $\mathcal{C}$.

For an object I of $\mathcal{C}$, the *slice category* $\mathcal{C}/I$ is a simple category-theoretic construction whose objects intuitively represent I-indexed families of classes. Formally, $\mathcal{C}/I$ is defined as the category whose objects are morphisms $X \longrightarrow I$ in $\mathcal{C}$, and whose morphisms from $X \longrightarrow I$ to $Y \longrightarrow I$ are maps $X \longrightarrow Y$ in $\mathcal{C}$ making the triangle commute. Intuitively, one thinks of the object $f: X \longrightarrow I$ in $\mathcal{C}/I$ as representing the family $\{f^{-1}(i)\}_{i \in I}$. Following this intuition, we use the notation $\{X_i\}_{i \in I}$ for an arbitrary object of $\mathcal{C}/I$. Given a map $h \in I \longrightarrow J$ in $\mathcal{C}$, one obtains two functors: a *reindexing* functor $h^*: \mathcal{C}/J \to \mathcal{C}/I$ that maps each $g: Y \longrightarrow J$ to the pullback of g along h; and a functor $\Sigma_h: \mathcal{C}/I \to \mathcal{C}/J$ that maps $f: X \longrightarrow I$ to $h \circ f: X \longrightarrow J$ (with the latter

[21] A *regular epi(morphism)* is a morphism that can be obtained as a coequalizer. Dually, a *regular mono(morphism)* is a morphism that can be obtained as an equalizer.

functor left adjoint to the former). Using the indexed class notation above, the action of these functors is more intuitively described as follows:

$$\begin{aligned} h^*(\{Y_j\}_{j\in J}) &= \{Y_{h(i)}\}_{i\in I} \\ \Sigma_h(\{X_i\}_{i\in I}) &= \{\textstyle\sum_{i\in h^{-1}(j)} X_i\}_{j\in J}\,. \end{aligned}$$

(The actions of the functors on morphisms are obvious.) Importantly, the assumed structure on $\mathcal{C}$ is preserved under slicing, i.e., for every I, the slice category $\mathcal{C}/I$ is a positive Heyting category and, for every $h\colon J \longrightarrow I$ the functor h^* preserves the positive Heyting category structure.

We now turn to the axioms needed for the notion of smallness. We shall assume that we have a collection $\mathcal{S}$ of objects of $\mathcal{C}$, designated as *small*, and satisfying the axioms expressing properties of smallness listed below. Furthermore, we shall assume that all this structure is preserved under slicing. In other words, for each slice category $\mathcal{C}/I$, we require an associated collection $\mathcal{S}_I$ of objects of $\mathcal{C}/I$, which are again designated as small, and which satisfy the same axioms on smallness listed below, and are preserved by reindexing functors. The non-category-theoretically-minded reader is encouraged to ignore the issue of slicing entirely, and simply think of the above data using the intuition that $\mathcal{S}$ is the collection of those classes that are sets, and $\mathcal{S}_I$ is the collection of families $\{A_i\}_{i\in I}$ in which A_i is a set for each $i \in I$. Indeed, we shall refer to a small object $\{A_i\}_{i\in I}$ in $\mathcal{C}/I$ as a *family of small objects*.

We impose five basic axioms on small objects, each named in accordance with the set-theoretic principle it corresponds to. The first four axioms express closure properties of smallness, and thus address the third law of sets from Section 3.2.

Restricted separation If B is small and $A \rightarrowtail B$ is a regular mono then A is small.

Replacement If A is small and $A \twoheadrightarrow B$ is regular epi then B is small.

Disjoint union If A is small and $\{B_x\}_{x\in A}$ is a family of small objects then $\sum_{x\in A} B_x$ is small.

Pairing The object $\mathbf{1}+\mathbf{1}$ is small.

The fifth axiom addresses the first and second laws of sets from Section 3.2. It ensures that small classes (i.e. sets) themselves appear as elements of certain (power)classes, and that these elements behave extensionally. In order to formulate this, we write *(I-indexed) family of subobjects of X* to mean a subobject $\{X_i\}_{i\in I}$ of the constant I-indexed family $\{X\}_{i\in I}$ (this constant family is the object of $\mathcal{C}/I$ given by the projection $X\times I \longrightarrow I$). Similarly, a *family of small subobjects of X* is a family $\{X_i\}_{i\in I}$ of subobjects of X that is itself a family of small objects.

Powerclass For every X there exists an object $\mathcal{P}(X)$ and family $\{M_A\}_{A\in\mathcal{P}(X)}$ of small subobjects of X such that, for every family $\{X_i\}_{i\in I}$ of small subobjects of X, there exists a unique $f\colon I \longrightarrow \mathcal{P}(X)$ such that $\{X_i\}_{i\in I}$ and $\{M_{f(i)}\}_{i\in I}$ are equal as families of subobjects of X.

For intuition here, think of $\mathcal{P}(X)$ as the class of all subsets of X, and think of M_A as the set of elements of the element $A \in X$ (i.e. as the set A itself). The family $\{M_A\}_{A\in\mathcal{P}(X)}$

is given by an object $\in_X \longrightarrow \mathcal{P}(X)$ of $\mathcal{C}/\mathcal{P}(X)$ which, as it is a family of subobjects of $\mathcal{P}(X)$, comes as a subobject $\in_X \rightarrowtail X \times \mathcal{P}(X)$ in $\mathcal{C}$. This subobject defines the membership relation on $\mathcal{P}(X)$ in $\mathcal{C}$.

Definition 3.5 *A category with basic class structure is given by a positive Heyting category together with a (stable under slicing) collection $\mathcal{S}$ of small objects satisfying the above axioms.*

The notion of basic class structure roughly embodies the closure properties on sets of the theory BCST$^-$ (precise results are given below).

We shall need two consequences of basic class structure below. The first is that every small object A of $\mathcal{C}$ is exponentiable in the sense that there is a natural function space object X^A in $\mathcal{C}$.[22] The second consequence is that the mapping $X \mapsto \mathcal{P}(X)$ on objects extends to an endofunctor on $\mathcal{C}$ whose action on morphisms takes $f: X \longrightarrow Y$ to the function $\mathcal{P}(f): \mathcal{P}(X) \longrightarrow \mathcal{P}(Y)$ that maps a subset A of X to its image under f, which is a subset of Y.

The following additional axioms on class structure exactly mirror the axioms we considered earlier as possible additions to BCST$^-$.[23]

Exponentiation (Exp) If A, B are small then so is B^A.

Powerset (Pow) If A is small then so is $\mathcal{P}(A)$.

Separation (Sep) If A is small and $B \rightarrowtail A$ is mono then B is small.

Infinity (Inf) The category $\mathcal{S}$, of small objects, has a natural numbers object.

Induction (Ind) $\mathcal{C}$ has a small natural numbers object.

Collection (Coll) The functor $\mathcal{P}(\cdot): \mathcal{C} \to \mathcal{C}$ preserves regular epis.

There is one aspect of sets and classes, as they are used in (constructive) set theory, which we have not implemented in the axioms we have so far considered for $\mathcal{C}$ and $\mathcal{S}$. In set theory, there is a distinguished largest class, namely the universe itself. This can be implemented in a category of classes by asking for $\mathcal{C}$ to have a universal object in the sense of [12].

Definition 3.6 *A universal object in $\mathcal{C}$ is an object U such that, for every object X, there is a distinguished mono $X \rightarrowtail U$.*

Given a category with basic class structure and universal object U, it is possible to interpret the first-order language of BCST$^-$ internally in $\mathcal{C}$ as expressing properties of U. In general, a formula ϕ with k free variables will be interpreted as a subobject $\phi \rightarrowtail U^k$ (where U^k is the k-fold product of U with itself). The unary predicate $\mathcal{S}(x)$ is interpreted as the distinguished mono $u: \mathcal{P}(U) \rightarrowtail U$ and the binary predicate $x \in y$ is interpreted as the composite

$$\in_U \rightarrowtail U \times \mathcal{P}(U) \overset{\mathsf{id}\times u}{\rightarrowtail} U \times U.$$

[22]Technically, A is *exponentiable* if the functor $A \times (\cdot): \mathcal{C} \to \mathcal{C}$ has a right adjoint.

[23]Again, these axioms are to be imposed on all slice categories $\mathcal{C}/I$ as well as $\mathcal{C}$ itself.

The connectives, quantifiers and equality predicate are simply interpreted by their own internal selves. We write $(C, U) \models \phi$ to mean that the subobject $\phi \rightarrowtail U^k$ is the whole of U^k (i.e. its representing mono is an isomorphism).

The theorem below shows that the assumed structure on our category-theoretic models $\mathcal{C}$ corresponds exactly to the constructive set theories considered in Sections 3.3–3.5.

Theorem 3.7 (Soundness and completeness) The following are equivalent for any ϕ.[24]

1. ϕ is a theorem of BCST$^-$ (plus any combination of Exp, Pow, Sep, Inf, Ind and Coll).
2. For all $\mathcal{C}$ with basic class structure (plus the same combination of Exp, Pow, Sep, Inf, Ind and Coll as in 1) and universal object U, it holds that $(C, U) \models \phi$.

Particular (but fully illustrative) cases of this result are proved in [12–14].

3.7 Categories of Sets

It is pleasing that all the constructive set theories we have considered have sound and complete classes of category-theoretic models that are easily axiomatized. The goal of this section is to show that, for certain constructive set theories, there are interesting examples of models that occur naturally in mathematics. Such natural models provide a source of possible applications of the associated constructive set theory, and thus potentially enable a mathematician to appreciate the value of the theory for non-ideological reasons.

Our approach to finding such models is to analyse the category-theoretic structure of categories of sets independently of the containing category of classes. First we state properties of the category of sets $\mathcal{S}$ which fall out easily from the assumed structure on $\mathcal{C}$.

Proposition 3.8 *Suppose $\mathcal{C}$ has basic class structure.*

1. *$\mathcal{S}$ is a Heyting pretopos.*[25]
2. *If* Exp *holds then $\mathcal{S}$ is locally cartesian closed.*
3. *If* Pow *holds then $\mathcal{S}$ is an elementary topos.*

Surprisingly, the above properties completely characterize the possible categories of sets in each case. The technology for demonstrating this has been developed in [13, 14, 24]. To a small category $\mathcal{E}$, one associates a category Idl($\mathcal{E}$) of 'ideals' over $\mathcal{E}$. This construction enjoys the following properties.[26]

Theorem 3.9 [13, 14, 24] Suppose that $\mathcal{E}$ is a Heyting pretopos.

[24] We have omitted spelling out how the additional constants $N, 0, s$ are interpreted in U in the case that the Inf and Ind axioms are considered. This is routine.

[25] A *Heyting pretopos* is a positive Heyting category in which all equivalence relations have effective quotients.

[26] For discussion on the appropriate meta-theory for proving the results in this section, see [13].

1. $\mathrm{Idl}(\mathcal{E})$ has basic class structure with $\mathcal{E}$ as its full subcategory of small objects, it has a universal object and it satisfies Coll.
2. If $\mathcal{E}$ is locally cartesian closed then $\mathrm{Idl}(\mathcal{E})$ satisfies Exp
3. If $\mathcal{E}$ is an elementary topos then $\mathrm{Idl}(\mathcal{E})$ satisfies Pow.

The proof of this theorem is distributed over the three references [13, 14, 24]. First, in Awodey *et al.* [13], $\mathrm{Idl}(\mathcal{E})$ was defined for an elementary topos $\mathcal{E}$ using a 'system of inclusions' on $\mathcal{E}$, and it was shown that $\mathrm{Idl}(\mathcal{E})$ has basic class structure and satisfies Pow and Coll. Later, in [24] (see also [23]), this construction was improved to avoid the system of inclusions, by defining $\mathrm{Idl}(\mathcal{E})$ to be the full subcategory of the presheaf category $[\mathcal{E}^{\mathrm{op}}, \mathbf{Set}]$ whose objects are directed colimits of diagrams of monos between representables.[27] That the improved construction also adapts to models of predicative set theories was realized by Awodey, Warren and the present author, who proved statement 2 of Theorem 3.9, after which Awodey and Warren extended the approach to obtain statement 1 [14].

Let us spell out the ramifications for constructive set theories of the results above. By Theorem 3.7, models of the impredicative theory

$$\mathrm{BCST} + \mathrm{Pow}\ (+\ \mathrm{Coll})$$

are given by categories with basic class structure satisfying Inf, Pow (and Coll). By Proposition 3.8 and Theorem 3.9, the categories of sets arising in such categories of classes are exactly the elementary toposes with natural numbers object (nno). Thus, the categories of sets that model BCST + Pow (+ Coll) are exactly the elementary toposes with nno. Elementary toposes are abundant and natural mathematical structures, hence the theory BCST + Pow (+ Coll) is of independent mathematical interest in being strongly tied to such categories. Indeed, this is the viewpoint developed in [13], where the theory BCST + Pow and its models are studied in detail.

An analogous story holds for the predicative theory

$$\mathrm{BCST} + \mathrm{Exp}\ (+\ \mathrm{Coll})\,.$$

For this theory, the associated categories of sets are exactly the locally cartesian closed (lcc) pretoposes with nno.[28] There are many interesting examples of lcc pretoposes that are not elementary toposes. For example, if one performs the usual construction that builds a realizability topos from a partial combinatory algebra (pca) [26], but does so using a *typed pca* in the sense of Longley then the resulting category is an lcc pretopos, but not necessarily a topos, see [27]. A particularly natural mathematical example of an lcc pretopos that is not an elementary topos is the 'exact completion' of the category of topological spaces [28]. The possibility of using BCST + Exp (+ Coll) to reason within such categories gives non-ideological mathematical motivation for the consideration of such predicative constructive set theories.

[27] In order to obtain a universal object, one needs to define a 'universe' U in $\mathrm{Idl}(\mathcal{E})$ and then restrict to the full subcategory of subobjects of U.

[28] Every lcc pretopos is indeed Heyting.

It should be noted that the above results characterizing categories of sets apply only to theories not involving the Ind and Sep axiom schemata. For the theories that do include these schemata, it is not clear whether it is possible to characterize the properties of the category of sets without involving the containing category of classes in the characterization. The difficulty is that the axiom schemata themselves relate properties of sets and classes, rather than stating closure properties of sets alone. Nevertheless, for the strongest theory

$$\text{BCST} + \text{Pow} + \text{Sep} + \text{Coll},$$

we can at least demonstrate that the theory has a rich collection of interesting mathematical models.

Theorem 3.10 [13] If $\mathcal{E}$ is a cocomplete topos or a realizability topos then $\mathcal{E}$ fully and faithfully embeds as the subcategory of small objects in a category with basic class structure satisfying Pow, Sep (hence Ind) and Coll.

We end the chapter with a challenge to find an analogue of Theorem 3.10 for the predicative theory BCST + Exp + Ind + Coll. In particular, we would like to see a naturally occurring mathematical example of a category of sets for this theory which is truly predicative in the sense that, even when considered classically, it cannot be used to model Pow or Sep. Even more interesting would be to find a naturally occurring truly predicative model of constructive set theories that include general mechanisms for inductive definition (including their associated full induction principles) of the sort considered in, e.g. [6,21,29].

Acknowledgements

This chapter presents a non-technical survey of recent results developed, in part, by the author in collaboration with Steve Awodey, Carsten Butz and Thomas Streicher [12,13], and further developed by Steve Awodey, Henrik Forssell and Michael Warren [14,24]. Explicit attributions are given in the text where appropriate. I would like to thank Peter Aczel, Carsten Butz, Nicola Gambino, Thomas Streicher and, especially, Steve Awodey for discussions on the subject of this paper.

This research was supported by an EPSRC Advanced Research Fellowship.

References

1. Bishop, E. (1967). *Foundations of Constructive Analysis*. McGraw-Hill, New York.
2. Myhill, J. (1975). Constructive set theory. *J. Symbolic Logic*, **40**, 347–382.
3. Friedman, H. (1977). Set theoretic foundations for constructive analysis. *Ann. of Math.*, **105**, 1–28.
4. Aczel, P. (1978). The type theoretic interpretation of constructive set theory. In: *Logic Colloquium '77*. Number 96 in Stud. Logic Foundations Math. pp. 55–66. North Holland.

5. Ščedrov, A. (1985). Intuitionistic set theory. In: *Harvey Friedman's Research on the Foundations of Mathematics*. Number 117 in Stud. Logic Foundations Math. pp. 257–284. North-Holland.
6. Aczel, P. and Rathjen, M. (2001). Notes on constructive set theory. Technical Report 40, 2000/2001. Mittag-Leffler Institute, Sweden.
7. Martin-Löf, P. (1984). *Intuitionistic Type Theory*. Studies in Proof Theory. Naples, Bibliopolis.
8. Seely, R. A. G. (1984). Locally cartesian closed categories and type theory. *Math. Proc. Cambridge Philos. Soc.*, **95**, 33–48.
9. Lambek, J. and Scott, P. J. (1986). *Introduction to Higher Order Categorical Logic*. Number 7 in Cambridge Studies in Advanced Mathematics. Cambridge University Press.
10. Fourman, M. P. (1980). Sheaf models for set theory. *J. Pure Appl. Algebra*, **19**, 91–101.
11. Joyal, A. and Moerdijk, I. (1995). *Algebraic Set Theory*. Number 220 in London Mathematical Society Lecture Note Series. Cambridge University Press,
12. Simpson, A. K. (1999). Elementary axioms for categories of classes. In: *Proc. 14th Annual IEEE Symposium on Logic in Computer Science*. pp. 77–85.
13. Awodey, S., Butz, C., Simpson, A. K., and Th. Streicher, (2004). *Relating Set Theories, Toposes and Categories of Classes*. In preparation.
14. Awodey, S. and Warren, M. (2005). Predicative Algebraic Set Theory. *Theory and Applications of Categories*, **15**, 1–39.
15. Shoenfield, J. R. (1977). Axioms of set theory. In *Handbook of Mathematical Logic*. Number 90 in Stud. Logic Foundations Math. North Holland, pp. 321–344.
16. Aczel, P. (1988). *Non-well-founded sets*. No. 14 in CSLI Lecture Notes. Stanford University.
17. Coquand, Th. and Huet, G. (1988). The calculus of constructions. *Inform. and Comput.*, **76**, 95–120.
18. Luo, Z. (1994). *Computation and Reasoning: A Type Theory for Computer Science*. No. 11 in International Series of Monographs on Computer Science. OUP.
19. Werner, B. (1997). Sets in types, types in sets. In: *Theoretical Aspects of Computer Science (Sendai, 1997)*. pp. 530–546. Number 1281 in Lecture Notes in Comput. Sci. Springer.
20. Gambino, N. (1999). Types and sets: a study of the jump to full impredicativity. Laurea Thesis, Department of Pure and Applied Mathematics, University of Padova.
21. Moerdijk, I. and Palmgren, E. (2002). Type theories, toposes and constructive set theory: predicative aspects of AST. *Ann. Pure Appl. Logic*, **114**, 155–201.
22. Butz, C. (2003). Bernays–Gödel type theory. *J. Pure Appl. Algebra*, **178**, 1–23.
23. Rummelhoff, I. (2004). Algebraic set theory. PhD thesis, University of Oslo, forthcoming,
24. Awodey, S. and Forssell, H. (2004). *Categorical Models of Intuitionistic Theories of Sets and Classes*. In preparation.

25. Johnstone, P. T. (2002). *Sketches of an Elephant: A Topos Theory Compendium. Vol. 1*. No. 43 in Oxford Logic Guides. Oxford University Press.
26. Hyland, J. M. E., Johnstone, P. T., and Pitts, A. M. (1980). Tripos theory. *Math. Proc. Camb. Phil. Soc.*, **88**, 205–232.
27. Lietz, P. and Streicher, Th. (2002). Impredicativity entails untypedness. *Math. Structures Comput. Sci.*, **12**, 335–347.
28. Birkedal, L., Carboni, A., Rosolini, G., and Scott, D. S. (1998). Type theory via exact categories. In: *Proc. 13th Annual IEEE Symposium on Logic in Computer Science*; pp. 188–198.
29. Aczel, P. (1986). The type theoretic interpretation of constructive set theory: inductive definitions. In: *Logic, methodology and philosophy of science, VII*. Number 114 in Stud. Logic Foundations Math. North-Holland.

4

PRESHEAF MODELS FOR CONSTRUCTIVE SET THEORIES

NICOLA GAMBINO

Abstract

We introduce a new kind of model for constructive set theories based on categories of presheaves. These models are a counterpart of the presheaf models for intuitionistic set theories defined by Dana Scott in the 1980s. We also show how presheaf models fit into the framework of algebraic set theory and sketch an application to an independence result.

4.1 Variable Sets in Foundations and Practice

Presheaves are of central importance both for the foundations and the practice of mathematics. The notion of a presheaf formalizes well the idea of a variable set, which is relevant in all the areas of mathematics concerned with the study of indexed families of objects [1]. One may then readily see how presheaves are of interest also in the foundations: both Cohen's forcing models for classical set theories and Kripke models for intuitionistic logic involve the idea of sets indexed by stages.

Constructive aspects start to emerge when one considers the internal logic of categories of presheaves. This logic, which does not include classical principles such as the law of the excluded middle, provides a useful language to manipulate objects and arrows, and can be used as an alternative to diagrammatic reasoning [2]. Furthermore, it is sufficiently expressive to allow the definitions of complex mathematical constructions. This aspect has led to important developments in the study of elementary toposes [3].

The main purpose of this chapter is to show how presheaves can be used to obtain models for constructive set theories [4, 5] analogous to those defined by Dana Scott for intuitionistic set theories [6]. In order to do so, we will have to overcome the challenges intrinsic to working with *generalized predicative* formal systems. By a generalized predicative formal system we mean here a system that is proof-theoretically reducible to Martin-Löf dependent type theories with W-types and universes [7, 8]. Generalized predicative systems typically contain axioms allowing generalized forms of inductive definitions [9] instead of proof-theoretically strong axioms such as Power Set.

Our development will focus on categories of classes rather than categories of sets as the starting point to define presheaves, thus assuming the perspective of algebraic set theory [10–14]. The main reason for this choice is that the properties of categories of sets do not always reflect directly the set-theoretical axioms adopted to define them. There are indeed axioms, such as Replacement, that do not express directly properties of sets, but regard the interaction between sets and classes. In categories of classes we can overcome this problem without loss, since sets can be isolated as special objects, those that are in some sense 'small'. One is then led to consider a notion of *small map* in such a way that the axioms of a set theory correspond directly to the axioms for small maps.

This approach has two main advantages. First, it allows us to give a homogeneous treatment of presheaf models of different set theories. Indeed, one of the initial motivations for the research described here was to investigate whether it was possible to generalize the presheaf models for intuitionistic set theories to constructive ones and present them in a uniform way. Secondly, we can show how Scott's presheaf models fit into the paradigm of algebraic set theory.

The study of presheaf and sheaf categories in the generalized predicative setting was initiated by Ieke Moerdijk and Erik Palmgren in [13, 15]. They introduced the notion of *stratified pseudo-topos* as a candidate for the notion of a *predicative topos*. A predicative topos should be a counterpart at the generalized predicative level of the notion of an elementary topos. They supported their axiomatization by showing how stratified pseudo-toposes support the construction of internal presheaves, in the sense that categories of internal presheaves in a stratified pseudo-topos are again stratified pseudo-toposes. They also proved that a stratified pseudo-topos can be used to define models of constructive set theories. The combination of these two facts comes close to exhibiting presheaf models for constructive set theories, but does not quite achieve it. This is because categories of classes arising in constructive set theories do not satisfy the axioms for a stratified pseudo-topos. For example, as pointed out in the context of intuitionistic set theories in [11], they fail to be *exact*, and, as we will discuss in Subsection 4.2, they do not have arbitrary *exponentials*. Even if they satisfied all the axioms for a stratified pseudo-topos, however, there would still be the problem of obtaining an explicit description of presheaf models, of developing their study, and of finding applications. All of these aspects will be considered here.

Related work on categories of classes in constructive set theories is also presented in [16]. The approach to the construction of categorical models for constructive set theories taken there is slightly different from the one assumed here, even if both follow the perspective of algebraic set theory. While in [16] categorical models for constructive set theories are defined exploiting the existence of a universal object, here they will be obtained via W-types and the assumption of a universal small map.

Let us conclude these introductory remarks with an overview of the contents of the chapter. In Section 4.2 we isolate axiomatically the structure on a category that is necessary to obtain a categorical model for a constructive set theory. We also discuss how the category of classes arising from a constructive set theory is an example of such a structure. That section also serves as an introduction to algebraic set theory. We then shift our attention to presheaves. Section 4.3 introduces the basic notions, describes the structure on the category of presheaves that is relevant for our study, and defines presheaf models. The study of these models via the so-called Kripke–Joyal semantics is presented in Section 4.4. We end the paper in Section 4.5 by sketching an application to an independence result. To be as self-contained as possible, we included background material on category theory and set theory.

4.2 Classes and Sets

4.2.1 *Set-theoretic axioms*

Set theories based on intuitionistic logic are formulated to provide an axiomatic basis to support the development of intuitionistic mathematics in set theory. Their axioms will be formulated here in an extension of first-order intuitionistic logic with equality, obtained by adding restricted quantifiers of the form $(\forall x \in a)$ and $(\exists x \in a)$ as primitive, and standard axioms for them. The membership relation can then be defined. A formula is said to be *restricted* if it contains only quantifiers that are restricted.

Classes provide a convenient notation to manipulate sets and formulas in mathematical practice. If ϕ is a formula with a free variable x and A is defined by $A =_{\text{def}} \{x \mid \phi\}$, we let $a \in A =_{\text{def}} \phi[a/x]$. Two classes are said to be *extensionally equal* if they have the same elements. Note that every set can be viewed as a class, and that proper classes cannot be considered as elements of other classes. Equality between sets is disciplined by the axiom of Extensionality, stating that two sets are equal if they are extensionally equal as classes.

The basic set existence axioms of Pairing, Union, and Infinity, familiar from classical set theory, simply assert that certain classes are sets. Using these, standard definitions allows us to introduce the forms of classes

$$A \times B, \quad A + B, \quad \textstyle\sum_{a \in A} B_a,$$

that denote binary cartesian products, binary disjoint unions, and indexed disjoint unions, respectively. There is a natural notion of function between classes, that generalizes the notion of function between sets. We write $f : A \to B$ to express that f is a function from A to B, and by this we mean that f is a subclass of $A \times B$ that is total and single-valued as a relation from A to B. Two functions are considered equal if they are extensionally equal as classes.

The set-theoretic universe, defined by

$$V =_{\text{def}} \{x \mid x = x\}, \tag{4.1}$$

cannot be asserted to be a set if we wish to avoid Russell's paradox. The interplay between classes and sets is specified further by other axioms. Restricted Separation is the axiom scheme asserting that for each set a and each restricted formula ϕ, the class $\{x \in a \mid \phi\}$ is a set. This weakening of the usual Full Separation axiom scheme is sufficient for many purposes in mathematical practice. In the absence of Full Separation, one needs to distinguish carefully between *subsets* and *subclasses* of a set. When considering two sets, we assume that functions between them are given as subsets, rather than subclasses, of the cartesian product of the domain and codomain.

The axiom of Exponentiation, originally introduced in [4], asserts that the class of functions between any two sets is again a set. Exponentiation is a consequence of Power Set that deserves to be isolated if one wishes to consider set theories that do not include Power Set. Strong Collection is the scheme

$$(\forall x \in a)(\exists y)\phi \to (\exists u)\big((\forall x \in a)(\exists y \in u)\phi \wedge (\forall y \in u)(\exists x \in a)\phi\big),$$

where a is a set and ϕ is an arbitrary formula. A consequence of Strong Collection is Replacement, asserting that for a function $f : A \to B$ between classes, if A is a set then so is its image. Strong Collection also allows us to have generalized forms of inductive definitions of classes [5, Chapter 5]. For instance, if A is a class and $(B_a \mid a \in A)$ is family of sets, we can form the associated class $W_{a \in A} B_a$ of *well-founded trees*. This is defined as the smallest class X such that if $a \in A$ and $t \in X^{B_a}$ then $(a, t) \in X$. Classes of well-founded trees are a set-theoretic counterpart of the well-ordering types introduced by Martin-Löf [7].

The axiom of Set Induction is used to view the set-theoretic universe as an inductively defined class. It asserts that the class V defined in (4.1) is the smallest class X such that, for all sets p, if $p \subseteq X$ then $p \in X$. Set Induction is relevant in our study since it allows us to see the set-theoretic universe as an initial algebra for an endofunctor on the category of classes, and this suggests a way to formulate a general notion of model for set theories with Set Induction, as we discuss in Subsection 4.2.3.

In the following, we will focus our attention on the constructive set theory CST whose axioms are Extensionality, Set Induction, Pairing, Union, Infinity, Restricted Separation, Exponentiation and Strong Collection. This is an extension of Myhill's original system [4], obtained by adding Strong Collection, and a subsystem of Aczel's system CZF, obtained by replacing Subset Collection with Exponentiation [5]. Robert Lubarsky has recently proved that Subset Collection is independent of Exponentiation, thus showing that CST is a proper subsystem of CZF [17]. The intuitionistic set theory IZF which is essentially an intuitionistic counterpart of classical Zermelo–Frankel set theory, is obtained from CST by adding Full Separation and Power Set [18]. Presheaf and sheaf models for IZF have been considered in [6, 19].

4.2.2 *Categories of classes*

We now associate to the constructive set theory CST a category, called **CST**, and study how the axioms of CST determine the properties of **CST**. The category **CST** is defined as having classes as objects and functions between them as arrows. In the study of the properties of this category, we follow [14].

The category **CST** has an obvious terminal object, given by the singleton class $1 =_{\text{def}} \{\emptyset\}$. Pullbacks in **CST** will now be defined explicitly. Given two arrows $u : X \to A$ and $f : B \to A$, we let $Y =_{\text{def}} \{(b, x) \in B \times X \mid fb = ux\}$. The required pullback diagram is then given by

$$\begin{array}{ccc} Y & \xrightarrow{g} & X \\ {\scriptstyle v}\downarrow & & \downarrow{\scriptstyle u} \\ B & \xrightarrow[f]{} & A \end{array}$$

where v and g are the first and second projection, respectively. Sometimes we will write $B \times_A X$ to denote the class Y defined above. The operation of pullback can be thought of as reindexing, or substitution: if one regards an arrow $u : X \to A$ as a family $(X_a \mid a \in A)$, where $X_a =_{\text{def}} \{x \in X \mid ux = a\}$, for $a \in A$, then, in the pullback diagram above, the family $(Y_b \mid b \in B)$ is isomorphic to the family $(X_{fb} \mid b \in B)$, obtained by reindexing the family $(X_a \mid a \in A)$ via the function f. Since **CST** has a terminal object and pullbacks, it has all finite limits [20, Section V.2].

For an arrow $f : B \to A$ we may consider the pullback of f along itself. This determines a diagram of the form $B \times_A B \rightrightarrows B$ that is called the *kernel pair* of f. Such a diagram determines an equivalence relation on B according to which $b, b' \in B$ are related if and only if $f(b) = f(b')$ holds. The quotient of B under this equivalence relation is isomorphic to the image of f in A, written $\mathrm{Im}(f)$ here, and we thus have a diagram of the form

$$B \times_A B \rightrightarrows B \longrightarrow \mathrm{Im}(f) \tag{4.2}$$

that is *exact* [10, Appendix B]. The possibility of defining quotients of this kind can be expressed abstractly by saying that we have *coequalizers of kernel pairs*. The arrow $B \to \mathrm{Im}(f)$ is obviously a surjection and it is easy to show that in **CST** every surjection is a *regular epimorphism*, i.e. that it fits into a diagram of the form in (4.2). Surjections are stable, in the sense that the pullback of a surjection is again a surjection. This discussion indicates that **CST** is a *regular category*, in the sense specified by the next definition.

Definition 4.1 *A category $\mathcal{E}$ is regular if it has finite limits, coequalizers of kernel pairs, and if regular epimorphisms (i.e. epimorphisms that arise as coequalizers of kernel pairs) are stable under pullback.*

The category **CST** also has *disjoint finite coproducts*, given by disjoint unions, and these are stable in the sense that they are preserved by pullbacks. In particular, the empty coproduct is given by $0 =_{\mathrm{def}} \emptyset$. The regular structure of **CST** is sufficient to define an adjunction $\exists_f \dashv \Delta_f$ between the functors

$$\mathit{Sub}\, A \underset{\exists_f}{\overset{\Delta_f}{\rightleftarrows}} \mathit{Sub}\, B$$

for any arrow $f : B \to A$, where we write $\mathit{Sub}\, X$ for the category of subclasses of a class X, with arrows given by inclusions. The functors Δ_f and $\exists_f$ are defined by letting, for $P \subseteq A$ and $Q \subseteq B$,

$$\Delta_f(P) =_{\mathrm{def}} \{b \in B \mid fb \in P\}, \quad \exists_f(Q) =_{\mathrm{def}} \{a \in A \mid (\exists b \in B_a)\, b \in Q\}.$$

This indicates that regular categories have enough structure to interpret a large fragment of first-order intuitionistic logic, which however does not include the universal quantifier [21, Section 4.4]. To interpret it, one requires *dual images*, i.e. the existence of a right adjoint $\forall_f : \mathit{Sub}\, B \to \mathit{Sub}\, A$ to the functor Δ_f, for every $f : B \to A$. These right adjoints exist in **CST** and can be defined by letting, for $Q \subseteq B$

$$\forall_f(Q) =_{\mathrm{def}} \{a \in A \mid (\forall b \in B_a)\, b \in Q\}.$$

So far, we have described some of the structure available on categories of classes. As pointed out in [11], categories of classes arising from intuitionistic or constructive set theories generally fail to be *exact*, in the sense that it is not possible to define quotients of arbitrary equivalence relations, but only of equivalence relations whose equivalence classes are sets.

Further structure on the category **CST** arises from the interplay between classes and sets, but does not seem to be directly expressible in terms of universal properties. For example, for two classes A and B, the functions from A to B can be collected into a class only if A is set, since elements of classes are required to be sets. This means that **CST** does not have arbitrary exponentials, but has exponentials of sets. More generally, if we have a set A and a family of classes $(B_a \mid a \in A)$ we can define the class $\prod_{a \in A} B_a$ whose elements are the functions f with domain A such that, for all $a \in A$, $f(a) \in B_a$.

To axiomatize this situation, one may introduce the notion of *small map*. The next definition introduces a natural notion of small map for the category **CST**.

Definition 4.2 *An arrow $u : X \to A$ in* **CST** *is said to be small if, for all a in A, the class $X_a =_{\mathrm{def}} \{x \mid ux = a\}$ is a set.*

Note that a set is just a class for which the canonical arrow into the singleton set 1 is small. In the next subsection, we will arrive at a definition of a categorical model for CST by isolating a group of axioms that are satisfied by the small maps in the regular category **CST**.

4.2.3 *Axioms for small maps*

In [10, 13] axioms for small maps were considered in the context of *pretoposes*, but they can be studied in the more general context of regular categories with stable disjoint coproducts. This is indeed the setting that we consider here, and from now on we let $\mathcal{E}$ be a regular category with stable disjoint coproducts. In order to introduce these axioms, it is convenient to recall some basic notions.

For $A \in \mathcal{E}$, one may define $\mathcal{E}/A$, the *slice category over* A, whose objects are arrows $u : X \to A$ of $\mathcal{E}$. For two objects $u : X \to A$, $u' : X' \to A$ of $\mathcal{E}/A$, an arrow in $\mathcal{E}/A$ between them is given by $v : X \to X'$ in $\mathcal{E}$ such that $u = u' v$ holds. In the category **CST** it is convenient to represent an object $X \to A$ of $\mathcal{E}/A$ as the family $(X_a \mid a \in A)$. The operation of pullback along $f : B \to A$ determines a functor $\Delta_f : \mathcal{E}/A \to \mathcal{E}/B$. This functor always has a left adjoint $\Sigma_f : \mathcal{E}/B \to \mathcal{E}/A$ defined by composition with f. The action of Σ_f can be described in **CST** using indexed disjoint unions.

For $f : B \to A$ in $\mathcal{E}$, if the pullback functor $\Delta_f : \mathcal{E}/A \to \mathcal{E}/B$ has a right adjoint $\Pi_f : \mathcal{E}/B \to \mathcal{E}/A$, we can define $P_f : \mathcal{E} \to \mathcal{E}$, the *generalized polynomial functor* associated to f, as the composite

$$\mathcal{E} \xrightarrow{\Delta_B} \mathcal{E}/B \xrightarrow{\Pi_f} \mathcal{E}/A \xrightarrow{\Sigma_A} \mathcal{E}$$

where we made use of the canonical arrows $A : A \to 1$ and $B : B \to 1$.

In the category **CST**, the functor Π_f can be defined if $f : B \to A$ is small. Identifying objects in the slice categories with families of classes, we define the functor Π_f by letting, for a family $(Y_b \mid b \in B)$ indexed by B,

$$\Pi_f (Y_b \mid b \in B) =_{\mathrm{def}} \left(\textstyle\prod_{b \in B_a} Y_b \mid a \in A\right). \tag{4.3}$$

We can then obtain the following explicit description of the generalized polynomial functor associated to an arrow $f : B \to A$. For a class X we have

$$P_f(X) = \textstyle\sum_{a \in A} X^{B_a} .$$

By its very definition, the class of well-founded trees $W_{a \in A} B_a$ is an initial algebra for this functor. This observation leads to define a general notion of well-founded tree in categories [15]. These are defined as initial algebras for generalized polynomial functors. Note that a natural numbers object can be characterized as an initial algebra for the polynomial functor associated to either of the two canonical arrows $1 \to 1 + 1$. For more on well-founded trees, see [15, 22].

The next definition isolates axioms for small maps corresponding to the properties of small maps for CST, as we will discuss after the definition. Our axioms are imposed on top of a standard group of axioms, those for a *class of open maps*, that are not recalled here. These are axioms (**A1**)–(**A7**) of [10, Section 1.1], and were first formulated in [23]. The axioms presented below are tailored to construct models of CST. Axioms (**S1**), (**S2**) were introduced in [10, Definition 1.1], while (**S3**) was considered in [14,23]. By a *small object* we mean an object for which the canonical map into the terminal object is small.

Definition 4.3 *Let $\mathcal{E}$ be a regular category with stable disjoint coproducts, and $\mathcal{S}$ a family of open maps. We say that $\mathcal{S}$ is a family of* CST-*small maps if the following axioms hold.*

- **(S1)** *If $f : B \to A$ is in $\mathcal{S}$, the pullback functor $\Delta_f : \mathcal{E}/A \to \mathcal{E}/B$ has a right adjoint $\Pi_f : \mathcal{E}/B \to \mathcal{E}/A$.*
- **(S2)** *There is a map $\pi : W \to V$ in $\mathcal{S}$ such that for any map $u : X \to A$ in $\mathcal{S}$ there exists a diagram of form*

$$\begin{array}{ccccc} X & \longleftarrow & Y & \longrightarrow & W \\ {\scriptstyle u}\downarrow & & \downarrow & & \downarrow{\scriptstyle \pi} \\ A & \underset{f}{\twoheadleftarrow} & B & \longrightarrow & V \end{array}$$

where $f : B \twoheadrightarrow A$ is an epimorphism and both squares are pullbacks.

- **(S3)** *For every $f : B \to A$, the canonical map $B \to B \times_A B$ is in $\mathcal{S}$.*
- **(S4)** *If $f : B \to A$ is in $\mathcal{S}$, then $\Pi_f : \mathcal{E}/B \to \mathcal{E}/A$ preserves smallness of maps.*
- **(S5)** *For every $f : B \to A$ in $\mathcal{S}$, the polynomial functor $P_f : \mathcal{E} \to \mathcal{E}$ has an initial algebra, whose underlying object is written W_f here. The natural numbers object is small.*
- **(S6)** *For every $f : B \to A$ in $\mathcal{S}$ there is an exact diagram*

$$E \rightrightarrows W_f \longrightarrow V_f$$

where $E \rightarrowtail W_f \times W_f$ is a small subobject such that

$$(\forall (a,t),\ (a',t') \in W_f)\ \Big(((a,t),(a',t')) \in E \leftrightarrow$$
$$(\forall x \in a')(\exists x' \in a')(tx, t'x') \in E \wedge (\forall x' \in a')(\exists x \in a)(tx, t'x') \in E \Big).$$

holds in the internal logic of $\mathcal{E}$.

We now wish to indicate how a category $\mathcal{E}$ with the properties of Definition 4.3 provides a categorical model for CST. The idea follows essentially from the sets-as-trees interpretation of constructive set theory [24–26].

First, consider the small map $\pi : W \to V$ of axiom (**S2**) and its associated well-founded tree W_π. In the category **CST**, a map satisfying axiom (**S2**) can be defined by letting $W =_{\text{def}} \{(x, y) \mid y \in x\}$, $V = \{x \mid x = x\}$, and taking π to be the first projection. Indeed, this map satisfies a stronger property than the one of axiom (**S2**) in that every small map in **CST** can be obtained simply as a pullback of it. The well-founded tree associated to this family is then defined as the smallest class X such that if a is a set and t is a function with domain a, then (a, t) is in X. Such a pair (a, t) may be thought of as a non-extensional set, given as the family $(tx \mid x \in a)$. The non-extensionality means for example that tx and tx' are considered distinct elements of the family even if $tx = tx'$. To obtain extensional sets, it is therefore necessary to take a quotient.

We define V_π to be the quotient of the well-founded tree W_π under the equivalence relation of axiom (**S6**). In **CST** this equivalence relation, denoted by $\cdot \simeq \cdot$, is such that for (a, t) and (b, s) in W_π, we have

$$(a,t) \simeq (b,s) \leftrightarrow (\forall x \in a)(\exists y \in b) tx \simeq sy \wedge (\forall y \in b)(\exists x \in a) tx \simeq sy .$$

The quotient of W_π under this equivalence relation is a class that is isomorphic to the cumulative hierarchy. In an arbitrary category $\mathcal{E}$, the object V_π is then a natural candidate to model CST. The canonical subobject

$$V_\pi \xrightarrow{\ =_{V_\pi}\ } V_\pi \times V_\pi$$

is used to interpret the equality relationship of the language of CST. The interpretation of the rest of the syntax is done as usual [21, Section 4.5], and restricted quantifiers are interpreted using appropriately small maps [13, Remark 3.8]. We do not need to define the interpretation of the membership relation, as we assumed that this is defined using equality and restricted quantifiers.

Theorem 4.4 (Soundness and completeness)

- If $\mathcal{E}$ is a regular category with disjoint coproducts and dual images, equipped with a class of CST-small maps, the object V_π of $\mathcal{E}$ is such that $(V_\pi, =_{V_\pi})$ is a model of CST.
- **CST** is a regular category with disjoint coproducts and dual images, equipped with a family of CST-maps.

Proof. The first part of the theorem follows essentially by Theorem 7.1 of [13]. Definition 4.3 was indeed introduced to isolate the essential elements of the proof necessary to prove the claim.

For the second part of the theorem, one uses the axioms of CST to verify the required conditions. The existence of the structure of a regular category with disjoint coproducts and dual images has already been discussed in Subsection 4.2.2. The functors required by (**S1**) are defined as in (4.3), and (**S2**) is a consequence of the definition of $\pi : W \to V$ given above. For (**S3**) one should recall that equality is a restricted formula, while for (**S4**) one uses Exponentiation and Restricted Separation. Finally, axioms (**S5**) and (**S6**) hold because we can define well-founded trees by induction and relations on them by double set-recursion [8], and the class of natural numbers is a set by Infinity. □

It is possible to have an alternative form of intuition about the cumulative hierarchy object V_π defined above. For a small map $f : B \to A$ in $\mathcal{E}$ let us consider the generalized polynomial functor $P_f : \mathcal{E} \to \mathcal{E}$ and recall that in **CST** this can be expressed as

$$P_f(X) = \textstyle\sum_{a \in A} X^{B_a}\ .$$

In the presence of sufficiently strong axioms for quotients, that are valid in the category of classes, one can then follow [13, Section 6] and define suitable quotients of $P_f(X)$ so as to determine a functor $\mathbb{P}_f : \mathcal{E} \to \mathcal{E}$. In the special case of the category **CST**, this is expressed as

$$\mathbb{P}_f(X) = \{\ p \mid p \subseteq X\ ,\ \ (\exists a \in A)(\exists t \in X^{B_a})\ p = \mathrm{Im}(t)\ \}\ .$$

In **CST**, the functor $\mathbb{P}_\pi$ associated to the arrow $\pi : W \to V$ defined above is exactly the *power-class operation*, defined by letting $\mathbb{P}(X) =_{\mathrm{def}} \{p \mid p \subseteq X\}$, where X is a class. Note that elements of $\mathbb{P}(X)$ are subsets of X, rather than subclasses, and that generally $\mathbb{P}(X)$ is a class, even when X is a set.

Just as the set-theoretic universe is an initial algebra for $\mathbb{P}$: **CST** $\to$ **CST** by the Set Induction axiom, the object V_π defined above is an initial algebra for the functor $\mathbb{P}_\pi : \mathcal{E} \to \mathcal{E}$, as proved in [13]. This point of view will be exploited in the next section, where we define models of CST in categories of presheaves. In order to do so, we recall a property that characterizes the functor $\mathbb{P}_\pi : \mathcal{E} \to \mathcal{E}$ from [10, 11]. We first need a definition.

Definition 4.5 *For $X, I \in \mathcal{E}$, we say that a subobject $R \rightarrowtail I \times X$ is an I-indexed family of small subobjects of X if $R \rightarrowtail I \times X \to I$ is a small map.*

The functor $\mathbb{P}_\pi : \mathcal{E} \to \mathcal{E}$ has the property (**P1**), expressed as in [11].

(P1) For every object X there is a $\mathbb{P}_\pi(X)$-indexed family of small subobjects of X, $\ni_X \rightarrowtail \mathbb{P}_\pi(X) \times X$, such that for all I-indexed families of small subobjects of X, $R \rightarrowtail I \times X$, there exists a unique map $\bar{R} : I \to \mathbb{P}_\pi(X)$ for which there is a pullback diagram of the form

$$\begin{array}{ccc} R & \longrightarrow & \ni_X \\ \downarrow & & \downarrow \\ I \times X & \xrightarrow[\bar{R} \times 1_X]{} & \mathbb{P}_\pi(X) \times X \end{array}$$

It is easy to verify that this property holds in **CST**. For a proof that the property (**P1**) holds in pretoposes, see [10, Section 1.3]. In [11] the property (**P1**) was introduced as part of a simplified axiomatization of categorical models for IZF. It would be of interest to formulate a simple axiom that allows us to derive both the existence of the quotient required by axiom (**S6**) and the definability of the functors $\mathbb{P}_\pi : \mathcal{E} \to \mathcal{E}$.

4.3 Presheaves

In this section, we work again with the constructive set theory CST and consider the category **CST** of classes arising from it. We will use **CST** to define new categories equipped with a class of small maps satisfying all the axioms of Definition 4.3. One could have considered, more generally, working internally in an arbitrary category $\mathcal{E}$ that satisfies the axioms of Definition 4.3 rather than with the explicitly defined category **CST**. The approach taken here, however, will be sufficient to show how Dana Scott's presheaf models [6] fit into the framework of algebraic set theory, and has the advantage of keeping the presentation quite simple.

4.3.1 *Basic definitions*

In this section, $\mathbb{C}$ is a fixed small category, in the sense that we assume that objects and arrows of $\mathbb{C}$ form a set in the constructive set theory CST. Objects and arrows are denoted with $a, b, c, \ldots$, and $f : b \to a$, $g : c \to b$, $h : d \to c, \ldots$, respectively. The identity arrow on an object a will be written $1_a : a \to a$ and the composite of f and g as above will be written $f\,g : c \to a$. For $a, b \in \mathbb{C}$, we let $[b, a]$ be the set of arrows from b to a. To help the intuition, one may think of an object a as a stage in a process, and of an arrow $f : b \to a$ in $\mathbb{C}$ as a transition from the stage a to the stage b.

The *opposite* of $\mathbb{C}$ is the category $\mathbb{C}^{\mathrm{op}}$ whose objects are the same of $\mathbb{C}$ and whose arrows are obtained by formally reversing the direction of the arrows of $\mathbb{C}$. A *presheaf* is a functor $\mathbb{C}^{\mathrm{op}} \to$ **CST**. Thus, a presheaf $X : \mathbb{C}^{\mathrm{op}} \to$ **CST** consists of a family of classes $X(a)$, for $a \in \mathbb{C}$, together with a family of functions

$$\begin{array}{ccc} X(a) \times [b, a] & \longrightarrow & X(b) \\ (x, f) & \longmapsto & x \cdot f \end{array}$$

that satisfy the equations

$$x \cdot 1_a = x\,, \quad (x \cdot f) \cdot g = x \cdot f\,g$$

for all $x \in X(a)$, and $f : b \to a$, $g : c \to b$. We may think of a presheaf X as a class varying through stages, and of the members of $X(a)$, for $a \in \mathbb{C}^{\mathrm{op}}$, as the elements of the variable class X at stage a. The function $X(a) \to X(b)$ determined by an arrow $f : b \to a$ can then be imagined as describing the evolution of the variable class X along the transition f.

As usual, we write $\widehat{\mathbb{C}}$ for the category of presheaves and natural transformations between them. If X and Y are presheaves, we say that a Y is a *subpresheaf* of X if, for all $a \in \mathbb{C}$, $Y(a) \subseteq X(a)$ holds. We now give a few examples of presheaves.

(i) For a class X, we define $X : \mathbb{C}^{\mathrm{op}} \to \mathbf{CST}$, the *constant* presheaf associated to X, as the functor mapping every object into X and any arrow into the identity map on X. The constant presheaf determined by 1 is a terminal object in $\widehat{\mathbb{C}}$.

(ii) For $a \in \mathbb{C}^{\mathrm{op}}$, we define $\mathbf{y}_a : \mathbb{C}^{\mathrm{op}} \to \mathbf{CST}$, the presheaf *represented by* a, by letting $\mathbf{y}_a(b) =_{\mathrm{def}} [b, a]$, for $b \in \mathbb{C}$. The actions are defined so that a pair $(f, g) \in \mathbf{y}_a(b) \times [c, b]$ is mapped into $f\,g \in \mathbf{y}_a(c)$.

(iii) Binary products, and disjoint binary coproducts are defined pointwise. For details, see [1].

The next definition shows how the category $\widehat{\mathbb{C}}$ inherits a notion of small map from **CST**. The definition of small map in **CST** was given in Definition 4.2.

Definition 4.6 *A natural transformation $\alpha : X \to Y$ is said to be small if, for all $a \in \mathbb{C}$, the function between classes $\alpha_a : X(a) \to Y(a)$ is a small map in* **CST**.

To define a presheaf model of CST, i.e. a model of CST whose underlying object is a presheaf, one may prove that $\widehat{\mathbb{C}}$ is a category satisfying all the hypotheses of the first part of Theorem 4.4 by combining the ideas in [10, Section IV§3] and in [13, 15]. We prefer, however, to give an explicit description of the model. This will be achieved by defining explicitly a power-presheaf functor that satisfies axiom (**P1**) and then isolating an initial algebra for it.

4.3.2 *Presheaf models*

Definition 4.7 *Let $X \in \widehat{\mathbb{C}}$. For $a \in \mathbb{C}$, we say that p is a presheaf subset of X at stage a if the following hold:*

(i) *p is a function with domain $\bigcup_{b \in \mathbb{C}} \mathbf{y}_a(b)$,*

(ii) *for all $b \in \mathbb{C}$ and $f : b \to a$, $p(f)$ is a set, and it holds that $p(f) \subseteq X(b)$,*

(iii) *for all $b,\ c \in \mathbb{C}$ and $f : b \to a$, $g : c \to b$, if $x \in p(f)$ then $x \cdot g \in p(f\,g)$.*

The next lemma states a useful property of presheaf subsets. Its proof is a direct consequence of Definition 4.7.

Lemma 4.8 *Let p be presheaf subset of X at stage a, for $X \in \widehat{\mathbb{C}}$ and $a \in \mathbb{C}$. We have $(\forall x \in X_a)\ x \in p(1_a)$ if and only if $(\forall b \in \mathbb{C})(\forall f \in \mathbf{y}_a(b))\ x \cdot f \in p(f)$.*

We begin to define the power-presheaf functor $\mathbb{P}_\pi : \widehat{\mathbb{C}} \to \widehat{\mathbb{C}}$ by letting

$$\mathbb{P}_\pi(X)(a) =_{\mathrm{def}} \{p \mid p \text{ presheaf subset of } X \text{ at stage } a\}.$$

for $X \in \widehat{\mathbb{C}}$ and $a \in \mathbb{C}$. To define the action

$$\begin{array}{ccc} \mathbb{P}_\pi(X)(a) \times [b,a] & \longrightarrow & \mathbb{P}_\pi(X)(b) \\ (\,p,\ f\,) & \longmapsto & p \cdot f \end{array}$$

we let $p \cdot f$ be the function mapping $g \in [c, b]$ into $p(f\,g)$. The functoriality required by the definition of presheaf follows directly. We are almost ready to verify the property of the power-presheaf functor stated in axiom (**P1**). We only need to exhibit, for every presheaf X, a $\mathbb{P}_\pi(X)$-indexed family of small subobjects of X that plays the role of a membership relation. For $X \in \widehat{\mathbb{C}}$ we define $\ni_X$ by letting

$$(\ni_X)(a) =_{\text{def}} \{(p, x) \in \mathbb{P}_\pi(X)(a) \times X(a) \mid x \in p(1_a)\}$$

for $a \in \mathbb{C}$. Observe that $\ni_X$ is a subpresheaf of $\mathbb{P}_\pi(X) \times X$ by Lemma 4.8 and that it is a $\mathbb{P}_\pi(X)$-indexed family of small subobjects of X.

Proposition 4.9 *The functor* $\mathbb{P}_\pi : \widehat{\mathbb{C}} \to \widehat{\mathbb{C}}$ *satisfies the property* (**P1**).

Proof. For an I-indexed family of small subobjects of X, $R \rightarrowtail I \times X$, we need to define a natural transformation $\bar{R} : I \to \mathbb{P}_\pi(X)$. For $a \in \mathbb{C}$, $i \in I(a)$ and $f : b \to a$ we define

$$\bar{R}_a(i)(f) =_{\text{def}} \{y \in X(b) \mid (i \cdot f, y) \in R(b)\},$$

where we assumed that R is a subpresheaf of $I \times X$ for simplicity. This definition determines a function $\bar{R}_a(i)$ that is a presheaf subset of X at stage a, so that $\bar{R}_a(i) \in \mathbb{P}_\pi(X)$. For $a \in \mathbb{C}$, we then have a function $\bar{R}_a : I(a) \to \mathbb{P}_\pi(X)(a)$, which gives us the components of the required natural transformation. □

Dana Scott's definition of a presheaf cumulative hierarchy in [6] is exactly an initial algebra for the power-presheaf functor. The next result recalls its characterization and states that the required presheaf can be defined in CST.

Theorem 4.10 *We can define a presheaf* V_π *such that, for* $a \in \mathbb{C}$, *it holds that* $s \in V_\pi(a)$ *if and only if the following conditions hold:*

- *s is a function with domain* $\bigcup_{b \in \mathbb{C}} \mathbf{y}_a(b)$,
- *for all* $f : b \to a$, *we have* $s(f) \in V_\pi(b)$,
- *for all* $f : b \to a$ *and* $g : c \to b$, *if* $t \in s(f)$ *then* $t \cdot g \in s(f\,g)$,

where, for $s \in V_\pi(a)$ *and* $f : b \to a$, *we let* $s \cdot f$ *be the function with domain* $\bigcup_{c \in \mathbb{C}} \mathbf{y}_b(c)$ *mapping* $g : c \to b$ *into* $s(f\,g)$.

Proof. The claim is a consequence of the possibility of defining classes by general forms of inductive definitions in CST. The details of the appropriate inductive definition are given in [27, Section 6.3], where it is also shown how V_π can be seen as the smallest presheaf that satisfies the requirements above. □

4.4 Kripke–Joyal Semantics

The presheaf V_π can be used to give a direct interpretation of all the axioms of CST using the Kripke–Joyal semantics [1, Section VI.6]. To do so, we need to fix some syntactic conventions. For $a \in \mathbb{C}$, we define the language $\mathcal{L}(a)$ to be the extension of the language $\mathcal{L}$ of CST with constants for elements of $V_\pi(a)$. As usual, we do not distinguish between elements of $V_\pi(a)$ and the constants added to the language $\mathcal{L}$ and use letters $s, t, r, \ldots$ for them. If ϕ is a formula of $\mathcal{L}(a)$ and $f : b \to a$ we define $\phi \cdot f$ to be the formula obtained from ϕ by leaving unchanged free variables and substituting each constant s appearing in ϕ with the constant $s \cdot f$. Observe that if ϕ is a sentence of $\mathcal{L}(a)$ then $\phi \cdot f$ is a sentence of $\mathcal{L}(b)$. The Kripke–Joyal semantics can then be defined by structural induction as in Table 4.1. Lemma 4.11 then states one of the expected properties of the semantics.

Lemma 4.11 (Monotonicity) Let $a \in \mathbb{C}$ and ϕ be a sentence of $\mathcal{L}(a)$. If $a \Vdash \phi$ then for all $b \in \mathbb{C}$ and all $f \in \mathbf{y}_a(b)$ it holds that $b \Vdash \phi \cdot f$.

Proof. The claim follows by structural induction on ϕ. □

To illustrate some of the properties of presheaf models, we investigate in more detail the interpretation of sentences. Let $a \in \mathbb{C}$. We say that a class P of arrows with codomain a is a *sieve* on a if for all $f : b \to a$ and $g : c \to b$ if $f \in P$ then $f\,g \in P$. We say that a sieve on a is a *set-sieve* if it is a set. Let $a,\ b \in \mathbb{C}$. For a set-sieve p on a and $f : b \to a$, define $p \cdot f =_{\text{def}} \{g \mid f\,g \in p\}$ and observe that $p \cdot f$ is a set-sieve on b. We can then define a presheaf Ω by letting, for $a \in \mathbb{C}$

$$\Omega_a =_{\text{def}} \{p \mid p \text{ set-sieve on } a\}.$$

Table 4.1 *Definition of the Kripke–Joyal semantics*

$$a \Vdash \bot =_{\text{def}} \bot$$

$$a \Vdash s = t =_{\text{def}} s = t$$

$$a \Vdash \phi \wedge \psi =_{\text{def}} (a \Vdash \phi) \wedge (a \Vdash \psi)$$

$$a \Vdash \phi \vee \psi =_{\text{def}} (a \Vdash \phi) \vee (a \Vdash \psi)$$

$$a \Vdash \phi \to \psi =_{\text{def}} (\forall b \in \mathbb{C})(\forall f \in \mathbf{y}_a(b))(b \Vdash \phi \cdot f \to b \Vdash \psi \cdot f)$$

$$a \Vdash (\exists x \in s)\phi =_{\text{def}} \big(\exists x \in s(1_a)\big)\; a \Vdash \phi$$

$$a \Vdash (\forall x \in s)\phi =_{\text{def}} (\forall b \in \mathbb{C})(\forall f \in \mathbf{y}_a(b))\,(\forall x \in s(f))\, b \Vdash \phi \cdot f$$

$$a \Vdash (\exists x)\phi =_{\text{def}} \big(\exists x \in V_\pi(a)\big)\; a \Vdash \phi$$

$$a \Vdash (\forall x)\phi =_{\text{def}} (\forall b \in \mathbb{C})(\forall f \in \mathbf{y}_a(b))\big(\forall x \in V_\pi(b)\big)\, b \Vdash \phi \cdot f$$

The next definition provides a link between the Kripke–Joyal semantics of sentences and sieves. For a sentence ϕ of $\mathcal{L}(a)$ define

$$[\![\phi]\!] =_{\text{def}} \bigcup_{b \in \mathbb{C}} \{f \in [b, a] \mid b \Vdash \phi \cdot f\}.$$

Proposition 4.12 *Let a in $\mathbb{C}_0$. Let ϕ be a sentence of $\mathcal{L}(a)$. The class $[\![\phi]\!]$ is a sieve on a, and if ϕ is restricted then $[\![\phi]\!]$ is a set-sieve on a.*

Proof. For the first claim, Lemma 4.11 gives the desired conclusion. For the second, use structural induction on ϕ, observing that the clauses defining the semantics of a restricted formula are themselves restricted. □

Theorem 4.13 $(V_\pi, =)$ *is a model of* CST.

Proof. The claim follows mainly from Theorem 6.17 in [27], apart from the validity of Exponentiation, which is a consequence of Theorem 7.1 and Theorem 9.6 of [13]. □

4.5 Conclusions

Since partially ordered sets are special small categories, presheaf models give as a special case extensions for constructive set theories of Kripke models for intuitionistic logic. In [28] Erik Palmgren applied a Kripke model construction to show an independence result for first-order intuitionistic logic. The result regards the notion of *pseudo-order* on a set. A binary relation $\cdot < \cdot$ on a set A is a pseudo-order if the following hold:

1. $\neg\big((a < b) \wedge (b < a)\big)$, for all $a, b \in A$
2. $\neg(a < b) \wedge \neg(b < a) \rightarrow a = b$, for all $a, b \in A$
3. $a < b \rightarrow \big((a < c) \vee (c < a)\big)$, for all $a, b, c \in A$

An example of pseudo-order is given by the strict order on Cauchy or Dedekind reals. Classically, every pseudo-order is a linear order and thus every two elements have a supremum. Palmgren's result shows that this is not the case in intuitionistic logic. For the proof, obtained using a Kripke model, see [28].

Proposition 4.14 (Palmgren) In intuitionistic first-order logic, the axioms of a pseudo-order do not imply that every two elements of a pseudo-order have a supremum.

It seems straightforward to generalize this result to an independence result for CST using presheaf models. Presheaf models offer, however, much more generality. One may indeed consider models, relevant for the study of the lambda-calculus [6], in which the category $\mathbb{C}$ is taken to be a monoid. Furthermore, it is possible to generalize presheaf models to sheaf models [10, 13, 27] which give variants of the double-negation translation suitable for constructive set theories. In this chapter, we set the ground for these promising applications.

Acknowledgements

I am most grateful to Dana Scott, who made available his notes on presheaf models. I also acknowledge useful discussions with Alex Simpson, Peter Aczel, Steve Awodey, Erik Palmgren, Benno van den Berg, Claire Kouwenhoven-Gentil, and Robert Lubarsky.

This research was supported by an EPSRC Postdoctoral Fellowship in Mathematics (GR/R95975/01).

References

1. MacLane, S. and Moerdijk, I. (1992). *Sheaves in Geometry and Logic*. Springer.
2. Pitts, A. M. (2000). Categorical Logic. In: *Handbook of Logic in Computer Science*, Vol. 5 (eds S. Abramsky, D. M. Gabbay, and T. S. E. Maibaum), Oxford University Press.
3. Joyal, A. and Tierney, M. (1984). An extension of the Galois theory of Grothendieck. *Mem. Amer. Math. Soc.*, **51**(309).
4. Myhill, J. R. (1975). Constructive Set Theory. *Journal of Symbolic Logic*, **40**(3), 347–382.
5. Aczel, P. and Rathjen, M. (2001). Notes on Constructive Set Theory. Technical Report 40. Mittag-Leffler Institute; Available from the web page *http://www.cs.man.ac.uk/petera/papers.html.*
6. Scott, D. S. (1985). Category-theoretic models for Intuitionistic Set Theory. Manuscript slides of a talk given at Carnagie-Mellon University.
7. Martin-Löf, P. (1984). *Intuitionistic Type Theory*. Bibliopolis.
8. Griffor, E. R. and Rathjen, M. (1994). The strength of some Martin-Löf type theories. *Archive for Math Logic*. **33** 347–385.
9. Aczel, P. (1977). An introduction to inductive definitions. In: *Handbook of Mathematical Logic* (ed J. Barwise), pp. 739–782. North-Holland.
10. Joyal, A. and Moerdijk, I. (1995). *Algebraic Set Theory*. Cambridge University Press.
11. Simpson, A. K. (1999). Elementary axioms for the category of classes. In: *14th Annual IEEE Symposium on Logic in Computer Science*. pp. 77–85. IEEE Press.
12. Butz, C. (2003). Bernays–Gödel type theory. *Journal of Pure and Applied Algebra*, **178**(1), 1–23.
13. Moerdijk, I. and Palmgren, E. (2002). Type theories, toposes and Constructive Set Theory: predicative aspects of AST. *Annals of Pure and Applied Logic*, **114**(1–3), 155–201.
14. Awodey, S., Butz, C., Streicher, T., and Simpson, A. K. (2003). Relating topos theory and set theory via categories of classes. Draft paper, available from the web page *http://www.andrew.cmu.edu/user/awodey.*
15. Moerdijk, I. and Palmgren, E. (2000). Wellfounded trees in categories. *Annals of Pure and Applied Logic*, **104**, 189–218.
16. Warren, M. (2004). Predicative categories of classes. Master's thesis. Carnagie Mellon University.
17. Lubarsky, R. S. Independence results around Constructive ZF. *Annals of Pure and Applied Logic*, **132**, 209–225.

18. Friedman, H. M. (1973). The consistency of classical set theory relative to a set theory with intuitionistic logic. *Journal of Symbolic Logic*, **38**, 315–319.
19. Fourman, M. P. (1980). Sheaf models for set theory. *Journal of Pure and Applied Algebra*, **19**, 91–101.
20. MacLane, S. (1971). *Categories for the working mathematician*. Springer.
21. Jacobs, B. (1999). *Categorical Logic and Type Theory*. North-Holland.
22. Gambino, N. and Hyland, M. (2004). Wellfounded trees and dependent polynomial functors. In: *Types for Proofs and Programs* (eds S. Berardi, M. Coppo, and F. Damiani), Vol. **3085** of Lecture Notes in Computer Science. Springer.
23. Joyal, A. and Moerdijk, I. (1994). A completeness theorem for open maps. *Annals of Pure and Applied Logic*, **70**, 51–86.
24. Aczel, P. (1978). The type theoretic interpretation of Constructive Set Theory. In: *Logic Colloquium '77* (eds nd L. Pacholski, A. MacIntyre, and J. Paris), pp. 55–66. North-Holland.
25. Aczel, P. (1982). The type theoretic interpretation of Constructive Set Theory: choice principles. In: *The L. E. J. Brouwer Centenary Symposium* (eds A. S. Troelstra, and D. van Dalen), pp. 1–40. North-Holland.
26. Aczel, P. (1986). The type theoretic interpretation of Constructive Set Theory: inductive definitions. In: *Logic, Methodology and Philosophy of Science VII* (eds Marcus R. Barcan, G. J. W. Dorn, and P. Weinegartner), pp. 17–49. North-Holland.
27. Gambino, N. (2002). Sheaf interpretations for generalised predicative intuitionistic systems. PhD thesis. University of Manchester.
28. Palmgren, E. (2003). Constructive completions of ordered sets, groups and fields. Technical Report U.U.D.M. Report 2003:5. Department of Mathematics, University of Uppsala.

5

UNIVERSES IN TOPOSES

THOMAS STREICHER

Abstract

We discuss a notion of *universe* in toposes which from a logical point of view gives rise to an extension of Higher Order Intuitionistic Arithmetic (HAH) that allows one to construct families of types in such a universe by structural recursion and to quantify over such families. Further, we show that (hierarchies of) such universes do exist in all sheaf and realizability toposes but neither in the free topos nor in the $V_{\omega+\omega}$ model of Zermelo set theory.

Though universes in **Set** are necessarily of strongly inaccessible cardinality it remains an open question whether toposes with a universe allow one to construct internal models of Intuitionistic Zermelo–Fraenkel set theory (IZF).

The background information about toposes and fibred categories as needed for our discussion in this chapter can be found e.g. in the fairly accessible sources [1–3].

5.1 Background and Motivation

It is commonly agreed on that elementary toposes with a natural numbers object (NNO) provide a concise and flexible notion of model for constructive Higher Order Arithmetic (HAH). Certainly, a lot of mathematics can be expressed within HAH. So what is the need then for set theory (ZFC) which is generally accepted as *the* foundation for mainstream mathematics? Well, ZFC is much stronger than HAH in the following respects:

(1) ZFC is based on classical logic whereas HAH is based on the weaker intuitionistic logic.

(2) ZFC postulates the axiom of choice whereas HAH does not.

(3) ZFC postulates the axiom of replacement which cannot even be formulated in HAH.[1]

The logic of toposes (with NNO) is inherently intuitionistic and in HAH the axiom of choice implies classical logic. Therefore, we have to give up (1) and (2) above when

[1] Notice, however, that the axiom of replacement obtains its full power only in the presence of the full separation scheme. In recent, yet unpublished work by S. Awodey, C. Butz, A. Simpson and T. Streicher [4] it has been shown that set theory with *bounded separation*, i.e. separation restricted to bounded formulas, but with replacement (and even strong collection) is equiconsistent to HAH as long as the underlying logic is intuitionistic. Otherwise the classical principle of excluded middle allows one to derive full separation from replacement.

In the context of this paper when we say replacement we mean the power of replacement together with full separation (although the latter does not make sense from a type-theoretic point of view!).

considering Grothendieck and realizability toposes as models of some kind of set theory. But what about (3), the axiom of replacement?

First of all notice that there are models of set theory without replacement but satisfying classical logic and choice, namely $V_{\omega+\omega}$.[2] On the other hand a lot of toposes, in particular Grothendieck toposes and realizability toposes, do model the axiom of replacement whereas in most cases they refute classical logic and the axiom of choice. More precisely, the above mentioned toposes model IZF, i.e. ZF with intuitionistic logic and the axiom of regularity reformulated as $\in$-induction.[3]

Though a large class of toposes validates IZF one still may complain that the formulation of IZF suffers from 'epsilonitis', i.e. that it 'implements' informal mathematics via the $\in$-relation rather than axiomatizing mathematical practice in terms of its basic notions. So one may ask *what is the mathematical relevance of the set-theoretic replacement axiom?* Maybe a set-theorist would answer 'for constructing ordinals greater than $\omega+\omega$' which, however, may seem a bit disappointing because most mathematics can be formulated without reference to transfinite ordinals.[4] Actually, what the axiom of replacement is mainly needed for in mathematical practice is to define families of sets indexed by some set I carrying some inductive structure as, typically, the set $\mathbb{N}$ of natural numbers. For example, most mathematicians would not hesitate to construct the sequence $(\mathcal{P}^n(\mathbb{N}))_{n\in\mathbb{N}}$ by (primitive) recursion over $\mathbb{N}$. Already in ZC, however, this is impossible because $\{\langle n, \mathcal{P}^n(\mathbb{N})\rangle \mid n\in\mathbb{N}\} \notin V_{\omega+\omega}$. Usually, in ZFC the sequence $(\mathcal{P}^n(\mathbb{N}))_{n\in\mathbb{N}}$ is constructed by applying the axiom of replacement to an appropriately defined class function from the set of natural numbers to the class of all sets. However, in a sense that does not properly reflect the mathematician's intuition who thinks of $(\mathcal{P}^n(\mathbb{N}))_{n\in\mathbb{N}}$ as a function f from $\mathbb{N}$ to sets defined recursively as $f(0) = \mathbb{N}$ and $f(n{+}1) = \mathcal{P}(f(n))$. There is, however, a 'little' problem, namely that the collection of all sets does not form a set but a proper class. Notice, however, that a posteriori the image of f does form a set as ensured by the replacement axiom. Notice, moreover, that the image of f is contained as a subset in the set $V_{\omega+\omega} = \bigcup_{n\in\mathbb{N}} \mathcal{P}^n(V_\omega)$ whose existence can be ensured again by the axiom of replacement in a similar way as above. Thus, if we had $V_{\omega+\omega}$ available as a set beforehand we could define f as a function from the set $\mathbb{N}$ to the set $V_{\omega+\omega}$ simply by primitive recursion. The distinguishing feature of $V_{\omega+\omega}$ is that it is closed under subsets, cartesian products, powersets (and, thus, also under exponentiation) and contains V_ω, the set of hereditarily finite sets, and, therefore, also $\mathbb{N}$ as elements. Actually, the set $V_{\omega+\omega}$ has even stronger closure properties, namely, that

(1) every element of $V_{\omega+\omega}$ is also a subset of $V_{\omega+\omega}$, i.e. $V_{\omega+\omega}$ is a so-called transitive set;

---

[2]But notice that $V_{\omega+\omega}$ validates ZC, i.e. ZFC *without* replacement, which, however, is still stronger than HAH as Z proves the consistency of HAH.

[3]The axiom of regularity and $\in$-induction are equivalent only classically as in IZF the principle of excluded middle follows from the axiom of regularity just as in HAH the principle of excluded middle follows from the least number principle.

[4]There are notable exceptions typically in the area of descriptive set theory as, e.g. *Borel determinacy* which is provable in ZF (as shown by D. A. Martin) but not in Z (as shown by H. Friedman). Even IZF does not decide Borel determinacy as it holds in **Set** but not in Hyland's effective topos **Eff**.

(2) $V_{\omega+\omega}$ with $\in$ restricted to it provides a model of Zermelo set theory (with choice), i.e. ZF(C), without the axiom of replacement.

Sets satisfying these two properties are called *Zermelo universes* and they are abundant because V_λ is a Zermelo universe for all limit ordinals $\lambda > \omega$. Thus, there are as many Zermelo universes as there are ordinals.

A Zermelo universe U satisfying the additional requirement that

(3) if f is a function with $\mathsf{dom}(f) \in U$ and $\mathsf{rng}(f) \subseteq U$ then $\mathsf{rng}(f) \in U$

is called a *Grothendieck universe* because it was A. Grothendieck who introduced this notion for the purpose of a convenient and flexible set-theoretic foundation of category theory. More precisely, he suggested to use ZFC together with the requirement that every set A be contained in some Grothendieck universe guaranteeing at least[5] an infinite sequence

$$U_0 \in U_1 \in \ldots U_{n-1} \in U_n \in U_{n+1} \in \ldots$$

of Grothendieck universes. As Grothendieck universes are transitive such a sequence is also *cumulative* in the sense that

$$U_0 \subseteq U_1 \subseteq \ldots U_{n-1} \subseteq U_n \subseteq U_{n+1} \subseteq \ldots$$

holds as well. One can show that V_λ is a Grothendieck universe if and only if λ is a *strongly inaccessible* cardinal, i.e. λ is an infinite regular cardinal with $2^\kappa < \lambda$ for all $\kappa < \lambda$.

Obviously, ZFC does not prove the existence of Grothendieck universes (or of strongly inaccessible cardinals) as otherwise ZFC could prove its own consistency (as a Grothendieck universe provides a small inner model of ZFC) which is impossible by Gödel's second incompleteness theorem.[6]

5.2 Universes in Toposes

We now define a notion of *universe in an (elementary) topos* that is stronger than the set-theoretic axiom of replacement but adapted to the 'spirit' of type theory and thus freed from 'epsilonitis'.

We do not claim any originality for the subsequent notion as it is inspired by categorical semantics (see, e.g. [5]) for an impredicative version of Martin-Löf's universes as can be found in the Extended Calculus of Constructions (see [6]). Categorical semantics of universes was anticipated by Jean Bénabou's influential paper [7] (from 1971!) introducing (among other important things) a notion of *topos internal to a topos*.[7]

---

[5] Actually, postulating choice for classes gives rise to a class function *Un* that assigns to every set a a Grothendieck universe $Un(a)$ with $a \in Un(a)$. Then by transfinitely iterating the function *Un* one obtains incredibly big hierarchies of Grothendieck universes.

[6] Notice, however, that the notion of Grothendieck universe is stronger than the notion of *small inner model* which is a Zermelo universe required to satisfy condition (3) above only for those f which are first order definable in the language of set theory!

[7] Alas, later work on categorical semantics of type theories usually does not refer to [7] but rather implicitly to some 'folklore' dating back to the early 1970s when [7] was written.

Definition 5.1 *A universe in a topos $\mathcal{E}$ is given by a class $\mathcal{S}$ of morphisms in $\mathcal{E}$ satisfying the following conditions:*

(1) *$\mathcal{S}$ is stable under pullbacks along morphisms in $\mathcal{E}$, i.e. for every pullback*

$$\begin{array}{ccc} B & \longrightarrow & A \\ {\scriptstyle b}\Big\downarrow & \lrcorner & \Big\downarrow{\scriptstyle a} \\ J & \xrightarrow[f]{} & I \end{array}$$

in $\mathcal{E}$ it holds that $b \in \mathcal{S}$ whenever $a \in \mathcal{S}$.

(2) *$\mathcal{S}$ contains all monos of $\mathcal{E}$.*

(3) *$\mathcal{S}$ is closed under composition, i.e. if $f : A \to I$ and $g : B \to A$ are in $\mathcal{S}$ then $\Sigma_f g = f \circ g \in \mathcal{S}$.*

(4) *$\mathcal{S}$ is closed under dependent products, i.e. if $f : A \to I$ and $g : B \to A$ are in $\mathcal{S}$ then $\Pi_f g \in \mathcal{S}$ (where Π_f is right adjoint to $f^* : \mathbb{E}/I \to \mathbb{E}/A$).*

(5) *In $\mathcal{S}$ there is a generic*[8] *morphism, i.e. a morphism $El : E \to U$ in $\mathcal{S}$ such that for every $a : A \to I$ in $\mathcal{S}$ we have*

$$\begin{array}{ccc} A & \longrightarrow & E \\ {\scriptstyle a}\Big\downarrow & \lrcorner & \Big\downarrow{\scriptstyle El} \\ I & \xrightarrow[f]{} & U \end{array}$$

for some morphism $f : I \to U$ in $\mathcal{E}$, i.e. $a \cong f^*El$.

A universe $\mathcal{S}$ is called impredicative iff $\Omega \to 1$ is in $\mathcal{S}$. If $\mathcal{E}$ has a natural numbers object N then we say that $\mathcal{S}$ contains N iff $N \to 1$ is in $\mathcal{S}$.

As already mentioned this notion of universe is inspired by a similar notion introduced by J. Bénabou in [7]. His main motivation was to provide a notion of *internal topos* capturing finite cardinals, i.e. Kuratowski finite sets with decidable equality, inside a topos $\mathcal{E}$ with natural numbers object N for which a generic morphism is provided by

$$k = \pi_2 : K = \{(i, n) \in N^2 \mid i < n\} \longrightarrow N.$$

The class $\mathbb{F}$ of all morphisms which can be obtained as pullback of the generic family k coincides with the class of families of finite cardinals. It satisfies conditions (1) and (3)–(5) of Definition 5.1. The monos in $\mathbb{F}$ are classified by $\mathsf{inl} : 1 \to 1{+}1$ and, accordingly, $\mathbb{F}$ contains all monos if and only if $\mathcal{E}$ is boolean. Thus, the logic of the internal topos of finite cardinals as given by $\mathbb{F}$ coincides with the logic of the ambient topos $\mathcal{E}$ if and only if $\mathcal{E}$ is boolean. On the other hand the morphism k is not only generic but also

[8] Notice that we do not require uniqueness of f, i.e. *El* is not a 'classifying' but only a 'generic' family for $\mathcal{S}$. Some authors use also the word 'weakly classifying' instead of 'generic'.

classifying[9] for $\mathbb{F}$ because in the internal logic of $\mathcal{E}$ from $K(n) \cong K(m)$ it follows that $n = m$.

Obviously, there are many possibilities for varying the notion of universe according to one's needs. The minimal notion is given by a class $\mathcal{S}$ of morphism satisfying condition (1) (which already entails that $\mathcal{S}$ is closed under isomorphism). Such classes $\mathcal{S}$ satisfying (1) coincide with full (and replete) subfibrations of the fundamental fibration $P_{\mathcal{E}} = \partial_1 : \mathcal{E}^{\mathbf{2}} \to \mathcal{E}$ of $\mathcal{E}$ and, therefore, are abundant in category theory and categorical logic, in particular, as they provide the correct fibrational generalization of the notion of full (and replete) subcategory. In semantics of dependent type theories such pullback stable classes $\mathcal{S}$ are called 'classes of display maps' for some internal collection of types (see e.g. [2, 5] for a more detailed treatment). In the *algebraic set theory* of Joyal and Moerdijk [8] they are called classes of 'small' maps and are thought of as 'families of sets indexed by a class'. In a sense the word 'small' is somewhat misleading here because for $a : A \to I$ in $\mathbb{F}$ and $m : A' \rightarrowtail A$ the composite $a \circ m$ need not be in $\mathbb{F}$ even for $I = 1$ because, constructively, finite cardinals are not closed under subsets. Already subterminals need not be finite cardinals and, accordingly, in general $\mathbb{F}$ will not satisfy condition (2) claiming that families of subterminals are 'small' families. But this phenomenon has nothing to do with finiteness *per se* as in some (non-standard) set theories subclasses of sets need not be sets themselves.[10]

Condition (3) says that (in each fibre) for a family of 'small' sets indexed over a 'small' set its sum (disjoint union) is small, too. From (1) and (3) it follows that for $a : A \to I$ and $b : B \to I$ the fibrewise product $a \times_I b : A \times_I B \to I$ is in $\mathcal{S}$, too. If, moreover, condition (2) is assumed then for $a : A \to I$ in $\mathcal{S}$ and arbitrary subobjects $m : A' \rightarrowtail A$ the composite $a \circ m$ is in $\mathcal{S}$, too. Thus, under assumption of (1), (2) and (3) for every object I in $\mathcal{E}$ the full subcategory $\mathcal{S}/I$ of $\mathcal{E}/I$ (on maps of $\mathcal{S}$ with codomain I) is finitely complete and inherits its finite limits from $\mathcal{E}/I$.[11]

Condition (4) says that the full subcategory as given by $\mathcal{S}$ is closed under dependent products and, therefore, under exponentiation. Under assumption of (1), (2) and (3) condition (4) is equivalent to the requirement that every $\mathcal{S}/I$ is closed under exponentiation in $\mathcal{E}/I$, i.e. that $\mathcal{S}/I$ is a full subcartesian closed category of $\mathcal{E}/I$.

---

[9] In [7] the more general notion of generic family was not considered. Probably because unique existence is 'more categorical' than mere existence and certainly because of the above example of finite cardinals. One might wonder whether one can always construct a classifying family from a generic one. However, this seems to be unlikely because it amounts to choosing representatives from isomorphism classes which is not only unnatural but also impossible constructively.

[10] In Gödel–Bernays–von Neumann class theory (GBN) every subclass of a set is guaranteed to be a set. But GBN guarantees only the existence of classes which are *first order definable* in the language of set theory. However, there is no reason why the intersection of a set with an *arbitrary non-standard* class should be a set in general. Consider, e.g. the class of *standard* elements of the set $\mathbb{N}$ of all natural numbers. Such phenomena lie at the heart of non-standard set/class theories like E. Nelson's Inner Set Theory, P. Vopenka's Alternative Set Theory or E. Gordon's Non-standard Class Theory. Non-standard Class Theory was investigated and developed by J. Bénabou in the early 1970s in quite some detail but, alas, never published.

[11] Actually, a weaker condition than (2) suffices for this purpose, namely that every *regular monomorphism* is in $\mathcal{S}$. Under assumption of (1) and (3) this weakening of (2) is equivalent to the requirement that $\mathcal{S}$ contains all isos and $fg \in \mathcal{S}$ implies $g \in \mathcal{S}$. Thus, it follows in particular that morphisms between small maps are small themselves.

Under assumption of conditions (1)–(4) condition (5) is equivalent to the requirement that the subfibration of $P_{\mathcal{E}}$ as given by $\mathcal{S} \hookrightarrow \mathcal{E}^{\mathbf{2}}$ is *equivalent*[12] to a *small* fibration. This still holds even if condition (2) is weakened to the requirement that for all $I \in \mathcal{E}$ the subcategory $\mathcal{S}/I$ of $\mathcal{E}/I$ is closed under equalizers. In general, a generic family need not be classifying as it can well happen that for distinct $f_1, f_2 : I \to U$ the families f_1^*El and f_2^*El are isomorphic as families over I. The family El is classifying iff in the internal logic of $\mathcal{E}$ it holds that $\forall a, b \in U\big(El(a) \cong El(b) \Rightarrow a =_U b\big)$. Notice that this requirement fails already when $\mathcal{E}$ is **Set** and $\mathcal{S}$ is the family $(a)_{a \in U}$ for some Grothendieck universe U in **Set** because U will contain an awful lot of distinct but equipollent sets. This, of course, can be overcome by restricting U to the cardinal numbers in U which, however, is only possible in the presence of the axiom of choice and in any case does not seem very natural.

One of the useful consequences of condition (5) is that the maps of $\mathcal{S}$ are closed under $+$ which can be seen as follows. Suppose $a : A \to I$ and $b : B \to J$ are maps in $\mathcal{S}$. By condition (5) there exist maps $f : I \to U$ and $g : J \to U$ with $a \cong f^*El$ and $b \cong g^*El$. Due to the extensivity properties of toposes we then have $a + b \cong f^*El + g^*El \cong [f, g]^*El \in \mathcal{S}$ as desired.

Notice that $\mathcal{S} = \mathrm{Mono}(\mathcal{E})$ is a class of maps satisfying conditions (1)–(5). Thus, a universe need not contain the terminal projection $\Omega_{\mathcal{E}} \to 1_{\mathcal{E}}$. However, if it does, i.e. if $\mathcal{S}$ is 'impredicative', then every $\mathcal{S}/I$ contains the object $\pi : I \times \Omega \to I$ and thus $\mathcal{S}/I$ is a subtopos of $\mathcal{E}/I$ in the sense that $\mathcal{S}/I \hookrightarrow \mathcal{E}/I$ is a logical functor. Moreover, for $\mathcal{E} = \mathbf{Set}$ the class $\mathbb{F}$ of families of finite cardinals gives rise to an impredicative universe which, however, does not contain N.

One of the useful consequences of impredicativity is that $\mathcal{S}$ is closed under quotients in the sense that if $a \circ e \in \mathcal{S}$ and e is epic then $a \in \mathcal{S}$. This follows immediately from the facts that for all objects I in $\mathcal{E}$ the inclusion $\mathcal{S}/I \hookrightarrow \mathcal{E}/I$ is logical and that in a topos for every epimorphism $e : A \twoheadrightarrow B$ the object B appears as subobject of $\mathcal{P}(A)$ via $e^{-1} : B \rightarrowtail \mathcal{P}(A)$.

Finally, we explain how condition (5) allows one (in the presence of the other conditions of Definition 5.1) to define families of sets in U via recursion over some index set (as e.g. the natural numbers object N) and to quantify over families of small sets (over a fixed index set).

Let $\pi_1, \pi_2 : U \times U \to U$ be the first and second projection, respectively. Then by condition (1) the maps π_1^*El and π_2^*El are in $\mathcal{S}$. From conditions (3) and (4) it follows that then the exponential $\pi_2^*El^{\pi_1^*El}$ is also in $\mathcal{S}$. By condition (5) there exists a map $fun : U \times U \to U$ such that $\pi_2^*El^{\pi_1^*El} \cong fun^*El$. Obviously, the map $fun : U \times U \to U$ *internalizes* the exponentiation of sets in U. We often write b^a as an abbreviation for $fun(a, b)$. If $\mathcal{S}$ is impredicative then there exists an $\omega : 1 \to U$ such that ω^*El is isomorphic to the terminal projection $!_\Omega : \Omega \to 1$. Then the map $pow = fun \circ \langle \omega \circ !_U, id_U \rangle : U \to U$ *internalizes* the powerset operation on U since $pow(a) = \omega^a$. If $\mathcal{S}$ contains N then there exists an $n : 1 \to U$ such that n^*El is isomorphic to $!_N$:

[12]This fibration need not itself be small as there need not be a classifying family, only a generic one. But it is equivalent to the small fibration arising from the internal category in $\mathcal{E}$ whose set of objects is given by U and whose family of morphisms is given by the exponential El^{El} in $\mathcal{E}/U$.

$N \to 1$. Due to the universal property of the NNO N there exists a unique map $p : N \to U$ with $p(0) = n$ and $p(k+1) = pow(p(k))$ for all $k \in N$. Obviously, the family p^*El provides the desired generalization of $(\mathcal{P}^k(\mathbb{N}))_{k\in\mathbb{N}}$ to toposes with a universe.

For every object A of $\mathcal{E}$ the exponential U^A exists. Thus, we can quantify over U^A, i.e. A-indexed families of small sets. As the family $\big(U^{El(a)}\big)_{a\in U}$ is given by the exponential $(!_U^* U)^{El}$ in $\mathcal{E}/U$ and every topos is in particular locally cartesian closed we have available in toposes with a universe also quantifications such as $\big(\forall a : U\big)\big(\forall b : U^{El(a)}\big)\big(\forall f : (\Pi x{:}El(a))El(b(x))\big) \ldots$, i.e. quantification over all sections of families of small sets indexed by small sets.

In particular, a topos with a universe $\mathcal{S}$ provides *internal quantification* over all sorts of small structures (where the notion of smallness is given by $\mathcal{S}$).

5.2.1 *Hierarchies of universes*

If one accepts one universe then there is no good reason why one shouldn't accept a further universe containing the previous one. The next definition makes precise what 'containing' actually means. For a similar definition in the context of (semantics of) dependent type theory see the Appendix of [5].

Definition 5.2 *Let $\mathcal{E}$ be a topos and $\mathcal{S}_1$ and $\mathcal{S}_2$ universes in $\mathcal{E}$. We say that $\mathcal{S}_1$ is included in $\mathcal{S}_2$ iff there is a generic family $El_1 : E_1 \to U_1$ for $\mathcal{S}_1$ such that both El_1 and $!_{U_1} : U_1 \to 1$ are maps in $\mathcal{S}_2$.*

Obviously $El_1 \in \mathcal{S}_2$ is equivalent to $\mathcal{S}_1 \subseteq \mathcal{S}_2$. But we also want that U_1 appears as an element of U_2, i.e. $!_{U_1} \cong u_1^* El_2$ for some global element $u_1 : 1 \to U_2$, which, obviously, is equivalent to the requirement that $!_{U_1} : U_1 \to 1$ is in $\mathcal{S}_2$.

We suggest that a reasonable notion of model for impredicative constructive mathematics is provided by a topos $\mathcal{E}$ with a natural numbers object N together with a sequence $(\mathcal{S}_n)_{n\in\mathbb{N}}$ of impredicative universes containing N such that every $\mathcal{S}_n$ is contained in $\mathcal{S}_{n+1}$ (in the sense of Definition 5.2 above). An appropriate internal language for such a structure is given by Z. Luo's *Extended Calculus of Constructions* (ECC), see [6], together with the Axiom of Unique Choice (AUC), extensionality for functions, the *propositional extensionality principle* $\forall p, q{\in}Prop\big((p{\Leftrightarrow}q) \Rightarrow p{=}q\big)$ and the principle of *proof-irrelevance* stating that propositional types, i.e. types in *Prop*, contain at most one element. We call this formal system ECCT as an acronym for *Extended Calculus of Constructions within a Topos*.

Determining the proof-theoretic strength of ECCT is an open problem which, however, seems to be fairly difficult for the following reasons. On the one hand in **Set** an impredicative universe containing N has at least strongly inaccessible cardinality (because it is infinite, regular and closed under $\mathcal{P}(-)$, i.e. $2^{(-)}$). Thus, in **Set** an infinite cumulative sequence of such universes requires the existence of infinitely many strongly inaccessible cardinals. For this reason one might expect that ECCT is as strong as IZF with an external[13] cumulative sequence of Grothendieck universes. Postulating the axiom

[13] ECCT does not prove the consistency of IZF together with the axiom that every set is an element of a Grothendieck universe. For this purpose one would have to extend ECCT with a further universe U_ω containing all U_n for $n < \omega$.

of choice (AC) this has been achieved by B. Werner in [9]. However, in toposes AC does not hold unless their logic is boolean. But in [8] A. Joyal and I. Moerdijk have constructed so-called initial ZF-algebras, i.e. internal models of IZF, from universes $\mathcal{S}$ which on the one hand are a little weaker than our notion (see the discussion in Section 5.4) but on the other hand are stronger in the sense that they validate the so-called type-theoretic *Collection Axiom*

$$\text{(CA)} \qquad (\forall X)(\forall A{:}U)(\forall e{:}X \to A)\ \mathrm{Epic}(e) \Rightarrow (\exists C{:}U)(\exists f{:}C \to X)\ \mathrm{Epic}(e \circ f)$$

which is needed for verifying that the initial ZF-algebra validates the set-theoretic replacement axiom and they have verified the existence of such universes for all Grothendieck and realizability toposes. Thus, alas, these comparatively well-known models cannot serve the purpose of disproving the claim that ECCT proves consistency of IZF. On the other hand one has got the impression that something like (CA) is needed for verifying that the initial ZF-algebra does actually validate the replacement axiom.

5.3 Existence of Universes in Toposes

After having introduced the notion of universe in a topos we now discuss the question of their existence. We are primarily interested in the existence of impredicative universes containing N and from now on refer to them simply as universes. Accordingly, we assume all toposes to have a NNO denoted as N.

First of all notice that generally in toposes universes need *not* exist. Consider for example the free topos $\mathcal{T}$ with a NNO N. If in $\mathcal{T}$ there existed a(n impredicative) universe (containing N) then HAH could prove its own consistency as the universe would allow one to construct a model for HAH inside $\mathcal{T}$ which is impossible by Gödel's second incompleteness theorem.

Quite expectedly, the situation is different for toposes whose construction depends on **Set** like realizability and Grothendieck toposes.

Let us first consider the somewhat simpler case of realizability toposes. Let $\mathcal{A}$ be a *partial combinatory algebra* (pca). Then the $\mathcal{P}(\mathcal{A})$-valued cumulative hierarchy $V^{(\mathcal{A})}$ is defined as

$$V^{(\mathcal{A})} = \bigcup_{\alpha \in \mathsf{Ord}} V_\alpha^{(\mathcal{A})}$$

where

$$V_\alpha^{(\mathcal{A})} = \bigcup_{\beta<\alpha} \mathcal{P}(\mathcal{A} \times V_\beta^{(\mathcal{A})})$$

is defined by transfinite recursion over α. Again by transfinite recursion one may define $\in$ and $=$ as $\mathcal{P}(\mathcal{A})$-valued binary predicates on $V^{(\mathcal{A})}$ as follows[14]

$$\begin{array}{lll} e \Vdash x \subseteq y & \text{iff} & \forall \langle a, z\rangle \in x.\ e{\cdot}a \Vdash z \in y \\ e \Vdash x = y & \text{iff} & \mathrm{pr}_1(e) \Vdash x \subseteq y \quad \text{and} \quad \mathrm{pr}_2(e) \Vdash y \subseteq x \\ e \Vdash x \in y & \text{iff} & \exists z \in V^{(\mathcal{A})}.\ \mathrm{pr}_1(e) \Vdash x = z \wedge \langle \mathrm{pr}_2(e), z\rangle \in y \end{array}$$

[14] We write $a{\cdot}b$ for 'a applied to b' and pr_1 and pr_2 for first and second projection w.r.t. the coding of pairs available in any pca.

where $e \Vdash x \in y$ and $e \Vdash x = y$ stand for $e \in [\![x \in y]\!]$ and $e \in [\![x = y]\!]$, respectively. One may show that this gives rise to a model for IZF as was done for the first Kleene algebra (of Gödel numbers for partial recursive functions) in McCarty's thesis [10] but his proof extends to arbitrary pca's without any further effort. It has recently been shown by A. Simpson and the author (see [4]) that the topos derived from this model of IZF is actually equivalent to RT($\mathcal{A}$), the realizability topos over $\mathcal{A}$. Now for strongly inaccessible cardinals κ one may easily show that $V_\kappa^{(\mathcal{A})}$ is a Grothendieck universe within $V^{(\mathcal{A})}$ and thus gives rise to a universe inside $V^{(\mathcal{A})}$.

For Grothendieck toposes the situation is somewhat more delicate. Suppose $\mathbf{U}$ is a Grothendieck universe in **Set**. Now if $\mathbb{C}$ is a category in $\mathbf{U}$ then this gives rise to a universe $\mathcal{U}$ inside the presheaf topos $\widehat{\mathbb{C}} = \mathbf{Set}^{\mathbb{C}^{\mathsf{op}}}$ which is defined as follows (see [11]). First recall that for every $A \in \widehat{\mathbb{C}}$ the slice category $\widehat{\mathbb{C}}/A$ is equivalent to $\widehat{\mathsf{Elts}(A)}$ (where $\mathsf{Elts}(A)$ is the category of elements of A obtained via the Grothendieck construction). We define a morphism $b : B \to A$ to be contained in $\mathcal{U}$ iff the corresponding presheaf (via $\widehat{\mathbb{C}}/A \simeq \widehat{\mathsf{Elts}(A)}$) is isomorphic to one factoring through $\mathbf{U}$, i.e., more explicitly, iff $b_I^{-1}(x)$ is isomorphic to a set in $\mathbf{U}$ for all $I \in \mathbb{C}$ and $x \in A(I)$. It is more or less straightforward to verify that $\mathcal{U}$ satisfies the conditions required for a universe. The only slightly delicate point is the existence of a morphism $El : E \to U$ generic for $\mathcal{U}$. We define U as the presheaf over $\mathbb{C}$ with $U(I) = \mathbf{U}^{(\mathbb{C}/I)^{\mathsf{op}}}$ and for $\alpha : J \to I$ in $\mathbb{C}$ we put $U(f) = f^* = \mathbf{U}^{(\mathbb{C}/f)^{\mathsf{op}}}$. We define the generic family $El : E \to U$ as the object of $\widehat{\mathbb{C}}/U$ corresponding (via $\widehat{\mathbb{C}}/U \simeq \widehat{\mathsf{Elts}(U)}$) to the presheaf $\mathrm{E} : \mathsf{Elts}(U)^{\mathsf{op}} \to \mathbf{U}$ which is defined as follows: with every object (I, A) in $\mathsf{Elts}(U)$ we associate the set $\mathrm{E}(I, A) = A(id_I)$ in $\mathbf{U}$ and with every morphism $f : (J, f^*A) \to (I, A)$ in $\mathsf{Elts}(U)$ we associate the map $\mathrm{E}(f) = A(f) : \mathrm{E}(I, A) \to \mathrm{E}(J, f^*A)$.

This construction for $\widehat{\mathbb{C}}$ extends to sheaf toposes $\mathcal{E} = \mathsf{Sh}(\mathbb{C}, \mathcal{J})$ for Grothendieck topologies $\mathcal{J}$ on $\mathbb{C}$ in the following way. Let $\mathcal{U}$ be the universe in $\widehat{\mathbb{C}}$ as constructed above. We define a universe $\mathcal{U}_\mathcal{E}$ in $\mathcal{E}$ as the intersection of $\mathcal{U}$ and $\mathcal{E} \subseteq \widehat{\mathbb{C}}$. The class $\mathcal{U}_\mathcal{E}$ consists of all maps of $\mathcal{E}$ that are isomorphic to some arrow $\mathsf{a}(f)$ where $f \in \mathcal{U}$ and $\mathsf{a} : \widehat{\mathbb{C}} \to \mathsf{Sh}(\mathbb{C}, \mathcal{J})$ is the sheafification functor left adjoint to the inclusion $\mathcal{E} = \mathsf{Sh}(\mathbb{C}, \mathcal{J}) \hookrightarrow \widehat{\mathbb{C}}$. (One readily checks that the image of $\mathcal{U}$ under a is contained in $\mathcal{U}$ (using the assumption that $\mathbb{C}$ is internal to $\mathbf{U}$) and therefore $\mathcal{U}_\mathcal{E}$ coincides with the image of $\mathcal{U}$ under a.) From this observation it follows that $\mathcal{U}_\mathcal{E}$ is closed under composition and as the sheafification functor a preserves finite limits it is immediate that $\mathcal{U}_\mathcal{E}$ is stable under pullbacks along arbitrary arrows in $\mathcal{E}$. Moreover, the map $\mathsf{a}(El)$ is generic for $\mathcal{U}_\mathcal{E}$ because El is generic for $\mathcal{U}$. As a preserves monos and $\mathcal{U}$ contains all monos of $\widehat{\mathbb{C}}$ it follows that $\mathcal{U}_\mathcal{E}$ contains all monos of $\mathcal{E}$. That $\mathcal{U}_\mathcal{E}$ satisfies condition (4) of Definition 5.1 can be seen as follows. Under the conditions (1), (2) and (3) (already established for $\mathcal{U}_\mathcal{E}$) condition (4) is equivalent to the requirement that for all $A \in \mathcal{E}$ the slice $\mathcal{U}_\mathcal{E}/A$ has exponentials inherited from $\mathcal{E}/A$. Suppose $b_1 : B_1 \to A$ and $b_2 : B_2 \to A$ are objects in $\mathcal{U}_\mathcal{E}/A$. Then their exponential $b_2^{b_1}$ taken in $\widehat{\mathbb{C}}/A$ stays within $\mathcal{U}$ as $\mathcal{U}$ is a universe in $\widehat{\mathbb{C}}$ and it stays within $\mathcal{E}/A$ as $\mathcal{E}/A$ is a subtopos[15] of $\widehat{\mathbb{C}}/A$ and subtoposes are closed under exponentiation. Thus $b_2^{b_1}$ is in $\mathcal{U}_\mathcal{E}/A$ concluding the argument that $\mathcal{U}_\mathcal{E}/A$ is closed under exponentiation taken in $\mathcal{E}/A$.

Next we show that the universe $\mathcal{U}_\mathcal{E}$ is impredicative. First notice that $\mathsf{a}(\top_{\widehat{\mathbb{C}}})$ is a generic mono for $\mathcal{E}$.[16] Thus, there exists a map $s : \Omega_\mathcal{E} \to \mathsf{a}(\Omega_{\widehat{\mathbb{C}}})$ with $\top_\mathcal{E} \cong s^*\mathsf{a}(\top_{\widehat{\mathbb{C}}})$. As there is also a map $p : \mathsf{a}(\Omega_{\widehat{\mathbb{C}}}) \to \Omega_\mathcal{E}$ with $\mathsf{a}(\top_{\widehat{\mathbb{C}}}) \cong p^*\top_\mathcal{E}$ it follows that $(p \circ s)^*\top_\mathcal{E} \cong \top_\mathcal{E}$. Thus, we have $p \circ s = id_{\Omega_\mathcal{E}}$ and, therefore, the map $s : \Omega_\mathcal{E} \to \mathsf{a}(\Omega_{\widehat{\mathbb{C}}})$ is a split mono. As $\mathcal{U}$ is impredicative it contains the terminal projection of $\Omega_{\widehat{\mathbb{C}}}$ and, accordingly, $\mathcal{U}_\mathcal{E}$ contains the terminal projection of $\mathsf{a}(\Omega_{\widehat{\mathbb{C}}})$. Thus, as $\mathcal{U}_\mathcal{E}$ is closed under subobjects in $\mathcal{E}$ it follows that $\mathcal{U}_\mathcal{E}$ contains also the terminal projection of $\Omega_\mathcal{E}$, i.e. that $\mathcal{U}_\mathcal{E}$ is impredicative.

As $\mathcal{U}_\mathcal{E}$ contains the terminal projection of $N_\mathcal{E} = \mathsf{a}(N)$, the NNO of $\mathcal{E}$, the universe $\mathcal{U}_\mathcal{E}$ contains $N_\mathcal{E}$.

5.4 Further Properties and Generalizations

It is a desirable property of a universe in a topos that for arbitrary families of types $a : A \to I$ the collection of those $i \in I$ with A_i small constitutes a subobject of I. The following mathematical precision of this informal idea is due to J. Bénabou[17] (see, e.g. [3]).

Definition 5.3 *A pullback stable class $\mathcal{S}$ of morphisms in a topos $\mathcal{E}$ is called definable if for every morphism $a : A \to I$ in $\mathcal{E}$ there is a subobject $m : I_0 \rightarrowtail I$ such that $m^*a \in \mathcal{S}$ and every $f : J \to I$ with $f^*a \in \mathcal{S}$ factors through m.*

At first sight in the presence of a generic family $El : E \to U$ for $\mathcal{S}$ definability of $\mathcal{S}$ seems to be evident by considering the subobject

$$I_0 = \{i \in I \mid \exists u \in U.\ A_i \cong El(u)\}$$

of I expressible in the internal language of $\mathcal{E}$. However, when unfolding the definition of I_0 following precisely the rules of Kripke–Joyal semantics (see [1]) one observes that I_0 is the *greatest* subobject m of I with $e^*m^*a \in \mathcal{S}$ for some epi e, i.e. $m^*a \in \mathcal{S}_\ell$ where $\mathcal{S}_\ell$, the so-called *stack completion of* $\mathcal{S}$, is the collection of all $b : B \to J$ with $e^*b \in \mathcal{S}$ for some epi $e : K \twoheadrightarrow J$. Again using Kripke–Joyal semantics one sees that $b : B \to J$ is in $\mathcal{S}_\ell$ iff $\forall j \in J.\exists u \in U.\ B_j \cong El(u)$ holds in $\mathcal{E}$. Notice that $b : B \to J$ might well be in $\mathcal{S}_\ell$ even if there is no $g : J \to U$ with $b \cong g^*El$ because every B_j may be isomorphic

[15] In the geometric sense because $\mathsf{a}_{/A} : \widehat{\mathbb{C}}/A \to \mathcal{E}/A$ is a finite limit-preserving left adjoint to the inclusion $\mathcal{E}/A \hookrightarrow \widehat{\mathbb{C}}/A$

[16] In general, sheafification does not preserve subobject classifiers. Actually, it does if and only if the corresponding Lawvere–Tierney topology j preserves implication in the sense that $j \circ \to\ = \ \to \circ (j \times j)$.

[17] He introduced the notion of definability for (full) subfibrations of *arbitrary* fibrations and not just for the particular case of (full) subfibrations of the fundamental fibration $\partial_1 : \mathcal{E}^2 \to \mathcal{E}$ of a topos $\mathcal{E}$.

to some $El(u)$ though one might not be able to *choose* such a $u \in U$ uniformly in $j \in J$. This discussion shows that S is definable in the sense of Bénabou already if S satisfies the following *descent property*: $a \in S$ whenever $e^*A \in S$ for some epi e. One easily shows that definability of S implies[18] the descent property for S. Thus, we have

Theorem 5.4 *A universe S in a topos $\mathcal{E}$ is definable in the sense of Bénabou if and only if S satisfies the descent property.*

A further characterization of definability of S can be found in [12]. By definability for every object A of $\mathcal{E}$ there exists a subobject $m : \mathcal{P}_S(A) \rightarrowtail \mathcal{P}(A)$ with $m^*(\in_A; \pi') \in S$, i.e.

$$
\begin{array}{ccc}
\in_A^S & \rightarrowtail & \in_A \\
\downarrow & \lrcorner & \downarrow \\
A \times \mathcal{P}_S(A) & \xrightarrow{A \times m} & A \times \mathcal{P}(A) \\
{\scriptstyle \pi'}\downarrow & \lrcorner & \downarrow{\scriptstyle \pi'} \\
\mathcal{P}_S(A) & \xrightarrow{\;\;m\;\;} & \mathcal{P}(A)
\end{array}
$$

with $\in_A^S; \pi' \in S$, such that for every $r : R \rightarrowtail A \times B$ with $r; \pi' \in S$ there exists a unique $\rho : B \to \mathcal{P}_S(A)$ with

$$
\begin{array}{ccc}
R & \longrightarrow & \in_A^S \\
{\scriptstyle r}\downarrow & \lrcorner & \downarrow \\
A \times B & \xrightarrow{A \times \rho} & A \times \mathcal{P}_S(A)
\end{array}
$$

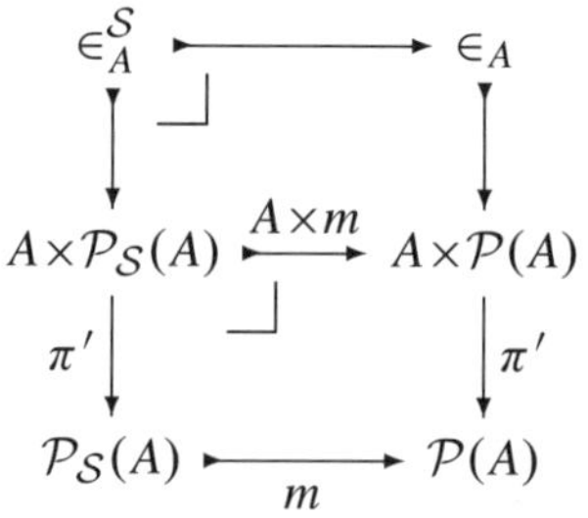

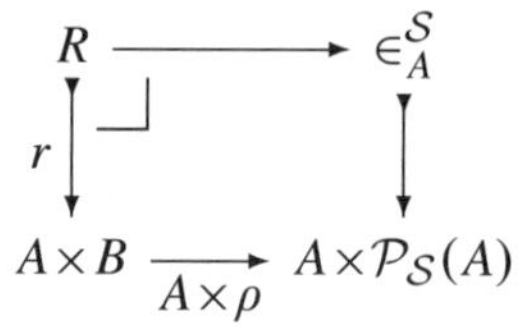

Obviously, the subobject $\mathcal{P}_S(A)$ consists of those subsets of A which are small in the sense of S. In [12] it has been shown that existence of such *small power objects* entails the descent property for S.[19]

---

[18]If $e^*a \in S$ for some epi $e : J \twoheadrightarrow I$ then e factors through some $m : I_0 \rightarrowtail I$ with $m^*a \in S$. But then m is an iso and, therefore, already $a \in S$.

[19]If $b \in S$ and

$$
\begin{array}{ccc}
B & \xrightarrow{e'} & A \\
{\scriptstyle b}\downarrow & \lrcorner & \downarrow{\scriptstyle a} \\
J & \xrightarrow{\;\;e\;\;} & I
\end{array}
$$

then $a \cong \phi^*(\in_A^S; \pi')$ where the classifying map $\phi : I \to \mathcal{P}_S(A)$ is given by

$$
x = \phi(i) \iff \exists j \in J. i = e(j) \wedge x = e'_!(b^{-1}[j])
$$

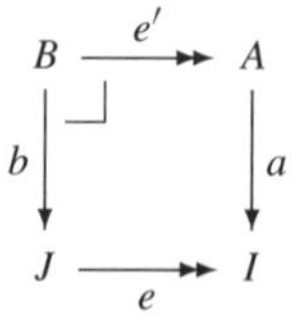

Currently it is not (yet) clear (to us) whether the universes introduced in Section 5.3 are definable. Though this seems to be very likely the case for Grothendieck toposes the question for realizability toposes seems to be much harder.[20]

However, instead of a universe $\mathcal{S}$ one might instead consider its stack completion $\mathcal{S}_\ell$ still satisfying conditions (1)–(4) of Definition 5.1 but instead of condition (5) only

(5.1) (descent)
for every $a : A \to I$ if $e^*a \in \mathcal{S}_\ell$ for some epi $e : J \twoheadrightarrow I$ then already $a \in \mathcal{S}_\ell$

(5.2) (weakly generic family)
there exists a map $El : E \to U$ in $\mathcal{S}_\ell$ such that for every $a : A \to I$ in $\mathcal{S}_\ell$ there is an epi $e : J \twoheadrightarrow I$ and $f : J \to U$ with $e^*a \cong f^*El$.

We leave it for future investigations to find out whether the stack completions $\mathcal{S}_\ell$ of the universes $\mathcal{S}$ constructed in Section 5.3 do validate the type-theoretic Comprehension Axiom (CA) discussed at end of Section 5.2.

5.5 Conclusions and Open Questions

We have introduced a (not too surprising) notion of (hierarchy of) universe(s) in toposes which we consider as an alternative to IZF. We think that the corresponding language ECCT being based on type theory is closer to mathematical practice than the first order language of IZF where everything is coded up in terms of $\in$. Nevertheless ECCT provides the possibility of defining families of types (in a universe) by recursion over the index set which is usually achieved by appeal to the set-theoretic replacement axiom in a much less direct way.

It is not clear, though very likely, that ECCT does not allow one to construct models for IZF without postulating further axioms like the type-theoretic Collection Axiom of [8]. Moreover, as far as we know one has not yet found 'mathematical', i.e. not meta-mathematical, statements expressible in the language of HAH which are not provable in HAH but are derivable in ECCT or IZF.[21] However, in [13] it has been shown that HAH does not allow one to prove the existence of solutions of domain equations (for quite general functors on domains) although IZF does and similarly so does ECCT. The reason is that solutions of domain equations arise as inverse limits of recursively defined families of domains, i.e. particular sets in the setting of [13]. It would be a pity

where $e'_!$ is the direct image map for e' and b^{-1} is the inverse image map for b, i.e.

$$\phi(i) = \{x \in A \mid \exists y \in B.\, e(b(y)) = i \wedge e'(y) = x\}\ .$$

[20] One would have to show for example that for the class $\mathcal{S}$ of small maps in the realizability topos **Eff** as considered in [8] there exists a generic family and not only a weakly generic one. But this is difficult as one doesn't even know the size of $\Gamma^{-1}(1)$, i.e. how many (up to isomorphism) objects $X \in$ **Eff** exist such that X has precisely one global element.

[21] However, recently (in June 2005) Harvey Friedman has informed me that he has found mathematically natural Π^0_1 sentences A derivable in ZF but not in classical HAH. Thus, by Friedman's $\neg\neg$-translation of ZF to IZF it follows that sentences are provable in IZF but not in HAH.

if this quite convincing example from applied mathematics were to remain the only one demonstrating the need for universes!

Acknowledgements

I am grateful for detailed comments by Alex Simpson and Bill Lawvere on an earlier draft where they pointed out to me that I underestimated the strength of universes in toposes and confounded Grothendieck universes with small inner models.

Furthermore, I want to thank Jean Bénabou for discussions about his seminal work on universes in [7] (which has a much more general aim than the notion of universes presented in this chapter) and for explaining to me his work on non-standard GBN class theory.

Finally, I thank Jaap van Oosten for reminding me that sheafification does not in general preserve subobject classifiers which I wrongly assumed in a previous version.

References

1. MacLane, S. and Moerdijk, I. (1994). *Sheaves in Geometry and Logic.* Springer.
2. Jacobs, B. (1999). *Categorical Logic and Type Theory.* North-Holland.
3. Streicher, T. *Fibred Categories à la Jean Bénabou.* Lecture notes (2003) available electronically at http:/www.mathematik.tu-darmstadt.de/~streicher/FIBR/FibLec.ps.gz.
4. Awodey, S., Butz, C., Simpson, A., and Streicher, T. (2004). *Relating set theories, toposes and categories of classes.* Paper in preparation.
5. Streicher, T. (1991). *Semantics of Type Theory.* Birkhäuser.
6. Luo, Z. (1994). *Computation and Reasoning. A Type Theory for Computer Science.* Oxford University Press.
7. Bénabou, J. (1973). *Problèmes dans les topos.* Lecture Notes of a Course from 1971 taken by J.-R. Roisin and published as Tech. Rep., Univ. Louvain-la-Neuve
8. Joyal, A. and Moerdijk, I. (1995). *Algebraic Set Theory.* London Mathematical Society Lecture Notes Series, **220**. Cambridge University Press.
9. Werner, B. *Sets in Types, Types in Sets.* Proc. of TACS'97, SLNCS1281.
10. McCarty, C. (1984). *Realizability and Recursive Mathematics.* PhD thesis, Oxford.
11. Hofmann, M. and Streicher, T. *Lifting Grothendieck Universes.* unpublished note available electronically at http://www.mathematik.tu-darmstadt.de/~streicher/NOTES/lift.dvi.gz.
12. Simpson, A. (1999). *Elementary axioms for categories of classes.* Proc. of LICS'99, pp. 77–85, IEEE Press.
13. Simpson, A. (2004). *Computational adequacy for recursive types in models of intuitionistic set theory. Annals of Pure and Applied Logic*, **130**, 207–275. (Special issue for selected pages from LICS 02, ed. Plotkin, G).

6

TOWARD A MINIMALIST FOUNDATION FOR CONSTRUCTIVE MATHEMATICS

MARIA EMILIA MAIETTI AND GIOVANNI SAMBIN

Abstract

The two main views in modern constructive mathematics, usually associated with constructive type theory and topos theory, are compatible with the classical view, but they are incompatible with each other, in a sense explained by some specific results which we briefly review. So it is desirable to design a common core which is compatible with all the theories in which mathematics has been developed, like Zermelo–Fraenkel set theory, topos theory, Martin-Löf's type theory, etc. and can be understood as it is by any mathematician, whatever foundation is adopted.

A requirement with increasing importance is that of developing mathematics in such a way that it can be formalized on a computer. This theoretically means that the foundation should obey the proofs-as-programs paradigm.

We claim that to satisfy both requirements it is necessary to use a minimal type theory mTT, which is obtained from Martin-Löf's type theory by relaxing the identification of propositions with sets. This ground type theory mTT is intensional and is needed for formalization. A 'toolbox' of extensional concepts, needed to do mathematics, is built on it. The common core is obtained at this level, by subtraction.

The underlying conceptual novelty is that one should give up to the expectation of an all-embracing foundation, also in the concrete sense that one needs a formal system living at two different levels of abstraction.

After abandoning the classical view and all its limits, a prospective constructivist is faced with the problem of choosing among a variety of views. They generally share the choice for intuitionistic logic, but they differ in mathematical principles.

The principal distinction is between two views. One maintains that the meaning of mathematics lies in its computational content, and thus keeps its formalization in a computer language in mind. It is usually associated with type theory; the axiom of choice then turns out to be valid and (hence) the powerset axiom is not accepted [29]. The other favours the mathematical structure beyond its particular presentations. It is usually expressed through category theory and often identified with topos theory. Extensionality is an essential feature, the powerset axiom is mostly accepted and hence the axiom of choice is not accepted as valid [23].

Both views are reasonable, well motivated and apparently cannot be dispensed with. It is however a matter of fact that they are incompatible, in the sense that by accepting both one is forced back to the classical view.

The same tension is revealed also as a technological challenge: while implementation of mathematics in a computer is intrinsically intensional (just think of the fact that a computer handles only expressions and no objects), mathematics itself needs to deal with objects independently of their presentations, and hence is extensional. Because of the first purpose, intensional type theory cannot be given up (it is actually used in most implementations), while an extensional foundation, like set theory or the theory of a generic topos, seems to be essential for the second (extensionality is assumed in all the actual approaches to mathematics). So again both visions seem to be indispensable; and yet they remain incompatible.

In our opinion, this incompatibility has to be faced openly. Communication between different approaches to mathematics clearly cannot be dismissed; and the urge to implement mathematics will certainly increase in the future. So it is necessary to identify a new foundational theory, which on the one hand permits us to keep communication alive and on the other hand can be used as a convenient basis to address the problem of implementation. Because of the incompatibility between the two general views, this theory is not to be obtained by union but by subtraction, that is by finding a common kernel. Thus we are pushed to abandon the widespread expectation of a single all-embracing foundation, which is good for all purposes. Actually, this is true in two different ways.

First, the purpose of implementing mathematics requires us to keep a clear-cut distinction between a ground level with an intensional type theory, which is necessary for implementation and actually is essentially an abstract functional programming language, and a more abstract level, in which one has only extensional concepts and constructs which are needed to develop mathematics. These extensional tools can be put on top of type theory in the spirit of a 'toolbox' (see the Preface of [1] and Section 3.2.1 of [2]). Thus any tool should be obtained according to the forget-restore principle: to reach extensionality, it is necessary to *forget* some information, like the algorithm to compute a function or the specific proof of a proposition; but to obtain implementation, one has to be able to *restore* such information at will. So the formal system must include independent treatment of two different levels of abstraction, as well as a systematic way to connect them. Typically, the axiom of choice is valid at the lower level, while it fails at the extensional level. Moreover, to be able to implement the toolbox, it is necessary to introduce proof-irrelevance for propositions. Thus propositions must be kept well distinct from sets. So a key step is to give a justification of logic independently of set theory, and we do this via the principle of reflection [3].

Second, keeping communication between different views alive requires the design of a common theory in which all mathematicians can believe, while leaving them free to keep their intended semantics in mind. By leaving out both the axiom of choice and the powerset axiom at the level of the toolbox, one obtains a common basis which is immediately understood as it is (that is, with no translations) and believed in by any mathematician. Thus, contrary to what is usually required on a foundational theory, our minimal theory cannot be the best possible description of a single semantics. Rather, it is designed to admit different, and actually mutually incompatible interpretations, or extensions.

After illustrating these technical and conceptual reasons, we present a specific formal system, called mTT. It is only a first proposal, and we are aware that it could be improved; still it works as a precise basis for a minimalist foundation. In fact, the toolbox developed over Martin-Löf type theory in [1] actually has our present mTT as the intended ground theory; to confirm this, one can observe that no real use is made of the axiom of choice. So, for example, since its beginning [4] all of formal topology, as long as it uses only extensional notions defined as 'tools' over type theory, has actually adopted—with informal rigour—the minimalist approach described here.

Awareness of the fact that all this could be expressed in a formal way, and moreover that it brings to compatibility with topos theory, comes from [5]. The two authors have discussed these topics since then; an anticipation of the conceptual content of the present paper was given in Section 3.1.2 of [2]. As a final remark, we recall that a minimalist foundation has aided the emergence of new and deeper structures, like those in the basic picture [6].

6.1 Arguments for a Minimal Constructive Type Theory

Here we present and discuss our arguments to show that some natural expectations are satisfied only by a minimalist foundation of constructive mathematics. In the next section, we propose a formal system which satisfies such expectations and we briefly discuss its connections with the literature.

6.1.1 *The computational view*

In the conception of Bishop [7–9] and of Martin-Löf [10], the meaning of constructive mathematics is given by its computational content. So every mathematical entity should be computable: sets must be inductively generated, functions must be given by an effective algorithm, etc. In other words, every object of mathematical reasoning must be expressible in and computable by a computer, which theoretically means a Turing machine. This can be done by defining a concept of computation for sets and functions beside that for number-theoretic functions. Then the mathematical reasoning itself is required to preserve the computational content. So the logic must be intuitionistic, and this is usually explained by saying that it satisfies Heyting semantics (or the so-called BHK interpretation) of propositions and of logical constants: every proposition can be identified with the set of its proofs, and every proof gives an effective method. We call this the computational view of constructivism.

The mathematics developed in such a computational way is by definition formalizable on a computer, in the strict sense that all the constructions and functions are computed by a computer. So one needs a foundational theory satisfying what we call the proofs-as-programs paradigm:

Proofs-as-programs. *A theory satisfies the* proofs-as-programs *paradigm if (i) all its set-theoretic constructors preserve computability, in the sense that their element constructors are given by effective methods so that all effective methods between natural numbers are computed by machine programs; (ii) its underlying logic satisfies Heyting semantics.*

This is a strict formulation of what it means for a foundation to be constructive in the computational sense. At the same time it is a way to express theoretically the requirement that the mathematics developed on it can be implemented on a computer. Even if we do not commit ourselves to any definition of computation between arbitrary sets (like in Martin-Löf type theory, Coq, PCF, etc.), this means that all mathematical concepts and proofs must admit a realizability interpretation, where the realizers are just the real programs of the Turing machine.

We can now see that if a theory satisfies proofs-as-programs and it is formulated as a many-sorted logic whose sorts include finite types like the type of functions $A \to B$ and whose logic includes first order logic, then it is consistent with the axiom of choice and the formal Church thesis. By the axiom of choice we mean the statement

$$AC: \quad (\forall x \in A)\,(\exists y \in B)\, R(x, y) \longrightarrow (\exists f \in A \to B)\,(\forall x \in A)\, R(x, f(x))$$

expressing the idea that from a proof of a specification $(\forall x \in A)\,(\exists y \in B)\, R(x, y)$ one can extract a function (or program) f which on the input $x \in A$ computes the output $f(x) \in B$ satisfying $R(x, f(x))$.

By the *formal* Church thesis we mean the statement expressing that all number-theoretic functions are programs:

$$CT: \quad (\forall f \in \mathbb{N} \to \mathbb{N})\,(\exists e \in \mathbb{N})\,(\forall x \in \mathbb{N})\,(\exists y \in \mathbb{N})\,(T(e, x, y)\,\&\, U(y) = f(x))$$

where $T(e, x, y)$ is the Kleene predicate expressing that y is the computation executed by the program with code number e on the input x and $U(y)$ is the output of the computation y.

Now, since a proofs-as-programs theory enjoys Heyting semantics, then it must be consistent with AC, since Heyting semantics validates AC because of the meaning given to the existential quantifier. Moreover, since we required that all effective methods between natural numbers are computed by programs, we deduce that the considered theory is also consistent with CT. Therefore, we conclude that *a many sorted theory satisfying proofs-as-programs must be consistent with the axiom of choice and the formal Church thesis, that is with AC+CT.*

A natural question to ask is what theories apt to develop mathematics satisfy the proofs-as-programs paradigm, and if there are any. First of all, one can observe immediately that the classical theory of sets ZF does not satisfy proofs-as-programs, given that not all number-theoretic functions are recursive. But not many other classical theories can satisfy the paradigm, since even many-sorted classical arithmetic fails to do it, as proved by the following (see for example [11]):

Proposition 6.1 *Let S be a many sorted theory of first-order classical logic, whose sorts include types of simply typed lambda calculus [12] like that of natural numbers $\mathbb{N}$, the product type $A \times B$ and arrow types like $A \to B$. If S satisfies the Peano axioms, the formal Church thesis and the axiom of choice (written by means of the arrow type), then it is inconsistent.*

Proof. Consider any predicate $P(x)$ and observe that

$$\forall x \in \mathbb{N}\; \exists y \in \mathbb{N} \;\; P(x) \longleftrightarrow y = 0$$

is valid thanks to the principle of excluded middle and recalling that $0 \neq 1$ is one of the Peano axioms: if $P(x)$ holds put $y = 0$ and, if not, put $y = 1$. Then, by applying CT to it there exists a recursive function f such that $P(x)$ iff $f(x) = 0$; that is, all predicates are recursive, which is not possible. □

This proof reveals that, in order to keep arithmetic and to realize the proofs-as-programs paradigm, and hence to be consistent with AC and CT, one possibility is to discharge the principle of excluded middle, as it seems to be the main cause of the inconsistency in the above proof. Therefore, from now on we only consider intuitionistic theories. Among them, we have intuitionistic set theories formulated in the style of Zermelo–Fraenkel set theories like IZF [13], Myhill's CST [14] or Aczel's CZF [15]. In the approach of category theory, we could also consider a generic elementary topos as a universe of sets in which to develop intuitionistic mathematics.

However, none of these proposals satisfies the proofs-as-programs paradigm. The main reason is that they validate extensional principles like the principle that two functions are equal if they have equal values. This can be seen well in the following proposition, where we prove that also extensional Martin-Löf's type theory does not satisfy proofs-as-programs. The proof is obtained by adapting that in [16] (page 498) of the fact that Heyting arithmetic with finite types HA^{ω} is inconsistent with CT+AC and function extensionality:

Proposition 6.2 *Martin-Löf's extensional type theory [17], which validates the axiom of choice, is inconsistent with the formal Church thesis.*

Proof. By applying AC on CT one gets a function $h \in (\mathbb{N} \to \mathbb{N}) \to \mathbb{N}$ such that

$$(*) \qquad (\forall f \in \mathbb{N} \to \mathbb{N})\ (\forall x \in \mathbb{N})\ (\exists y \in \mathbb{N})\ \ (T(h(f), x, y)\ \&\ U(y) = f(x)).$$

By extensionality of functions, that is the fact that $\lambda x.f(x) =_{\mathbb{N}\to\mathbb{N}} \lambda x.g(x)$ is derivable if and only if $\forall x \in \mathbb{N}\ f(x) =_{\mathbb{N}} g(x)$, one can prove that h is injective: from $(*)$ and $h(\lambda x.f(x)) =_{\mathbb{N}} h(\lambda x.g(x))$ one has $\forall x \in \mathbb{N}\ f(x) =_{\mathbb{N}} g(x)$ and hence by extensionality also $\lambda x.f(x) =_{\mathbb{N}\to\mathbb{N}} \lambda x.g(x)$. Since $\lambda x.f(x) =_{\mathbb{N}\to\mathbb{N}} \lambda x.g(x)$ obviously implies $h(\lambda x.f(x)) = h(\lambda x.g(x))$ because h is a function, $\lambda x.f(x) = \lambda x.g(x)$ turns out to be equivalent to $h(\lambda x.f(x)) = h(\lambda x.g(x))$. Since equality of natural numbers is decidable, then extensional equality of recursive functions would also be decidable, which it is not. Hence, the theory is inconsistent. □

The main cause of the inconsistency in the above proof seems to be the principle of extensionality of functions, which is obtained by means of the extensional equality. So to keep consistency with AC+CT one possibility is to consider a type theory with an *intensional* equality, like Martin-Löf's intensional type theory in [18, 19]. From now on we call it M-L type theory.

It is proposed as a full scale approach to a constructive (intuitionistic and predicative) mathematics. It satisfies the proofs-as-programs paradigm since it is a functional programming language which satisfies all the desirable properties, like type-checking, needed for computation. Moreover, it can be seen as a rigorous and general description of the identification of propositions with sets (propositions-as-sets interpretation), of which Heyting semantics is a consequence. So validity of AC follows from the interpretation of quantifiers.

All of this is very satisfactory. So why should one look for something different? There are some good reasons to do that.

6.1.2 *Intrinsic reasons: intensionality vs. extensionality*

Even assuming wholeheartedly the computational view, there are some good reasons to look for a modification of M-L type theory. We begin here with some intrinsic reasons.

Of course, our type theory should be sufficient for the working mathematician to develop mathematics on it, and not only to formalize it. A peculiarity which is common to all mathematics is its extensionality: sets and subsets with the same elements are equal, functions giving the same values are equal, etc. However, as seen in Proposition 6.2, adding extensionality to a theory makes it inconsistent with proof-as-programs, and hence unsuitable to formalize mathematics. This is a matter of fact that cannot be ignored. Extensional aspects, like extensional equality as in [17], simply cannot be required directly at the level of the ground type theory, if this has to satisfy the proofs-as-programs paradigm.

There is a way out, and it seems unavoidable to us: to develop mathematics in a type theory, one must have both a ground intensional level, which satisfies proofs-as-programs and where formalization of mathematics is implemented, and an extensional level, where mathematics is actually developed. These two levels must be kept carefully distinct, but of course they must also be linked in a clear and systematic way. We do not see them as two theories, but as two connected levels of abstraction in the same theory, as we now explain.

Mathematics is commonly conceived as dealing with 'objects', which are independent of their different presentations. This is just a way to say that it is extensional. So the higher, extensional level should be obtained from the lower, intensional one by leaving out those details of the presentations which are not necessary to determine a mathematical 'object', that is by 'forgetting' some information. For instance, as most often in mathematics, one is interested only in the provability or truth of a proposition, and not in the specific form of its proofs (proof-irrelevance). Therefore it is convenient to use a judgement of the form $A\ true$ (see [17]) for a proposition A with the meaning that A is provable. This requirement implies that in the ground type theory, on which we abstract, propositions should be formally well distinct from sets. In fact, proof-irrelevance must act on only propositions; it simply would make no sense on sets.

The design of the extensional level over type theory has to be done by following a certain discipline. We propose that it should be achieved by building a suitable collection of (extensional) mathematical concepts, or tools; from now on, we call it the *toolbox*, as in [1].

It is then in toolbox that mathematics can be developed. But one of our requirements was that mathematics should be formalizable in a computer. To this aim, one usually devises a computational model of the extensional theory. We don't need to do this for our toolbox of extensional concepts. In fact, if all tools are introduced according to what has been called the *forget–restore* principle (for a detailed explanation, see the Preface of [1] and Section 3.2.1 of [2]), then their implementability at the ground level is automatically preserved. In fact, *restore* says that the computational content is fully under control (and thus *forget* was correct) since it can be restored whenever wished. Thus the tool can be brought back to the ground theory. In other words, the forget–restore principle says that only those tools can be introduced for which implementation is known.

Another example of extensional theory apt to develop mathematics, and that admits a constructive interpretation in type theory, is Aczel's system CZF of (constructive) set theory. However, its interpretation in M-L type theory is global and static (technically, it uses type theory as a meta-language to construct a model of CZF, and this requires W-types and universes). So implementation has to be studied case by case. Instead, in our approach, the link between toolbox and our ground type theory is local and dynamic (it can be seen in the same moment mathematics is done), and the computational content is systematically preserved (which is the reason for keeping proof-terms of propositions). A precise and practical difference is hence that all the mathematics which is developed in toolbox is automatically formalizable.

The fact that to formalize mathematics a theory with two different levels is needed helps to clarify the debate about the validity of the axiom of choice. As recently emphasized by Martin-Löf, one must distinguish between an 'intensional' AC, which holds constructively simply by the BHK interpretation of quantifiers, and an 'extensional' AC, which simply fails (constructively).[1]

So also when developing mathematics on the basis of type theory, either one specifies that mathematics is meant to be 'intensional' (which is not done in practice), or one carefully justifies each application of AC, by showing that the choice function respects the equality involved (there are sets, like the natural numbers, in which the generation of elements automatically produces 'objects', so intensional and extensional equality coincide and AC is justified). In other words, indiscriminate use of AC is admissible only in a classical axiomatic set theory, like ZFC.

Using toolbox guarantees that this problem is solved. In fact, because of Proposition 6.2 (and the wish of compatibility, see the next subsection), we need to block the validity of AC at the level of toolbox; this is done by keeping propositions distinct from sets, and by adopting a weak notion of existential quantifier. Of course, it remains true that the corresponding formulation of AC for sets, which becomes just the distributivity of Σ over Π, is valid in the ground theory.

6.1.3 *Extrinsic reasons: compatibility*

There are also other good reasons to consider a variant of M-L type theory, and these are extrinsic. We firmly believe that even if one decides to develop mathematics in the computational way, one should not ignore all the mathematics already developed and the important results achieved in other conceptions, like Zermelo–Fraenkel set theory and topos theory. It would be very strange to develop mathematics in different ways and with no communication between them.

[1] In his talk entitled *100 years of Zermelo's axiom of choice: what was the problem with it?* given at TYPES '04, Jouy-en-Josas (France), 15-18 Dec. 2004, he also stressed that the debate among mathematicians about the acceptance of AC should be revisited with this distinction in mind.

Reaching mutual communication is not just an ideal wish, there are strong and concrete reasons that one can experience in the actual development of mathematics. A common base would allow for a more systematic transfer of techniques and results than what is possible today between two approaches to the same topic (for example, point-free topology), one developed in topos theory (theory of locales) and the other in type theory (formal topology). Ideally, one would like to develop constructive mathematics in a theory which is compatible with existing foundations and can be understood and used by most mathematicians.

To reach mutual communication with existing relevant theories, besides proofs-as-programs, we add also the following condition to our desiderata:

Compatibility. *The theory should be* compatible *with known theories such as classical set theory and the theory of a generic topos [20, 21], Martin-Löf's intensional type theory [18] and the Calculus of Construction [22], and hence it should be* minimal *with respect to such more expressive existing theories.*

Note that the above condition implies compatibility also with all the set theories that are interpretable in M-L type theory, as CZF [15].

To reach compatibility with classical set theory, ZFC, we certainly need to abandon the internal validity of the formal Church thesis, whilst we can allow the internal validity of the axiom of choice. Giving up CT is not so painful since CT is not necessarily valid in Heyting semantics.

However, we also have to give up the internal validity of the axiom of choice if we want to reach compatibility also with the theory of a generic topos. The reason is that a topos satisfying the internal axiom of choice is boolean [23], that is it enjoys classical logic. Hence, a topos with a natural numbers object is inconsistent with AC+CT, because classical arithmetic is already inconsistent with such principles. The same applies to intuitionistic set theories in the style of Zermelo–Fraenkel like IZF [13], Myhill's CST [14] or Aczel's CZF [15].

The fact that topos theory is inconsistent with AC+CT is certainly due to AC, given that there are examples of toposes, like the effective topos [24], that validate CT. The reason for such incompatibility is well visible in the proof of the following proposition (first given in [25]) in the context of a set theory validating the axiom of choice and some other simple axioms (see also [26, 27]):

Proposition 6.3 *Let $\mathcal{S}$ be a set theory in a many sorted language with the empty set axiom, pair set axiom, comprehension axiom, set extensionality axioms and the axiom of choice. Then $\mathcal{S}$ validates the principle of excluded middle.*

Proof. Let P be any formula of the set theory. Let $0 \equiv \emptyset$ and $1 \equiv \{\emptyset\}$. Consider the sets

$$V_o \equiv \{x \in \{0, 1\} \mid x = 0 \vee P\} \qquad V_1 \equiv \{x \in \{0, 1\} \mid x = 1 \vee P\}.$$

Note that we can apply the axiom of choice on

$$(\forall z \in \{V_o, V_1\})\ (\exists y \in \{0, 1\})\ (y \in z)$$

which is always valid, to get a function f from $\{V_o, V_1\}$ to $\{0, 1\}$. Through this function, which must give equal values on extensionally equal sets, we can decide whether P holds or not by comparing the values $f(V_o)$ and $f(V_1)$. □

This proof reveals that extensionality is constructively incompatible with the axiom of choice. Indeed, the choice function does not need to respect set extensionality since a proof of $(\forall x \in A)\,(\exists y \in B)\,R(x, y)$ can be done *intensionally* by associating a b in B to any $a \in A$ without necessarily respecting a given equivalence relation on A.

Extensionality is an intrinsic aspect of topos theory. Indeed, the internal type theory of a topos formulated in the style of Martin-Löf's type theory is extensional because of the presence both of the extensional propositional equality and of extensional powersets [28]. This makes the axiom of choice become what in type theory—or better in the extensional theory of setoids built upon type theory—is recognized as the *extensional* axiom of choice, since the choice function must be extensional by definition.

Hence a type theorist to be understood by a topos theorist has to give up the axiom of choice that the topos theorist automatically understands as the extensional one. This difficulty of expressing intensional concepts in topos theory is indeed a limit and a possible source of misunderstanding, given that the extensional axiom of choice is by no means valid in Heyting semantics (and in general), contrary to the intensional axiom of choice which certainly is.

Finally, to make our proofs-as-programs theory compatible with M-L type theory we must certainly give up extensional powersets. In fact dropping extensional propositional equality is not enough to keep constructive compatibility with AC if we keep extensional powersets as described in [29].[2]

Therefore we conclude that to satisfy the compatibility condition our proofs-as-programs theory has to avoid the internal validity of CT, of AC and the presence of extensional powersets.

6.1.4 *Conceptual reasons: against reductionism*

Two further reasons of a more conceptual nature speak in favour of a minimalist approach, since it avoids the reduction of logic and of geometrical intuition to computation.

The explanation of logic via the propositions-as-sets interpretation evidently makes logic strongly dependent on set theory. But while it seems legitimate to reduce mathematics to computation, it is hard to see reasons why this should apply also to logic. In fact, it seems clear that logic should be applicable also to fields not (yet) treated mathematically, and thus it is quite natural that its justification should not require any prior justification of set theory. So one reason to modify M-L type theory is to obtain a distinction of logic from set theory, that is of deductions from computations.

A second reason is that the computational view underlying M-L type theory is so strict that it clashes with some kind of spatial or geometric intuition. For instance, one would like to find a precise formulation of Brouwer's notion of choice sequence [30],

[2] There it is proved that an extension of M-L type theory [18, 19] with *extensional* powersets, in which subsets are represented by propositional functions, necessarily validates the principle of excluded middle.

but this appears to be impossible in M-L type theory unless it is modified to reach control over the use of AC (see Section 3.2.6 of [2]).

6.2 A Proposal for a Minimal Type Theory

After giving some good motivations in favour, we now have to show that a minimal theory is possible. We agree with Bishop and Martin-Löf that AC is a direct consequence of the explanation of intuitionistic logic as in the BHK interpretation, or in the propositions-as-sets view. Given that AC is to be abandoned, we have to abandon propositions-as-sets too and thus we must explain logic in a different way, independently of proof-terms and of set theory in general.

We believe that logical constants (connectives and quantifiers) can be explained in a satisfactory way via the *principle of reflection* (see [3]), which relies only on the notion of truth of a proposition, possibly depending on an arbitrary element of a domain. In this way logic becomes independent of full set theory. This is a delicate point. For instance, also the explanation of logical constants given by Martin-Löf in [31], following Prawitz's inversion principle, does not make explicit use of proof-terms. But one central idea there is that introduction rules define the meaning, and elimination rules are completely determined by them. In our opinion, this idea has to be abandoned too, otherwise the propositions-as-sets interpretation is implicitly present (a fact which goes together with validity of AC, which in our approach fails).

6.2.1 *Logic explained via the principle of reflection*

The principle of reflection says that the meaning of a logical constant is given by its definitional equation, which requires the logical constant to be the reflection in the object language of a link between judgements (assertions of truth of a proposition) at the meta-language. The formal rules of inference, defining the logical constant explicitly, are derived from its definitional equation. This is done for all connectives in [3], where it is shown that they are all explained by using only two meta-linguistic links, *and* and *yields* (corresponding to comma , and turnstile $\vdash$ in a sequent calculus). Moreover, the derivations of inference rules all follow the same pattern. We here show that also the quantifiers $\forall$ and $\exists$ can be explained in the same way. With respect to connectives, one needs the additional notion of propositional function depending on an element of a domain.

The informal version of the definitional equation for $\forall$ is: $\Gamma \vdash (\forall x \in D)A(x)$ iff *for every* $d \in D, \Gamma \vdash A(d)$. A more rigorous formulation is obtained by replacing the right member with $\Gamma, z \in D \vdash A(z)$, on the assumption that Γ does not depend on z:

$$\Gamma \vdash (\forall x \in D)A(x) \quad \text{iff} \quad \Gamma, z \in D \vdash A(z).$$

So to explain the meaning of $\forall$ one first has to understand the meaning of $\Gamma, z \in D \vdash A(z)$, that is, the fact that one can replace z with any element d of D and obtain that Γ true yields $A(d)$ true. In the explanation of $\forall$ given in the BHK interpretation, one has to understand this and *moreover* that a method is given to transform d and a proof of Γ

into a proof of $A(d)$. We do not need here the addition of the method, because only the truth of propositions (with no proof-term) is involved.

Of course, knowing that D is a set in the sense of type theory, that is, with rules to generate all its elements, is certainly sufficient to give a clear explanation of the meaning of $\Gamma, z \in D \vdash A(z)$. It does not seem, however, to be also necessary; in fact, it seems that one can be in the position of knowing $\Gamma \vdash A(d)$ whenever $d \in D$ is known, however this information may be obtained. For example, knowing that d is a human being, one understands that d is mortal, even if one is not able to give fixed rules to generate all human beings as a set.

The solution of the definitional equation for quantifiers follows the same pattern as for connectives. We prefer to skip the details in the case of $\forall$ and give them only for $\exists$, because it is a bit less intuitive. The informal version of the definitional equation for $\exists$ is: $\Gamma, (\exists x \in D)A(x) \vdash \Delta$ iff *for every* $d \in D$, $\Gamma, A(d) \vdash \Delta$. It is a bit puzzling at first that $\exists$ at object level is explained as the reflection of what looks as a universal quantification at the meta-level, but actually this is explained by observing that *for every* $d \in D$ is reflected at the left of $\vdash$, that is on assumptions. Then the definitional equation can be seen as an expression of the standard meaning of $\exists$: to know that from the pure existence of an element satisfying A one can conclude Δ means precisely to know that one can conclude Δ from $A(d)$, whatever the element d is such that $A(d)$ holds.

Here too, the precise formulation of the definitional equation for $\exists$ is obtained by replacing *for every* $d \in D$, $\Gamma, A(z) \vdash \Delta$ with $\Gamma, z \in D, A(z) \vdash \Delta$, under the condition that Γ, Δ do not depend on z:

$$\Gamma, (\exists x \in D)A(x) \vdash \Delta \quad \text{iff} \quad \Gamma, z \in D, A(z) \vdash \Delta .$$

The formation rule is just one direction of the definitional equation:

$\exists$-formation

$$\frac{\Gamma, z \in D, A(z) \vdash \Delta}{\Gamma, (\exists x \in D)A(x) \vdash \Delta}$$

under the condition that Γ, Δ do not depend on z. The other direction can also be written as a rule; it is called 'implicit reflection' since it can be seen as information on $\exists$ given implicitly:

$\exists$-implicit reflection:

$$\frac{\Gamma, (\exists x \in D)A(x) \vdash \Delta}{\Gamma, z \in D, A(z) \vdash \Delta} .$$

One can think of it as a wish which is to be satisfied by finding a proper rule (that is, one in which $\exists$ does not appear in the premises) which is equivalent to it. This is achieved as follows. First, the premise of $\exists$-implicit reflection is made trivially valid by choosing $\Gamma = \emptyset$ and $\Delta = (\exists x \in D)A(x)$. So one obtains:

$\exists$-axiom

$$z \in D, A(z) \vdash (\exists x \in D)A(x).$$

This is closer to the standard explanation of $\exists$, since it says that if one has any element of D on which A holds, then one can conclude $(\exists x \in D)A(x)$. The proper rule of reflection is now obtained by composing the axiom with derivations of the two components $z \in D$ and $A(z)$:

∃-explicit reflection:

$$\frac{\Gamma \vdash z \in D \quad \Gamma' \vdash A(z)}{\Gamma, \Gamma' \vdash (\exists x \in D)A(x)}\,.$$

In detail, from the premise $\Gamma \vdash z \in D$ and the ∃-axiom, by cut one obtains $\Gamma, A(z) \vdash (\exists x \in D)A(x)$ and hence by composing also with $\Gamma' \vdash A(z)$ one obtains the conclusion $\Gamma, \Gamma' \vdash (\exists x \in D)A(x)$.

Now it is easy to prove that actually explicit reflection is equivalent to implicit reflection. In fact, choosing $\Gamma = z \in D$ and $\Gamma' = A(z)$, ∃-explicit reflection gives the ∃-axiom, and ∃-implicit reflection is obtained from the ∃-axiom by cutting $(\exists x \in D)A(x)$. Hence ∃-formation and ∃-explicit reflection together are equivalent to the definitional equation, which is thus solved.

It is still an open problem to see how the principle of reflection should be extended to include a treatment of proof-terms.

6.2.2 *A formal system for minimal type theory*

After explaining logic by means of the reflection principle, we want to embed it into a set theory that satisfies our requirements. We make a proposal of such a theory and we call it mTT, for Minimal Type Theory. Here, we just give a brief description of its rules and the motivations behind its design, and we collect in the appendix all the formal rules.

Our type theory is obtained by embedding intuitionistic predicate logic into a constructive theory of sets as M-L type theory [18]. However we do not want to do that by a meta-translation of propositions into sets, but we simply add rules forming propositions to those forming sets. In particular, to build sets we use the four kinds of judgements in [18]:

$$A\ set\ [\Gamma] \qquad A = B\ set\ [\Gamma] \qquad a \in A\ set\ [\Gamma] \qquad a = b \in A\ set\ [\Gamma]$$

that is the judgements about set formation and their terms, equality between sets and equality between terms of the same set. As specific set constructors we take those of [18], like the empty set, lists, dependent products, disjoint sums and strong indexed sums.

Then, to build propositions we use new kinds of judgement saying that something is a proposition and when two propositions are equal:

$$A\ prop\ [\Gamma] \qquad A = B\ prop\ [\Gamma].$$

mTT includes the formation of propositions like falsum, universal and existential quantifications, implication, disjunction, conjunction and an intensional propositional equality, with their terms and equalities. Their elimination rules act only on propositions and hence they are more restrictive than the corresponding rules in M-L type theory.

Considering that the rules of logic are obtained by following the reflection principle in a sequent calculus style, the best would be to formulate all the system in a sequent calculus style, including set theory. But, being more familiar with type theories formulated in natural deduction style, we leave this problem to future work and we express logic too in the natural deduction style.

Since we want our theory to satisfy the proofs-as-programs paradigm, we must give all the inference rules for propositions with explicit treatment of how proofs are constructed; formally, we must provide the rules with proof-terms. In other words, we identify a proposition with the set if its proofs. This might at first seem to contradict our view that logic is to be independent of set theory. However, this choice is just forced on us if we want to keep proofs as programs: the only way we have to 'teach a computer' how to deal with deductions is to identify them with some kind of computations. From the conceptual point of view, this move is less artificial than it looks at first: while it certainly remains odd to assume that a proposition like 'All human beings are mortal' is identified with the set of its proofs, this is justified when we restrict our attention to mathematical propositions, and moreover we want to express them and their proofs within a formal system. After all, requiring that also propositions are defined by inductive rules is a way of expressing internally how the proofs in a formal system, like a sequent calculus LJ for intuitionistic logic, are defined inductively.

To express formally that a proposition is identified with the set of its proofs, and so that it is a set, we add the rule:

prop-into-set

$$ps)\ \frac{A\ prop}{A\ set}.$$

We also need to add the rule saying that if $A = B\ prop$ then $A = B\ set$. Since we think of propositions as special sets, like small sets in M-L type theory, we do not introduce new kinds of judgements to express the formation of a proof of a proposition and the definitional equality between proofs, like for example $a \in A\ prop\ [\Gamma]$ and $a = b \in A\ prop\ [\Gamma]$. If we did that, then the *prop-into-set* rule should be enriched with the rule saying that if $a \in A\ prop\ [\Gamma]$ then $a \in A\ set\ [\Gamma]$ and also conversely if $A\ prop\ [\Gamma]$ and $a \in A\ set\ [\Gamma]$ then $a \in A\ prop\ [\Gamma]$, where the converse expresses the fact that the elements of propositions are only their proof-terms. Since we choose to avoid these extra kinds of judgement about proof-terms, then we can reduce contexts to lists of assumptions of variables varying on sets. Therefore, the rules to form contexts in mTT are simply the usual ones of M-L type theory:

$$1C)\ \frac{}{\emptyset\quad cont} \qquad 2C)\ \frac{\Gamma\quad cont \qquad A\ set\ [\Gamma]}{\Gamma, x \in A\quad cont}\ (x \in A \notin \Gamma)\,.$$

Then, to express the fact that equality is an equivalence relation, we add all the inference rules that express reflexivity, symmetry and transitivity of equality between sets, propositions and their terms. Moreover, to express the fact that equality preserves the typing of a term we add the set equality rule

$$seteq)\ \frac{a \in A\,[\Gamma] \quad A = B\; set\,[\Gamma]}{a \in B\,[\Gamma]}\ .$$

Of course, we add the rule of assumption of variables

$$var)\ \frac{\Gamma, x \in A, \Delta \quad cont}{x \in A\,[\Gamma, x \in A, \Delta]}$$

saying that if $\Gamma, x \in A, \Delta$ is a context then x is an element of A under that context. By this formulation of variable assumption the structural rules of weakening, substitution and of a suitable exchange turn out to be derivable.

The formulation of the *prop-into-set* rule we present resembles the formation of universes *à la* Russell (see [17]). This observation reminds us that we could also formulate the *prop-into-set* rule *à la* Tarski, saying that if A is a proposition then $T(A)$ is its set of proofs together with an encoding of proofs of the proposition A into proof-terms of $T(A)$, under the encoding of the propositional context, and with also a decoding of the proof-terms of $T(A)$ into proofs of A, under the corresponding contexts.

The formulation *à la* Tarski of the *prop-into-set* rule corresponds to the rule in the version of the Calculus of Constructions in [22] expressing that a proposition $\phi : Prop$ can be turned into the type of its proofs $T(\phi)\ type$ by associating the proofs of the dependent product type to those of the universal quantification. Then, the main difference between our type theory and the Calculus of Constructions is that our type theory is predicative also on propositions and hence that propositions are defined inductively with proof-terms.

The presence of proof-terms is necessary from a computational point of view to get a type theory for which type-checking can be performed.

Proof-terms are also crucial to implement a toolbox to deal with extensional concepts, as that in [1]. For example, proof-terms are needed to implement the notion of function between subsets. Recall that a subset is represented in [1] by a propositional function $U(a)\ prop\ [a \in S]$. Then the difficulty of implementing a function between subsets is due to the fact that it is a partial function between the underlying sets, and a partial function is not definable directly as a proof-term in a type theory where all functions are total. However, we can solve this difficulty by using the indexed sum of a set with a propositional function, namely $\Sigma_{a \in S} U(a)$ for the propositional function $U(a)\ prop\ [a \in S]$. And it is the *prop-into-set* rule that makes the set $\Sigma_{a \in S} U(a)$ well formed by turning $U(a)\ prop\ [a \in S]$ into $U(a)\ set\ [a \in S]$.

Another important role of the *prop-into-set* rule is to allow proofs by induction of a proposition depending on an inductive set. For example, consider a proposition $A(x)\ prop\ [x \in \mathbb{N}]$ depending on the set of natural numbers (see the appendix for the rules[3]). Since the proposition $A(x)\ prop\ [x \in \mathbb{N}]$ is also a family of sets $A(x)\ set\ [x \in \mathbb{N}]$, by means of the *prop-into-set* rule we can use the elimination rule of the

[3]Note that we can represent $\mathbb{N} = List(N_1)$ as the set of lists on the singleton set which is in turn represented as $N_1 \equiv List(N_0)$.

natural numbers to derive $El_{List}(a, l, n) \in A(n)$ given the proof-terms $a \in A(0)$ and $l(x, y) \in A(s(x))$ $[x \in \mathbb{N}, y \in A(x)]$. Hence, we do not need to add the specific induction principles for propositions singularly, as in [32].

Considering that in developing mathematics one is interested only in the provability of a propositions and not in its proofs, one could be unsatisfied with the introduction of proof-terms for propositions. This is certainly a legitimate request, but to be implemented not at the level of type theory but at the level of toolbox. This is why in toolbox to talk about the validity of a proposition one can use an additional kind of judgement (introduced in [17]) of the form $A\ true$ for a proposition A, with the meaning that the proposition A is true. However, according to the introduction in [1] and Section 3.2.1 in [2], one must be careful when using proof-irrelevance of propositions since this is a derived property depending on the underlying type theory. Indeed, it is acceptable to forget the information of proof-terms that is necessary to give a presentation of an extensional object, like for example a function between subsets, only when we do not lose the possibility of restoring it, that is when we do not destroy the computation necessary to implement the concept itself.

Whilst a proposition is identified with the set of its proofs, in the resulting type theory the separation between logic and set theory remains clearly visible. In fact, while the elimination rules of sets follow the inversion principle (see principle (1) on page 8 of [17]), the elimination rules of propositions do not. Indeed, the sets of proofs of mathematical propositions originate from the reflection principle and hence the elimination rule of a proposition refers only to propositions. In other words to generate the proofs of propositions we do not follow the principle that the introduction rules determine the elimination rules, but we simply give a different justification of the rules of logic with respect to those of set theory.

To clarify this point, recall that also Martin-Löf in [31] gives an explanation of logic independently from its interpretation in set theory, but still using Heyting semantics. Then, to develop mathematics he interprets logic into set theory; it is here that he interprets the quantifier $\exists$ into the set-theoretic constructor Σ, probably—our guess—because they have the same introduction rules. But we may conceive of different indexed sum constructors with the same introduction rules and different elimination rules, such as the strong indexed sum in [18, 19] and Howard's weak indexed sum in [33], which we report here:

Weak indexed sum set

$$\text{F-}\Sigma^{w})\ \frac{C(x)\ set\ \ [x \in B]}{\Sigma^{w}_{x\in B}C(x)\ set} \qquad \text{I-}\Sigma^{w})\ \frac{b \in B \quad c \in C(b)}{< b,_{w} c > \in \Sigma^{w}_{x\in B}C(x)}$$

$$\text{E-}\Sigma^{w})\ \frac{\begin{array}{l} M\ set \\ d \in \Sigma^{w}_{x\in B}C(x) \quad m(x,y) \in M\ [x \in B, y \in C(x)] \end{array}}{El_{\Sigma^{w}}(d, m) \in M}$$

$$\text{C-}\Sigma^{w})\ \frac{\begin{array}{l} M\ set \\ b \in B \quad c \in C(b) \quad m(x,y) \in M\ [x \in B, y \in C(x)] \end{array}}{El_{\Sigma^{w}}(< b,_{w} c >, m) = m(b,c) \in M}.$$

Observe that the elimination rule of the weak indexed sum is restricted to sets not depending on the weak indexed sum itself. Hence, the weak indexed sum solves a definitional equation of weaker complexity than the one of the strong indexed sum.

Our rules for the existential quantifier in mTT (see the appendix) resemble those of Howard's weak indexed sum, but they are applied only to propositions. This is a crucial restriction to get a type theory where the axiom of choice (and probably also the axiom of unique choice) is no longer valid. In fact, at first glance it might appear that an alternative proposal for a minimal type theory could be the extension of M-L type theory with Howard's weak indexed sum, to be used to interpret the existential quantification. But it turns out that, when considered as propositions, weak indexed sum sets are provably equivalent to strong ones, as first noted by Luo [34]. Hence, interpreting the existential quantifier as the weak indexed sum would not change the theorems of the system and in particular the validity of choice principles.

In mTT we are able to avoid such an equivalence between the strong indexed sum set and the existential quantifier since the existential quantifier solves an even weaker definitional equation than the one for Howard's weak indexed sum, because, after distinguishing propositions and sets, its elimination rule can be applied only to propositions and not to generic sets.

An analysis of the connections between the sets of propositional proofs and the set-theoretic operations on sets of proofs is left to future research (for example, for propositions A, B we have that if $A \to B$ is provable then $\Pi_A B$ is not empty and conversely, but this connection does not seem to follow between $A \vee B$ and $A + B$ as well as between $\exists_{x \in A} B$ and $\Sigma_{x \in A} B$ with A simply a set, as just observed).

Whilst mTT satisfies proofs-as-programs informally, we think that a mathematical model of this type theory where AC+CT are valid could be built starting from the results in [35].

A categorical semantics of mTT, for which mTT provides an internal language, is an open question because of its intensional aspect. Certainly we can design extensional categorical models, considering that for many extensional type theories we know categorical semantics having them as an internal language [28].

Thanks to its genesis, mTT can be seen easily embeddable in M-L type theory in [18, 19] and in the theory of a generic topos, simply because they are obtained by the addition of some further principle. In this strong sense, mTT is compatible with both, and hence also with classical set theory, and this explains our adjective minimal.

Translation of mTT into Martin-Löf's set theory. To translate mTT into M-L type theory it is enough to translate propositions into the corresponding sets following Martin-Löf's propositions-as-sets translation, since all the proofs made with propositions become proofs of the corresponding sets. This also shows that the lambda calculus underlying mTT is strongly normalizing as the one underlying M-L type theory.

Translation of mTT in the internal theory of a topos. Consider the typed calculus of toposes with a natural numbers object introduced in [5, 28] written in the style of extensional Martin-Löf's type theory in [17]. In such a calculus we can also define the falsum type, the disjoint sum type and the quotient type on any equivalence relation. In particular any type can be turned into a type with at most one proof, called *mono type*, by quotienting it over the total relation.

Then, we translate all the sets of mTT into the corresponding types of the internal typed calculus of toposes and the propositions into mono types by induction. In particular, the intensional equality proposition is translated into the extensional equality type, which is mono. Note that the *prop-into-set* rule is valid by definition since propositions are translated as special types.

Translation of mTT into the Calculus of Constructions. We could translate mTT in a version of the Calculus of Constructions [22] where we assume that the types include the inductive sets of M-L type theory. Then, we translate sets into the corresponding types and propositions into the small types of proofs of the form $T(\phi)$ for a proposition ϕ by induction.

Toolbox for mTT. A toolbox for subsets over mTT is ready in [1]. Whilst this was built for M-L type theory, it can be thought of as built for mTT, since it is based on truth judgements $A\ true$ for propositions whose logical rules coincide with those of mTT. This might be surprising considering that the mTT existential quantifier is no longer identified with the strong indexed sum and it does not make AC valid. But note that if one consider the rules for the existential quantifier at the level of truth judgements, one can see that the same rules can be obtained also starting from different type constructors. Indeed, the strong indexed sum and Howard's weak one (via the propositions-as-sets interpretation) produce the same rules after suppressing proof-terms, namely those of the existential quantifier of intuitionistic logic which also coincide with the rules of mTT existential quantifier. Hence all these constructors cannot be distinguished at the pure level of logical judgements. As a consequence, the axiom of choice does not follow from the meaning of the logical constants expressed in terms of truth judgements (see [36, 37]). Also the validity of the axiom of unique choice is not automatic in toolbox since one can distinguish a single valued relation from a program that computes it. Several other tools, like setoids, finite subsets, etc. are also needed, but are not yet formally completed.

Related work. The logic-enriched type theory by P. Aczel and N. Gambino in [32] is a theory for constructive mathematics compatible with topos theory and classical set theory. However this is closer to a many-sorted logic approach, while the one we propose here is closer to the type-theoretic approach, since we have proof-terms for propositions in order to be able to perform type-checking. Moreover, by our embedding of propositions into sets we have induction principles for propositions on the various set-theoretic constructions built into it.

In order to reproduce topos theory within M-L type theory, in [38] a predicative notion of topos is introduced. This notion has clearly an extensional character, as that of topos, since it is based on a pretopos whose internal dependent type theory is extensional (see [28]) in the sense of extensional type theory [17].

A comparison between the weak and strong existential quantifiers was also considered in the framework of the Calculus of Constructions in [22, 39] or to represent data types in programming languages in [40].

6.3 Some Visible Benefits

It is encouraging to observe that the foundational attitude underlying minimal type theory, which apparently was born for reasons of compatibility and communication, actually has a specific identity and an intrinsic mathematical interest. Our claim that minimal type theory, together with toolbox, can be used to develop constructive mathematics is proved by the fact that most of formal topology has already been developed over it, though informally (necessarily so, since the underlying formal system is given here for the first time). In fact, formal topology is developed based on toolbox in [1], which is meaningful only if propositions are kept distinct from sets and which can be implemented in mTT; this is confirmed also by the fact that in most of formal topology no use is made of choice principles.

In our opinion, some other benefits of the minimalistic attitude are even more interesting. In mathematics, the idea of dealing at the same time with a variety of different interpretations is not new. It is just the attitude of abstract algebra, and of the modern axiomatic method in general. What we propose is to do the same at the level of foundations. Like what happened with abstract algebra, adopting weaker foundational assumptions has stimulated the emergence of new, stronger structures. Actually, the real and deep reason for choosing a new foundation is that by using it some *new* mathematics is developed (see [2], Section 1.3.2). The whole new field called the basic picture [2, 6], a new mathematical structure which underlies and at the same time generalizes constructive topology, has sprung out of an informal but rigorous adherence to a minimalist foundation.

The basic picture is patently a piece of mathematics, since it is extensional, it is very simple technically and can be understood as it is (with no translation) both by a type theorist and by a topos theorist. This means that any mathematician can understand and take advantage of the new mathematical concepts arising in constructive mathematics. Its main conceptual novelty lies in a specific mathematical treatment of notions with existential character. So the well known notions of inclusion $\subseteq$ between subsets, of open subset, of cover $\lhd$ generated by induction, etc., are now accompanied by their dual notions of overlap $\between$ between subsets, of closed subset, of positivity relation $\ltimes$ generated by coinduction [41], etc.

Finally, since the axiom of choice does not hold, the notion of formal point is not reduced to be lawlike, and thus the possibility of conceiving choice sequences as formal points remains open (see [2], Section 3.2.6).

6.4 Open Questions and Further Work

A natural question is to ask what known type theories satisfy our proofs-as-programs paradigm as we specified. In particular we have open questions such as:

Is there a realizability model testifying that Martin-Löf's intensional type theory in [18, 19] is consistent with CT? We strongly believe that Martin-Löf's intensional type theory is consistent with CT and on the other hand we hope to see a realizability model of it to get a mathematical proof of that belief. This model would also be a model of mTT and hence a proof of its consistency with AC+CT.

Is second order many-sorted (on finite types including function types) intuitionistic logic, or the Calculus of Constructions consistent with propositional AC+CT? In other words, does some version of the Calculus of Constructions satisfy our proofs-as-programs paradigm? We are very curious to know a proof of its consistency with AC+CT.

Acknowledgements

The first author thanks Enrico Martino for fruitful discussions on the definition of a proofs-as-programs theory. We thank Thierry Coquand, Giovanni Curi, Ferruccio Guidi, Per Martin-Löf, Eike Ritter, Bas Spitters, Thomas Streicher, Silvio Valentini for their comments on the topics of this chapter.

References

1. Sambin, G. and Valentini, S. (1998). Building up a toolbox for Martin-Löf's type theory: subset theory. In: *Twenty-five years of constructive type theory, Proceedings of a Congress held in Venice, October 1995* (ed. Sambin, G. and Smith, J.), pp. 221–244. Oxford U. P.
2. Sambin, G. (2003). Some points in formal topology. *Theoretical Computer Science*, **305**(1–3), 347–408.
3. Sambin, G., Battilotti, G., and Faggian, C. (2000). Basic logic: reflection, symmetry, visibility. *J Symbolic Logic*, **65**(3), 979–1013.
4. Sambin, G. (1987). Intuitionistic formal spaces – a first communication. *Mathematical logic and its applications* pp. 187–204.
5. Maietti, M. E. (1998, February). The type theory of categorical universes. PhD thesis. University of Padova.
6. Sambin G. (with the collaboration of Gebellato, S. and Martin-Löf, P. and Capretta, V.), (2003, May). The basic picture; Preprint n. 08, Dipartimento di Matematica, Università di Padova.
7. Bishop, E. (1967). *Foundations of Constructive Analysis*. McGraw-Hill, New York.
8. Bishop, E. (1985). Schizophrenia in contemporary mathematics. In: *Errett Bishop: reflections on him and his research (San Diego, Calif., 1983)*. (ed. M. Rosenblatt). Volume. 39 of Contemporary Mathematics. Amer. Math. Soc. pp. 1–32.
9. Bishop, E. and Bridges, D. (1985). *Constructive analysis. Grundlehren der Mathematischen Wissenschaften [Fundamental Principles of Mathematical Sciences]*. Volume. 279. Springer-Verlag, Berlin.
10. Martin-Löf, P. (1970). *Notes on Constructive Mathematics*. Almqvist & Wiksell.
11. Coquand, T. (1997). Computational content of classical logic. In: *Semantics and Logics of Computation (Cambridge, 1995)*. Volume. 14 of *Publ. Newton Inst.*, pp. 33–78. Cambridge University Press, Cambridge.
12. Girard J.-Y. (with the collaboration of Lafont, Y and Taylor, P), (1989). *Proofs and Types*, Volume. 7 of *Cambridge Tracts in Theoretical Computer Science*. Cambridge University Press, Cambridge.

13. Scedrov, A. (1985). Intuitionistic set theory. In: *Harvey Friedman's research on the foundations of mathematics*. Volume. 117 of *Stud. Logic Found. Math.*, pp. 257–284. North-Holland, Amsterdam.
14. Myhill, J. 1975 Constructive set theory. *J Symbolic Logic*, **40**, 347–383.
15. Aczel, P. and Rathjen, M. Notes on constructive set theory. Mittag-Leffler Technical Report No.40, 2000/2001.
16. Troelstra, A. and Dalen, D. van. (1988) *Constructivism in mathematics. An Introduction. Vol. II.* Volume 123 of *Studies in Logic and the Foundations of Mathematics*. North-Holland, Amsterdam.
17. Martin-Löf, P. (1984). *Intuitionistic Type Theory. Notes by G. Sambin of a series of lectures given in Padua, June 1980.* Bibliopolis, Naples.
18. Nordström, B., Peterson, K., and Smith, J. (1990). *Programming in Martin Löf's Type Theory*. Clarendon Press, Oxford.
19. Martin-Löf, P. (1998). An intuitionistic theory of types. In: *Twenty five years of Constructive Type Theory* (ed. Sambin G. and Smith, J.), pp. 127–172. Oxford Science Publications.
20. Johnstone, P. T. (2002*b*). *Sketches of an Elephant: A Topos Theory Compendium.* Vol. 2. Volume. 44, *Oxford Logic Guides*. The Clarendon Press, Oxford University Press, New York.
21. Johnstone, P. T. (2002*a*). *Sketches of an elephant: a topos theory compendium.* Vol. 1. Volume. 43, of Oxford Logic Guides. The Clarendon Press, Oxford University Press, New York.
22. Coquand, T. (1990). Metamathematical investigation of a calculus of constructions. In: *Logic in Computer Science*. (ed. P. Odifreddi). pp. 91–122. Academic Press.
23. Diaconescu, R. (1975). Axiom of choice and complementation. *Proc. Amer. Math. Soc.*, **51**, 176–178.
24. Hyland, J. M. E. (1982). The effective topos. In: *The L.E.J. Brouwer Centenary Symposium (Noordwijkerhout, 1981)*. Volume. 110 of *Stud. Logic Foundations Math.*, pp. 165–216. North-Holland, Amsterdam-New York.
25. Goodman, N. Myhill, J. (1978). Choice implies excluded middle. *Z Math Logik Grundlag Math*, **24**, 461.
26. Bell, J. L. (1988). *Toposes and Local Set Theories: An Introduction*. Oxford: Claredon Press.
27. Bell, J. (1997). Zorn's lemma and complete Boolean algebras in intuitionistic type theories. *J Symbolic Logic*, **62**, 1265–1279.
28. Maietti, M. E. (2001). Modular correspondence between dependent type theories and categorical universes. *Math Struct Comput Sci*, to appear.
29. Maietti, M. E. and Valentini, S. (1999). Can you add powersets to Martin-Löf intuitionistic type theory? *Mathematical Logic Quarterly*, **45**, 521–532.
30. Brouwer, L. E. J. (1975). *Collected Works* I. North Holland;
31. Martin-Löf, P. (1985). On the meanings of the logical constants and the justifications of the logical laws. In: *Proceedings of the conference on mathematical logic*

(Siena, 1983/1984). Volume. 2, pp. 203–281. Reprinted in: *Nordic J. Philosophical Logic*, 1 (1996), no. 1, pages 11–60.
32. Aczel, P. and Gambino, N. (2002). Collection principles in dependent type theory. In: *Types for Proofs and Programs (Durham, 2000)*. Volume. 2277 of *Lecture Notes in Comput. Sci.*, pp. 1–23. Springer, Berlin.
33. Howard, W. A. (1980). The formulae-as-types notion of construction. In: *To H. B. Curry: Essays on Combinatory Logic, Lambda Calculus and Formalism*. pp. 480–490. Academic Press, London-New York.
34. Luo, Z. (1994). *Computation and Reasoning. A Type Theory for Computer Science.*, Volume. 11 of *International Series of Monographs on Computer Science*. The Clarendon Press, Oxford University Press, New York.
35. Hyland, J. M. E. (2002). Variations on realizability: realizing the propositional axiom of choice. *Math Structures Comput Sci.*, **12**, 295–317.
36. Swaen, M. D. G. (1992). A characterization of ML in many-sorted arithmetic with conditional application. *J Symbolic Logic*, **57**, 924–953.
37. Swaen, M. D. G. (1991). The logic of first order intuitionistic type theory with weak sigma-elimination. *J Symbolic Logic*, **56**, 467–483.
38. Moerdijk, I. and Palmgren, E. (2002). Type theories, toposes and constructive set theory: predicative aspects of ast. *Annals of Pure and Applied Logic*, **114**(1-3), 155–201.
39. Coquand, T. (1990). On the analogy between propositions and types. In: *Logical Foundations of Functional Programming*, University of Texas at Austin Year of Programming, p. 393. Addison-Wesley, Reading, MA. Originally in LNCS 242.
40. Mitchell, J. C. and Plotkin, G. D. (1988). Abstract types have existential type. *ACM Transactions Programming Languages and Systems*, **10**(3), 470–502.
41. Martin-Löf, P. and Sambin, G. (2005). Generating positivity by coinduction. [6]

6.5 Appendix: The System mTT

We present here the inference rules for sets and propositions in mTT. For brevity, in presenting formal rules we omit the corresponding equality rules, as in [17]. Moreover, the piece of context common to all judgements involved in a rule is omitted. The typed variables appearing in a context are meant to be added to the implicit context as the last one. We also recall that the contexts are made of assumptions on sets only, and that we have the rule:

Prop-into-set

$$ps)\ \frac{A\ prop}{A\ set}$$

The underlying set theory of mTT is Martin-Löf's constructive set theory [18]. Hence, it includes the following sets:

Empty set

F-Em) $N_0\ set$ E-Em) $\dfrac{a \in N_0 \quad A(x)\ set\ [x \in N_0]}{\mathsf{emp_o}(a) \in A(a)}$

Strong Indexed Sum set

$F-\Sigma)\ \dfrac{C(x)\ set\ \ [x \in B]}{\Sigma_{x \in B} C(x)\ set}$ I-Σ) $\dfrac{b \in B \quad c \in C(b)}{< b, c > \in \Sigma_{x \in B} C(x)}$

E-Σ) $\dfrac{\begin{array}{l} M(z)\ set\ [z \in \Sigma_{x \in B} C(x)] \\ d \in \Sigma_{x \in B} C(x) \quad m(x, y) \in M(< x, y >)\ [x \in B, y \in C(x)] \end{array}}{El_{\Sigma}(d, m) \in M(d)}$

C-Σ) $\dfrac{\begin{array}{l} M(z)\ set\ [z \in \Sigma_{x \in B} C(x)] \\ b \in B \quad c \in C(b) \quad m(x, y) \in M(< x, y >)\ [x \in B, y \in C(x)] \end{array}}{El_{\Sigma}(< b, c >, m) = m(b, c) \in M(< b, c >)}$

Disjoint Sum set

$F-+)\ \dfrac{B\ set \quad C\ set}{B + C\ set}$ I_1-+) $\dfrac{b \in B}{\mathsf{inl}(b) \in B + C}$ I_2-+) $\dfrac{c \in C}{\mathsf{inr}(c) \in B + C}$

E-+) $\dfrac{\begin{array}{l} A(z)\ set\ [z \in B + C] \\ w \in B + C \quad a_B(x) \in A(\mathsf{inl}(x))\ [x \in B] \quad a_C(y) \in A(\mathsf{inr}(y))\ [y \in C] \end{array}}{El_{+}(w, a_B, a_C) \in A(w)}$

C_1-+) $\dfrac{\begin{array}{l} A(z)\ set\ [z \in B + C] \\ b \in B \quad a_B(x) \in A(\mathsf{inl}(x))\ [x \in B] \quad a_C(y) \in A(\mathsf{inr}(y))\ [y \in C] \end{array}}{El_{+}(\mathsf{inl}(b), a_B, a_C) = a_B(b) \in A(\mathsf{inl}(c))}$

C_2-+) $\dfrac{\begin{array}{l} A(z)\ set\ [z \in B + C] \\ c \in C \quad a_B(x) \in A(\mathsf{inl}(x))\ [x \in B] \quad a_C(y) \in A(\mathsf{inr}(y))\ [y \in C] \end{array}}{El_{+}(\mathsf{inr}(c), a_B, a_C) = a_C(c) \in A(\mathsf{inr}(c))}$

List set

F-list) $\dfrac{C\ set}{List(C)\ set}$ I_1-list) $\epsilon \in List(C)$ I_2-list) $\dfrac{s \in List(C) \quad c \in C}{\mathsf{cons}(s, c) \in List(C)}$

E-list) $\dfrac{\begin{array}{l} L(z)\ set\ [z \in List(C)] \quad s \in List(C) \quad a \in L(\epsilon) \\ l(x, y, z) \in L(\mathsf{cons}(x, y))\ [x \in List(C), y \in C, z \in L(x)] \end{array}}{El_{List}(a, l, s) \in L(s)}$

$$\text{C}_1\text{-list)}\ \frac{L(z)\ set\ [z \in List(C)] \quad a \in L(\epsilon) \qquad l(x,y,z) \in L(\mathsf{cons}(x,y))\ [x \in List(C), y \in C, z \in L(x)]}{El_{List}(a,l,\epsilon) = a \in L(\epsilon)}$$

$$\text{C}_2\text{-list)}\ \frac{L(z)\ set\ [z \in List(C)] \quad s \in List(C) \quad c \in C \quad a \in L(\epsilon) \qquad l(x,y,z) \in L(\mathsf{cons}(x,y))\ [x \in List(C), y \in C, z \in L(x)]}{El_{List}(a,l,\mathsf{cons}(s,c)) = l(s,c,El_{List}(a,l,s)) \in L(\mathsf{cons}(s,c))}$$

Dependent Product set

$$\text{F-}\Pi\ \frac{C(x)\ set\ [x \in B]}{\Pi_{x \in B} C(x)\ set} \qquad \text{I-}\Pi\ \frac{c \in C(x)[x \in B]}{\lambda x^B.c \in \Pi_{x \in B} C(x)}$$

$$\text{E-}\Pi\ \frac{b \in B \quad f \in \Pi_{x \in B} C(x)}{\mathsf{Ap}(f,b) \in C(b)} \quad \beta\text{C-}\Pi\ \frac{b \in B \quad c(x) \in C(x)[x \in B]}{\mathsf{Ap}(\lambda x^B.c(x), b) = c(b) \in C(b)}$$

Then, mTT includes the following propositions:

Falsum

$$\text{F-Fs)}\ \bot\ prop \qquad \text{E-Fs)}\ \frac{a \in \bot \quad A\ prop}{\mathsf{r_o}(a) \in A}$$

Existential quantification

$$F - \exists)\ \frac{C(x)\ prop\ [x \in B]}{\exists_{x \in B} C(x)\ prop} \qquad \text{I-}\exists)\ \frac{b \in B \quad c \in C(b)}{< b,_{\exists} c > \in \exists_{x \in B} C(x)}$$

$$\text{E-}\exists)\ \frac{M\ prop \qquad d \in \exists_{x \in B} C(x) \quad m(x,y) \in M\ [x \in B, y \in C(x)]}{El_{\exists}(d,m) \in M}$$

$$\text{C-}\exists)\ \frac{M\ prop \qquad b \in B \quad c \in C(b) \quad m(x,y) \in M\ [x \in B, y \in C(x)]}{El_{\exists}(< b,_{\exists} c >, m) = m(b,c) \in M}$$

Disjunction

$$F - \vee)\ \frac{B\ prop \quad C\ prop}{B \vee V\ prop} \quad \text{I}_1 - \vee)\ \frac{b \in B}{\mathsf{inl}_{\vee}(b) \in B \vee C} \quad \text{I}_2 - \vee)\ \frac{c \in C}{\mathsf{inr}_{\vee}(c) \in B \vee C}$$

$$\text{E-}\vee)\ \frac{A\ prop \qquad w \in B \vee C \quad a_B(x) \in A\ [x \in B] \quad a_C(y) \in A\ [y \in C]}{El_{\vee}(w, a_B, a_C) \in A}$$

$$\text{C}_1\text{-}\vee)\quad \frac{A\ prop \quad b \in B \quad a_B(x) \in A\ [x \in B] \quad a_C(y) \in A\ [y \in C]}{El_\vee(\mathsf{inl}_\vee(b), a_B, a_C) = a_B(b) \in A}$$

$$\text{C}_2\text{-}\vee)\quad \frac{A\ prop \quad c \in C \quad a_B(x) \in A\ [x \in B] \quad a_C(y) \in A\ [y \in C]}{El_\vee(\mathsf{inr}_\vee(c), a_B, a_C) = a_C(c) \in A}$$

Implication

$$\text{F-}\to\quad \frac{B\ prop \qquad C\ prop}{B \to C\ prop}$$

$$\text{I-}\to\quad \frac{B\ prop \qquad c(x) \in C\ [x \in B]}{\lambda_\to x^B.c(x) \in B \to C} \qquad \text{E-}\to\quad \frac{b \in B \quad f \in B \to C}{\mathsf{Ap}_\to(f, b) \in C}$$

$$\beta\text{C-}\to\quad \frac{B\ prop \qquad b \in B \quad c \in C\ [x \in B]}{\mathsf{Ap}_\to(\lambda_\to x^B.c, b) = c(b) \in C}$$

Universal quantification

$$\text{F-}\forall\quad \frac{C(x)\ prop\ [x \in B]}{\forall_{x \in B} C(x)\ prop} \qquad \text{I-}\forall\quad \frac{c(x) \in C(x)\ [x \in B]}{\lambda_\forall x^B.c(x) \in \forall_{x \in B} C(x)}$$

$$\text{E-}\forall\quad \frac{b \in B \quad f \in \forall_{x \in B} C(x)}{\mathsf{Ap}_\forall(f, b) \in C(b)} \qquad \beta\text{C-}\forall\quad \frac{b \in B \quad c(x) \in C(x)\ [x \in B]}{\mathsf{Ap}_\forall(\lambda_\forall x^B.c(x), b) = c(b) \in C(b)}$$

Propositional equality

$$\text{Id})\quad \frac{A\ set \quad a \in A \quad b \in A}{\mathsf{Id}(A, a, b)\ prop} \qquad \text{I-Eq})\quad \frac{a \in A}{\mathsf{id}_\mathsf{A}(a) \in \mathsf{Id}(A, a, a)}$$

$$\text{E-Eq})\quad \frac{C(x, y)\ prop\ [x : A, y \in A] \quad a \in A \quad b \in A \quad p \in \mathsf{Id}(A, a, b) \quad c(x) \in C(x, x)\ [x \in A]}{El_\mathsf{Id}(p, (x)c(x)) \in C(a, b)}$$

$$\text{C-Eq})\quad \frac{C(x, y)\ prop\ [x : A, y \in A] \quad a \in A \quad c(x) \in C(x, x)\ [x \in A]}{El_\mathsf{Id}(\mathsf{id}_A(a), (x)c(x)) = c(a) \in C(a, a)}$$

Conjunction

$$F - \wedge)\quad \frac{B\ prop \qquad C\ prop}{B \wedge C\ prop} \qquad \text{I-}\wedge)\quad \frac{b \in B \quad c \in C}{< b,_\wedge c > \in B \wedge C}$$

$$\text{E}_1\text{-}\wedge\quad \frac{d \in B \wedge C}{\pi_1^B(d) \in B} \qquad \text{E}_2\text{-}\wedge\quad \frac{d \in B \wedge C}{\pi_2^C(d) \in C}$$

$$\beta_1\ \text{C-}\wedge\quad \frac{b \in B \quad c \in C}{\pi_1^B(< b,_\wedge c >) = b \in B} \qquad \beta_2\ \text{C-}\wedge\quad \frac{b \in B \quad c \in C}{\pi_2^C(< b,_\wedge c >) = c \in C}$$

7

INTERACTIVE PROGRAMS AND WEAKLY FINAL COALGEBRAS IN DEPENDENT TYPE THEORY

PETER HANCOCK AND ANTON SETZER

Abstract

We reconsider the representation of interactive programs in dependent type theory, proposed by the authors in earlier papers. The basis of this approach is monadic I/O as used in Haskell. We consider two versions: in the first the interface with the real world is fixed, while in the second the potential interactions can depend on the history of previous interactions. We consider also both client and server programs that run on opposite sides of an interface. Whereas in previous versions the type of interactive programs was introduced in an ad hoc way, it is here defined as a weakly final coalgebra for a general form of polynomial functor. We give formation/introduction/elimination/equality rules for these coalgebras. Finally we study the relationship of the corresponding rules with guarded induction. We show that the introduction rules are nothing but a slightly restricted form of guarded induction. However, the form in which we write guarded induction is not recursive equations (which would break normalization—we show that type checking becomes undecidable), but instead involves an elimination operator in a crucial way.

7.1 Introduction

Naïvely conceived, programs developed in dependent type theory are not interactive. They are functions that receive one or more arguments as input, and return one value as output or result. This view of program execution as consisting of a single step of interaction is perhaps appropriate for batch programming, prevalent in the 1960s and 1970s. At that time a job was submitted to the computer, typically consisting of some numerical computation on prepared data, and the results printed or stored in a file.

Nowadays one expects programs to be interactive. A running program should receive input from external devices (e.g. keyboard, mouse, network or sensors), and in response send output to external devices (e.g. display, sound card, network, actuators), and this cycle should repeat over and over again, perhaps forever.

The chief interest of dependent type theory for programming is not merely that it is a programming language, but rather that it is a framework for specifying and reasoning about programs. It is therefore necessary to understand how to develop interactive programs in dependent type theory. We hope to use it to develop verified interactive programs.

In this article we explore one approach to the representation of interactive programs in dependent type theory. This approach takes as its basis the concept of 'monadic

IO' [1] used in functional programming. We shall see that in dependent type theory, besides non-dependent interactive programs in which the interface between the user and the real world is fixed as in ordinary functional programming, there is a natural notion of state-dependent interactive programs, in which the interface changes over time. The representation is based on a structure identified by Petersson and Synek in [2].

We shall see that the concept of an interactive program is closely connected with the representation of weakly final coalgebras for certain specific functors. This notion will then be generalized to coalgebras for general polynomial functors.

We will then suggest an extension of Martin-Löf type theory by rules for weakly final coalgebras. There is work on encoding weakly final coalgebras in standard intensional Martin-Löf type theory. However, because it seems that reasoning about final coalgebras will be important in the future, we believe that it is natural to have them as 'first class', directly represented objects, as given by our rules. We shall see that a restricted form of guarded induction (where there is exactly one constructor on the right hand side and where reference can be made only to the function one is defining and not to elements of the coalgebra defined previously) is in exact correspondence with coiteration. Finally we will show that bisimulation is a state-dependent weakly final coalgebra.

7.1.1 *Other approaches to interactive programs in dependent type theory*

As pointed out to us by Peter Dybjer, in a certain sense one can use an expression with an algebraic data type as an interactive program. First one brings the expression to constructor form. (The reductions considered by Martin-Löf reduce a term only to weak head normal form.) Then one can 'peel away' the constructor, choose one of its operands, and reduce it further. To use the expression as an interactive program, one associates with each constructor some action upon the world, and with each response or output forthcoming from that action a selector that chooses an operand of the constructor. For instance, $2 + 3$ reduces to $\mathrm{S}(2 + 2)$, and one can then decide to investigate the argument $2 + 2$ further and find that $2 + 2$ reduces to $\mathrm{S}(2 + 1)$, etc. These successive reductions gives rise to a sequence of (trivial) interactions: handing over a coin to a shop-keeper for example, and receiving an acknowledgement. Or, if one defines $B : \{0, 1\} \to \mathrm{Set}$, $B\ 0 = \emptyset$, $B\ 1 = \mathbb{N}$, $C := \mathrm{W}x : \{0, 1\}.B\ x$, and starts with an element $c : C$, then c reduces to the form $\sup(a, f)$. In the case $a = 1$ one can apply f to an externally given natural number in order to obtain another element of C, and so on; in the case $a = 0$ no response is possible, and the process comes to an end. Note that the process of interpreting a constructor as a command or action, peeling them away and using the response to select an operand with which to continue is not an operation *within* type theory, but an extra-mathematical application *of* type theory. However, in order to obtain strong normalization and therefore decidable type checking one usually requires that types are well-founded, entailing that such a sequence of interactions will necessarily terminate eventually with some constructor without operands. So non-terminating sequences of interactions are impossible. For this reason, if we are not content merely to model terminating interactive programs, we need to consider coalgebras rather than algebras.

7.1.2 *Related work*

H. Geuvers has introduced in [3] rules for inductive and coinductive type corresponding to corecursion in the context of the simply typed λ-calculus and in the context of system F. He showed that the resulting systems are strongly normalizing. E. Gimenéz [4] (see as well the book on Coq [5], Chapter 13, for an exposition of the coalgebraic data types in Coq, which are based on the work by Gimenéz) has studied guarded recursion for weakly final coalgebras and a corresponding general recursive scheme for initial algebras in the context of Coq. He showed that the definable functions are extensionally the same as those definable by the rules given by Geuvers. However, interactive programs are not studied in their work, nor do they investigate in depth the formation/introduction/elimination/equality rules in the context of Martin-Löf type theory. What is not obvious in the work by Gimenéz is that not only can guarded induction be interpreted using the rules for weakly final coalgebras, but in fact the rules for weakly final coalgebras are exactly those arising from a slightly restricted form of guarded induction. Furthermore, the syntax used by Gimenéz seems to suggest that when one introduces recursive functions by guarded induction, it is only lazy evaluation which prevents their complete reduction. On the other hand, when looking at the rules one realizes that this is not the case, and the evaluation of these functions is driven by applying case distinction to an element of the coalgebra, corresponding to our elim-function discussed below.

The problem of representing final coalgebras in type theory was addressed in the special case of Aczel's non-well-founded sets by Lindström in [6], who gave a representation using an inverse-limit construction that requires an extensional form of type theory. Markus Michelbrink is working on an encoding of weakly final coalgebras in standard intensional Martin-Löf type theory, i.e. on introducing sets representing the weakly final coalgebras and functions corresponding to those given by the introduction and elimination rules such that the equalities given by the rules hold w.r.t. bisimilarity rather than definitional equality.

7.1.3 *Notations and type theory used*

In this chapter we work in standard Martin-Löf type theory, based on the logical framework with both dependent products and function types. Apart from sets introduced by rules added to type theory in this chapter, we use the constructs of the logical framework including Set, finite sets, the set of natural numbers, the disjoint union of sets and the identity type for forming new types.

We write $(x : A) \to B$ for the type of dependent functions f, where f takes for its argument an $a : A$ and returns an element $f(a)$ of type $B[x := a]$. This type is the logical framework version of the dependent function type denoted by $\Pi x : A.B$; the difference is that for $(x : A) \to B$ the η-rule is postulated at the level of judgemental equality, whereas for $\Pi x : A.B$ it holds at the level of propositional equality. We write

$\lambda x.s$ for the function f taking argument a and returning $s[x := a]$. If for $x : A$ we have $s : B$, then $\lambda x.s : ((x : A) \to B)$. We write $\lambda x, y.s$ for $\lambda x.\lambda y.s$. We write $f(a)$ for application of f to a, $f(a, b)$ for the application of f to a and b, and similarly for longer sequences of applications.

We write $(x : A) \times B$ for the dependent product. The elements of this type are pairs $\langle a, b\rangle$ where $a : A$ and $b : B[x := a]$. We write $\pi_0(a)$ and $\pi_1(a)$ for the first and second projection of an element of this type. $(x : A) \times B$ is the logical framework version of the type $\Sigma x : A.B$; again the difference is that with $(x : A) \times B$ we postulate the η-rule at the level of judgemental equality, whereas with $\Sigma x : A.B$ it holds rather at the level of propositional equality.

We write $(x : A, y : B) \to C$ for $(x : A) \to ((y : B) \to C)$, and $(x : A) \times (y : B) \times C$ for $(x : A) \times ((y : B) \times C))$, similarly for longer chains of types. Sometimes we assign a variable to the last set in a product, e.g. $(x : A) \times (y : B) \times (z : C)$ although z is never used.

We omit from products and function types variables which are not used (e.g. $(x : A, B, z : C) \to D$ instead of $(x : A, y : B, z : C) \to D$, if C, D don't depend on y), and write $A \to B$ and $A \times B$ instead of $(x : A) \to B$ and $(x : A) \times B$, respectively, where B does not depend on x. Furthermore we write $\langle a, b, c\rangle$ for $\langle a, \langle b, c\rangle\rangle$, and similarly for longer sequences.

If A, B : Set then $A + B$: Set is the disjoint union of A and B with constructors $\text{inl} : A \to (A + B)$ and $\text{inr} : B \to (A + B)$. If $f : A \to C$ and $g : B \to C$, then we define $[f, g] : (A + B) \to C$ as the function such that $[f, g](\text{inl}(a)) = f(a)$, $[f, g](\text{inr}(b)) = g(b)$. Further we usually write $A_0 + A_1 + \cdots + A_m$ instead of $A_0 + (A_1 + \cdots + (A_{m-1} + A_m))$. When referring to this type, we write in_i^m for the injection from A_i into $A_0 + \cdots + A_m$.

We have the set $\mathbf{1}$: Set, which has sole element $* : \mathbf{1}$, and assume the η-rule for $\mathbf{1}$, so we have that if $x : \mathbf{1}$, $x = * : \mathbf{1}$. Furthermore we have the empty set $\emptyset$: Set, with no elements and elimination rule $\text{efq}_A : (x : \emptyset) \to (A\ x)$ for any function $A : \emptyset \to \text{Set}$. We usually omit the index A of efq.

We frequently refer to the set $\mathbf{1}+C$ (in Haskell called Maybe(C)), and in connection with this set, write inl instead of inl$(*)$.

When carrying out proofs, for convenience we usually work in extensional type theory, although many proofs can be carried out in intensional type theory. We write $\text{Id}(A, a, b)$ for the equality type expressing equality of $a : A$ and $b : A$. The canonical element of $\text{Id}(A, a, a)$ will be called $\text{refl}_A(a)$. When the overhead is not too great, we make basic definitions in intensional type theory. Then we use J for the transfer principle derived from the elimination rule, where $\text{J} : (C : A \to \text{Set}, a : A, b : A, x : \text{Id}(A, a, b), C(a)) \to C(b)$, and $\text{J}(C, a, a, \text{refl}_A(a), c) = c$.

We denote the set with two elements $*_0$ and $*_1$ by $\mathbf{2}$.

Apart from the type constructions above, we have one additional type, the type of small types called Set. Elements of Set are types. Set will be closed under all type constructions mentioned in this section (including the function type and product), except for Set itself.

7.2 Non-Dependent Interactive Programs

We have studied two main approaches taken in functional programming languages that allow interactive programs to be written:[1]

- constants whose evaluation has side effects;
- the IO-monad.

Constants with side effects are used for instance in ML and Lisp. One cannot use this approach in dependent type theory without imposing restrictions on the language, because in dependent type theory expressions are evaluated during type checking. For example, if $a, a' : A$, then the term $\lambda B, x.x$ is of type $(B : A \to \text{Set}) \to B\ a \to B\ a'$ if and only if a and a' are equal elements of type A, which is to say that a and a' evaluate to the same normal form. If there were constants with side effects, the evaluation of a might trigger interactions with the real world. The type correctness of the program might (bizarrely) depend on the results of these interactions.

The idea underlying the IO-monad, as it is used in Haskell, is that a program is a static, mathematical structure that can be used to determine the next interaction, on the basis of previous interactions. Performing an interaction is an external, or extra-mathematical operation, carried out in a loop. Suppose that zero or more interactions have already been performed, and responses to those interactions have been received. Then, using this structure, the next interaction is calculated and performed in the real world. Once a response is obtained the loop is repeated. This was the approach taken in our articles [9–11], and we repeat the key ideas in the following, following mainly [10].

In [10], an atomic interaction starts with the interactive program issuing a command in the real world (e.g. to write a character to the screen, or to return a code of the next key pressed by the user). In response to a command the real world returns an answer. For example if the command was to write a character on the screen, the answer is an acknowledgement message; if the command was to get a key pressed, the answer is a code of the key. Once the answer is obtained, the atomic interaction is finished, and the program continues with the next atomic interaction.

In type theory, we can represent the set of commands as a set $\mathrm{C} : \text{Set}$. The set of responses that can be returned to a command $c : \mathrm{C}$ is represented as a set $\mathrm{R}(c) : \text{Set}$. The interface of an interactive program with the real world is therefore represented by a pair $\langle \mathrm{C}, \mathrm{R} \rangle$, which is an element of

$$\text{Interface}^{\text{nondep}} := (C : \text{Set}) \times (R : C \to \text{Set}).$$

(In [10] we used the terminology 'world' instead of 'interface'. We now think that the new terminology is more appropriate.) In the following, when referring to non-dependent programs, we assume a fixed interface $\langle \mathrm{C}, \mathrm{R} \rangle$.

[1] For an overview of I/O in functional programming, see [8].

7.2.1 *The set* IO

To run an interactive program p appropriate to the interface $\langle \mathrm{C}, \mathrm{R} \rangle$, we need the following ingredients.

- We need to determine from p the command $c : \mathrm{C}$ to be issued next, by calculating the normal form of p.
- For every possible response $r : \mathrm{R}(c)$ to c we need to determine a continuation program q. When the atomic interaction initiated by issuing the command c is complete, the interactive program should continue with the interactive program q.

Let $\mathrm{IO} : \mathrm{Set}$ be the set of interactive programs—we will see below how to actually introduce this set and the associated functions elim, Coiter in type theory. Note that we suppress here the dependence of IO on the interface. (The same will apply to other operations like elim.) Then we need a function $c : \mathrm{IO} \to \mathrm{C}$ determining the command to issue and a function $\mathrm{next} : (p : \mathrm{IO}, \mathrm{R}(c(p))) \to \mathrm{IO}$ that determines the next program from the response. We can combine both ingredients into one function:

$$\mathrm{elim} : \mathrm{IO} \to ((c : \mathrm{C}) \times (\mathrm{R}(c) \to \mathrm{IO})).$$

Define

$$\mathrm{F} : \mathrm{Set} \to \mathrm{Set}\,, \qquad \mathrm{F}(X) := (c : \mathrm{C}) \times (\mathrm{R}(c) \to X).$$

Then

$$\mathrm{elim} : \mathrm{IO} \to \mathrm{F}(\mathrm{IO}).$$

If we have $p : \mathrm{IO}$, then execution of the program proceeds as follows. First, we compute $\mathrm{elim}(p) = \langle c, f \rangle$, and issue the command c. When we have obtained a response $r : \mathrm{R}(c)$ from the real world, we compute the new program $f(r) : \mathrm{IO}$. This cycle with its two phases of computation and interaction is repeated with $f(r)$, i.e. we compute $\mathrm{elim}(f(r))$, issue the relevant command, receive a response which we use to determine the next element of IO to be performed, and so forth.

The process terminates if and when one reaches an element $p : \mathrm{IO}$ which has associated with it a command c such that $\mathrm{R}(c) = \emptyset$. This means that no response by the real world is possible. It can also happen that the program 'hangs', or waits forever for a response because the real world never provides a response to a command, although there are possible responses.

Note that execution of interactive programs need not terminate. Consider for instance an editor or word-processor. There is no a priori bound on the length of an editing session. On the other hand, there are situations in which one wants to enforce termination of interactive programs. Consider for instance a program that writes a file to a disk. There will be several interactions with the disk, during which blocks of data are written to different sectors on the disk and information about their location is stored in the directory structure of the disk. In this case one expects that this process terminates after a certain amount of time, so it is natural to demand that only finitely many interactions are possible. In general many functions of an operating system, especially those controlling interactions with hardware, are of this kind.

7.2.2 *Introduction of elements of* IO

One could describe an interactive program as a labelled tree: the nodes are labelled by commands $c : \mathrm{C}$ and a node with label c has immediate subtrees indexed by $r : \mathrm{R}(c)$. When performing the corresponding program, one would start by issuing the command at the root. Then one would, depending on the response of the real world r, move to the subtree with index r, issue the command which is given as label of that node, and having received response r', move to the r'th subtree and so forth.

In type theory, it turns out to be technically simpler to omit two properties of trees, firstly that each node is reached at most once, and secondly that each node is reached at least once. If one omits these two conditions, then an interactive program is introduced by

- an $X : \mathrm{Set}$, corresponding to the nodes of the tree,
- a function which associates with each node $x : X$ the command $c : \mathrm{C}$ to be issued when control has reached that node and for every $r : \mathrm{R}(c)$ the next node, from which the program should continue having received the response r,
- the initial node of the tree $x : X$, with which the program starts.

This means that elements are introduced by a triple $\langle X, f, x\rangle$ where $X : \mathrm{Set}$, $f : X \to ((c : \mathrm{C}) \times (\mathrm{R}(c) \to X))$ and $x : X$. Note that $f : X \to \mathrm{F}(X)$. So the introduction rule for IO is that we have a constructor

$$\mathrm{Coiter} : (X : \mathrm{Set}, f : X \to \mathrm{F}(X)) \to X \to \mathrm{IO}.$$

(The name Coiter, which stands for coiteration, will be explained later. The principle of coiteration is well-known in the area of coalgebra theory.)

7.2.3 *Bisimilarity*

Two programs $p, q : \mathrm{IO}$ behave in the same way, if firstly they issue the same command, and secondly when supplied with the same response, they continue with programs which again issue the same command, and so on. The equivalence relation which holds between programs that behave in the same way is (as is well known) bisimilarity.

In our setting, bisimilarity can be defined as follows: a bisimulation relation is a relation $B \subseteq \mathrm{IO} \times \mathrm{IO}$, such that for every $p, p' : \mathrm{IO}$, if $B(p, p')$ holds and $\mathrm{elim}(p) = \langle c, n\rangle$ and $\mathrm{elim}(p') = \langle c', n'\rangle$, then there exists a proof $cc' : \mathrm{Id}(\mathrm{C}, c, c')$ and for $r : \mathrm{R}(c)$ we have $B(n(r), n'(r'))$, where $r' : \mathrm{R}(c')$ is obtained from $r : \mathrm{R}(c)$ using the transfer principle, i.e. $r' = \mathrm{J}(\lambda d.\mathrm{R}(d), c, c', cc', r)$.

If there is a bisimulation relation between p and p', then p and p' obviously exhibit the same behaviour. Conversely, if p and p' behave in the same way, then one can obtain a bisimulation relation, namely the one which identifies q and q' if and only if q is a descendent of p and q' is the corresponding descendant of p'. Therefore two interactive programs p and p' behave in the same way if and only if there exists a bisimulation relation B such that $B(p, p')$ holds.

Let B be the union of all bisimulation relations. Then B is called bisimilarity. It is a bisimulation relation, and moreover it is the largest one, since it contains any other bisimulation relation. We write $p \approx p'$ for $\mathrm{B}(p, p')$, and will show below how to define $\approx$ in type theory.

7.2.4 *Equalities and weakly final coalgebras*

When we introduce IO in type theory, we want the following equality to hold. Assume $X : \mathrm{Set}$, $f : X \to \mathrm{F}(X)$ and $x : X$. Assume $f(x) = \langle c, g\rangle$ where $g : \mathrm{R}(c) \to X$. Then $\mathrm{elim}(\mathrm{Coiter}(X, f, x)) = \langle c, \lambda r.\mathrm{Coiter}(X, f, g(r))\rangle$. In other words, if an element x in X has associated with it a command c and a function that for any $r : \mathrm{R}(c)$ returns x_r, then the corresponding IO-program should have associated with it the same command c and, depending on r, should return the program associated with the next node $x_r : X$, which is $\mathrm{Coiter}(X, f, x_r)$.

We can extend F to a functor Set $\to$ Set, whose action on morphisms $f : X \to Y$ gives a function $\mathrm{F}(f) : \mathrm{F}(X) \to \mathrm{F}(Y)$, where $\mathrm{F}(f, \langle c, g\rangle) := \langle c, \lambda r.f(g(r))\rangle$. The functor laws will however only be provable using extensional equality.

With this extension, we can see that IO together with elim will be a weakly final coalgebra for $\mathrm{F} : \mathrm{Set} \to \mathrm{Set}$.[2]

That (IO, elim) is a coalgebra means that $\mathrm{elim} : \mathrm{IO} \to \mathrm{F}(\mathrm{IO})$. That it is weakly final means that for every other coalgebra (X, f), where $X : \mathrm{Set}$ and $f : X \to \mathrm{F}(X)$, there exists an arrow $\mathrm{Coiter}(X, f) : X \to \mathrm{IO}$ such that $\mathrm{elim} \circ \mathrm{Coiter}(X, f) = \mathrm{F}(\mathrm{Coiter}(X, f)) \circ f$:

$$
\begin{array}{ccc}
X & \xrightarrow{\;f\;} & \mathrm{F}(X) \\
{\scriptstyle \mathrm{Coiter}(X,f)}\downarrow & & \downarrow{\scriptstyle \mathrm{F}(\mathrm{Coiter}(X,f))} \\
\mathrm{IO} & \xrightarrow[\;\mathrm{elim}\;]{} & \mathrm{F}(\mathrm{IO})
\end{array}
$$

We do not demand uniqueness of the arrow $\mathrm{Coiter}(X, f)$. If we had uniqueness of this arrow, then (IO, elim) would be a final coalgebra for F. We don't know whether there are rules which can be considered as a formulation of the existence of final coalgebras in intensional dependent type theory—the usual principles imply that bisimilarity is equality, which implies extensionality of the equality on $\mathrm{N} \to \mathrm{N}$.

[2]Note that IO emerges here as a coalgebra rather than an algebra. This is natural, since what we actually need for running such programs is the function elim. One could introduce instead IO as an F-algebra (which can be introduced by the Petersson–Synek trees [2]) and use the fact that under some weak initiality condition, an F-algebra is as well an F-coalgebra. Then IO is an inductive-recursive definition and therefore part of a standard extension of Martin-Löf type theory. However, unless one uses non-well-founded type theory, one would not be able to introduce non-well-founded elements of IO.

Continuation-passing I/O (see [8], Section 7.6 for an excellent description) represents IO as an algebraic type. If one distinguishes between algebras and coalgebras, this could be considered to be a coalgebra. That type is very close to our definition of IO.

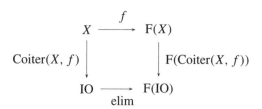

7.3 Dependent Interactive Programs

7.3.1 *Dependent interactive programs*

In the preceding section we haven't fully exploited the power of dependent types. With dependent types, it is possible to vary the set of commands available at different times. A typical example would be a program interacting through several windows. Once the program has opened a new window, it can interact with it (e.g. read the user input to this window, or write to that window). After closing the window, such interaction is no longer possible. Another example might be the switching on and off of a printer. After switching it on we can print, whereas when the printer is switched off we can no longer print. Sometimes, the commands available depend on responses of the environment to our commands. For instance, if we try to open a network connection, we either get a success message—then we can communicate via the new channel created—or a failure message—then we can't communicate.

A very general situation can be modelled by having a set $\mathrm{S} : \mathrm{Set}$ of states of the system. Depending on $s : \mathrm{S}$ we have a set of commands $\mathrm{C}(s) : \mathrm{Set}$. For every $s : \mathrm{S}$ and $c : \mathrm{C}(s)$ we have a set of responses $\mathrm{R}(s, c)$ to this command. After a response to a command is received the system reaches a new state, so we have, depending on $s : \mathrm{S}, c : \mathrm{C}(s)$ and $r : \mathrm{R}(s, c)$, a next state $\mathrm{n}(s, c, r)$ of the system. A dependent interface consists of these four components,

$$\begin{aligned} \mathrm{S} &: \mathrm{Set} \\ \mathrm{C} &: \mathrm{S} \to \mathrm{Set} \\ \mathrm{R} &: (s : \mathrm{S}, c : \mathrm{C}(s)) \to \mathrm{Set} \\ \mathrm{n} &: (s : \mathrm{S}, c : \mathrm{C}(s), r : \mathrm{R}(s, c)) \to \mathrm{S}. \end{aligned}$$

So the set of dependent interfaces is

$$\begin{aligned} \mathrm{Interface}^{\mathrm{dep}} := {} & (S : \mathrm{Set}) \\ & \times (C : S \to \mathrm{Set}) \\ & \times (R : (s : S, c : C(s)) \to \mathrm{Set}) \\ & \times ((s : S, c : C(s), r : R(s, c)) \to S). \end{aligned}$$

7.3.2 *Programs for dependent interfaces*

As with non-dependent interfaces, we require for $\langle \mathrm{S}, \mathrm{C}, \mathrm{R}, \mathrm{n} \rangle : \mathrm{Interface}$, that we have a set of interactive programs $\mathrm{IO}(s) : \mathrm{Set}$ for every $s : \mathrm{S}$. $\mathrm{IO}(s)$ should be the set of interactive programs starting in state s. In order to be able to perform an interactive program $p : \mathrm{IO}(s)$, we need to determine the command $c : \mathrm{C}(s)$ to be issued, and a function which for every $r : \mathrm{R}(s, c)$ returns a program to be performed after this response, starting in state $\mathrm{n}(s, c, r)$. That program is therefore an element of $\mathrm{IO}(\mathrm{n}(s, c, r))$. Let elim be the function which determines c and the next program. If we define

$$\mathrm{F} : (S \to \mathrm{Set}) \to (S \to \mathrm{Set}),\ \mathrm{F}(X, s) := (c : \mathrm{C}(s)) \times ((r : \mathrm{R}(s, c)) \to X(\mathrm{n}(s, c, r)))$$

then we obtain

$$\begin{aligned} \mathrm{IO} &: S \to \mathrm{Set}, \\ \mathrm{elim} &: (s : \mathrm{S}) \to \mathrm{IO}(s) \to \mathrm{F}(\mathrm{IO}, s). \end{aligned}$$

The introduction of elements of $\mathrm{IO}(s)$ is similar to the case of non-dependent interfaces. Instead of one set X of nodes, as in the non-dependent case, the introduction of an interactive program now requires for every state s a set of nodes $X(s)$, i.e. an S-indexed set $X : \mathrm{S} \to \mathrm{Set}$. For every $s : \mathrm{S}$ and $x : X(s)$ we need to determine from $p : \mathrm{IO}(s)$ the command $c : \mathrm{C}(s)$ to be issued and for $r : \mathrm{R}(s, c)$ the next node of type $X(\mathrm{n}(s, c, r))$ with which the program continues. As before, these two functions can be summarized by one function $f : (s : \mathrm{S}) \to X(s) \to \mathrm{F}(X, s)$. Further we need an initial node $x_0 : X(s)$. So we have the following introduction rule for IO:

$$\mathrm{Coiter} : (X : \mathrm{S} \to \mathrm{Set}, f : (s : \mathrm{S}) \to X(s) \to \mathrm{F}(X, s), s : \mathrm{S}) \to X(s) \to \mathrm{IO}(s).$$

7.3.3 *Weakly final coalgebras on* S → Set

As in the case of non-dependent interactive programs, we require an equality rule to hold. Assume $X : \mathrm{S} \to \mathrm{Set}$, $f : (s : \mathrm{S}) \to X(s) \to \mathrm{F}(X, s)$, $s : \mathrm{S}$, and $x : X(s)$. Assume $f(s, x) = \langle c, g \rangle$ where $g : \mathrm{R}(s, c) \to X(\mathrm{n}(s, c, r))$. Then

$$\mathrm{elim}(s, \mathrm{Coiter}(X, f, s, x)) = \langle c, \lambda r.\mathrm{Coiter}(X, f, \mathrm{n}(s, c, r), g(r)) \rangle.$$

Again we can extend F to an endofunctor on the presheaf category $\mathrm{S} \to \mathrm{Set}$, by having as morphism part for $X, Y : \mathrm{S} \to \mathrm{Set}$ and $f : (s : \mathrm{S}) \to X(s) \to Y(s)$ the function $\mathrm{F}(f) : (s : \mathrm{S}) \to \mathrm{F}(X, s) \to \mathrm{F}(Y, s)$, where $\mathrm{F}(f, s, \langle c, g \rangle) := \langle c, \lambda r.f(\mathrm{n}(s, c, r), g(r)) \rangle$. As before the functor laws can be proved only with respect to extensional equality.

With this extension, we can see that IO will be a weakly final coalgebra for F: we have $\mathrm{elim} : (s : \mathrm{S}) \to \mathrm{IO}(s) \to \mathrm{F}(\mathrm{IO}, s)$ and for every $X : \mathrm{S} \to \mathrm{Set}$ and $f : (s : \mathrm{S}) \to X(s) \to \mathrm{F}(X, s)$ there exists an arrow $\mathrm{Coiter}(X, f) : (s : \mathrm{S}) \to X(s) \to \mathrm{IO}(s)$ such that $\mathrm{elim} \circ \mathrm{Coiter}(X, f) = \mathrm{F}(\mathrm{Coiter}(X, f)) \circ f$ holds with respect to composition in the presheaf category (where $f \circ g := \lambda s, x.f(s, g(s, x))$):

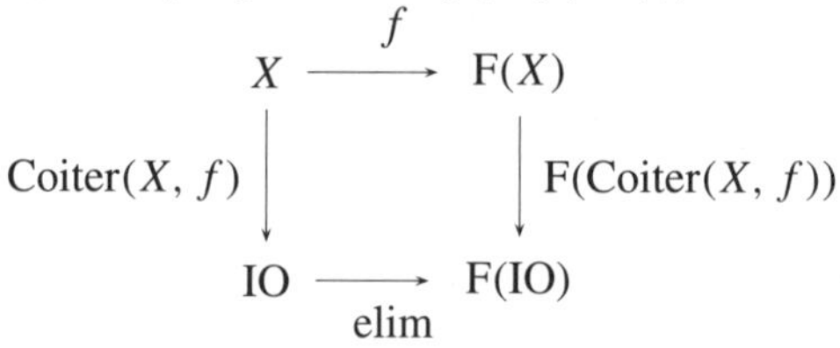

7.3.4 *Bisimilarity*

The definition of bisimilarity between non-state-dependent interactive programs extends directly to state-dependent interactive programs. A relation $B' : (s : \mathrm{S}) \to \mathrm{IO}(s) \to \mathrm{IO}(s) \to \mathrm{Set}$ is a bisimulation relation, if for $s : \mathrm{S}$ and $p, p' : \mathrm{IO}(s)$ such that $B'(s, p, p')$ we have that if $\mathrm{elim}(p) = \langle c, g \rangle$ and $\mathrm{elim}(p') = \langle c', g' \rangle$, then there exists a $cc' : \mathrm{Id}(\mathrm{C}(s), c, c')$ and for $r : \mathrm{R}(s, c)$ we have that $B'(\mathrm{n}(s, c, r), g(r), g''(r))$ holds, where $g'' : (r : \mathrm{R}(s, c)) \to \mathrm{IO}(\mathrm{n}(s, c, r))$ is obtained from $g' : (r : \mathrm{R}(s, c')) \to \mathrm{IO}(\mathrm{n}(s, c', r))$ by using the transfer principle and cc' (this short definition, which can be used with intensional equality, is due to M. Michelbrink). Bisimilarity B is now the largest such relation, i.e. the union of all bisimulation relations, and one can easily see that B is in fact a bisimulation relation. As before one can easily see that two programs

$p, q : \mathrm{IO}(s)$ behave in the same way if and only if $\mathrm{B}(s, p, q)$ holds. We write $p \approx_s q$ for $\mathrm{B}(s, p, q)$.

7.4 Server-Side Programs and Generalization to Polynomial Functors

7.4.1 *Server-side programs*

What we have described above are in a sense 'client-side' programs: the program issues a command and gets back a response from the other side of the interface. There are as well server-side programs, in which the program receives commands to which it returns responses.

An example is a user interface. Currently, the standard way for defining a user interface is that one first places components like buttons, text boxes and labels in the screen. Then one associates with certain components event handlers, which are functions that take as argument an event (e.g. the event of clicking the mouse on a button; the event will encode certain data about this, e.g. the coordinates of the mouse click, or flags indicating whether it was a single or double click). The event handler usually doesn't return an answer, but when it is executed, a side effect will take place.

This model of a GUI corresponds to a server-side program: the program waits for a command, e.g. a mouse click event associated with a button. Depending on that event, it performs one or more interactions with the window manager, the database and possibly other systems. Once these interactions are finished the program is waiting for the next event.

Server-side programs correspond to the same definition of the set of interactive programs as above, but with respect to a different functor F, namely in case of non-dependent interfaces

$$\mathrm{F}(X) = (c : \mathrm{C}) \to ((r : \mathrm{R}(c)) \times X)$$

and in case of dependent interfaces

$$\mathrm{F}(X, s) = (c : \mathrm{C}(s)) \to ((r : \mathrm{R}(s, c)) \times X(\mathrm{n}(s, c, r))).$$

Let us write in the following F^{∞} for the set of interactive programs corresponding to functor F. In the non-dependent case we need $\mathrm{F}^{\infty} : \mathrm{Set}$ and a function $\mathrm{elim} : \mathrm{F}^{\infty} \to \mathrm{F}(\mathrm{F}^{\infty})$, which, depending on a program and a command $c : \mathrm{C}$ from the outside world, determines the response of the interactive program to it and the next interactive program to be performed. In the dependent version, we need an S-indexed set $\mathrm{F}^{\infty} : \mathrm{S} \to \mathrm{Set}$ of interactive programs, and a function $\mathrm{elim} : (s : \mathrm{S}) \to \mathrm{F}^{\infty}(s) \to \mathrm{F}(\mathrm{F}^{\infty}, s)$, which determines for every $s : \mathrm{S}$, $p : \mathrm{F}^{\infty}(s)$ and every command $c : \mathrm{C}(s)$ from the outside world a response $r : \mathrm{R}(s, c)$ and a program $p' : \mathrm{F}^{\infty}(\mathrm{n}(s, c, r))$, the program continues with.

7.4.2 *Generalisation: polynomial functors on families of sets*

We have seen the need to introduce sets $F^{\infty} : \mathrm{S} \to \mathrm{Set}$ such that there exists $\mathrm{elim} : (s : \mathrm{S}) \to F^{\infty}(s) \to F(F^{\infty}, s)$ and constructors

$$\mathrm{Coiter} : (X : \mathrm{S} \to \mathrm{Set}) \to ((s : \mathrm{S}) \to X(s) \to F(X, s)) \to (s : \mathrm{S}) \to X(s) \to F^{\infty}(s)$$

where we used two kinds of endofunctors on S $\to$ Set, namely $F = \lambda X, s.(c : \mathrm{C}(s)) \times (\mathrm{R}(s, c) \to X(\mathrm{n}(s, c, r)))$ and $F = \lambda X, s.(c : \mathrm{C}(s)) \to (\mathrm{R}(s, c) \times X(\mathrm{n}(s, c, r)))$. Note that endofunctors of the first kind are more general than those of the second kind: functors of the second kind are by the axiom of choice equivalent to functors of the first kind, whereas the other direction does not always hold.

We can generalize the above to general polynomial functors $F : (\mathrm{S} \to \mathrm{Set}) \to (\mathrm{S} \to \mathrm{Set})$, which are essentially of the form $\lambda X.(c : \mathrm{C}(s)) \to ((r : \mathrm{R}(s, c)) \times ((d : \mathrm{D}(s, r, c)) \to \cdots$. All these functors are strictly positive, and we will in a later article extend the set of polynomial functors to a set of strictly positive ones.

The definition of the set of polynomial functors (which is a dependent form of the usual definition of polynomial functors and extends for instance [12]) is as follows: first we define inductively the set of polynomial functors $F : (\mathrm{S} \to \mathrm{Set}) \to \mathrm{Set}$:

- (*Projection.*) For $s : \mathrm{S}$, $\lambda X.X(s)$ is a polynomial functor.
- (*The constant functor.*) If $A : \mathrm{Set}$, then $\lambda X.A$ is a polynomial functor.
- If $A : \mathrm{Set}$ and for $a : A$ $F_a : (\mathrm{S} \to \mathrm{Set}) \to \mathrm{Set}$ is a polynomial functor, so is $\lambda X.(a : A) \to F_a(X)$.
- If $A : \mathrm{Set}$ and for $a : A$ $F_a : (\mathrm{S} \to \mathrm{Set}) \to \mathrm{Set}$ is a polynomial functor, then $\lambda X.(a : A) \times F_a(X)$ is a polynomial functor.
- If F, F' are polynomial functors $(\mathrm{S} \to \mathrm{Set}) \to \mathrm{Set}$, so is $\lambda X, s.F(X, s) + F'(X, s)$.

(The last case could be reduced to the previous cases by defining it as $\lambda X.s.(x : \mathbf{2}) \times G_x(X, s)$ where $G_{*_0}(X, s) = F(X, s)$, $G_{*_1}(X, s) = F'(X, s)$.)

If for $s : \mathrm{S}$, $F_s : (\mathrm{S} \to \mathrm{Set}) \to \mathrm{Set}$ is a polynomial functor, then $\lambda X, s.F_s(X)$ is a polynomial functor $(\mathrm{S} \to \mathrm{Set}) \to \mathrm{S} \to \mathrm{Set}$.

Polynomial functors $F : \mathrm{Set} \to \mathrm{Set}$ are inductively defined in the same way as polynomial functors $F : (\mathrm{S} \to \mathrm{Set}) \to \mathrm{Set}$, except the clause for projection is replaced by the following:

- (*Identity*). $\lambda X.X$ is a polynomial functor.

It is an easy exercise to introduce for polynomial functors F the morphism part, i.e. for $f : (s : \mathrm{S}) \to X(s) \to Y(s)$ a function $F(f) : (s : \mathrm{S}) \to F(X, s) \to F(Y, s)$. However to show that the functor-laws hold on the category of pre-sheaves S $\to$ Set requires extensional equality.

7.4.3 *Equivalents of polynomial functors*

In the presence of extensional type theory, one can show that each polynomial functor $F : (\mathrm{S} \to \mathrm{Set}) \to (\mathrm{S} \to \mathrm{Set})$ is equivalent to a functor G of the form $G(X, s) = (c : \mathrm{C}(s)) \times ((r : \mathrm{R}(s, c)) \to X(\mathrm{n}(s, c, r)))$ for some C, R, n, i.e. there exists a natural equivalence $f : F \dot{\to} G$. (A similar result for a weaker (non-dependent) version of polynomial functors was shown in [12].) We show first that every polynomial functor $F : (\mathrm{S} \to \mathrm{Set}) \to \mathrm{Set}$ is equivalent to a functor $\lambda X.(c : \mathrm{C}) \times ((r : \mathrm{R}(c)) \to X(\mathrm{n}(c, r)))$ for some $\mathrm{C} : \mathrm{Set}$, $\mathrm{R} : \mathrm{C} \to \mathrm{Set}$ and $\mathrm{n} : (c : \mathrm{C}) \to \mathrm{R}(c) \to \mathrm{S}$:

- In case of $F(X) = X(s)$ we define $\mathrm{C} := \mathbf{1}$, $\mathrm{R}(c) := \mathbf{1}$, $\mathrm{n}(c, r) := s$. The corresponding functor $G = \lambda X.\mathbf{1} \times (\mathbf{1} \to X(s))$ is easily seen to be equivalent to F.
- In case of $F(X) = A$ for A : Set we define $\mathrm{C} := A$, $\mathrm{R}(c) := \emptyset$, $\mathrm{n}(c, r) := \mathrm{efq}(r)$. Using extensional equality, one can easily see that the corresponding functor $G = \lambda X.A \times ((r : \emptyset) \to X(\mathrm{efq}(r)))$ is equivalent to F.
- In case of $F(X) = (a : A) \to F_a(X)$, and $F_a(X)$ being equivalent to $(c : C'(a)) \times ((r : R'(a, c)) \to X(n'(a, c, r)))$, let $\mathrm{C} := ((a : A) \to C'(a))$, $\mathrm{R}(c) := (a : A) \times R'(a, c(a))$, $\mathrm{n}(c, \langle a, r\rangle) := n'(a, c(a), r)$. The corresponding functor $G = \lambda X.(c : (a : A) \to C'(a)) \times ((r' : (a : A) \times (r : R'(a, c(a)))) \to X(\mathrm{n}(c, r')))$ is equivalent to $\lambda X.(c : (a : A) \to C'(a)) \times ((a : A) \to (r : R'(a, c(a))) \to X(n'(a, c(a), r)))$, which by the axiom of choice is equivalent to $\lambda X.(a : A) \to ((c : C'(a)) \times ((r : R'(a, c)) \to X(n'(a, c, r))))$, which is equivalent to F.
- In case of $F(X) = (a : A) \times F_a(X)$ and $F_a(X)$ being equivalent to $(c : C'(a)) \times ((r : R'(a, c)) \to X(n'(a, c, r)))$, let $\mathrm{C} := (a : A) \times C'(a)$, $\mathrm{R}(\langle a, c\rangle) := R'(a, c)$ and $\mathrm{n}(\langle a, c\rangle, r) := n'(a, c, r)$. The corresponding functor $G = \lambda X.(b : ((a : A) \times C'(a))) \times ((r : \mathrm{R}(b)) \to X(\mathrm{n}(b, r)))$ is equivalent to $\lambda X.(a : A) \times ((c : C'(a)) \times ((r : R'(a, c)) \to X(n'(a, c, r))))$, which is equivalent to F.
- In case of $F(X) = F_0(X) + F_1(X)$ and $F_i(X)$ being equivalent to $(c : C_i) \times ((r : R_i(c)) \to X(n_i(c, r)))$, let $\mathrm{C} := C_0 + C_1$ and $\mathrm{R}(\mathrm{inl}(c)) := R_0(c)$, $\mathrm{R}(\mathrm{inr}(c)) := R_1(c)$, $\mathrm{n}(\mathrm{inl}(c), r) := n_0(c, r)$, $\mathrm{n}(\mathrm{inr}(c), r) := n_1(c, r)$. The corresponding functor $G = \lambda X.(c : \mathrm{C}) \times ((r : \mathrm{R}(c)) \to X(\mathrm{n}(c, r)))$ is equivalent to $\lambda X.((c : C_0) \times ((r : R_0(c)) \to X(n_0(c, r)))) + ((c : C_1) \times ((r : R_1(c)) \to X(n_1(c, r))))$, which is equivalent to F.

If now $F : (\mathrm{S} \to \mathrm{Set}) \to (\mathrm{S} \to \mathrm{Set})$ is polynomial and $F(X, s) = F_s(X)$ where F_s is equivalent to $G_s := \lambda X.(c : \mathrm{C}(s)) \times ((r : \mathrm{R}(s, c) \to X(\mathrm{n}(s, c, r)))$, then F is equivalent to $\lambda X, s.G_s(X)$, which is of the desired form.

7.4.4 *The natural numbers, conatural numbers, iteration, and coiteration*

The natural numbers can be introduced as the initial algebra of the polynomial functor $\mathrm{F} : \mathrm{Set} \to \mathrm{Set}$, where $\mathrm{F}(X) := \mathbf{1} + X$. That N is an F-algebra means that we have a constructor intro : $\mathrm{F}(\mathrm{N}) \to \mathrm{N}$. The relationship to the usual constructors 0 and S of the natural numbers is that $\mathrm{intro}(\mathrm{inl}) = 0$, $\mathrm{intro}(\mathrm{inr}(n)) = \mathrm{S}(n)$. That N is a weakly initial algebra with respect to F means that if A : Set and $f : A \to (\mathbf{1} + A)$, then there exists a function $\mathrm{Iter}'(A, f) : \mathrm{N} \to A$ such that $\mathrm{Iter}'(A, f) \circ \mathrm{intro} = f \circ \mathrm{F}(\mathrm{Iter}'(A, f))$. If one specializes the equalities for $\mathrm{Iter}'(A, f)$ to $\mathrm{inl} = 0$ and $\mathrm{inr}(n) = \mathrm{S}(n)$ one obtains $\mathrm{Iter}'(A, f, 0) = f(\mathrm{inl})$ and $\mathrm{Iter}'(A, f, \mathrm{S}(n)) = f(\mathrm{inr}(\mathrm{Iter}'(A, f, n)))$. If one replaces the argument f in Iter′ by $n := f(\mathrm{inl})$ and $g := f \circ \mathrm{inr}$, one obtains a function $\mathrm{Iter} : (A : \mathrm{Set}) \to A \to (A \to A) \to \mathrm{N} \to A$ such that $\mathrm{Iter}(A, a, g, 0) = a$ and $\mathrm{Iter}(A, a, g, \mathrm{S}(n)) = g(\mathrm{Iter}(A, a, g, n))$, i.e. $\mathrm{Iter}(A, a, g, n) = g^n(a)$. Therefore Iter is the principle of *iteration*. N is not only a weakly initial algebra but an initial algebra, which means that $\mathrm{Iter}'(a, f)$ (or equivalently $\mathrm{Iter}(a, f)$) is the only function fulfilling the above mentioned equation. This is not guaranteed by the principle of iteration alone. In type theory it *is* guaranteed by the principle of induction—using induction one can show that if g is any other function s.t. $g \circ \text{intro} = f \circ \text{F}(g)$, then $(n : \text{N}) \to \text{Id}(A, g(n), \text{Iter}'(a, f, n))$.

The weakly final coalgebra for F introduces the conatural numbers N^∞. So we have $\text{elim} : \text{N}^\infty \to (\mathbf{1} + \text{N}^\infty)$. Let us write 0_A for $\text{inl} : \mathbf{1} + A$, $\text{S}_A(a)$ for $\text{inr}(a) : \mathbf{1} + A$, where $a : A$, and let us omit the subscript A in case $A = \text{N}^\infty$. Then the elimination rule means that for every element of $n : \text{N}^\infty$ we have that $\text{elim}(n) = 0$ or $\text{elim}(n) = \text{S}(m)$ for some $m : \text{N}^\infty$.

The existence of Coiter means that, if we have $A : \text{Set}$ and $g : A \to (\mathbf{1} + A)$, then there exists a function $\text{Coiter}(A, g) : A \to \text{N}^\infty$ such that

$$\text{elim} \circ \text{Coiter}(A, g) = \text{F}(\text{Coiter}(A, g)) \circ g.$$

This means: If $a : A$ and $g(a) = 0_A$, then $\text{elim}(\text{Coiter}(A, g, a)) = 0$. If $g(a) = \text{S}_A(a')$, then $\text{elim}(\text{Coiter}(A, g, a)) = \text{S}(\text{Coiter}(A, g, a'))$. Coiter is the dual of Iter, and here called *coiteration*. Note that if one defines $n := \text{Coiter}(\mathbf{1}, \lambda x.\text{inr}(*))$, we get $\text{elim}(n) = \text{S}(n)$, so N^∞ contains infinite conatural numbers.

7.5 Coiteration in Dependent Type Theory

The standard rules for dependent type theory allow us to introduce inductively defined sets, which correspond to (weakly) initial algebras. Coalgebraic types are not represented in a direct way. Markus Michelbrink is working on modelling state-dependent coalgebras in intensional type theory. At the time of writing this chapter it seems that he has succeeded, although the proof is complex, and has yet to be verified. Even when his approach is finally accepted, it will still be rather complicated to carry out proofs about coalgebras in this way. Furthermore, if one models interactive programs in this way, it would probably be rather inefficient to actually execute such programs.

The usual approach in dependent type theory is to introduce new types directly as first class citizens rather than reducing them using complicated methods to already existing types. That's what one needs in programming in general: a rich type structure rather than a minimal one, that allows one to program without having to carry out a complicated encoding.

In the same way we think it is the right approach to extend type theory by new rules for weakly final coalgebras. Of course one needs to show that such an extension is consistent, and we will do so in a future article by developing a PER module (in [13] a set theoretic model for final coalgebras was developed).

All rules will depend on a polynomial function $\text{F} : \text{S} \to \text{Set}$, which we suppress. In fact, it requires a derivation to show that F is a polynomial functor, which means that all rules have additional premises which derive that F is a polynomial functor. A complete set of rules for deriving polynomial functor would require the introduction of a data type of such functors analogous to the data type of inductive-recursive definitions introduced in [14]. However, such a theory would lie outside the scope of the present article. For the moment it suffices to restrict the theory to functors of the form

$F = \lambda X, s.(c : C(s)) \times ((r : R(s, c)) \rightarrow X(n(s, c, r)))$. That F is a polynomial functor is guaranteed by $\langle S, C, R, n \rangle$: Interface.

From the considerations in the previous section we obtain the following rules for weakly final coalgebras (these rules are well-known in the area of coalgebra theory, but have to our knowledge not yet been discussed in the context of Martin-Löf type theory):

Formation Rule:

$$\frac{s : S}{F^{\infty}(s) : \text{Set}}$$

Introduction Rule:

$$\frac{A : S \rightarrow \text{Set} \qquad f : (s : S) \rightarrow A(s) \rightarrow F(A, s) \qquad s : S \qquad x : A(s)}{\text{Coiter}(A, f, s, x) : F^{\infty}(s)}$$

Elimination Rule:

$$\frac{s : S \qquad p : F^{\infty}(s)}{\text{elim}(s, p) : F(F^{\infty}, s)}$$

Equality Rule:

$$\text{elim}(s, \text{Coiter}(A, f, s, x)) = F(\text{Coiter}(A, f), s, f(x)).$$

Note that, in case $F(X, s) = (c : C(s)) \times ((r : R(s, c)) \rightarrow X(n(s, c, r)))$ we have that if $f(x) = \langle c, g \rangle$ then $\text{elim}(s, \text{Coiter}(A, f, s, x)) = \langle c, \lambda r.\text{Coiter}(A, f, s, g(r)) \rangle$. So we can write the equality in this case as follows.

$$\begin{aligned} \text{elim}(s, \text{Coiter}(A, f, s, x)) = \text{case } & (f(x)) \text{ of} \\ & \langle c, g \rangle \rightarrow \langle c, \lambda r.\text{Coiter}(A, f, s, g(r)) \rangle \,. \end{aligned}$$

7.5.1 *Non-dependent version*

In the case $S = \mathbf{1}$ we have by the η-rule that $S \rightarrow$ Set and Set are isomorphic. Instead of using this isomorphism, it is more convenient to add special rules for non-dependent polynomial functors F : Set $\rightarrow$ Set, which are as follows:

Formation Rule:

$$F^{\infty} : \text{Set}$$

Introduction Rule:

$$\frac{A : \text{Set} \qquad f : A \rightarrow F(A) \qquad x : A}{\text{Coiter}(A, f, x) : F^{\infty}}$$

Elimination Rule:

$$\frac{p : F^{\infty}}{\text{elim}(p) : F(F^{\infty})}$$

Equality Rule:

$$\text{elim}(\text{Coiter}(A, f, x)) = F(\text{Coiter}(A, f), f(x)).$$

7.5.2 *Inductive data types vs. coalgebras*

If we compare the above rules with the rules for inductive data types like the natural numbers or the W-type, we observe the following:

- With inductive data types, the introduction rules are 'simple': they don't refer to all sets. On the other hand, the elimination rules are complex and refer universally to all sets (e.g. induction on N can have any set as result type).
- For coalgebras, the elimination rules are 'simple', and don't refer to arbitrary sets. However the introduction rules are complex, and can refer existentially to arbitrary sets.

This duality is in the nature of things. In the case of inductive data types, we form the least set closed under various operations. What 'closed' means is given by introduction rules, but is described in a simple way. The real power of these types lies in the stipulation that we have the *least* such closed set, and this requires an induction principle referring to all sets.

In case of coalgebras, we form the largest set which fulfils a certain elimination principle. The elimination principle corresponds to a simple elimination rule. The strength comes from the fact that we have the *largest* such set and that requires reference to arbitrary sets.

7.6 Guarded Induction

7.6.1 *Coiter and guarded induction*

Coiteration can be read as a recursion principle. In order to make clear what we mean by this, let us consider the weakly final F-coalgebra IO for the functor $\mathrm{F} = \lambda X.(c : \mathrm{C}) \times (\mathrm{R}(c) \to X) : \mathrm{Set} \to \mathrm{Set}$. A function $f : X \to (c : \mathrm{C}) \times (\mathrm{R}(c) \to X)$ can be split into two function $c : X \to \mathrm{C}$ and $\mathrm{next} : (x : X) \to \mathrm{R}(c(x)) \to X$. Then the rules for $\mathrm{Coiter}(X, f)$ express that for functions c and next as above there exists a function $g : X \to \mathrm{IO}$ (defined as $\mathrm{Coiter}(X, f)$) such that for $x : X$ we have

$$\mathrm{elim}(g(x)) = \langle c(x), \lambda r.g(\mathrm{next}(x, r))\rangle.$$

If one thinks of C, R as an interface of an interactive system, this can be read as: the interactive program $g(x)$ is defined recursively by determining for every $x : X$ a command $c(x)$ and then the continuation function that handles a response to this command, defined in terms of g itself.

In coalgebra theory one often discusses the introduction of coalgebras A by definitions of the form

$$A = \mathrm{codata}\ C_0(\cdots) \mid \cdots \mid C_n(\cdots),$$

where C_i are constructors, and the arguments of the constructors may refer to A itself at strictly positive positions, as well as to previously defined sets and set constructors. For example the conatural numbers N^∞, the set of streams of values of type A, and the set of interactive programs can be introduced by the definitions

$$\begin{aligned}
\mathrm{N}^{\infty} &= \text{codata } 0 \mid \mathrm{S}(n : \mathrm{N}^{\infty}), \\
\mathrm{Stream}(A) &= \text{codata cons}(a : A, l : \mathrm{Stream}(A)), \\
\mathrm{IO} &= \text{codata do}(c : \mathrm{C}, \text{next} : \mathrm{R}(c) \to \mathrm{IO}).
\end{aligned}$$

With this point of view one might be tempted to reread the above recursion equation for g as

$$g(x) = \mathrm{do}(c(x), \lambda r.g(\mathrm{next}(x, r))).$$

However, were we to allow definitions of that kind, we would immediately get non-terminating programs (set, for instance, $g : \mathbf{1} \to \mathrm{IO}$ by $g(x) = \mathrm{do}(c, \lambda r.g(*))$). With the original equation $\mathrm{elim}(g(x)) = \langle c(x), \lambda r.g(\mathrm{next}(x, r))\rangle$, termination is maintained because the elimination constant elim must be applied to $g(x)$ to obtain a reducible expression. (The example just given reads $\mathrm{elim}(g(x)) = \langle c, \lambda r.g(*)\rangle$, which is unproblematic.)

The principle for defining $g(x) = \mathrm{do}(c(x), \lambda r.g(\mathrm{next}(x, r)))$ is a simple case of guarded induction [15] (see as well the work [4] by Gimenéz; its relationship to the current work was discussed in the introduction). The idea of guarded induction is that one can define elements of such a codata set recursively, as long as every reference in the right hand side of the definition to the function we are defining recursively is 'guarded' by at least one constructor. So one can define $f : \mathbf{1} \to \mathrm{N}^{\infty}$ ('infinity') by $f(x) = \mathrm{S}(f(x))$, one can define $f : \mathrm{N}^{\infty} \to \mathrm{N}^{\infty}$ (the successor of a conatural) by $f(x) = \mathrm{S}(x)$ (without recursion), and one can define $f : \mathrm{N} \to \mathrm{Stream}(\mathrm{N})$ by $f(n) = \mathrm{cons}(n, f(n + 1))$. However, one cannot define $f : \mathbf{1} \to \mathrm{N}^{\infty}$ by $f(x) = f(x)$, since in this equation the recursion is not guarded.

Guarded induction is unproblematic in the context of lazy functional programming, as long as one doesn't need to test for equality, since there reduction to weak head normal form suffices. In dependent type theory, type checking depends on the decidability of equality of terms (see the example at the beginning of Section 2), and we have the following theorem.

Theorem 7.1 *In intensional Martin-Löf type theory extended by the principle of guarded recursion for streams in its original form equality of terms and therefore type checking is undecidable.*

Proof. Define, depending on $g_i : \mathbb{N} \to \mathbb{N}$, the functions $f_i : \mathbb{N} \to \mathrm{Stream}$ $(i = 0, 1)$ by guarded recursion as $f_i(n) = \mathrm{cons}(g_i(n), f_i(n + 1))$. Then $f_0(0) = f_1(0)$ if and only if g_0 and g_1 are extensionally equal, which is undecidable. □

If one instead reads guarded induction as a definition of $\mathrm{elim}(g(x)) = \cdots$ rather than $g(x) = \cdots$, we obtain an unproblematic principle, and one can see that a restricted form of guarded induction is, when viewed in this form, equivalent to coiteration:

The right hand side of the codata definition

$$A = \text{codata } C_0(x_0 : A_{0,0}, \ldots, x_{m_0} : A_{0,m_0}) \mid \cdots \mid C_l(x_l : A_{l,0}, \ldots, x_{m_l} : A_{l,m_l})$$

can be read as a polynomial functor $\mathrm{F} : \mathrm{Set} \to \mathrm{Set}$,

$$\mathrm{F}(X) = \mathrm{F}_0(X) + \cdots + \mathrm{F}_l(X)$$

where

$$\mathrm{F}_i(X) = ((x_0 : A'_{i,0}(X)) \times \cdots \times (x_{m_0} : A'_{i,m_0}(X)))$$

and $A'_{i,j}(X)$ is obtained from $A_{i,j}$ by replacing A by X. Note that $A_{i,j}$ either does not depend on X, so the functor $A'_{i,j}$ is constant, or $A_{i,j}$ is of the form $(y_0 : D_0) \to \cdots \to (y_k : D_k) \to A$, in which case

$$A'_{i,j} = \lambda X.(y_0 : D_0) \to \cdots \to (y_k : D_k) \to X,$$

which is a polynomial functor.

Let C_i be the injection from $\mathrm{F}_i(\mathrm{F}^\infty)$ into $\mathrm{F}(\mathrm{F}^\infty)$. Then for every element of F^∞ we have $\mathrm{elim}(a) = C_i(a)$ for some i and $a : \mathrm{F}_i(\mathrm{F}^\infty)$.

Now the principle of coiteration means that if we have a set A and a function $f : A \to \mathrm{F}(A)$ then we get a function $g : A \to \mathrm{F}^\infty$ such that $\mathrm{elim}(g(a)) = \mathrm{F}(g)(f(a))$. This reads:

$$\text{if } f(a) = C_i(\langle a_0, \ldots, a_k \rangle), \text{ then } \mathrm{elim}(g(a)) = C_i(\langle a'_0, \ldots, a'_k \rangle),$$

where the a'_i are defined as follows: $a'_i = a_i$ if $A_{i,j}(X)$ does not depend on X, and $a'_i = \lambda y_0, \ldots, y_k.g(a_i(y_0, \ldots, y_k))$ if $A_{i,j}(X) = (y_0 : D_0) \to \cdots \to (y_k : D_k) \to X$.

We can reread this as follows. We can define a function $f : A \to \mathrm{F}^\infty$ by defining $f(a)$ for $a : A$ as some $C_i(\langle b_0, \ldots, b_k \rangle)$ where b_k refers, when an element of F^∞ is needed, to f applied to any other element of A. This corresponds to a restricted form of guarded induction, where the right hand side of the recursion has exactly one constructor, and one never refers to F^∞ but only to f applied to some other arguments.

Let us consider now the definition of an **indexed codata** definition, i.e.

$$A_0(x : B_0) = \text{codata } C_{0,0}(\cdots) \mid \cdots \mid C_{0,m_0}(\cdots)$$
$$\cdots$$
$$A_l(x : B_l) = \text{codata } C_{l,0}(\cdots) \mid \cdots \mid C_{l,m_l}(\cdots),$$

where the arguments of $C_{i,j}$ refer to A_i at strictly positive positions. Let $B := B_0 + \cdots + B_l$. Then the above can be read as the definition of a B-indexed weakly final coalgebra $A = \mathrm{F}^\infty : B \to \mathrm{Set}$ for a suitable polynomial functor F, which is introduced in a similar way as before. The analogy between guarded induction and iteration is as before, except that one defines now a function $f : (b : B, C(b)) \to \mathrm{F}^\infty(b)$ recursively by defining, in case $b = \mathrm{in}_i^l(b')$ $\mathrm{elim}(b', f(b', c)) = C_{i,j}(\langle c_0, \cdots, c_k \rangle)$, where c_k can refer (and has to refer), in case an element of $\mathrm{F}^\infty(b'')$ is needed, to an element $f(b'', c')$ for some c'. So dependent coalgebras correspond to indexed codata definitions.

7.6.2 *Bisimilarity as a state-dependent coalgebra*

Bisimilarity in case of the functor $\mathrm{F}(X, s) = (c : \mathrm{C}(s)) \times (\mathrm{R}(s, c) \to X(\mathrm{n}(s, c, r)))$ can be considered as a weakly final coalgebra over the index set

$$(s : \mathrm{S}) \times \mathrm{F}^{\infty}(s) \times \mathrm{F}^{\infty}(s).$$

The condition for a bisimulation relation B as introduced above is that, if $B(s, p, p')$ holds, and $\mathrm{elim}(p) = \langle c, g\rangle$ and $\mathrm{elim}(p') = \langle c', g'\rangle$, then we have $cc' : \mathrm{Id}(\mathrm{C}(s), c(s), c'(s))$ and for $r : \mathrm{R}(s, c(s))$ we have $B(\mathrm{n}(s, c, r), g(r), g''(r))$, where g'' was obtained from g' using cc'. We can now define the polynomial functor (we curry the arguments for convenience)

$$
\begin{aligned}
&\mathrm{G} : ((s : \mathrm{S}) \to \mathrm{F}^{\infty}(s) \to \mathrm{F}^{\infty}(s) \to \mathrm{Set}) \to (s : \mathrm{S}) \to \mathrm{F}^{\infty}(s) \to \mathrm{F}^{\infty}(s) \to \mathrm{Set}\\
&\mathrm{G}(X, s, p, p') = \text{case } \mathrm{elim}(p) \text{ of}\\
&\qquad \langle c, g\rangle \to \text{case } \mathrm{elim}(p') \text{ of}\\
&\qquad\qquad \langle c', g'\rangle \to (cc' : \mathrm{Id}(\mathrm{C}(s), c, c')) \times\\
&\qquad\qquad\qquad ((r : \mathrm{R}(s, c)) \to X(\mathrm{n}(s, c, r), g(r), g''(r)))
\end{aligned}
$$

with g'' defined as above (using cc'). Then

$$\mathrm{elim} : (s : \mathrm{S}) \to (p, p' : \mathrm{F}^{\infty}(s)) \to \mathrm{G}^{\infty}(s, p, p') \to \mathrm{G}(\mathrm{G}^{\infty}, s, p, p')$$

expresses that $\mathrm{G}^{\infty}(s, p, p')$ is a bisimulation relation. Further the principle of coiteration means that if we have

$$B : (s : \mathrm{S}) \to \mathrm{F}^{\infty}(s) \to \mathrm{F}^{\infty}(s) \to \mathrm{Set}$$

and

$$f : (s : \mathrm{S}, p, p' : \mathrm{F}^{\infty}(s)) \to B(s, p, p') \to \mathrm{G}(B, s, p, p')$$

then

$$\mathrm{Coiter}(B, f) : (s : \mathrm{S}, p, p' : \mathrm{F}^{\infty}(s)) \to B(s, p, p') \to \mathrm{G}^{\infty}(s, p, p').$$

The existence of f means that B is a bisimulation relation, and $\mathrm{Coiter}(B, f)$ means that B is contained in G^{∞}. So Coiter expresses that G^{∞} contains any bisimulation relation. The introduction and equality rules together express therefore that G^{∞} is the largest bisimulation relation, i.e. bisimilarity. So the above rules allow us to introduce bisimulation in type theory in a direct way, and one can use guarded induction as a proof principle for carrying out proofs about properties of bisimulation.

7.6.3 *Normalization*

It seems that the normalization proof by Geuvers [3] carries over to the intensional version of type theory used in this chapter, and that therefore intensional type theory with the rules for state-dependent coalgebras is normalizing. If one had guarded induction, normalization would fail. A counterexample is $f : \mathbf{1} \to \mathrm{N}^{\infty}$, $f(*) = \mathrm{S}(f(*))$. Translating the guarded induction principle used here back into our rules, we obtain a function

$f := \mathrm{Coiter}(\mathbf{1}, g)$, where $g := \lambda x.\mathrm{inr}(x) : \mathbf{1} \to (\mathbf{1} + \mathbf{1})$. Note that $f(*)$ is already in normal form. The recursion is carried out only when one applies elim to $f(*)$, and we then obtain $\mathrm{elim}(f(*)) = \mathrm{S}(f(*))$, where the right hand side is again in normal form. So evaluation of full recursion is inhibited, since one needs to supply one application of elim in order to trigger a one step reduction of f.

7.7 Conclusion

We have introduced one approach to the representation of interactive programs in dependent type theory, and seen that it gives rise to weakly final coalgebras for polynomial functors. We have investigated rules for final coalgebras that correspond to coiteration, and shown why they correspond to a certain form of guarded induction, namely the definition of functions $g : A \to \mathrm{F}^{\infty}$ by equations $\mathrm{elim}(g(x)) = C_i(t_0, \ldots, t_k)$ where the terms t_i can (and in fact have to) refer, if an element of F^{∞} is required, to g itself.

The story is far from complete. The next step would be to introduce rules which correspond to a more general form of guarded induction in which we can refer directly to previously introduced elements of F^{∞} in the terms t_i, where an element of F^{∞} is required. Those rules will express the principle of corecursion, rather than merely coiteration. A further extension that remains to be formulated is a principle which, when considered as guarded induction, allows further uses of the constructors for the coalgebra in the terms t_i. This would allow for instance the definition of a function from N into streams of natural numbers such that $\mathrm{elim}(f(n)) = \mathrm{cons}(n, \mathrm{cons}(n, f(n+1)))$.

We haven't yet given a PER model for the rules (however, in [13] a set theoretic model is given). We haven't explored in full the relationship between guarded induction and the monad (in [13] it is shown that one obtains a monad up to bisimulation). Further we haven't yet interpreted non-state-dependent coalgebras in ordinary type theory. All this will be presented in a future article.

Acknowledgements

This work was supported by Nuffield Foundation, grant ref. NAL/00303/G and EPSRC grant GR/S30450/01.

References

1. Moggi, E. (1989). Computational lambda-calculus and monads. In: *Proceedings of the Logic in Computer Science Conference*.
2. Kent Petersson, K. and Synek, D. (1989). A set constructor for inductive sets in Martin-Löf's type theory. In: *Category theory and computer science (Manchester, 1989)*, pp. 128–140. LNCS 389, Springer.
3. Geuvers, H. (1992). Inductive and coinductive types with iteration and recursion. In: *Informal proceedings of the 1992 workshop on Types for Proofs and Programs, Bastad 1992, Sweden*, (eds B. Nordström, K. Petersson, and G. Plotkin), pp. 183–207.

4. Gimenéz, E. (1994). Codifying guarded definitions with recursive schemes. In *Proceedings of the 1994 Workshop on Types for Proofs and Programs*, pp. 39–59. LNCS No. 996.
5. Bertot, Y. and Castéran, P. (2004). *Interactive Theorem Proving and Program Development*. Springer.
6. Lindström, I. (1989). A construction of non-well-founded sets within Martin-Löf's type theory. *Journal of Symbolic Logic*, **54**(1), 57–64.
7. Nordström, B., Petersson, K., and Smith, J. M. (1990). *Programming in Martin-Löf's Type Theory: An Introduction.* Clarendon Press, Oxford.
8. Gordon, A. (1994). *Functional programming and Input/Output.* Distinguished Dissertations in Computer Science. Cambridge University Press.
9. Hancock, P. and Setzer, A. The IO monad in dependent type theory. In *Electronic proceedings of the workshop on dependent types in programming, Göteborg, 27–28 March 1999.* Available via *http://www.md.chalmers.se/Cs/Research/Semantics/APPSEM/dtp99.html.*
10. Hancock, P. and Setzer, A. (2000). Interactive programs in dependent type theory. In: *Computer Science Logic. 14th international workshop, CSL 2000*, (eds P. Clote and Schwichtenberg H.), Volume 1862 of *Springer Lecture Notes in Computer Science*, pp. 317–331.
11. Hancock, P. and Setzer, A. Specifying interactions with dependent types. In *Workshop on subtyping and dependent types in programming, Portugal, 7 July 2000.* Electronic proceedings, available via *http://www-sop.inria.fr/oasis/DTP00/Proceedings/proceedings.html.*
12. Dybjer, P. (1997). Representing inductively defined sets by wellorderings in Martin-Löf's type theory. *Theoret. Comput. Sci.*, **176**(1–2), 329–335.
13. Michelbrink, M. and Setzer, A. (2005). State dependent IO-monads in type theory. Proceedings of the CTCS'04, Electronic Notes in Theoretical Computer Science, **122**. pp. 127–146. Elsevier.
14. Dybjer, P. and Setzer, A. (2003). Induction-recursion and initial algebras. *Annals of Pure and Applied Logic*, **124**, 1–47.
15. Coquand, T. (1994). Infinite objects in type theory. In: *Types for Proofs and Programs, International Workshop TYPES'93, Nijmegen, The Netherlands, May 24-28, 1993, Selected Papers*, (eds H. Barendregt and T. Nipkow), volume 806 of *Lecture Notes in Computer Science*, pp. 62–78. Springer.
16. Altenkirch, T. (2001). Representations of first order function types as terminal coalgebras. In: *Typed Lambda Calculi and Applications, TLCA 2001*, number 2044 in Lecture Notes in Computer Science, pp. 8–21.
17. Hallnäs, L. (1987). An intensional characterization of the largest bisimulation. *Theoretical Computer Science*, **53**, 335–343.
18. Hancock, P. (2000). Ordinals and interactive programs. PhD thesis, LFCS, University of Edinburgh.
19. Hancock, P. and Hyvernat, P. (2004). Programming as applied basic topology. Submitted, available via *http://homepages.inf.ed.ac.uk/v1phanc1/chat.html.*

20. Jacobs, B. (1999). *Categorical Logic and Type Theory*. Studies in Logic and the Foundations of Mathematics **141**, North Holland, Elsevier.
21. Martin-Löf, P. (1984). *Intuitionistic Type Theory*, volume 1 of *Studies in Proof Theory: Lecture Notes*. Bibliopolis, Napoli.
22. Milner, R. (1989). *Communication and Concurrency*. Prentice Hall.
23. Moggi, E. (1991). Notions of computation and monads. *Information and Computation*, **93**(1), 55–92.
24. Peyton Jones, S. L. and Wadler, P. (1993, January). Imperative functional programming. In: *20'th ACM Symposium on Principles of Programming Languages*, Charlotte, North Carolina.
25. Wadler, P. (1994). Monads for functional programming. In: *Program Design Calculi*, volume 118 of *NATO ASI series, Series F: Computer and System Sciences*. (ed M. Broy), Springer Verlag.

8

APPLICATIONS OF INDUCTIVE DEFINITIONS AND CHOICE PRINCIPLES TO PROGRAM SYNTHESIS

ULRICH BERGER AND MONIKA SEISENBERGER

Abstract

We describe two methods of extracting constructive content from classical proofs, focusing on theorems involving infinite sequences and nonconstructive choice principles. The first method removes any reference to infinite sequences and transforms the theorem into a system of inductive definitions, the other applies a combination of Gödel's negative- and Friedman's A-translation. Both approaches are explained by means of a case study on Higman's lemma and its well-known classical proof due to Nash-Williams. We also discuss some proof-theoretic optimizations that were crucial for the formalization and implementation of this work in the interactive proof system Minlog.

8.1 Introduction

This paper is concerned with the problem of finding constructive content in nonconstructive mathematical theorems. By the constructive content of a theorem we mean:

- possibly, a constructive reformulation of the statement of the theorem;
- a constructive proof corresponding to the given non-constructive proof;
- a program extracted from the proof that computes witnesses of existential statements in the theorem.

In the paper we will:

- describe two different methods of transforming a non-constructive proof into a constructive proof and critically assess them from a constructive point of view;
- sketch how to extract programs from these proofs and discuss the nature of these programs;
- discuss proof-theoretic optimizations which are of independent interest and seem to be necessary in order to make these methods feasible for the implementation of larger case studies in an interactive theorem prover.

The overall aim of the paper is to give insight into possible constructive aspects hidden in non-constructive theorems and describe—in a non-technical way—how these aspects can be brought to light and computationally exploited. The results presented here summarize work done by the authors over the past few years [1–5] largely as part of the Minlog group at the University of Munich [6].

We will focus on theorems about infinite sequences as they typically occur in analysis and infinitary combinatorics. Our running example and main case study will be Higman's lemma [7][1] and its classical proof due to Nash-Williams [10] which we will briefly describe now. Higman's lemma is concerned with *well-quasiorders* (wqos), that is, binary relations $(A, \leq_A)$ on a set A of letters with the property that each infinite sequence $(a_i)_{i\in\omega}$ of letters is 'good', i.e. there are indices $i < j$ such that $a_i \leq_A a_j$. (The simplest examples of wqos are finite sets with equality.) From $(A, \leq_A)$ one can derive a binary relation $(A^*, \leq_{A^*})$ on the set A^* of (finite) words over the alphabet A where $v \leq_{A^*} w$ means that v is *embeddable* into w, i.e. v can be obtained from w by deleting some letters or replacing them with ones which are smaller with respect to $\leq_A$.

Higman's lemma *If $(A, \leq_A)$ is a well-quasiorder, so is $(A^*, \leq_{A^*})$.*

Proof. [Nash-Williams, 1963] We focus on the details relevant for later discussion. Our assumption is that every infinite sequence of letters is good. We have to prove that every infinite sequence of words is good. That a bad sequence of words, i.e. a sequence that is not good, is impossible is an immediate consequence of the following two facts:

1. For every bad sequence $(w_n)_{n\in\omega}$ of words there exists a bad sequence $(w'_n)_{n\in\omega}$ which is lexicographically smaller, i.e. there is an index n_0, such that $w'_i = w_i$ for all $i < n_0$ and w'_{n_0} is a proper initial segment of w_{n_0}.
2. If there exists a bad sequence of words, then there also exists a minimal bad sequence of words (with respect to the above lexicographic ordering).

Proof of 1. If $(w_n)_{n\in\omega}$ is bad, then all w_n must be non-empty, i.e. of the form $w_n = v_n * a_n$ (a_n being the last letter of w_n). As $(A, \leq_A)$ is a wqo, the sequence $(a_n)_{n\in\omega}$ must contain a 'monotone' subsequence $a_{n_0} \leq_A a_{n_1} \leq_A \dots$ with strictly increasing indices n_i. (Such a sequence can be obtained by applying a non-constructive choice principle we—omit the details). Now, the sequence $w_0, \dots, w_{n_0-1}, v_{n_0}, v_{n_1}, \dots$ is lexicographically smaller and also bad, since, if it were good, then so would be $(w_n)_{n\in\omega}$.

Proof of 2. Assume there is a bad sequence of words. We define a minimal bad sequence of words by repeated non-constructive choices: if words $w_0, \dots, w_{n-1}$ have already been chosen such that they begin a bad sequence, we choose a shortest word w_n such that $w_0, \dots, w_{n-1}, w_n$ begins bad sequence as well. Clearly, the resulting sequence $(w_n)_{n\in\omega}$ is bad and lexicographically minimal. □

The aspect of Nash-Williams' proof which is most problematic from a constructive point of view is not so much the use of classical logic, but the definition of infinite sequences in a constructively unacceptable way. One way around this problem is to reformulate the statement of the theorem in such a way that any reference to infinite objects is avoided. In Section 8.2 we will do this, following an approach of Coquand and Fridlender [11] by expressing the property of being a well-quasiorder in terms of

[1] Higman's lemma is used, for example, in term rewriting theory for termination proofs [8, 9].

an inductive definition. In contrast to [11] we allow an arbitrary alphabet A with a decidable well-quasiorder.

In section 8.3 we take another approach: we leave the statement of Higman's lemma as it is, but apply a combination of Gödel's negative- and Friedman's A-translation [12] to constructivize the proof. This idea goes back to Constable and Murthy [13] who formalized Higman's lemma in a system of second order arithmetic representing infinite sequences by their graphs and proving the necessary choice principles by impredicative comprehension and classical logic [14]. Our formalization differs from theirs in that we work in a finite type system where infinite sequences are available as objects type $\mathbb{N} \to \rho$. This has the advantage that the formalization of Nash-Williams' proof is technically simpler, however we have to include the non-constructive choice principles as axioms. Applying negative- and A-translation we obtain a proof which is constructive modulo the translated choice principles. These translated principles can be reduced to a relativized version of bar induction, a principle whose constructive status is controversial. Nevertheless it is possible to extract an executable program that computes for any infinite sequence of words indices $i < j$ witnessing that it is good. However, to show the termination of this program one again needs bar induction.

Analysing the programs extracted from the respective proofs we will see that both use recursion along certain well-founded trees which, however, are given in very different ways. The program extracted from the inductive proof works—as expected—on inductively defined trees, while the program from the A-translated proof recurs on the well-founded tree given by a total continuous functional of type two.

We will also discuss some proof-theoretic optimizations which were crucial for fully formalizing and implementing this case study. The logical and inductive rules of the proof calculus are modified in order to be able to discard redundant parts of the programs already at extraction time, and the combined Gödel/A-translation is optimized with the effect that extracted programs have much lower types and a simpler control structure.

Proof-theoretically, both methods yield only suboptimal results as Higman's lemma is in fact provable in ACA_0, a system that is conservative over arithmetic [15, 16]; see also [17–21]. The interest in our methods lies in the fact that they both yield proofs and programs that make constructive use of the crucial ideas in Nash-Williams' non-constructive proof, whereas the proof in [15] does not seem to be related to Nash-Williams' proof. The same applies to another inductive proof given by Fridlender [22] which is based on an intuitionistic proof of Veldman (published in [23]). Veldman's proof does not require decidability of $\leq_A$ and uses the idea of Higman's original proof. The proof-theoretic strength of a form of the minimal bad sequence argument that is sufficiently general for Nash-Williams' proof has been analysed by Marcone [24].

8.2 Inductive Definitions

In this section we describe how Nash-Williams' classical proof of Higman's lemma given in the introduction may be transformed into a constructive inductive argument. The crucial idea, due to Coquand and Fridlender [11], is to reformulate the notion of a well-quasiorder without referring to infinite sequences, but using a (generalized) inductive definition instead. For a finite sequence $as := [a_0, \ldots, a_{n-1}]$ of elements in $(A, \leq_A)$, let $\mathsf{Good}\, as$ be the property that there are $i < j < n$ such that $a_i \leq_A a_j$ (hence, an infinite sequence is good iff all its finite initial segments are). Define a predicate $\mathsf{Bar}_A \subseteq A^*$ inductively by

$$\frac{\mathsf{Good}\, as}{\mathsf{Bar}\, as} \qquad \frac{\forall a\, \mathsf{Bar}\, as * a}{\mathsf{Bar}\, as}.$$

Classically, $\mathsf{Bar}\, as$ means that every infinite sequence starting with as is good. Hence, classically, $(A, \leq_A)$ is a wqo iff $\mathsf{Bar}\,[\,]$ holds. The 'if' part of this equivalence can be proven easily using the induction principle for Bar

$$\frac{\forall as\,(\mathsf{Good}\, as \to B(as)) \qquad \forall as\,(\forall a\, B(as * a) \to B(as))}{\forall as\,(\mathsf{Bar}\, as \to B(as))}$$

with $B(as)$:= 'every infinite sequence extending as is good'. The converse direction requires a non-constructive choice principle which will be discussed in detail in the next section.

Inductive formulation of Higman's lemma $\mathsf{Bar}_A\,[\,] \to \mathsf{Bar}_{A^*}\,[\,]$.

In the following we sketch a constructive proof of this reformulation of Higman's lemma that makes use of the essential ideas in Nash-Williams' proof. For simplicity we first restrict ourselves to a finite alphabet A (well-quasiordered by equality). Our goal is to prove $\mathsf{Bar}_{A^*}\,[\,]$. In the first part of the classical proof one defines from a bad sequence $(w_n)_{n<\omega}$ a lexicographically smaller bad sequence of the form $w_0, \ldots, w_{n_0-1}, v_{n_0}, v_{n_1}, \ldots$. We mimic this on the level of finite sequences by defining a relation $<_{\mathsf{NW}}$ (NW for Nash-Williams) such that $vs <_{\mathsf{NW}} ws$ holds iff all words in ws are non-empty and vs is obtained from ws by the following process. Take a letter a that occurs as the last letter of some word in ws and scan through ws from left to right, chopping off the last letter of the current word if this letter is a, respectively deleting the current word if it does not end with a but some word ending with a has been encountered before. The contrapositive of the first part of the classical proof, 'if there is no bad sequence which is lexicographically smaller than $(w_n)_{n\in\omega}$, then $(w_n)_{n\in\omega}$ is good', corresponds now to

$$\forall ws\,(\forall vs(vs <_{\mathsf{NW}} ws \to \mathsf{Bar}_{A^*}\, vs) \to \mathsf{Bar}_{A^*}\, ws\,). \tag{+}$$

This formula immediately entails $\mathsf{Bar}_{A^*}\,[\,]$, since for $ws = [\,]$ the premise of (+) trivially holds ($[\,]$ has no $<_{\mathsf{NW}}$-predecessors). The main idea for proving (+) is to introduce an inductive predicate Bars on A^{***} and a function $\mathsf{Folder} : A^{**} \to A^{***}$ such that

$\mathsf{Bars}\,(\mathsf{Folder}\,ws)$ is equivalent to $\forall vs(vs <_{\mathsf{NW}} ws \to \mathsf{Bar}_{A^*}\,vs)$. The inductive definition of Bars parallels Bar:

$$\frac{\mathsf{Good}\,vs_i}{\mathsf{Bars}\,[vs_0,\ldots,vs_{n-1}]} \qquad \frac{\forall w\,\mathsf{Bars}\,[vs_0,\ldots,vs_i * w,\ldots vs_{n-1}]}{\mathsf{Bars}\,[vs_0,\ldots,vs_{n-1}]}.$$

The statement $\forall ws.\mathsf{Bars}\,(\mathsf{Folder}\,ws) \to \mathsf{Bar}_{A^*}\,ws$, which is equivalent to (+), can be proven by main induction on the number of letters that do not occur as an end letter in ws and side induction on $\mathsf{Bars}\,(\mathsf{Folder}\,ws)$. More precisely, given ws with $\mathsf{Bars}\,(\mathsf{Folder}\,ws)$ and the respective induction hypotheses one shows $\forall w\,\mathsf{Bar}_{A^*}\,ws * w$ by structural induction on w, and then concludes $\mathsf{Bar}_{A^*}\,ws$ with the second introduction rule for Bar. This second side induction corresponds roughly to the definition of the minimal bad sequence in the second part of the classical proof. To prove the induction step the decidability of equality on A ($\leq_A$ in the general case) is used.

An inductive proof of Higman's lemma for an arbitrary well-quasiordered alphabet—not only a finite one—is given in [1] (see also [5] for an earlier version). In this case, the induction on the number of letters that do not occur as last letters becomes an induction on the predicate Bar_A. Furthermore, the sequence structure, given by the predicate Folder, is replaced by a structure involving trees.

8.2.1 *Formalization and program extraction*

We formalized and implemented the inductive proof of Higman's lemma for a finite alphabet in the Minlog system. In order to obtain interesting computational content we proved from $\mathsf{Bar}_{A^*}\,[\,]$ the original statement of Higman's lemma, $\forall (w_n)_{n\in\omega}\,\exists i,j.\,i < j \wedge w_i \leq_{A^*} w_j$, using induction on Bar_{A^*}, and extracted a program from the latter proof. For the special case $A = \{0,1\}$ this may be found in the Minlog repository (www.minlog-system.de). The resulting program contains three nested recursions corresponding to the inductions in the proof. Its size as a Minlog term is surprisingly small—just about 50 lines—which is mainly due to refinements of the logic resulting in an optimized program extraction process. In order to explain these refinements, we need some basic facts about *realizability*, the proof-theoretic method underlying program extraction.

Realizability translates formulas into (possibly higher order) types essentially by removing all dependencies of formulas from objects. For example, a universal formula $\forall x^\rho A(x)$ translates into a function type $\rho \to \sigma$ (where A translates to σ) and an inductive predicate, as for example $\mathsf{Bar}\,as$, translates into a type of wellfounded trees. The logical rules are translated into natural operations on these types. For example, $\forall$-introduction and elimination translate into λ-abstraction and application, and introduction and elimination rules for an inductive definition become constructors and recursion operators for trees. Hence, extracted programs are higher type functional programs with iteration constructs restricted to (terminating) structural recursion.

The first refinement is achieved by distinguishing between two types of universal quantifiers, the usual quantifier $\forall$ and a 'non-computational' quantifier $\forall^{\mathsf{nc}}$. The meaning of $\forall^{\mathsf{nc}} x^\rho A(x)$ is roughly '$A(x)$ has been established for all x with a proof whose computational content does not depend on x'. Consequently one does not assign to $\forall^{\mathsf{nc}} x^{\rho} A(x)$ the type $\rho \to \sigma$, but just σ. Technically, the computational independence of a proof from an object variable is handled by a strengthened variable condition (see also [1, 25]).

The second refinement concerns the distinction between inductive definitions with and without computational content. For instance it is convenient to formalize the embeddability relation between words by means of an inductive definition,

$$\frac{}{[\,] \leq^* [\,]} \qquad \frac{v \leq^* w}{v \leq^* w * a} \qquad \frac{v \leq^* w}{v * a \leq^* w * a},$$

even though this relation is decidable and could therefore be represented by a boolean function. Normally, such an inductive definition would introduce an inductive data type into the program, but, if $\leq^*$ is declared as an inductive predicate without computational content, no extra data type is introduced. Of course, the introduction of inductive predicates without computational content is subject to extra conditions that ensure the soundness of the system. The distinction between logical constructs with and without computational content is similar to the distinction Set/Prop in intuitionistic type theory (see, e.g. [26]), but seems to be more flexible.

8.3 Classical Dependent Choice

Now we show how to extract computational content directly from Nash-Williams' proof. Recall that in this proof one derives a contradiction from the assumption that a given sequence of words, let us call it f, is bad. We use Gödel's negative translation combined with the Friedman's A-translation [12] to transform this classical proof into a constructive proof. For convenience we will in the following call this translation simply A-translation although we mean in fact the combination of Gödel- and A-translation. Hence, for a given existential formula A, the A-translation of a formula B, written B^A, is obtained by double negating all atomic and existential formulas (where $\neg C$ is defined as $C \to \bot$) and replacing $\bot$ (falsity) by the formula A. In the case of Higman's lemma, A will be the formula expressing that f is good, i.e.

$$A :\equiv \exists i, j.\, i < j \wedge f(i) \leq^* f(j).$$

Under this translation all axioms concerning classical logic become intuitionistically provable, and instances of mathematical principles like induction are translated into (different) instances of the same principle. Most importantly, the (false) assumption that f is bad is translated into an intuitionistically provable formula. Altogether one obtains an intuitionistic proof of the translation of $\bot$, i.e. A, from which a program computing indices i, j with $f(i) \leq^* f(j)$ can be extracted. However this is not quite the full story since we did not say how to deal with the non-constructive choices that occur in Nash-Williams' proof. The kind of choices used there are captured by the following scheme of *dependent choice*

$$\textbf{DC} \quad B([\,]) \wedge \forall xs(B(xs) \to \exists x\, B(xs * x)) \to \exists g \forall n\, B(\bar{g}n)$$

where $\bar{g}n := [g(0), \ldots, g(n-1)]$. In Nash-Williams' proof this scheme is used, for example, with the predicate $B(xs) :\equiv$ 'xs is a list of words that can be extended to an infinite bad sequence of words and if xs is non-empty then the last word in xs is as short as possible'. The A-translation of **DC** is (logically equivalent to)

$$\mathbf{DC}^A \quad \text{Hyp}_1 \wedge \text{Hyp}_2 \wedge \text{Hyp}_3 \to A$$

where

$$\begin{aligned} \text{Hyp}_1 &\equiv B([\,])^A \\ \text{Hyp}_2 &\equiv \forall xs(B(xs)^A \wedge (\exists x\, B(xs * x)^A \to A) \to A) \\ \text{Hyp}_3 &\equiv \exists g \forall n\, B(\bar{g}n)^A \to A. \end{aligned}$$

Unlike induction on natural numbers, **DC** does *not* prove intuitionistically its A-translation. However, $\mathbf{DC}^A$ can be reduced intuitionistically to *relativized bar induction* [27, 28] (a.k.a. extended bar induction or bar induction on species) a scheme whose constructive status is at least debatable [29, 30]. We do not go into this reduction here, but instead informally explain how to directly interpret $\mathbf{DC}^A$ computationally in terms of realizability.

The idea of the following interpretation of $\mathbf{DC}^A$ is due to Berardi, Bezem and Coquand [31]. In order to realize $\mathbf{DC}^A$ we assume we are given realizers G_1, G_2, G_3 of the hypotheses $\text{Hyp}_1, \text{Hyp}_2, \text{Hyp}_3$ respectively. We have to compute a realizer of A. We use G_3. So, we need to compute some function g and a realizer h of $\forall n\, B(\bar{g}n)^A$. We compute g and h in stages. Suppose we have already computed finite approximations xs and ys to g and h, i.e. $xs = [x_0, \ldots, x_{n-1}]$, $ys = [y_0, \ldots, y_n]$ and y_i realizes $B([x_0, \ldots, x_{i-1}])^A$ for $i \le n$ (we get started by setting $y_0 := G_1$). We run G_3 with the arguments g and h defined by

$$g(i) = \begin{cases} x_i & \text{if } i < n \\ 0 & \text{if } i \ge n \end{cases} \qquad\qquad h(i) = \begin{cases} y_i & \text{if } i \le n \\ \text{see below} & \text{if } i > n \end{cases}$$

If G_3 happens not to query h at any $i > n$, then we are done. If it does, we compute a realizer $h(i)$ of $B(xs * 0 * \ldots * 0)^A$ (with 0 repeated $i - n$ times) using a realizer of $A \to B(xs * 0 * \ldots * 0)^A$ (since B^A is logically equivalent to a formula of the form $C \to A$ such a realizer trivially exists). So, we are back to the task of computing a realizer of A. This time we use the realizer G_2. We apply G_2 to our xs and y_n (we assumed y_n realizes $B(xs)^A$) and some realizer of the formula $\exists x\, B(xs * x)^A \to A$ which we compute as follows: given x and a realizer y of $B(xs * x)^A$ we need to compute a realizer of A. We do this by recurring to our main process, but now with the larger approximations $xs * x$ and $ys * y$. Luckily, each branch of the whole computation will eventually terminate because G_3 is *continuous* and therefore will query its arguments at only finitely many values in order to compute a result.

The latter assumption can be justified by a model of realizability in which all functions are continuous. Such a model can, for example, be constructed from the total elements of the Scott/Ershov hierarchy of partial continuous functionals over the flat domains of partial booleans and partial natural numbers [2, 30, 32, 33]. It is important to note that models where all functionals are computable—like, for example, HEO [30], or, more generally, the effective topos [34]—cannot be used here, since in order for the argument given above to be valid the sequences g and h approximated by the xs and ys have to be (possibly non-computable) free choice sequences (a related phenomenon is the fact that the unit interval of recursive reals is not compact).

The recursive process described above can be written more formally as follows. A realizer of A is computed as $\Phi([], [G1])$ where for arguments xs, ys with lengths n and $n+1$, respectively (other arguments are uninteresting), $\Phi(xs, ys)$ is recursively defined by

$$\Phi(xs, ys) = G_1\left(\tilde{xs}, \lambda i.\begin{cases} y_i & \text{if } i \leq n \\ E(G_2(xs, y_n, (\lambda xs, ys.\Phi(xs * x, ys * y)))) & \text{if } i > n \end{cases}\right)$$

where $\tilde{xs} := \lambda i.\mathsf{if}\, i < |xs|\, \mathsf{then}\, x_i\, \mathsf{else}\, 0^\rho$ and E is the (trivial) realizer of $A \to B(xs * 0 * \ldots * 0)^A$. By some simple technical manipulations (like coding xs and ys into one sequence, etc.) this can be primitive recursively reduced to the following higher type recursion scheme

$$\textbf{MBR} \quad \Psi(xs) = Y\left(\lambda i.\begin{cases} x_i & \text{if } i < |xs| \\ H(xs, \lambda x.\Psi(xs * x)) & \text{if } i \geq |xs| \end{cases}\right)$$

where xs varies over finite sequences of some type ρ and the equation is of some type ν that does not contain function types (in the example of Higman's lemma ν is the type of pairs of integers). In [2] the functional Ψ is called *modified bar recursion* and it is shown that Spector's bar recursion [35], which is the scheme

$$\textbf{SBR} \quad \Psi(xs) = \begin{cases} G(xs) & \text{if } Y(\tilde{xs}) < |xs| \\ H(xs, \lambda x.\Psi(xs * x)) & \text{if } Y(\tilde{xs}) \geq |xs| \end{cases}$$

can be primitive recursively defined from **MBR**, but not vice versa. Berardi, Bezem and Coquand [31] proved the correctness of the above sketched computational interpretation of **DC** using a special realizability interpretation based on infinite terms. Oliva and the first author [2] showed that Kreisel's modified realizability [36] together with Plotkin's adequacy theorem [37] can be used instead (thus avoiding infinite terms, the role of which is taken over by the Scott/Ershov model of partial continuous functionals).

In our case study we worked with a realizer of $\mathbf{DC}^A$ which is defined from **MBR** [1]. It would be even more direct (and probably technically simpler) to work with the realizer of the minimal bad sequence argument in the form of an open induction principle [28, 38] instead of reducing open induction to (classical) dependent choice.

8.3.1 *Proof-theoretic and computational optimizations*

As explained above the A-translation replaces every atomic formula C by $(C \to A) \to A$ where A is an existential formula. This has the effect that higher types and many case distinctions come up in the extracted program which may lead to complex and inefficient code. In [3] a refined A-translation is introduced that minimizes double negations and hence reduces these negative effects. These refinements are implemented in Minlog and we have tested them in our case study.

Another improvement is specific to Minlog's implementation of normalization by evaluation [39]. We introduced the functional Ψ in Minlog as a program constant together with a rewrite rule corresponding to **MBR**. Normalization by evaluation means that in order to normalize a term, it is evaluated as a functional (Scheme) program where **MBR** is interpreted as a recursive higher type procedure that is defined according to the rewrite rule for **MBR**. This normalization procedure is rather fast. When running (i.e. normalizing) programs containing **MBR** one, however, observes a certain inefficiency which can be explained by the fact that if (in **MBR**) Y calls its argument at different values $\geq |xs|$, the expression $H(xs, \lambda x.\Psi(xs*x))$ (which does not depend on i) is evaluated repeatedly. An obvious method to avoid this inefficiency is to equip the argument of Y with an internal memory that stores the value of the expression $H(xs, \lambda x.\Psi(xs * x))$ after it has been computed for the first time. We have done this and gained a considerable speed-up.

8.4 Conclusion

In this chapter we discussed two methods of constructivizing classical proofs. We applied them to the classical Nash-Williams proof of Higman's lemma and extracted two different programs. Both methods and the case study are implemented in Minlog (www.minlog-system.de). The main computational principle used in the extracted programs is recursion on well-founded trees. However, while in the program extracted from the inductive proof the trees are inductively generated as the elements of an inductive data type, in the program obtained from the A-translated Nash-Williams' proof well-founded trees are given by continuous functionals of type two. Although both forms of well-founded recursion are known to be of different strength in general ([30, 35], Appendix by J. Zucker), it is possible that the particular instances used here are in some way equivalent. It was our hope that by analysing the extracted programs such an equivalence could be revealed. Unfortunately, the program corresponding to the second version is still too complex to permit such an analysis, although it is considerably shorter than the program extracted by Murthy [14]. It also remains unclear how our programs are related to those extracted by Murthy [14] and Herbelin [40]. Since Higman's lemma is provable in a system which is conservative over first-order arithmetic, we know that well-founded recursion is not needed at all, but Gödel primitive recursion (of type $\mathbb{N} \to \mathbb{N}$ if the alphabet A is finite) would suffice. Such a program could be extracted in principle from an implementation of, e.g. the proof by Schütte and Simpson [15], but nobody has so far undertaken such an implementation.

The inductive approach presented in this paper may be carried out in any theorem prover supporting program extraction from inductive definitions (see, for instance, [41] for an implementation of Higman's lemma in Isabelle and [22] for a formalization of a different proof in Alf). The second approach is more specific to the Minlog system, since it requires an implementation of the refined A-translation which does not seem to be available in other systems. In particular, the optimizations via memoization directly rely on Minlog's normalization procedure.

The program extraction from Nash-Williams' classical proof via A-translation described in Section 8.3 could be extended in a straightforward way to a corresponding classical proof of Kruskal's theorem, even in its strong form with gap condition [42]. The latter would be interesting because then we could extract a program from a theorem for which no constructive proof is known so far. On the other hand, the inductive method of Section 8.2 seems to be much harder to generalize, since, unlike the A-translated proof, the inductive proof is not obtained by a mechanical translation process, but rather by picking up the essential ideas and transforming them into a constructive argument. At present, it is an open problem to find a corresponding inductive proof for Kruskal's theorem.

Another interesting problem is whether *strong normalization*, i.e. termination of *every* reduction sequence, holds for suitable variants of **MBR** or **SBR** (the versions given above are obviously not strongly normalizing since the 'else' branch of the case analysis can always be rewritten). For example, **MBR** could be reformulated by replacing the conditional expression in the right hand side of its defining equation by a call of an auxiliary constant Ψ' with an extra (boolean) argument in order to force evaluation of the test $k < |xs|$ before the subterm $\Phi ygh(xs * x)$ may be further reduced:

$$\begin{aligned}\Psi(xs) &= Y(\lambda i.\Psi'(i, xs, i < |xs|)) \\ \Psi'(i, xs, \mathsf{T}) &= x_i \\ \Psi'(i, xs, \mathsf{F}) &= H(xs, \lambda x.\Psi(xs * x)).\end{aligned}$$

Proving strong normalization for recursion schemes of this kind is the subject of ongoing research.

References

1. Seisenberger, M. (2003). On the Constructive Content of Proofs. PhD thesis, University of Munich.
2. Berger, U. and Oliva, P. Modified Barrecursion and Classical Dependent Choice. To appear in Lecture Notes in Logic, Springer.
3. Berger, U. Buchholz, W. and Schwichtenberg, H. (2002). Refined program extraction from classical proofs. *Annals of Pure and Applied Logic*, **114**, 3–25.
4. Berger, U. Schwichtenberg, H. and Seisenberger, M. (2001). The Warshall Algorithm and Dickson's Lemma: Two Examples of Realistic Program Extraction. *Journal of Automated Reasoning*, **26**, 205–221.
5. Seisenberger, M. (2001). An inductive version of Nash-Williams' minimal-bad-sequence argument for Higman's lemma. In: *Types for Proofs and Programs*, P. Callaghan, et al., Lecture Notes in Computer Science **2277**, Springer.
6. Benl, H. Berger, U. Schwichtenberg, H. Seisenberger, M. and Zuber, W. (1998). Proof theory at work: Program development in the Minlog system. In: *Automated Deduction – A Basis for Applications II*, (ed W. Bibel and P. H. Schmitt), pp. 41–71. Kluwer, Dordrecht.
7. Higman, G. (1952). Ordering by divisibility in abstract algebras. *Proceedings of the London Mathematical Society (3)*, **2**(7), 326–336.
8. Cichon, E. A. and Tahhan Bittar, E. (1994). Ordinal recursive bounds for Higman's theorem. *Theoretical Computer Science*, **201**, 63–84.

9. Touzet, H. (2002). A characterisation of multiply recursive functions with Higman's lemma. *Information and Computation*, **178**, 534–544.
10. Nash-Williams, C. St. J. A. (1963). On well-quasi-ordering finite trees. *Proc. Cambridge Phil. Soc.*, **59**, 833–835.
11. Coquand, T. and Fridlender, D. (1994). A proof of Higman's lemma by structural induction. Chalmers University.
12. Friedman, H. (1978). Classically and intuitionistically provably recursive functions. In: *Higher Set Theory*, (eds D. Scott, G. Müller), Lecture Notes in Mathematrics **669**. pp. 21–28, Springer.
13. Constable, R. and Murthy, C. (1991). Finding computational content in classical proofs. In: *Logical Frameworks*, (eds G. Huet and G. Plotkin), pp. 341–362, Cambridge University Press.
14. Murthy, C. R. (1990). Extracting Constructive Content from Classical Proofs. PhD thesis, Cornell University Ithaca, New York.
15. Schütte, K. and Simpson, S. G. (1985). Ein in der reinen Zahlentheorie unbeweisbarer Satz über endliche Folgen von natürlichen Zahlen. *Archiv für Mathematische Logik und Grundlagenforschung*, **25**, 75–89.
16. Girard, J.-Y. (1987). *Proof theory and complexity*. Bibliopolis, Naples.
17. Simpson, S. G. (1988). Ordinal numbers and the Hilbert Basis Theorem. *J. Symbolic Logic*, **53**, 961–974.
18. de Jongh D. and Parikh R. (1977). Well partial orderings and their order types. *Indagationes Mathematicae*, **39**, 195–207.
19. Schmidt D. (1979). Well-orderings and their maximal order types. Habilitationsschrift, Mathematisches Institut der Universität Heidelberg.
20. Murthy, C. R. and Russell, J. R. (1990). A constructive proof of Higman's lemma. In: *Proceedings of the Fifth Symposium on Logic in Computer Science*, 257–267.
21. Richman, F. and Stolzenberg, G. (1993). Well quasi-ordered sets. *Advances in Mathematics*, **97**, 145–153.
22. Fridlender, D. (1997). Higman's Lemma in Type Theory. PhD thesis, Chalmers University of Technology and University of Göteburg.
23. Veldman, W. (2004). An intuitionistic proof of Kruskal's theorem. *Archive for Mathematical Logic*, **43**, 215–264.
24. Marcone, A. (1996). On the logical strength of Nash-Williams' theorem on transfinite sequences. In: *Logic: from Foundations to Applications; European logic colloquium*, (eds W. Hodges, M. Hyland, C. Steinhorn, and J. Truss), pp. 327–351.
25. Berger, U. (1993). Program extraction from normalization proofs. In *Typed Lambda Calculi and Applications*, (eds M. Bezem and J.F. Groote) volume 664 of *Lecture Notes in Computer Science*, pp. 91–106. Springer Verlag, Berlin, Heidelberg, New York.
26. Paulin-Mohring, C. and Werner, B. (1993). Synthesis of ML programs in the system Coq. *Journal of Symbolic Computation*, **15**, 607–640.
27. Coquand, T. (1991). Constructive topology and combinators. In *Constructivity in Computer Science*, Lecture Notes in Computer Science **613**, pp. 159–164.

28. Berger, U. (2004). A computational interpretation of open induction. *Proc 19th IEEE Symp. Logic in Computer Science.*
29. Luckhardt, H. Extensional Gödel functional interpretation – a consistency proof of classical analysis. *Lecture Notes in Mathematics*, **306**, Springer, 1973.
30. Troelstra, A. S. (1973). Metamathematical Investigation of Intuitionistic Arithmetic and Analysis, *Lecture Notes in Mathematics* **344**, Springer.
31. Berardi, S. Bezem, M. and Coquand, T. (1998). On the computational content of the axiom of choice. *The Journal of Symbolic Logic*, **63**(2), 600–622.
32. Scott, D. S. (1970). Outline of a mathematical theory of computation. *4th Annual Princeton Conference on Information Sciences and Systems*, pp. 169–176.
33. Ershov, Y. (1977). Model *C* of partial continuous functionals. In: *Logic Colloquium 1976*, (eds R. Gandy and M. Hyland), pp. 455–467, North Holland.
34. Hyland, M. (1982). The effective topos. In: *The L.E.J. Brouwer Centenary Symposium*, (eds A.S. Troelstra and D. Van Dalen), pp. 165–216. North Holland.
35. Spector, C. (1962). Provably recursive functionals of analysis: a consistency proof of analysis by an extension of principles in current intuitionistic mathematics. In: *Recursive function theory*, (ed F. D. E. Dekker), pp. 1–27, North–Holland.
36. Kreisel, G. (1959). Interpretation of analysis by means of constructive functionals of finite types. *Constructivity in Mathematics*, (ed A. Heyting), 101–128. North Holland.
37. Plotkin, G. D. (1977). LCF considered as a programming language. *Theoretical Computer Science*, **4**, 223–255.
38. Coquand, T. (1997). *A Note on the Open Induction Principle*. Chalmers University.
39. Berger, U. Eberl, M. and Schwichtenberg, H. (1998). Normalization by evaluation. In: *Prospects for Hardware Foundations*, (eds B. Möller and J.V. Tucker), Lecture Notes in Computer Science **1546**, pp. 117–137. Springer.
40. Herbelin, H. (1994). A program from an A-translated impredicative proof of Higman's lemma. *http://coq.inria.fr/contribs/higman.html.*
41. Berghofer, S. (2004). A constructive proof of Higman's Lemma in Isabelle. In: *Types for Proofs and Programs* (TYPES 2003), (eds S. Berardi and M. Coppo), Lecture Notes in Computer Science **3085**, Springer.
42. Simpson, S. G. (1985). Nonprovability of certain combinatorial properties of finite trees. In: *Harvey Friedman's Research on the Foundations of Mathematics*, L.A. Harrington, et al., 87–117, North–Holland.
43. Gödel, K. (1958). Über eine bisher noch nicht benützte Erweiterung des finiten Standpunkts. *Dialectica*, **12**, 280–287.

9

THE DUALITY OF CLASSICAL AND CONSTRUCTIVE NOTIONS AND PROOFS

SARA NEGRI AND JAN VON PLATO

Abstract

The method of converting mathematical axioms into rules of sequent calculus reveals a perfect duality between classical and constructive basic notions, such as equality and apartness, and between the respective rules for these notions. Derivations with the mathematical rules of a constructive theory are specular duals of corresponding classical derivations. The class of geometric theories is among those convertible into rules and the duality defines a new class of 'cogeometric' theories. Examples of such theories are projective and affine geometry.

The logical rules of classical sequent calculus are invertible, which has for quantifier-free theories the effect that logical rules in derivations can be permuted to apply after the mathematical rules. In the case of mathematical rules involving variable conditions, this separation of logic does not always hold, because quantifier rules may fail to permute down. A sufficient condition for the permutability of mathematical rules is determined and applied to give an extension of Herbrand's theorem from universal to geometric and cogeometric theories.

9.1 Introduction

A constructive approach to the real numbers uses the apartness of two real numbers as a basic relation. The axioms for this relation, written $a \neq b$, are as follows:

AP1. $\sim a \neq a$,
AP2. $a \neq b \supset a \neq c \vee b \neq c$.

Substituting a for c in AP2, we get $a \neq b \supset a \neq a \vee b \neq a$, so that symmetry of apartness follows by AP1. Equality is a defined notion:

EQDEF $a = b \equiv \sim a \neq b$.

By AP1, equality is reflexive. By the contraposition of symmetry of apartness, we have also symmetry of equality. By AP2 and symmetry of apartness, we have $a \neq b \supset a \neq c \vee c \neq b$, so contraposition gives transitivity of equality.

If instead of the constructively motivated notion of apartness we take equality as a basic notion, with its standard properties of reflexivity, symmetry, and transitivity, apartness can be defined by

APDEF $a \neq b \equiv \sim a = b$.

Irreflexivity and symmetry of apartness follow. For the 'splitting' property of an apartness $a \neq b$ into two cases $a \neq c \vee b \neq c$, the contraposition of transitivity of equality gives

$$\sim a = b \supset \sim (a = c \ \& \ c = b).$$

To distribute negation inside the conjunction, classical logic is needed.

The above game with classical and constructive notions can be carried further. In [1] the basic relations of constructive elementary geometry were treated. (Incidentally, apartness relations were already used in geometry in Heyting's doctoral dissertation of 1925, see [2].) The parallelism of two lines is a classical basic relation, and its constructive counterpart is the 'convergence' of two lines l and m, written $l \nparallel m$. The axioms are as for the apartness relation above.

The intuition with constructive basic notions is that the classical notions such as equality are 'infinitely precise', whereas apartness, if it obtains, can be verified by a finite computation. Something of this intuition can be seen already in Brouwer's first ideas [15] on the topic of apartness relations of 1924, where it is required that the set of objects considered be continuous. This was certainly the intention with Brouwer's constructive real numbers and with Heyting's constructive synthetic geometry. A set is defined as discrete if it has a decidable equality relation, otherwise it is continuous. The constructive interpretation of the law of excluded middle for equality, $a = b \vee \sim a = b$, is precisely that the basic set of objects is discrete. With such sets, it makes no difference which relations are used as basic, the constructive or classical ones, as the axioms are interderivable. (Incidentally, we have here an argument against the creation of a special "intuitionistic notation," such as *a#b*, parallel to the standard classical one. Such extra symbolism turns out redundant in the discrete case. All we need is to slash the standard symbols for relations.)

In a contribution to the previous Venice conference of 1999 [3] the constructivization of elementary axiomatics was extended to lattice theory. It then seemed that proofs that use apartness relations would be harder to find than corresponding classical proofs (see especially Theorem 7.1 and the discussion on p. 196). It has turned out, however, that there is an automatic bridge between classical and constructive notions and proofs. The matter is best seen on a formal level if for the representation of proofs Gentzen's sequent calculus is used. The method we shall apply was found in connection with a proof-theoretical investigation of apartness relations [4], and generalized to theories with universal axioms in [5] (a later paper that appeared earlier). The duality of classical and constructive notions and proofs was used first in a study of order relations, in [6].

9.2 From Mathematical Axioms to Mathematical Rules

Sequents are expressions of the form $\Gamma \to \Delta$ in which Γ and Δ are finite lists of formulas with order disregarded (i.e. finite multisets). The reading is that Δ gives the possible (open) cases that are derivable under the (open) assumptions Γ. In logical symbolism, with $\&\Gamma$ the conjunction of formulas in Γ and $\vee\Delta$ the disjunction of formulas in Δ, the sequent $\Gamma \to \Delta$ expresses the derivability of the formula $\&\Gamma \supset \vee\Delta$.

Table 9.1 *The logical rules of classical sequent calculus*

$$\frac{A, B, \Gamma \to \Delta}{A\&B, \Gamma \to \Delta}L\& \qquad \frac{\Gamma \to \Delta, A \quad \Gamma \to \Delta, B}{\Gamma \to \Delta, A\&B}R\&$$

$$\frac{A, \Gamma \to \Delta \quad B, \Gamma \to \Delta}{A \vee B, \Gamma \to \Delta}L\vee \qquad \frac{\Gamma \to \Delta, A, B}{\Gamma \to \Delta, A \vee B}R\vee$$

$$\frac{\Gamma \to \Delta, A \quad B, \Gamma \to \Delta}{A \supset B, \Gamma \to \Delta}L\supset \qquad \frac{A, \Gamma \to \Delta, B}{\Gamma \to \Delta, A \supset B}R\supset$$

$$\frac{\Gamma \to \Delta, A}{\sim A, \Gamma \to \Delta}L\sim \qquad \frac{A, \Gamma \to \Delta}{\Gamma \to \Delta, \sim A}R\sim$$

The logical rules of sequent calculus show how assumptions in the left, antecedent part, and cases in the right, succedent part of a sequent can be modified by the logical operations. We show only the rules for the connectives in Table 9.1.

We observe that the rule pairs $L\& - R\vee$, $L\vee - R\&$, and $L\sim - R\sim$ display a left–right mirror image duality.

The most remarkable property of these rules of classical logic is their invertibility: If a sequent of any of the forms given in a conclusion of a rule is derivable, the sequents we find in the corresponding premisses are also derivable. The other way, from the premisses to the conclusion is licensed by the rules themselves. Therefore, given a sequent $\Gamma \to \Delta$, we can decompose its formulas and get simpler sequents that together are equiderivable with the given sequent, until we arrive at topsequents of a derivation tree in which there is nothing to decompose left. If each of these leaves is an initial sequent, one that has a common atomic formula (atom) on both sides of the arrow, the endsequent $\Gamma \to \Delta$ is derivable by the rules given in Table 9.1, otherwise it is underivable.

Many axiom systems consist of universal formulas or, equivalently, of formulas of propositional logic with free parameters in the atoms. Each such formula is (at least classically) equivalent to a finite number of implications of the form

$$P_1\&\ldots\&P_m \supset Q_1 \vee \ldots \vee Q_n \tag{9.1}$$

with P_i, Q_j atoms and $m, n \geq 0$. Special cases are $m = 0$, with (9.1) reduced to $Q_1 \vee \ldots \vee Q_n$, and $n = 0$, with (9.1) reduced to $\sim(P_1\&\ldots\&P_m)$. Formula (9.1) can be converted into a sequent calculus rule in two ways: One is based on the idea that if each of the Q_j, together with other assumptions Γ, is sufficient to derive a number of cases Δ, as expressed formally by the sequent $Q_j, \Gamma \to \Delta$, then the P_i together are sufficient. We have the schematic rule

$$\frac{Q_1, \Gamma \to \Delta \quad \ldots \quad Q_n, \Gamma \to \Delta}{P_1, \ldots, P_m, \Gamma \to \Delta}\textit{L-rule}. \tag{9.2}$$

A dual scheme says that if each of the P_i follows as a case from some assumptions Γ, then the Q_j follow as cases:

$$\frac{\Gamma \to \Delta, P_1 \quad \dots \quad \Gamma \to \Delta, P_m}{\Gamma \to \Delta, Q_1, \dots, Q_n} \textit{R-rule}. \tag{9.3}$$

The rules that act on the left, or antecedent, part of sequents are best seen in a root-first order. Assume we are given a system of such left rules and assume that $\Gamma \to \Delta$ contains only atoms. Now try matching $\Gamma \to \Delta$ as a conclusion to the rules. Whenever there is a match, we have premisses that each get one additional atom in the antecedent. Thus, in the end, we obtain the deductive closure of Γ relative to the rules, with branchings into several possible closures each time there is more than one premiss. The succedent Δ remains untouched by the rules.

One special case merits attention, namely that of $n = 1$. We can then limit the rules to have just one formula in the succedent. A sequent $\Gamma \to P$ is derivable if and only if P belongs to the deductive closure of Γ relative to the rules.

Two more details need to be added before we can make an overall statement:

1. The formulas $P_1, \dots, P_m$ in the assumption part of the conclusion of scheme (9.2) have to be repeated in each of the premisses, and dually for $Q_1, \dots, Q_n$ in scheme (9.3). The intuitive justification is that if $P_1, \dots, P_m$ are among the assumptions in the endsequent, they can be permitted as assumptions anywhere else, and dually for $Q_1, \dots, Q_n$.
2. It can happen that instantiation of free parameters in atoms produces a duplication (two identical atoms) in the conclusion of a rule instance, say

$$P_1, \dots, P, P, \dots, P_m, \Gamma \to \Delta.$$

   By condition 1, each premiss has the duplication. We now require that the rule with the duplication P, P *contracted* into a single P is added to the system of rules. For each axiom system, there is only a bounded number of possible cases of contracted rules to be added, very often none at all.

Detailed explanations of the need for conditions 1 and 2 can be found in [5, 7].

Let us assume we are given a system Ax with a finite number of axioms of the form (9.1). Let HAx be the axiomatic system Ax together with a standard axiomatic system of classical logic. Let G3* be the Gentzen system G3 of Table 9.1 extended with the (left or right) sequent calculus rules and their contracted forms as determined by the axioms Ax. We have:

Theorem 9.1 T *he system* G3* *is complete, i.e.* $\Gamma \to \Delta$ *is derivable in* G3* *if and only if* $\&\Gamma \supset \vee\Delta$ *is derivable in* HAx.

A proof can be found in [5, 7]. It is easily seen, by the invertibility of the logical rules of G3, that instances of logical rules permute down relative to the mathematical rules. Therefore we can consider the mathematical rules in isolation from the logical ones.

9.3 Derivations in Left and Right Rule Systems

We shall show the duality of classical and constructive notions and proofs through examples that are easily seen to be representative of the general situation. Consider the theory of apartness. Its two axioms convert into the system of left rules

$$\frac{}{a \neq a, \Gamma \to \Delta}{\scriptstyle Irref} \qquad \frac{a \neq c, a \neq b, \Gamma \to \Delta \quad b \neq c, a \neq b, \Gamma \to \Delta}{a \neq b, \Gamma \to \Delta}{\scriptstyle Split.}$$

Symmetry of apartness is expressed by the sequent $\to a \neq b \supset b \neq a$ and has the derivation

$$\cfrac{\cfrac{\cfrac{}{a \neq a, a \neq b \to b \neq a}{\scriptstyle Irref} \quad b \neq a, a \neq b \to b \neq a}{a \neq b \to b \neq a}{\scriptstyle Split.}}{\to a \neq b \supset b \neq a}{\scriptstyle R\supset} \tag{9.4}$$

Now take rules *Irref* and *Split* and move all atoms to the other side by rule $R\sim$ of classical sequent calculus. Next write $a = b$ for $\sim a \neq b$, etc. The result can be written as the two rules for equality

$$\frac{}{\Gamma \to \Delta, a = a}{\scriptstyle Ref} \qquad \frac{\Gamma \to \Delta, a = b, a = c \quad \Gamma \to \Delta, a = b, b = c}{\Gamma \to \Delta, a = b}{\scriptstyle ETr.}$$

Here *ETr* stands for 'Euclidean transitivity', from the way transitivity is expressed by Euclid.

With our example derivation (9.4), switch atoms on the left and right sides of the arrow, erase the slashes, and change the rule names to get

$$\cfrac{\cfrac{\cfrac{}{b = a \to a = b, a = a}{\scriptstyle Ref} \quad b = a \to a = b, b = a}{b = a \to a = b}{\scriptstyle ETr.}}{\to b = a \supset a = b}{\scriptstyle R\supset} \tag{9.5}$$

The sequents in the mathematical part of derivation (9.5) are perfect mirror images of those in derivation (9.4).

Next we convert the two axioms of an apartness relation into a system of right rules:

$$\frac{\Gamma \to \Delta, a \neq a}{\Gamma \to \Delta}{\scriptstyle Irref} \qquad \frac{\Gamma \to \Delta, a \neq c, b \neq c, a \neq b}{\Gamma \to \Delta, a \neq c, a \neq b}{\scriptstyle Split.}$$

The symmetry of apartness now has the derivation

$$\cfrac{\cfrac{\cfrac{a \neq b \to b \neq a, a \neq a, a \neq b}{a \neq b \to b \neq a, a \neq a}{\scriptstyle Split.}}{a \neq b \to b \neq a}{\scriptstyle Irref}}{\to a \neq b \supset b \neq a}{\scriptstyle R\supset.} \tag{9.6}$$

The mirror image left rules for equality are

$$\frac{a=a,\Gamma\rightarrow\Delta}{\Gamma\rightarrow\Delta}\,Ref \qquad \frac{a=b,a=c,b=c,\Gamma\rightarrow\Delta}{a=c,b=c,\Gamma\rightarrow\Delta}\,ETr.$$

Symmetry is derived by the mirror image of derivation (9.6):

$$\cfrac{\cfrac{\cfrac{a=b,a=a,b=a\rightarrow a=b}{a=a,b=a\rightarrow a=b}\,ETr}{b=a\rightarrow a=b}\,Ref}{\rightarrow b=a\supset a=b}\,R\supset. \tag{9.7}$$

There are thus two kinds of systems of rules of equality, and the same for apartness. Euclidean equality has axioms that are Harrop (or Horn clause) formulas, i.e. have no disjunctions in their positive parts. As a consequence, derivations with the two rules of this theory are linear, with just one premiss. Also the mirror image right theory of apartness has linear derivations. It could be called a 'co-Harrop' theory, with axioms that have no conjunctions in their negative parts. (See [7] for the notions of Harrop formula and theory, and positive and negative parts of formulas.)

The above examples of rules and derivations are fully representative of the general situation: We can take the left rule scheme (9.2) and convert it into a right rule scheme (9.3) in exactly the same way as in the examples, with a change in the basic notions from constructive to classical or the other way around. The question remains what, if anything, is gained by the constructivization of classical elementary axiomatic theories; combinatorially, for each derivation in a constructive system of rules, there is a dual classical derivation and vice versa.

9.4 Geometric and Cogeometric Axioms and Rules

It is possible to extend the left and right rule schemes (9.2) and (9.3) by allowing in their active formulas the occurrence of free variables subject to a variable condition, and thus to express the role of quantifiers in a 'logic-free' way.

The left rule scheme takes the form

$$\frac{\overline{Q}_1(y_1/x_1),\overline{P},\Gamma\rightarrow\Delta \quad \dots \quad \overline{Q}_n(y_n/x_n),\overline{P},\Gamma\rightarrow\Delta}{\overline{P},\Gamma\rightarrow\Delta}\,GRS \tag{9.8}$$

where $\overline{Q}_j$ and $\overline{P}$ indicate the multisets of atomic formulas $Q_{j1},\dots Q_{jk_j}$ and $P_1,\dots,P_m$, respectively, and the eigenvariables $y_1,\dots,y_n$ of the premisses satisfy the condition of not having free occurrence in the conclusion of the scheme. This variable condition is the same as the one for the rules $R\forall$ and $L\exists$ of first-order logic.

A rule scheme of the form (9.8) expresses as a rule an axiom, called a *geometric axiom*, of the form

$$\forall \overline{z}(P_1 \& \ldots \& P_m \supset \exists x_1 M_1 \vee \ldots \vee \exists x_n M_n)$$

in which M_j is the conjunction of the multiset $\overline{Q}_j$, that is, $Q_{j_1} \& \ldots \& Q_{jk_j}$. Finite conjunctions of geometric axioms lead to the class of formulas, called *geometric implications*, that are sentences of the form

$$\forall \overline{z}(A \supset B)$$

in which A and B are formulas not containing $\supset$ or $\forall$.

Geometric theories are theories axiomatized by geometric implications. Interest in the study of geometric theories has arisen from different areas of logic and mathematics. In topos theory, geometric formulas are characterized as the fragment of first-order logic preserved by geometric morphisms. The preservation property extends beyond first-order logic to existential fixed-point formulas (cf. [8]). The nature of geometric logic as the logic of finite observations has been emphasized in localic approaches to constructive topology (cf. [9]). Geometric theories can be treated, as any other theory, by the addition of Hilbert-style axioms to a logical proof system, but axiomatic systems are hard to analyse proof-theoretically. In [10], geometric theories are presented through suitable rules extending (intuitionistic) natural deduction. A proof of normalization for the extensions thus obtained is given and the systems applied in a systematic study of the proof theory of intuitionistic modal logic; in [11], the so-called method of *dynamical proof*, establishing the derivability of one atom from a finite set of atoms, is applied to certain specific geometric theories such as the theory of algebraically closed fields. In [12], sequent rules in the form of the geometric rule scheme above have been introduced, permitting the extension of structural proof analysis from the first-order classical and intuitionistic systems of sequent calculus G3c and G3im to geometric theories. In particular, this method permits us to present geometric theories as contraction- and cut-free sequent systems and to obtain what is undoubtedly the simplest possible proof of *Barr's theorem*: If a geometric implication is provable classically in a geometric theory, then it is provable intuitionistically. The proof of this conservativity result consists in noting that a classical cut-free proof of a geometric implication is already an intuitionistic proof in sequent systems with rules for geometric theories.

We observe that Barr's theorem is not a characterization of the intuitionistic fragment of geometric theories, because we can go beyond geometric implications and maintain the conservativity result. First, following Dragalin's suggestion (cf. Section 3.7.3 in [13]) we can modify the intuitionistic left rule for implication by admitting a multisuccedent conclusion in the left premiss

$$\frac{A \supset B, \Gamma \to \Delta, A \quad B, \Gamma \to \Delta}{A \supset B, \Gamma \to \Delta} L\supset.$$

Rule $L\supset$ of the classical calculus (without $A \supset B$ in the left premiss) is then admissible in the modified intuitionistic calculus, thus the difference between the intuitionistic and classical sequent systems is confined to rules $R\supset$ and $R\forall$. An operational definition of formulas for which the conservativity of classical derivations holds can be given: If a formula is derivable classically in a geometric theory and the derivation contains no steps of $R\supset$ and $R\forall$ with a non-empty context in the premiss, then the derivation is an intuitionistic derivation. However, this is an empty characterization, stating nothing but that 'an intuitionistic derivation is an intuitionistic derivation'. A characterization in terms of the form of the formulas alone, not of their derivations, would be desirable. There are classes of formulas, such as geometric implications, the form of which forces the derivation to be of the stated kind. The same is true, for example, if the formula does not contain in its positive part implications or universally quantified formulas as components of a disjunction. Even so, there are still formulas outside the mentioned classes for which the conservativity holds.

Examples of geometric theories include the theory of real-closed fields, Robinson arithmetic, and constructive affine geometry. In order to obtain a geometric axiomatization some care is needed when formulating the axioms. For instance, the axiom stating the existence of inverses of non-zero elements in the theory of fields,

$$\sim x = 0 \supset \exists y\, x \cdot y = 1$$

is not geometric as it contains an implication the antecedent of which is an implication $(x = 0 \supset \bot)$, but it can be replaced by the geometric axiom

$$x = 0 \vee \exists y\, x \cdot y = 1$$

that can be given as a rule following the geometric rule scheme,

$$\frac{x = 0, \Gamma \rightarrow \Delta \quad x \cdot y = 1, \Gamma \rightarrow \Delta}{\Gamma \rightarrow \Delta}\text{L-}inv$$

with the variable condition y not free in Γ, Δ.

Alternatively, we can take inequality $\neq$ as the primitive relation and turn L-*inv* into the following right rule with the same variable condition on y,

$$\frac{\Gamma \rightarrow \Delta, x \neq 0 \quad \Gamma \rightarrow \Delta, x \cdot y \neq 1}{\Gamma \rightarrow \Delta}\text{R-}inv$$

corresponding to the axiom $\sim \forall y(x \neq 0 \;\&\; x \cdot y \neq 1)$.

All the other axioms for fields and real-closed fields can be given in terms of right rules for the primitive relation of inequality.

A similar transformation can be done with the axioms of constructive affine geometry. These axioms, presented in [1], are based on the primitive notions of distinct points, $a \neq b$, distinct lines, $l \neq m$, convergent lines, $l \not\parallel m$, and of a point outside a line, $a \notin l$, and on the constructions of a line $ln(a, b)$ connecting two distinct points a and b, and of a point $pt(l, m)$ obtained as the intersection of two convergent lines l and m.

We observed [12] that the extension with the axiom stating the existence of three non-collinear points,

$$\exists x \exists y \exists z (x \neq y \; \& \; z \notin ln(x, y))$$

maintains the theory geometric as the axiom corresponds to the following instance of the geometric rule scheme

$$\frac{x \neq y, z \notin l(x, y), \Gamma \rightarrow \Delta}{\Gamma \rightarrow \Delta}$$

with the variable condition x, y, z not free in Γ, Δ.

If the axiomatization is instead based on the primitive relation of equality of points, equality of lines, parallelism of lines, and incidence a point with a line, the axiom becomes

$$\exists x \exists y \exists z (\sim x = y \; \& \; \sim z \in ln(x, y))$$

which is no longer a geometric implication. Thus, as concluded in [12], 'classical geometry is not a geometric theory'. However, the axiom can be given in the form of the right rule, with the condition x, y, z not free in Γ, Δ,

$$\frac{\Gamma \rightarrow \Delta, x = y, z \in l(x, y)}{\Gamma \rightarrow \Delta}$$

by which

$$\sim \forall x \forall y \forall z (x = y \vee z \in ln(x, y))$$

is derivable. All the other axioms can also be uniformly presented as right rules for the primitive relations $a = b$, $l = m$, $l \parallel m$, and $a \in l$.

The above examples illustrate a general result:

Theorem 9.2 *Let* T *be a geometric theory based on the primitive relations* R_i, *with rules following the geometric rule scheme* GRS, *and let* T′ *be the theory obtained by formulating the axioms in terms of the dual relations* R_i'. *Then a contraction- and cut-free system for the theory* T′ *is obtained by turning all the instances of* GRS *into the form*

$$\frac{\Gamma \rightarrow \Delta, \overline{P}', \overline{Q}_1'(y_1/x_1) \quad \ldots \quad \Gamma \rightarrow \Delta, \overline{P}', \overline{Q}_n'(y_n/x_n)}{\Gamma \rightarrow \Delta, \overline{P}'} \textit{co-GRS}$$

in which the apices indicate the atoms transformed in terms of the dual relations R_i'.

We can ask what kinds of axioms are captured by the scheme *co-*GRS. Clearly, the scheme is interderivable with an axiom of the form

$$\forall \overline{z} (\forall x_1 M_1' \& \ldots \& \forall x_n M_n' \supset P_1' \vee \ldots \vee P_m') \qquad \textit{co-GA}$$

in which $M_j' \equiv Q_{j1}' \vee \ldots \vee Q_{jk_j}'$.

It is easy to verify that any formula of the form

$$\forall \overline{z}(A \supset B)$$

with A and B formulas not containing $\supset$ or $\exists$, can be brought to a canonical form consisting of conjunctions of formulas of the form given by *co-GA*. Formulas A, B not containing $\supset$ or $\exists$ will be called *cogeometric* and the implication $A \supset B$ a *cogeometric implication*. A theory axiomatized by cogeometric implications will be called a *cogeometric theory*. Classical projective and affine geometry with the axiom of non-collinearity included constitute examples of cogeometric theories.

The above examples have shown how the duality between geometric and cogeometric theories can be used for changing the primitive notions in the sequent formulation of a theory. Meta-theoretical results can be imported from one theory to its dual by exploiting the symmetry of their associated sequent calculi.

In [7], an extension of Herbrand's theorem to universal theories is presented. We recall the statement:

Theorem 9.3 (Herbrand's theorem for universal theories) *Let* T *be a theory with a finite number of purely universal axioms and let* G3cT *be the sequent system obtained by turning the theory into a system of mathematical rules. If the sequent* $\rightarrow \forall x \exists y_1 \ldots \exists y_k A$, *with A quantifier free, is derivable in* G3cT, *then there are terms* t_{i_j} *with* $i \leqslant n, j \leqslant k$ *such that*

$$\rightarrow \bigvee_{i=1}^{n} A(t_{i_1}/y_1, \ldots, t_{i_k}/y_k)$$

is derivable in G3cT.

Clearly, the theorem does not extend to geometric theories. In fact, if $\exists x P$ is an axiom of the theory T, then $\rightarrow \exists x P$ is derivable in G3cT but there is no finite disjunction such that $\rightarrow P(t_1) \vee \ldots \vee P(t_n)$ would be derivable in G3cT.

The crucial ingredient in the proof of Herbrand's theorem is the possibility to assume a derivation in which the quantifier rules come last. In first-order logic and in universal theories, this is unproblematic. With mathematical rules involving variable conditions, like the geometric or the cogeometric rule scheme, the quantifier rules cannot in general be permuted last in a derivation. Suppose we have a derivation containing the steps

$$\dfrac{\dfrac{\overline{Q}_1(y_1/x_1), \overline{P}, \Gamma \rightarrow \Delta, \exists x A, A(t/x)}{\overline{Q}_1(y_1/x_1), \overline{P}, \Gamma \rightarrow \Delta, \exists x A}{\scriptstyle R\exists} \quad \ldots \quad \overline{Q}_n(y_n/x_n), \overline{P}, \Gamma \rightarrow \Delta, \exists x A}{\overline{P}, \Gamma \rightarrow \Delta, \exists x A}{\scriptstyle GRS}.$$

If the term t contains the variable y_1, the permutation of $R\exists$ to below GRS fails because the variable condition for a correct application of GRS would no longer be satisfied. This is the exact structural reason for the failure of Herbrand's theorem for existential theories. We can nevertheless impose an additional hypothesis that makes the permutation possible. The hypothesis ensures that a fresh variable substitution *limited to the atoms* $\overline{Q}_1$ is possible.

Lemma 9.4 *Let* T *be a geometric theory and let* G3cT *be the sequent system obtained by turning the theory into a system of left mathematical rules. Suppose that the sequent* $\overline{Q}_i(y_i/x_i), \overline{P}, \Gamma \to \Delta, A(t/x)$ *is derivable in* G3cT*, that* y_i *is not free in* Γ, Δ*, and that no atom* $\overline{Q}_i$ *occurs positively in* A*. Then* $\overline{Q}_i(z/x_i), \overline{P}, \Gamma \to \Delta, A(t/x)$ *is derivable for an arbitrary fresh variable* z*.*

Proof. Consider the initial sequents in a derivation of the given sequent. By the assumptions that y_i not occur free in Δ and that no atom among the $\overline{Q}_i$ be in the positive part of A, it follows that the principal atoms of the axioms are not among the $\overline{Q}_i$. Thus, after the substitution of the variable y_i with a fresh variable z in the atoms $\overline{Q}_i(y_i/x_i)$, the leaves of the tree remain initial sequents, and the logical steps remain correct because the atoms in $\overline{Q}_i$ are never principal in logical rules. Since z is a fresh variable, also the instances of the geometric rule scheme remain correct, thus the substitution produces a derivation of $\overline{Q}_i(z/x_i), \overline{P}, \Gamma \to \Delta, A(t/x)$ in G3cT. □

By the lemma, under the additional hypothesis of non-occurrence of the atoms $\overline{Q}_i$ in positive parts of A, we can assume a derivation in which the mathematical rules come first, followed by propositional rules, followed by a linear part consisting of quantifier rules. The rest of the proof of Herbrand's theorem is then a routine matter. Thus we have:

Theorem 9.5 (Herbrand's theorem for geometric theories) *Let* T *be a geometric theory and let* G3cT *be the sequent system obtained by turning the theory into a system of mathematical rules following the geometric rule scheme GRS. If the sequent* $\to \forall x \exists y_1 \dots \exists y_k A$*, with* A *quantifier-free, is derivable in* G3cT *and no atom* $\overline{Q}_i$ *occurs positively in* A*, then there are terms* t_{i_j} *with* $i \leqslant n$*,* $j \leqslant k$ *such that*

$$\to \bigvee_{i=1}^{n} A(t_{i_1}/y_1, \dots, t_{i_k}/y_k)$$

is derivable in G3cT.

By exploiting the symmetry between a left and a right rule system we obtain the corresponding results for cogeometric theories.

Lemma 9.6 *Let* T *be a cogeometric theory and let* G3cT *be the sequent system obtained by turning the theory into a system of right mathematical rules. Suppose the sequent* $\Gamma \to \Delta, \overline{Q}_i(y_i/x_i), \overline{P}, A(t/x)$ *is derivable in* G3cT*,* y_i *is not free in* Γ, Δ*, and no atom* $\overline{Q}_i$ *occurs negatively in* A*. Then* $\Gamma \to \Delta, \overline{Q}_i(z/x_i), \overline{P}, A(t/x)$ *is derivable for an arbitrary fresh variable* z*.*

Theorem 9.7 (Herbrand's theorem for cogeometric theories) *Let* T *be a cogeometric theory and let* G3cT *be the sequent system obtained by turning the theory into a system of right non-logical rules following the cogeometric rule scheme co-GRS. If the sequent* $\to \forall x \exists y_1 \dots \exists y_k A$*, with* A *quantifier-free, is derivable in* G3cT *and no atom* $\overline{Q}_i$ *occurs* *negatively in A, then there are terms t_{i_j} with $i \leqslant n$, $j \leqslant k$ such that*

$$\rightarrow \bigvee_{i=1}^{n} A(t_{i_1}/y_1, \ldots, t_{i_k}/y_k)$$

is derivable in G3cT.

9.5 Duality of Dependent Types and Degenerate Cases

The axiomatization of elementary geometry with constructive basic notions leads in a natural way to *dependent typing*: A formula with a constructed line $ln(a, b)$, such as the incidence axiom $a \in ln(a, b)$, is well-formed only if the condition of non-degeneracy $a \neq b$ is satisfied. In a first-order formulation, incidence axioms with conditions of non-degeneracy can be given as implications, as in [1]. For projective geometry, we have

$$a \neq b \supset {\sim} a \notin ln(a, b), \quad a \neq b \supset {\sim} b \notin ln(a, b),$$

and similarly for intersection points. The corresponding left rule for the first axiom is the zero-premiss rule

$$\overline{a \neq b, a \notin ln(a, b), \Gamma \rightarrow \Delta}^{\,Inc.}$$

By the duality of left and right rules, we have for the classical notions of equality and incidence the rule

$$\overline{\Gamma \rightarrow \Delta, a = b, a \in ln(a, b)}^{\,Inc.}$$

Thus, the incidence axioms for connecting lines in a classical formulation are

$$a = b \vee a \in ln(a, b), \quad a = b \vee b \in ln(a, b),$$

and similarly for the rest of the incidence axioms. The *degenerate cases* $a = b$ in these axioms are the classical duals of dependent typings in constructive geometry. The phenomenon is quite general; similar observations could be made about the condition for the inverse operation.

The use of constructions seems to be necessary for the conversion of mathematical axioms into systems of cut-free rules, be it a system based on classical or constructive notions. To see this, we formulate elementary geometry as a *relational theory* with existential axioms in place of constructions, as in

$$\forall x \forall y \exists z (x \in z \;\&\; y \in z).$$

(The sorts of the variables are determined from their places in the incidence relation: x and y points, z a line.) Next, uniqueness axioms are added, such as

$$\forall x \forall y \forall z \forall v (x \in z \;\&\; y \in z \;\&\; x \in v \;\&\; y \in v \supset z = v).$$

As mentioned above, it is possible to formulate geometry, the axiom of non-collinearity included, either as a constructive geometric theory, or as a classical cogeometric theory.

This result refers to a formulation with geometric constructions. With a relational formulation, a comparison of the form of the existential axioms that replace constructions with the form of the axiom of non-collinearity leads instead to the following result:

If non-collinearity is formulated as a geometric implication, the existence axioms are cogeometric; if the existence axioms instead are geometric, non-collinearity is cogeometric.

There is thus a fundamental incompatibility in both approaches, but it can be overcome through the use of constructions. This phenomenon is quite general and is met in, for example, lattice theory (as in [14]), and in field theory.

References

1. von Plato, J. (1995). The axioms of constructive geometry. *Annals of Pure and Applied Logic*, vol. 76, pp. 169–200.
2. Heyting, A. (1927). Zur intuitionistischen Axiomatik der projektiven Geometrie, *Mathematische Annalen*, vol. 98, pp. 491–538.
3. von Plato, J. (2001). Positive lattices. In *Reuniting the Antipodes*, (eds P. Schuster et al.,) pp. 185–197, Kluwer, Dordrecht 2001.
4. Negri, S. (1999). Sequent calculus proof theory of intuitionistic apartness and order relations, *Archive for Mathematical Logic*, vol. 38, pp. 521–547.
5. Negri, S. and J. von Plato (1998). Cut elimination in the presence of axioms, *The Bulletin of Symbolic Logic*, vol. 4, pp. 418–435.
6. Negri, S., J. von Plato, and T. Coquand (2004). Proof-theoretical analysis of order relations. *Archive for Mathematical Logic*, vol. 43, pp. 297–309.
7. Negri, S. and J. von Plato (2001). *Structural Proof Theory*. Cambridge University Press.
8. Blass, A. (1988). Topoi and computation. *Bulletin of the European Association for Theoretical Computer Science*, no. 36, pp. 57–65.
9. Vickers, S. (1989). *Topology via Logic*, Cambridge University Press.
10. Simpson, A. (1994). Proof Theory and Semantics of Intuitionistic Modal Logic. PhD thesis, School of Informatics, University of Edinburgh.
11. Coste, M., H. Lombardi, and M.-F. Roy (2001). Dynamical method in algebra: Effective Nullstellensätze, *Annals of Pure and Applied Logic*, vol. 111, pp. 203–256.
12. Negri, S. (2003). Contraction-free sequent calculi for geometric theories with an application to Barr's theorem, *Archive for Mathematical Logic*, vol. 42, pp. 389–401.
13. Troelstra, A. and H. Schwichtenberg (2000). *Basic Proof Theory*. 2nd edn, Cambridge University Press.
14. Negri, S. and J. von Plato (2004). Proof systems for lattice theory, *Mathematical Structures in Computer Science*, vol. 14, pp. 507–526.
15. Brouwer, L. (1924). Intuitionistische Zerlegung mathematischer Grundbegriffe, as reprinted in Brouwer's *Collected Works*, vol. 1, pp. 275–280.

PART II

Practice

10

CONTINUITY ON THE REAL LINE AND IN FORMAL SPACES

ERIK PALMGREN

Abstract

We show that formal topology provides a good framework for constructive topology, which on the one hand agrees with Bishop's theory for the real line, and on the other hand avoids the well-known difficulties of this theory with respect to composition of maps and the reciprocal function.

10.1 Introduction

As is well known, Brouwer introduced his axioms for intuitionism in order to regain central results about continuity. A notable example is the classical theorem that every real-valued continuous function on a finite, closed interval is uniformly continuous. His proof relied on the covering compactness of such intervals, which in turn was derived from the axioms using only constructive principles. In fact, the axiom known as the fan theorem (classically equivalent to König's lemma) suffices. The special axioms were avoided altogether in Bishop's development of constructive analysis, a development which is consistent with classical mathematics, as well as recursive mathematics. Bishop simply modified the definition of continuous function on the real numbers to mean: *uniformly continuous* on each finite and closed interval. This was a very successful step. However, it may also lead to difficulties, when going beyond metric spaces. If X is a general space with a topology given by a neighbourhood basis, the composition of two continuous functions

$$\mathbb{R} \to X \to \mathbb{R},$$

need not be a continuous function. That a class of topological spaces form a category seems to be a minimal requirement, for instance in the theory of manifolds, or in algebraic topology. Though little emphasized, the continuous functions of the category of locales, or formal spaces, agrees with Bishop's definition of continuous function on real numbers. Proving this within the framework of (Bishop) constructive mathematics is the purpose of the present chapter (Theorems 10.5 and 10.10). In addition, we show that the reciprocal map is included in the category (Section 10.5). With formal spaces it is thus not necessary to adopt the fan theorem as an axiom, for instance, to get a good category. For a discussion of this proposal, and some earlier constructions of categories of spaces addressing these problems, we refer to [1] and [2]. For a recursive version of formal topology and related results there we refer to [3].

Locale theory developed from the wealth of work in topos theory (see [4]), without regard for being a constructive alternative to Brouwer's solution. On the other hand, this was the explicit purpose in Martin-Löf's doctoral thesis [5], and he showed how to regain covering compactness of the Cantor space by inductively generating the covers in a constructive manner. This was the precursor to the theory of formal spaces as developed by Martin-Löf and Sambin.

10.2 Point-Free Topology

The fundamental principle of formal spaces is to work primarily with the basic neighbourhoods, and their relation with respect to covering. For instance in the case of real numbers, these can be open intervals (a, b) with rational endpoints (see Section 10.3). It is often essential that covers can be generated inductively, rather than merely being defined in terms of points. We refer to [6] and [7] for a general background on formal topology, and briefly recall some basic definitions.

Definition 10.1 *Let S be a set, and let $\lhd$ be a relation between elements of S and subsets of S, i.e. $\lhd \subseteq S \times \mathcal{P}(S)$. Extend $\lhd$ to a relation between subsets by letting $U \lhd V$ if and only if $a \lhd V$ for all $a \in U$. For a preorder $(X, \leq)$ and a subset $U \subseteq X$, its downwards closure $U_{\leq}$ consists of those $x \in X$ such that $x \leq y$ for some $y \in U$. Write $a_{\leq}$ for $\{a\}_{\leq}$.*

Definition 10.2 *A formal topology $\mathcal{S}$ is a preordered set $S = (S, \leq)$ (of so-called basic neighbourhoods) together with a relation $\lhd \subseteq S \times \mathcal{P}(S)$, the covering relation, satisfying the four conditions*

(R) $a \in U$ *implies* $a \lhd U$, (L) $a \lhd U$, $a \lhd V$ *implies* $a \lhd U_{\leq} \cap V_{\leq}$,
(T) $a \lhd U$, $U \lhd V$ *implies* $a \lhd V$, (E) $a \leq b$ *implies* $a \lhd \{b\}$.

A point is an inhabited subset $\alpha \subseteq S$ which is filtering with respect to $\leq$, and such that $U \cap \alpha$ is non-void, whenever $a \lhd U$ for some $a \in \alpha$. The collection of points is denoted $\mathrm{Pt}(\mathcal{S})$.

We use the term *formal space* interchangeably with *formal topology.*

10.3 Continuous Mappings

A continuous mapping between formal topologies is a certain relation between their basic neighbourhoods. The fundamental example of such a mapping is the one associated with a continuous function $f : \mathbb{R} \to \mathbb{R}$:

$$(a, b)\, A_f\, (c, d) \iff f(a, b) \subseteq (c, d).$$

See Theorem 10.5 below.

To define the general concept we introduce some notation. For a relation $R \subseteq S \times T$ the *inverse image of $V \subseteq T$ under the relation R* is, as usual,

$$R^{-1}[V] =_{\text{def}} \{a \in S : (\exists b \in V)\, a\, R\, b\}.$$

Thus for instance

$$A_f^{-1}\{(c,d)\} = \{(a,b) : f(a,b) \subseteq (c,d)\}.$$

Notice that, in general, $R^{-1}[U] \subseteq R^{-1}[V]$ whenever $U \subseteq V$, and

$$R^{-1}[\cup_{i \in I} U_i] = \cup_{i \in I} R^{-1}[U_i].$$

The relation R is naturally extended to subsets as follows. For $U \subseteq S$, let $U\, R\, b$ mean $(\forall u \in U)\, u R b$, and for $V \subseteq T$, we let $a\, R\, V$ mean $a \lhd R^{-1}[V]$.

Definition 10.3 *Let $\mathcal{S} = (S, \leq, \lhd)$ and $\mathcal{T} = (T, \leq', \lhd')$ be formal topologies. A relation $R \subseteq S \times T$ is a continuous mapping from $\mathcal{S}$ to $\mathcal{T}$ (and we write $R : \mathcal{S} \to \mathcal{T}$) if*

(A1) *$a\, R\, b$, $b \lhd' V$ implies $a\, R\, V$,*

(A2) *$a \lhd U$, $U\, R\, b$, implies $a\, R\, b$,*

(A3) *$a\, R\, T$, for all $a \in S$,*

(A4) *$a\, R\, V$, $a\, R\, W$ implies $a\, R\, (V_{\leq'} \cap W_{\leq'})$.*

Remark 1 *Note that by $b \lhd' \{b\}$, (A1) and (A2)*

$$\{a\}\, R\, b \Longleftrightarrow a\, R\, b \Longleftrightarrow a \lhd R^{-1}\{b\} \Longleftrightarrow a\, R\, \{b\}.$$

A continuous mapping R thus satisfies

$$a\, R\, b,\, b \lhd' \{b'\} \Longrightarrow a\, R\, b'.$$

Moreover (A4) may be replaced by the condition

(A4′) *$a\, R\, b$, $a\, R\, c \Longrightarrow a\, R\, (b_{\leq'} \cap c_{\leq'})$.*

The next properties are useful for checking closure under composition. Denote by $\tilde{U} = \{a : a \lhd U\}$, the saturation of U in the topology.

Proposition 10.4 *Let $R : \mathcal{S} \to \mathcal{T}$ be a continuous mapping. Then:*

(i) *$U \lhd V$ implies $R^{-1}[U] \lhd R^{-1}[V]$,*

(ii) *$b\, R\, U$ iff $b\, R\, \tilde{U}$,*

(iii) *$R^{-1}[U]^{\sim} = R^{-1}[\tilde{U}]^{\sim}$.* □

Let **FTop** be the following category of formal topologies and continuous mappings. For a formal topology $\mathcal{S} = (S, \leq, \lhd)$ we define a continuous mapping $I : \mathcal{S} \to \mathcal{S}$ (the identity) by

$$a I b \Longleftrightarrow a \lhd \{b\}.$$

For continuous mappings, $R_1 : \mathcal{S}_1 \to \mathcal{S}_2$ and $R_2 : \mathcal{S}_2 \to \mathcal{S}_3$, between formal spaces, define the composition

$$a(R_2 \circ R_1)b \Longleftrightarrow a \lhd R_1^{-1}[R_2^{-1}\{b\}].$$

This is a continuous mapping $(R_2 \circ R_1) : \mathcal{S}_1 \to \mathcal{S}_3$. The category is not locally small, within any known predicative meta-theory.

Let **Nbhd** be the category of (large) neighbourhood spaces and point-wise continuous functions. Only set-based spaces are considered in [8]. Then we have a functor $\mathrm{Pt} : \mathbf{FTop} \to \mathbf{Nbhd}$ given by

$$\mathrm{Pt}(R)(\alpha) = \{y \in T : (\exists x \in \alpha)\, x\, R\, y\}$$

for $R : \mathcal{S} \to \mathcal{T}$. For a basic neighbourhood a of the formal topology $\mathcal{S}$,

$$a^* =_{\mathrm{def}} \{\alpha \in \mathrm{Pt}(\mathcal{S}) : a \in \alpha\}$$

is a basic open of the neighbourhood space $\mathrm{Pt}(\mathcal{S})$. Moreover,

$$a\, R\, b \Longrightarrow \mathrm{Pt}(R)[a^*] \subseteq b^*. \tag{10.1}$$

Remark 2 *Sambin [6] defines continuous mappings in a slightly different way. We can most easily explain it by removing axiom (A2) and introducing an equivalence* $\sim$ *of mappings* $\mathcal{S} \to \mathcal{T}$ *as follows*

$$F \sim G \Longleftrightarrow \overline{F} = \overline{G}.$$

Here $\overline{H}$ *is the relation given by* $a\, \overline{H}\, b$ *iff* $a \lhd H^{-1}b$. *Composition is ordinary composition of relations. Moreover, relations on* $S \times T$ *are written as functions* $T \to \mathcal{P}(S)$. *The resulting category* **FTop′** *is categorically equivalent to* **FTop**, *via the forgetful functor* **FTop** $\to$ **FTop′** *and the reverse functor given by* $F \mapsto \overline{F}$. *An advantage of* **FTop′** *is the simple definition of composition, which does not involve the cover relation.*

Remark 3 *Instead of continuous mapping the term approximable mapping is often used to emphasize the generalization of the corresponding notion in Scott's domain theory.*

10.4 Functions on Real Numbers

We recall a standard construction of the formal space $\mathcal{R}$ of real numbers. Here we follow [9], but omit the positivity predicate and use a smaller set of neighbourhoods. The basic neighbourhoods of $\mathcal{R}$ are $\{(a, b) \in \mathbb{Q}^2 : a < b\}$ given the inclusion order (as intervals), denoted by $\leq$. The cover $\lhd$ is generated by

(G1) $(a, b) \lhd \{(a', b') : a < a' < b' < b\}$ for all $a < b$,

(G2) $(a, b) \lhd \{(a, c), (d, b)\}$ for all $a < d < c < b$.

Recall that this means that $\lhd$ is the smallest covering relation satisfying (G1) and (G2). The points $\mathrm{Pt}(\mathcal{R})$ of $\mathcal{R}$ form a structure isomorphic to the Cauchy reals $\mathbb{R}$ (see, e.g. [7]), via $\bar{\ } : \mathbb{R} \to \mathrm{Pt}(\mathcal{R})$ given by

$$\overline{x} = \{(a, b) \in \mathbb{Q}^2 : a < x < b\}.$$

The points are ordered as follows

$$\alpha < \beta \iff_{\text{def}} \exists (a, b) \in \alpha\ \exists (c, d) \in \beta\ b < c,$$
$$\alpha \leq \beta \iff_{\text{def}} \neg \beta < \alpha.$$

In Bishop's constructive analysis a function $f : \mathbb{R} \to \mathbb{R}$ is defined to be continuous iff it is uniformly continuous on each compact interval (closed and finite interval). Each such function gives rise to a continuous mapping $A_f : \mathcal{R} \to \mathcal{R}$ given by

$$(a, b)\, A_f\, (c, d) \iff f(a, b) \subseteq (c, d).$$

Theorem 10.5 *A continuous function $f : \mathbb{R} \to \mathbb{R}$ is represented by the continuous mapping A_f in the sense that $g = \mathrm{Pt}(A_f) : \mathrm{Pt}(\mathcal{R}) \to \mathrm{Pt}(\mathcal{R})$ satisfies $g(\overline{x}) = \overline{f(x)}$ for all $x \in \mathbb{R}$.*

Proof. The proof that A_f is continuous is straightforward for (A2)–(A4), noting that for (A3) we use that the image of a compact interval under f is bounded. Property (A1) is equivalent to

$$I \lhd U \implies (\forall J \in \mathcal{R})(J\, A_f\, I \Rightarrow J \lhd A_f^{-1}[U]). \tag{10.2}$$

This is proved by 'induction on $\lhd$'. Denoting the right hand side of (10.2) by $I\ K\ U$, this amounts to proving that K satisfies the axioms of a cover relation and the generators of $\mathcal{R}$. Since $\lhd$ is the least such relation, the statement (10.2) then follows. All these axioms are essentially straightforward to check, but let us note how to verify (G2), since this uses uniform continuity. We have to show

$$(u, v)\ K\ \{(u, c), (d, v)\}, \tag{10.3}$$

for $u < d < c < v$. Suppose $f(a, b) \subseteq (u, v)$. Let $\varepsilon = c - d$, and let $(a_1, b_1), \ldots, (a_n, b_n)$ be a cover of (a, b), with $a \leq a_k < b_k \leq b$, so finely meshed that for all k

$$x, y \in (a_k, b_k) \implies |f(x) - f(y)| < \varepsilon/3. \tag{10.4}$$

Put $d' = d + \varepsilon/3$ and $c' = c - \varepsilon/3$. Let k be arbitrary and take some $x \in (a_k, b_k)$. Since $d' < c'$ we have by cotransitivity of the order that $d' < f(x)$ or $f(x) < c'$. If $d' < f(x)$, then $f(a_k, b_k) \subseteq (d, v)$ by (10.4). On the other hand, if $f(x) < c'$, then $f(a_k, b_k) \subseteq (u, c)$ again by (10.4). This means that

$$\{(a_1, b_1), \ldots, (a_n, b_n)\} \subseteq A_f^{-1}\{(u, c), (d, v)\}.$$

The set on the left hand side was assumed to cover (a, b), so we are done proving (10.3).

For $x \in \mathbb{R}$ we have by definition of g

$$g(\bar{x}) = \{(c, d) : (\exists(a, b) \in \bar{x})\ f(a, b) \subseteq (c, d)\}.$$

Thus $(c, d) \in g(\bar{x})$ iff there are a, b with $a < x < b$ and $f(a, b) \subseteq (c, d)$. By the continuity of f the latter is equivalent to $c < f(x) < d$, i.e. $(c, d) \in \overline{f(x)}$. □

Next we show that every continuous mapping $\mathcal{R} \to \mathcal{R}$ gives rise to a continuous function in the sense of Bishop. We use the Heine–Borel theorem of [9].

For a formal topology $S = (S, \leq, \lhd)$ and a subset U of S define the *closed subspace topology* $S_U = (S, \leq, \lhd_U)$ (intuitively on the complement of the union of U) by defining a new cover relation

$$a \lhd_U V \Longleftrightarrow a \lhd U \cup V.$$

(For a general discussion of closed and open sublocales we refer to [4].)

Then $E : S_U \to S$ given by

$$a\, E\, b \Longleftrightarrow a \lhd_U \{b\}$$

is a continuous monomorphism in the category of formal spaces. In fact

$$a \lhd_U E^{-1}[V] \Leftrightarrow a \lhd_U V.$$

Let $i = \mathrm{Pt}(E)$. Then $i(\alpha) = \alpha$, for any $\alpha \in \mathrm{Pt}(S_U)$. Moreover, for $\beta \in \mathrm{Pt}(S)$,

$$\beta \in \mathrm{Pt}(S_U) \Leftrightarrow \beta \cap U = \emptyset.$$

We will be interested in the special case when $S = \mathcal{R}$ and when U is the set of basic intervals bounded away from $[\alpha, \beta]$, more precisely,

$$C(\alpha, \beta) = \{(a, b) :\ \bar{b} < \alpha \text{ or } \beta < \bar{a}\},$$

where $\alpha < \beta$ are some given points of $\mathcal{R}$. Any $\gamma \in \mathrm{Pt}(\mathcal{R})$ satisfies

$$\gamma \in \mathrm{Pt}(\mathcal{R}_{C(\alpha,\beta)}) \Leftrightarrow \gamma \cap C(\alpha, \beta) = \emptyset \Leftrightarrow \alpha \leq \gamma \leq \beta.$$

Thus we take $\mathcal{R}_{C(\alpha,\beta)}$ to be the formal space for the closed interval $[\alpha, \beta]$. Denote it by $\mathcal{I}(\alpha, \beta)$.

The Heine–Borel theorem of [9] goes through, without any important changes, for the version of formal spaces we have used here. They used an auxiliary relation $\lhd_{\text{fin}}$ (suggested by Thierry Coquand) to show the result

$$(a, b) \lhd V \Leftrightarrow (\forall u, v)(a < u < v < b \Rightarrow (u, v) \lhd_{\text{fin}} V).$$

Here $\lhd_{\text{fin}}$ is the cover relation generated by (G2) only. It satisfies the following important property

Lemma 10.6 *If* $(a, b) \lhd_{\text{fin}} U$ *then there is a finite* $U_0 \subseteq U$ *such that*

$$(a, b) \subseteq \cup U_0 \tag{10.5}$$

as intervals.

Proof. Define the relation $(a, b) \lhd_{\text{extfin}} U$ to hold iff there is some finite $U_0 \subseteq U$ satisfying (10.5). Note that since the endpoints of intervals in U_0 are rational it does not matter whether the intervals are considered as subsets of rational or real numbers. In fact (10.5) is decidable. The relation $\lhd_{\text{extfin}}$ clearly satisfies (G2), (R) and (E). (L) can easily be checked for (10.5) over the rational numbers. As for (T) there is only a finite choice principle involved. We conclude that $\lhd_{\text{fin}}$ is smaller than $\lhd_{\text{extfin}}$, thereby proving the lemma. □

For a finite set V_0, the relation $(a, b) \lhd_{\text{fin}} V_0$ is thus equivalent to (a, b) being covered by $\cup V_0$ (as intervals). The coherence property of $\lhd_{\text{fin}}$ then follows.

Theorem 10.7 (Cederquist and Negri [9])
For $\alpha < \beta$, suppose $\mathcal{I}(\alpha, \beta) \lhd_{C(\alpha,\beta)} V$. Then

(i) $\mathcal{I}(\alpha, \beta) \lhd_{C(\alpha,\beta)} V_0$ for some finite $V_0 \subseteq V$.
(ii) $(r, s) \lhd_{\text{fin}} C(\alpha, \beta) \cup V$, where $\bar{r} < \alpha < \beta < \bar{s}$.

To prove Theorem 10.10 below we need furthermore a simple version of Lebesgue's lemma.

Lemma 10.8 *If* $I_1, \ldots, I_n$ *are open intervals with rational endpoints whose union* S *is an interval, then there is rational* $\delta > 0$ *such that for every pair of real numbers* $x, y \in S$, *where* $|x - y| < \delta$, *there is some* k *with* $x, y \in I_k$. □

The δ is called the *Lebesgue number* of the covering.
Here then is a strengthening of the conclusion of Theorem 10.7(i).

Corollary 10.9 *For* $\alpha < \beta$, *suppose* $\mathcal{I}(\alpha, \beta) \lhd_{C(\alpha,\beta)} V$. *Then* $\mathcal{I}(\alpha, \beta) \lhd_{C(\alpha,\beta)} V_0$ *for some finite* $V_0 \subseteq V$ *and* $\cup V_0$ *is an interval.*

Proof. Write $C = C(\alpha, \beta)$. Theorem 10.7(ii) and the coherence of $\lhd_{\text{fin}}$ gives a finite

$$W_0 = \{I_1, \ldots, I_n\} \subseteq C \cup V, \tag{10.6}$$

and $\bar{r} < \alpha < \beta < \bar{s}$ such that

$$(r, s) \lhd_{\text{fin}} W_0.$$

We may assume that $\cup W_0$ is an interval (otherwise we may remove certain intervals outside (r, s)). By the Lebesgue lemma there is a positive rational δ so that for all real $x, y \in \cup W_0$,

$$|x - y| < \delta \Longrightarrow (\exists k)\, x, y \in I_k.$$

Now $\alpha \in \cup W_0$. Pick r' with $r < r' < \alpha$ and $\alpha - r < \delta$. Hence by the lemma there is some k_α such that $r', \alpha \in I_{k_\alpha}$. Similarly there is $\beta < s' < s$ and k_β with $\beta, s' \in I_{k_\beta}$. By (10.6) we find a finite choice function $f : \{1, \ldots, n\} \to \{1, 2\}$ so that

(i) $f(k) = 1 \Rightarrow I_k \in C$,
(ii) $f(k) = 2 \Rightarrow I_k \in V$.

Now form $V_0 = \{I_k : f(k) = 2\} \subseteq V$. The equality between basic neighbourhood's is decidable so the subset V_0 is finite as well. Since $\alpha \in I_{k_\alpha}$, $f(k_\alpha) = 2$. Similarly $f(k_\beta) = 2$. We now claim that

$$(r', s') \lhd_{\text{fin}} V_0.$$

By density, pick rational numbers $u \in I_{k_\alpha}$, $v \in I_{k_\beta}$ with $\alpha < u$ and $v < \beta$. If $v < u$ the claim is clear. If $u < v$, then for any rational $q \in (u, v)$ there is some k with $q \in I_k$. For such k, $f(k) = 1$ is impossible by the definition of C. Thus $f(k) = 2$ and the claim follows. Again we may assume that $\cup V_0$ is an interval. Finally, since C covers what is outside the interval $[\alpha, \beta]$ we have

$$\mathcal{I}(\alpha, \beta) \lhd_C V_0.$$ □

Now the following converse of Theorem 10.5 is fairly easy

Theorem 10.10 *Let $\alpha < \beta \in \mathrm{Pt}(\mathcal{R})$. If $G : \mathcal{I}(\alpha, \beta) \to \mathcal{R}$ is a continuous mapping, then*

$$g = \mathrm{Pt}(G) : \mathrm{Pt}(\mathcal{I}(\alpha, \beta)) \to \mathrm{Pt}(\mathcal{R})$$

is uniformly continuous.

Proof. Write $C = C(\alpha, \beta)$ and $\mathcal{I} = \mathcal{I}(\alpha, \beta)$. Let ε be a positive rational number. The intervals $I_n = (n\varepsilon/2, n\varepsilon/2 + \varepsilon)$, $n \in \mathbb{Z}$, form a cover S_ε of $\mathcal{R}$. Then since $\mathcal{I} \lhd_C G^{-1}[\mathcal{R}]$, we get

$$\mathcal{I} \lhd_C G^{-1}[S_\varepsilon].$$

By Corollary 10.9 there is a finite

$$V_0 = \{(a_1, b_1), \ldots, (a_n, b_n)\} \subseteq G^{-1}[S_\varepsilon]$$

with $\mathcal{I} \lhd_C V_0$. For each $k = 1, \ldots, n$ there is thus some $(c_k, d_k) \in S_\varepsilon$ with

$$(a_k, b_k)\, G\, (c_k, d_k). \tag{10.7}$$

Now $\cup V_0$ is an interval, so let $\delta > 0$ be the Lebesgue number of V_0. Suppose that $\gamma, \gamma' \in \mathrm{Pt}(\mathcal{I})$ and $|\gamma - \gamma'| < \delta$. Take any $(a, b) \in \gamma$. Then $(a, b) \lhd_C V_0$, so there is some $(a', b') \in (C \cup V_0) \cap \gamma$. Since $C \cap \gamma = \emptyset$, we have in fact $(a', b') \in V_0$. Hence $\gamma \in \cup V_0$. Similarly $\gamma' \in \cup V_0$. Thus by the Lebesgue lemma, $\gamma, \gamma' \in (a_k, b_k)$ for some k. By (10.7) we have

$$(c_k, d_k) \in g(\gamma) \cap g(\gamma'),$$

and thus $|g(\gamma) - g(\gamma')| < \varepsilon$. This proves that g is uniformly continuous. □

By an analogous argument as above: suppose that $G : \mathcal{I}(\alpha, \beta) \to \mathcal{X}$ is continuous, and U is a covering of the space $\mathcal{X}$. Then there is a finite cover $\{(a_1, b_1), \ldots, (a_n, b_n)\}$

of the interval, and a finite list of neighbourhoods $z_1, \ldots, z_n$ in U, with the property that for each point α in the interval (a_i, b_i), the function value $g(\alpha)$ is in the neighbourhood z_i. This is an often used device in homotopy theory.

Since the formal real topology $\mathcal{R}$ is regular, we have as a special case of Theorem 4.3 in [10].

Lemma 10.11 *Any two continuous mappings $F, G : \mathcal{A} \to \mathcal{R}$ with $F \subseteq G$ are equal.*

□

We identify $\mathbb{R}$ and $\mathrm{Pt}(\mathcal{R})$ and establish the following bijective correspondence of Bishop continuous functions on $\mathbb{R}$ and continuous mappings on $\mathcal{R}$ in the formal sense. Consider a continuous mapping $F : \mathcal{R} \to \mathcal{R}$. Then the composition with $E_{\alpha,\beta} : \mathcal{I}(\alpha, \beta) \to \mathcal{R}$ is continuous, and so

$$\mathrm{Pt}(F \circ E_{\alpha,\beta}) = \mathrm{Pt}(F) \circ \mathrm{Pt}(E_{\alpha,\beta}),$$

is uniformly continuous by Theorem 10.10. Hence $f = \mathrm{Pt}(F)$ is uniformly continuous on every compact interval, i.e. continuous in the sense of Bishop. Moreover, $A_f \subseteq F$ by (10.1). Thus $A_f = F$, by Lemma 10.11 and Theorem 10.5, and we are back to the continuous mapping. Conversely, if $f : \mathbb{R} \to \mathbb{R}$ is Bishop continuous, then by Theorem 10.5, $A_f : \mathcal{R} \to \mathcal{R}$ is a continuous mapping, and moreover $\mathrm{Pt}(A_f) = f$.

10.5 Open Subspaces and the Reciprocal Map

We show that the reciprocal map is a continuous mapping, so that any composition with it is again continuous.

Consider an arbitrary formal topology $X = (X, \leq, \lhd)$ and a set of neighbourhoods $G \subseteq X$ which is *downwards closed,* i.e. $G_{\leq} = G$. Define the *open subspace topology* $X^G = (X, \leq', \lhd')$ by letting

$$a \lhd' U \Longleftrightarrow_{\mathrm{def}} a_{\leq} \cap G \lhd U$$

and $a \leq' b$ iff $a \lhd' \{b\}$. These relations can be seen to extend $\lhd$ and $\leq$, respectively.

Theorem 10.12 *For a formal topology X and a downwards closed set of basic neighbourhoods G,*

(a) X^G is a formal topology,

(b) $\alpha \in \mathrm{Pt}(X^G)$ iff $\alpha \cap G$ is inhabited and $\alpha \in \mathrm{Pt}(X)$. □

Consider now the formal topology of real numbers $\mathcal{R} = (\mathcal{R}, \leq, \lhd)$. The set of neighbourhoods bounded away from 0

$$G = \{(a, b) \in \mathcal{R} : 0 < a \text{ or } b < 0\}$$

is downwards closed. $\mathcal{R}^G$ is the formal topology of *off-zero real numbers*. For $u = (a, b) \in G$ define the *reciprocal interval*

$$u^{-1} = (b^{-1}, a^{-1}).$$

Note that the operation is monotone with respect to $\leq$. The *reciprocal* is a continuous mapping I from $\mathcal{R}^G$ to $\mathcal{R}$. It is defined by

$$u\, I\, v \Longleftrightarrow u \in G \;\&\; u^{-1} \subseteq v.$$

A key lemma in the proof that I is indeed continuous is the following:

Lemma 10.13 *If $u \lhd U$, then $v \lhd' I^{-1}U$, for all $v \in G$ with $v\, I\, u$.*

Proof. By induction on $\lhd$. □

Then it is straightforward to check that for all $x \in \mathbb{R}$ apart from 0,

$$\mathrm{Pt}(I)(\overline{x}) = \overline{x^{-1}},$$

which proves I to represent the reciprocal function.

Acknowledgements

The present chapter is an elaboration of some statements of the author during the panel discussion of the workshop *From Sets and Types to Topology and Analysis* in Venice, May 12–16, 2003. I am grateful to Peter Schuster for pointing out some difficulties with the reciprocal map in other approaches of topology.

The author is supported by a grant from the Swedish Research Council (VR).

References

1. Waaldijk, F. (1998). Modern Intuitionistic Topology. PhD Thesis, Nijmegen.
2. Schuster, P. (2003). *What is continuity, constructively?* Preprint.
3. Sigstam, I. (1995). Formal spaces and their effective presentations. *Arch. Math. Logic*, **34**, 211–246.
4. Johnstone, P. T. (1982). *Stone Spaces.* Cambridge University Press.
5. Martin-Löf, P. (1970). *Notes on Constructive Mathematics.* Almqvist and Wiksell, Stockholm.
6. Sambin, G. (1987). Intuitionistic formal spaces—a first communication. In: *Mathematical Logic and its Applications,* (ed. D. Skordev), pp. 187–204. Plenum Press.
7. Negri, S. and Soravia, D. (1999). The continuum as a formal space. *Arch. Math. Logic,* **38**, (1999), no. 7, 423–447.
8. Bishop, E. and Bridges, D. S. (1985). *Constructive Analysis.* Springer.
9. Cederquist, J. and Negri, S. (1995). A constructive proof of the Heine-Borel covering theorem for formal reals. *Types for proofs and programs (Torino, 1995),* pp. 62–75, Lecture Notes in Comput. Sci. **1158**, Springer, Berlin, 1996.

10. Palmgren, E. Predicativity problems in point-free topology, In: *Proceedings of the Annual European Summer Meeting of the Association for Symbol Logic*, (eds V. Stoltenberg-Hansen and J. Väänänen), Helsinki, Finland, 2003, Lecture Notes in Logic 24, ASL (to appear).
11. Coquand, T. (1996). Formal topology with posets. Preprint.
12. Fourman, M. P. and Grayson, R. J. (1982). Formal spaces. In: *The L.E.J. Brouwer Centenary Symposium* (eds A. S. Troelstra, and D. van Dalen), pp. 107–122. North-Holland.
13. Palmgren, E. (2002). Maximal and partial points in formal topology. *Annals of Pure and Applied Logic,* to appear.
14. Palmgren, E. (2004). Regular universes and formal spaces. *Annals of Pure and Applied Logic,* to appear.
15. Sambin, G. (2003). Some points in formal topology. *Theoretical Computer Science,* **305**, 347–408.

11

SEPARATION PROPERTIES IN CONSTRUCTIVE TOPOLOGY

PETER ACZEL AND CHRISTOPHER FOX

Abstract

We formulate the notion of a constructive topological space in the constructive set theory CZF. For each $i = 0, 1, 2, 3$ we formulate three classically equivalent constructive versions of the T_i separation properties; some implications between these are proved, and counterexamples are given to the converse implications. Using a notion of upper real we introduce a variant notion of metric space and examine its relationship with the various separation axioms. Motivated by ideas of Bridges and Vita we also consider some separation properties relative to an inequality.

The point-free notion of a formal topology has been developed in the setting of constructive type theory. We introduce this notion in our constructive set theory setting along with a notion of sober constructive topological space and observe that the formal points of a formal topology form a sober constructive topological space. We also introduce a new, constructively weaker, notion of sobriety based on the work of Sambin et al. on the basic picture. In contrast to the standard definition of sobriety, it can be shown constructively that every T_2 space is weakly sober.

11.1 Introduction

Until the 1970s there was only a limited focus on the general notion of a topological space in constructive mathematics. An exception was the work of Anne Troelstra on intuitionistic general topology, starting with his thesis [9]. Most attention was paid to metric space notions both in intuitionistic analysis and in Bishop style constructive analysis. But in later years, because of the development of topos theory, the study of sheaf models and work on point-free topology, including work on formal topology its more predicative version, the notions of general topology in constructive mathematics have received more attention. The papers [5, 6] built on Troelstra's work to study general topological notions in sheaf models. In point-free topology the formal points of a locale form a (sober) topological space and it can be of interest to investigate how the properties of such a space of points relate to the properties of the locale it comes from. In formal topology the formal points of a formal topology still generally form a 'topological space' in some sense. But care is needed because in the more predicative context of formal topology the points do not always form a set. The limited aim of this chapter is to present a fairly systematic survey of the main separation properties that topological spaces can have in constructive mathematics. These are perhaps the first kinds of properties to consider when moving from the study of metric spaces to the study of general topological spaces.

Because we want to relate our work to work in formal topology, but also be compatible with work in sheaf models we choose to present our definitions and results in the setting of constructive set theory. More specifically the work can be formalized in the formal system CZF of constructive ZF, see [1], but it should not be necessary for the reader to have a precise working knowledge of the axioms. It should suffice for the reader to be aware that CZF is a subsystem of ZF that uses intuitionistic rather than classical logic and does not have the powerset axiom and does not have the full separation scheme, but only the restricted separation scheme. This is the version of the scheme which only entails that $\{x \in a \mid \phi\}$ is a set when a is a set and ϕ is a restricted formula, i.e. a formula in which all quantifiers are restricted to sets. In CZF there is the axiom scheme of subset collection whose main consequence is Myhill's exponentiation axiom, which expresses that the class of all functions $f : a \to b$ from a set a to a set b form a set b^a.

We use standard set theoretic class terminology. Given a class X we write $Pow(X)$ for the *powerclass of* X; i.e. the class of all subsets of the class X. Even when X is a set we are not able to take $Pow(X)$ to be a set. In fact the Powerset axiom is a consequence in CZF of the assumption that just the powerclass of a singleton set is a set. By adding the Powerset axiom and the full separation scheme to CZF we obtain the much stronger theory IZF of Intuitionistic ZF. This theory has the same proof theoretic strength as ZF.

Some useful classically true statements that cannot be proved even in IZF are REM, $wREM$ and LPO. REM is the principle of restricted excluded middle; i.e. the scheme that $\phi \vee \neg\phi$ for all restricted formulae. $wREM$ is the principle of weak restricted excluded middle; i.e. the scheme that $\neg\neg\phi \vee \neg\phi$ for all restricted formulae. LPO is the Limited Principle of Omniscience. These principles may be expressed as the following axioms.

REM: $\forall p \in Pow(\{0\})[p = 1 \vee \neg(p = 1)]$.
$wREM$: $\forall p \in Pow(\{0\})[\neg\neg(p = 1) \vee \neg(p = 1)]$.
LPO: $(\forall f \in 2^{\mathbb{N}})[(\exists n \in \mathbb{N})(f(n) = 1) \vee (\forall n \in \mathbb{N})(f(n) = 0)]$.

In Sections 11.2 and 11.3 we introduce the notions of constructive topological space (ct-space for short) and formal topology that we will use and give the construction of the ct-space of formal points of a formal topology. In Section 11.4 we introduce three versions of the classical T_i separation properties and develop the relationships between them. In Section 11.5 we relate these notions to some notions developed in [3]. In Section 11.6 we consider a generalised notion of metric space. In section 11.7 we show that the ct-space of formal points of a regular formal topology is $T_3^\sharp$. We end in Section 11.8 by considering the notion of a sober ct-space.

11.2 Constructive Topological Spaces (ct-spaces)

We formulate a notion of topological space, the notion of constructive topological space (ct-space) below, that is general enough to include the spaces of formal points of a formal topology, but restricted enough so that in IZF we get essentially the usual notion.

In order to be general enough we have to allow the points to form a class that might not be a set. In order to be restricted enough we assume certain smallness conditions. In particular we assume as part of the structure a set indexed base for the opens.

A *constructive topological space (*ct*-space)* consists of a class X of the points of the space together with a set S and a class $\Vdash\ \subseteq X \times S$ such that, letting $B_a = \{x \in X \mid x \Vdash a\}$ for each $a \in S$, $B_U = \bigcup_{a \in U} B_a$ for each $U \in Pow(S)$ and $\alpha_x = \{a \in S \mid x \Vdash a\}$ for each $x \in X$, the following conditions hold.

CS1 $X = B_S$,

CS2 If $a_1, a_2 \in S$ then $B_{a_1} \cap B_{a_2} \subseteq B_U$, where $U = \{a \in S \mid B_a \subseteq B_{a_1} \cap B_{a_2}\}$.

CS3 For each $x \in X$ the classes α_x and $\{y \in X \mid \alpha_y = \alpha_x\}$ are sets.

In a ct-space a class Y of points is *open* if $Y = B_U$ for some subclass U of S. So in a ct-space it is the open classes that form a 'topology' in something like the standard sense. But as part of the structure of a ct-space there is the base of open classes, B_a for $a \in S$, indexed by the set S.

A ct-space X is *small* if its class of points is a set. Then $\Vdash = \bigcup_{x \in X}\{x\} \times \alpha_x$ is a set so that each B_a is also a set. So a small ct-space is essentially just a topological space in the usual sense, except that the open sets form a class and, as part of the structure, there is a set-indexed base for the topology.

Proposition 11.1 *[IZF] Every* ct*-space is small.*

Proof. Given a ct-space, in IZF $Pow(S)$ is a set by the Powerset Axiom and so

$$Q = \{\ U \in Pow(S) \ \mid\ (\exists x \in X)(U = \alpha_x)\ \}$$

is a set by the full Separation Scheme. For each $U \in Q$ let $f(U) = \{y \in X \mid \alpha_y = U\}$. This is a set because $U = \alpha_x$ for some $x \in X$ and so $f(U) = \{y \in X \mid \alpha_y = \alpha_x\}$. Now observe that $X = \bigcup_{U \in Q} f(U)$, which is a set by the Replacement scheme and the Union axiom. □

We call a class F of points *closed* if $cl(F) \subseteq F$, where $cl(F)$ is the class of limit points of F. Here a *limit point* of F is a point x such that for every $a \in \alpha_x$ the class B_a has a point in common with F. Note that F is closed iff $F = \mathsf{rest}\ U$ for some subclass U of S, where $\mathsf{rest}\ U = \{y \in X \mid \alpha_y \subseteq U\}$. There is another notion that is also of interest. Call F *strongly closed* if it is the complement of an open class. Any strongly closed class is closed.

Proposition 11.2 *F is strongly closed iff $wcl(F) \subseteq F$, where $wcl(F)$ is the class of weak limit points of F. Here a point x is a weak limit point of F if for every $a \in \alpha_x$ the class B_a is not disjoint from F.*

Proof. Let $U_F = \bigcap_{y \in F}(S \backslash \alpha_y)$. Then $F \cap B_{U_F} = \emptyset$ and it is easy to check that x is a weak limit point of F iff $x \notin B_{U_F}$. It follows that every weak limit point of F is in F iff F is the complement of B_{U_F}. Thus the implication from right to left is proved. For the reverse implication assume that F is the complement of B_U. It follows that $U \subseteq U_F$ and so if x is a weak limit point of F then $x \notin B_U$ and so $x \in F$. Thus F is strongly closed. □

11.3 The Space of Formal Points of a Formal Topology

11.3.1 *Local collections*

We describe a simple general method for constructing **ct**-spaces that includes as a special case the construction of the **ct**-space of formal points of a formal topology. We first give a convenient representation of T_0 **ct**-spaces up to isomorphism. As defined in the next section, a **ct**-space is T_0 if, for all points x, y

$$[\ \alpha_x = \alpha_y \ \Rightarrow\ x = y\ ].$$

A class X of inhabited subsets of a set S is called a *local collection* of subsets of S if, for all $x \in X$ and all $a_1, a_2 \in x$ there is $a \in x$ such that

$$\forall y \in X[\ a \in y \ \Rightarrow\ a_1, a_2 \in y\ ].$$

Observe that if X is a local collection of subsets of a set S then it is the class of points of a T_0 **ct**-space where

$$\Vdash\ =\ \{\ (x, a) \in X \times S\ \mid\ a \in x\ \}.$$

Conversely every **ct**-space X determines a local collection $X' = \{\alpha_x \mid x \in X\}$. Moreover when X is T_0 then the map $x \mapsto \alpha_x$ is an isomorphism $X \cong X'$.

We now describe the general method to construct local collections that generalizes the construction of a **ct**-space from a formal topology. Let R be any binary relation on the set S. Call an inhabited subset x of S an *R-point* if the following hold.

1. If $a_1, a_2 \in x$ then for some $a \in x$, aRa_1 and aRa_2.
2. If bRa then $[\ b \in x\ \Rightarrow\ a \in x\ ]$.

Any class of R-points forms a local collection. Conversely each local collection X is a class of R_X-points, where

$$R_X = \{\ (a, b) \in S \times S\ \mid\ \forall x \in X[\ a \in x\ \Rightarrow\ b \in x\ ]\ \}.$$

Note that R_X is always reflexive, transitive.

11.3.2 *The notion of a formal topology*

We will use the following version of the notion of a formal topology. A *formal topology* on a set S is an operator $\mathcal{A} : Pow(S) \to Pow(S)$ such that for all sets $U, V \subseteq S$,

1. $U \subseteq \mathcal{A}U$,
2. $U \subseteq \mathcal{A}V \Rightarrow \mathcal{A}U \subseteq \mathcal{A}V$,
3. $\mathcal{A}U \cap \mathcal{A}V \subseteq \mathcal{A}(U \downarrow \cap V \downarrow)$,

where $a \leq b \Leftrightarrow a \in \mathcal{A}\{b\}$ for $a, b \in S$ and $U \downarrow = \{a \in S \mid \exists b \in U\ a \leq b\}$ for sets $U \subseteq S$. Usually these axioms are given in terms of the relation $\triangleleft$ where

$$a \triangleleft U \Leftrightarrow a \in \mathcal{A}U$$

for $a \in S$ and $U \in Pow(S)$. Note that given a small ct-space we may obtain a formal topology on S by defining, for each set $U \in Pow(S)$,

$$\mathcal{A}U = \{a \in S \mid B_a \subseteq B_U\}.$$

We need the ct-space to be small in order to ensure that $\mathcal{A}U$ is a set.

11.3.3 *The* ct-*space of formal points*

Given a formal topology $\mathcal{A}$ on S a *formal point* of the formal topology is a $\leq$-point x such that for all subsets U of S

$$x \between \mathcal{A}U \Rightarrow x \between U,$$

where, for classes A, B we write $A \between B$ if A, B have a common element. By definition these form a class of $\leq$-points and hence a local collection and so form a T_0 ct-space.

11.4 Constructive Separation Properties

In this section we start by giving the constructive formulations of the T_i separation properties, for $i = 0, 1, 2, 3$ used in [5] and then go on to give the classically equivalent stronger versions $T_i^\sharp$, also considered there. At the same time we also introduce a new family of intermediate separation properties, T_i^+ that were inspired by a study of [3].

11.4.1 *The separation properties for $i = 0, 1, 2$*

The standard classical formulations of these separation properties is usually in terms of separating via open sets a pair of distinct points. It seems that a correct constructive formulation of these separation properties instead expresses that inseparable points are equal using the following notions of inseparability. For points x, y of a ct-space X we define $x \sim_i y$ as follows.

- $x \sim_0 y$ iff $\alpha_x = \alpha_y$
- $x \sim_1 y$ iff $\alpha_x \subseteq \alpha_y$
- $x \sim_2 y$ iff $(\forall a \in \alpha_x)(\forall b \in \alpha_y)(B_a \between B_b)$.

We now state our first family of separation properties

- For $i = 0, 1, 2$ the ct-space X is T_i if, for all points x, y

$$(x \sim_i y) \Rightarrow (x = y).$$

The original classical formulation of T_i for $i = 0, 1, 2$ has the form that for all points x, y

$$\neg(x = y) \Rightarrow (x \not\sim_i y),$$

where

- $x \neq_0 y$ iff $(\exists a \in \alpha_x)(a \notin \alpha_y) \vee (\exists b \in \alpha_y)(b \notin \alpha_x)$
- $x \neq_1 y$ iff $(\exists a \in \alpha_x)(a \notin \alpha_y)$
- $x \neq_2 y$ iff $(\exists a \in \alpha_x)(\exists b \in \alpha_y)[B_a \cap B_b = \emptyset]$.

The classical T_i do not seem to be of constructive interest. Instead we consider the following classically equivalent notions.

For $i = 0, 1, 2$, a ct-space X is T_i^+ if, for all points x, y

$$\neg(x \neq_i y) \Rightarrow (x = y).$$

The following facts may be worth noting.

Proposition 11.3

1. *A* ct-*space is classical* T_1 *iff the complement of every point singleton is open.*
2. *A* ct-*space is* T_1 *iff every point singleton is closed.*
3. *A* ct-*space is* T_1^+ *iff every point singleton is strongly closed.*

We get a stronger family of separation notions as follows. For $i = 0, 1, 2$ a ct-space is a $T_i^\sharp$ ct-*space* if it is T_0 and, for all points x, y

$$(\forall a \in \alpha_x)[\ y \neq_i x \ \vee \ a \in \alpha_y\ ].$$

11.4.2 *Notions of regular space*

We formulate notions $R, R^+, R^\sharp$ corresponding to the classical notion of a regular space. A standard formulation of the classical notion states that a topological space is classical regular if whenever given a closed set and a point not in it there are two disjoint open sets the first including the closed set and the second containing the point. Let X be a ct-space. Our constructive versions replace the quantification over closed sets not containing the point by a quantification over (basic) open neighborhoods of the point.

- The space X is R if

$$\forall x \in X \forall a \in \alpha_x \exists b \in \alpha_x [cl(B_b) \subseteq B_a].$$

- The space X is R^+ if

$$\forall x \in X \forall a \in \alpha_x \exists b \in \alpha_x [wcl(B_b) \subseteq B_a].$$

- The space X is a $R^\sharp$ if

$$\forall x \in X \forall a \in \alpha_x \exists b \in \alpha_x [X \subseteq B_b^* \cup B_a],$$

where, for any class Z of points, Z^* is the largest open class disjoint from Z; i.e. it is B_U where $U = \{c \in S \mid B_c \cap Z = \emptyset\}$.

The space X is a T_3 *(*T_3^+ *or* $T_3^\sharp$*)* ct-*space* if it is T_0 and R (R^+ or $R^\sharp$ respectively).

11.4.3 *Some implications*

Clearly for all $x, y \in X$, $(x \sim_0 y) \Rightarrow (x \sim_1 y) \Rightarrow (x \sim_2 y)$, so $T_2 \Rightarrow T_1 \Rightarrow T_0$. Also $(x \neq_2 y) \Rightarrow (x \neq_1 y) \Rightarrow (x \neq_0 y)$, so $T_2^+ \Rightarrow T_1^+ \Rightarrow T_0^+$ and $T_2^\sharp \Rightarrow T_1^\sharp \Rightarrow T_0^\sharp$.

The implications $T_i^+ \Rightarrow T_i$ are easy to prove for $i = 0, 1, 2, 3$, as are the implications $T_i^\sharp \Rightarrow T_i^+$ for $i = 0, 2, 3$. $T_1^\sharp \Rightarrow T_1^+$ requires a little more work because the relation $\neq_1$ is not symmetric.

Proposition 11.4 *If X is $T_1^\sharp$ then X is T_1^+.*

Proof. Since X is $T_1^\sharp$, it must be $T_0^\sharp$ and hence T_0^+. We will show that $(\forall x, y \in X)[x \neq_0 y \rightarrow x \neq_1 y]$ so that X is T_1^+.

Suppose that $x \neq_0 y$. If $a \in \alpha_x$ and $a \notin \alpha_y$ then $x \neq_1 y$, so assume instead that there is an $a \in \alpha_y$ such that $a \notin \alpha_x$. Since X is $T_1^\sharp$ either $x \neq_1 y$ or $a \in \alpha_y$, but the latter is false so $x \neq_1 y$. □

Proposition 11.5 *If X is T_3 (T_3^+ or $T_3^\sharp$) then X is T_2 (T_2^+ or $T_2^\sharp$ respectively).*

Proof. If X is T_0 then observe the following characterizations.

- X is T_2 iff $(\forall x \in X)(\forall a \in \alpha_x)[\{y \in X \mid x \sim_2 y\} \subseteq B_a]$.
- X is T_2^+ iff $(\forall x \in X)(\forall a \in \alpha_x)[\{y \in X \mid \neg(x \neq_2 y)\} \subseteq B_a]$.
- X is $T_2^\sharp$ iff $(\forall x \in X)(\forall a \in \alpha_x)[X \subseteq \{y \in X \mid x \neq_2 y\} \cup B_a]$.

In view of the definitions of T_3, T_3^+ and $T_3^\sharp$ it suffices to show that if $x \in X$ and $a, b \in \alpha_x$ then the following hold.

- $\{y \in X \mid x \sim_2 y\} \subseteq cl(B_b)$,
- $\{y \in X \mid \neg(x \neq_2 y)\} \subseteq wcl(B_b)$,
- $B_b^* \subseteq \{y \in X \mid x \neq_2 y\}$.

These are straightforward to check. □

11.4.4 *Summary*

We have the following implications between the separation axioms:

$$\begin{array}{ccccccc} T_3^\sharp & \Longrightarrow & T_2^\sharp & \Longrightarrow & T_1^\sharp & \Longrightarrow & T_0^\sharp \\ \Downarrow & & \Downarrow & & \Downarrow & & \Downarrow \\ T_3^+ & \Longrightarrow & T_2^+ & \Longrightarrow & T_1^+ & \Longrightarrow & T_0^+ \\ \Downarrow & & \Downarrow & & \Downarrow & & \Downarrow \\ T_3 & \Longrightarrow & T_2 & \Longrightarrow & T_1 & \Longrightarrow & T_0 \end{array}$$

In the next subsection we show that none of the converses of these implications hold constructively.

11.4.5 *Some counterexamples*

We first show how some of the standard counterexamples from [8] can be constructed in CZF, and then consider the discrete topology on a set, which can provide weak counterexamples to the implications $T_i \Rightarrow T_i^+$ and $T_i^+ \Rightarrow T_i^\sharp$.

Example 11.6 Odd–even topology on the natural numbers ($R^\sharp$ but not T_0).

Let $X = \mathbb{N}$, $S = \mathbb{N}$ and $B_a = \{2a, 2a+1\}$ for each $a \in S$. The sets $(B_a)_{a \in S}$ form a partition of X, and it follows that X is a small **ct**-space.

X is $R^\sharp$: Let $x \in X$ and $a \in \alpha_x$. If we take $b = a$ then $B_a \cup B_b^* = B_a \cup (X \backslash B_a) = X$ since equality on $\mathbb{N}$ is decidable.

X is not T_0: $\alpha_0 = \alpha_1$ but $\neg(0 = 1)$, so X is not T_0. □

Example 11.7 Two-point[1] Sierpinski space ($T_0^\sharp$ but not T_1)

Let $X = \{0, 1\}$, $S = \{0, 1\}$, $B_0 = X$ and $B_1 = \{1\}$. Then X is a small **ct**-space.

X is T_0: Suppose that $\alpha_x = \alpha_y$. Then $x \in B_1$ iff $y \in B_1$, and hence $x = y$ by the decidability of equality on $\{0, 1\}$.

X is $T_0^\sharp$: Suppose that $a \in \alpha_x$ and $y \in X$. Either $a \in \alpha_y$ or $\neg(a \in \alpha_y)$. If $\neg(a \in \alpha_y)$ then $y \neq_0 x$.

X is not T_1: $\alpha_0 \subseteq \alpha_1$ but $\neg(0 = 1)$. □

Example 11.8 'Cofinite' topology on the natural numbers ($T_1^\sharp$ but not T_2).

Let $X = \mathbb{N}$, $S = \mathbb{N} \times \mathbb{N}$ and $B_{(a,b)} = \{x \in X \mid (x = a) \vee (b \leq x)\}$. Using the decidability of $=$ and $\leq$ on $\mathbb{N}$, it is straightforward to check that the intersection of two basic open sets is open, so X is a small **ct**-space. Note that although this space is classically the same as the finite complement topology, not every open class will be cofinite constructively.

X is T_0: Suppose that $x, y \in X$ and $\alpha_x = \alpha_y$. Let $a = (x, \max\{x, y\} + 1)$. Then $x \in B_a$, and hence $y \in B_a$, so $x = y$.

X is $T_1^\sharp$: Now suppose that $(a, b) \in \alpha_x$ and $y \in X$. Either $x = y$ or $\neg(x = y)$. If $x = y$ then $(a, b) \in \alpha_y$, and if $\neg(x = y)$ then $(y, \max\{x, y\} + 1) \in \alpha_y \backslash \alpha_x$ so $x \neq_1 y$.

X is not T_2: Any two basic open sets have an inhabited intersection. □

Example 11.9 Irrational slope topology ($T_2^\sharp$ but not R).

In this example we will use the Dedekind reals, as defined in [1]. Given two real numbers x and y, write $x \neq y$ for $(x < y) \vee (y < x)$ (x is *apart* from y), and as in [2] we define an *irrational number* to be a real number which is apart from every rational number.

---

[1] This is not the same as the usual constructive version of the Sierpinski space, whose points are the elements of the class $Pow(0)$. The powerclass Sierpinski space is not T_0^+ constructively.

Choose an arbitrary irrational number θ (for example $\sqrt{2}$). Let $X = \mathbb{Q} \times \mathbb{Q}^{\geq 0}$, $S = \mathbb{Q} \times \mathbb{Q}^{\geq 0} \times \mathbb{Q}^{>0}$ and

$$B_{(a,b,r)} = \{(a, b)\} \cup (N_r(a - b\theta) \times \{0\}) \cup (N_r(a + b\theta) \times \{0\})$$

where $N_r(\zeta) = \{z \in \mathbb{Q} \mid |z - \zeta| < r\}$.

The intersection of two basic open sets is open:
Suppose that $(x, y) \in B_{(a,b,r)} \cap B_{(c,d,s)}$. If $y > 0$ then $(x, y) = (a, b) = (c, d)$, so $(x, y) \in B_{(x,y,\min\{r,s\})}$. If $y = 0$ then there is a $t \in \mathbb{Q}^{>0}$ such that $N_t(x) \times \{0\} \subseteq B_{(a,b,r)} \cap B_{(c,d,s)}$, and hence $(x, y) \in B_{(t,0)} \subseteq B_{(a,b,r)} \cap B_{(c,d,s)}$.

So X is a small ct-space.

X is T_0: Suppose that $\alpha_{(x,y)} = \alpha_{(x',y')}$. If $y > 0$ and $y' > 0$ then $(x, y, 1) \in \alpha_{(x,y)} = \alpha_{(x',y')}$, so $(x', y') = (x, y)$. If $y = 0$, then $(x, y, r) \in \alpha_{(x',y')}$ for all $r \in \mathbb{Q}^{>0}$, so $y' = 0$ and $|x - x'| < r$ for all $r \in \mathbb{Q}^{>0}$, and so $x = x'$. Similarly if $y' = 0$ then $y = 0$ and $x = x'$.

X is $T_2^\sharp$: Suppose that $(a, b, r) \in \alpha_{(x,y)}$ and $(x', y') \in S$. Then $((x, y) = (x', y'))$ is decidable. If $(x, y) = (x', y')$, then $(x, y, r) \in \alpha_{(x',y')}$. Suppose that $\neg((x, y) = (x', y'))$. Since θ is irrational, we have $\theta \neq \frac{x-x'}{y'-y}$, so $(y' - y)\theta \neq x - x'$, and hence $x + y\theta \neq x' + y'\theta$. Similarly it can be shown that $x + y\theta \neq x' - y'\theta$, $x - y\theta \neq x' + y'\theta$ and $x - y\theta \neq x' - y'\theta$. So we can choose an $r \in \mathbb{Q}^{>0}$ such that $N_r(x - y\theta) \cup N_r(x + y\theta)$ is disjoint from $N_r(x' - y'\theta) \cup N_r(s' + y'\theta)$. $(x, y) \in B_{(x,y,r)}$, $(x', y') \in B_{(x',y',r)}$ and $B_{(x,y,r)} \cap B_{(x',y',r)} = \emptyset$, so $(x', y') \neq_2 (x, y)$.

***X* is not *R*:** Suppose that X is R. Let $x = (0, 1)$ and $a = (0, 1, 1) \in \alpha_x$. There must be a $b \in \alpha_x$ such that $cl(B_b) \subseteq B_a$. $b = (0, 1, r)$ for some $r \in \mathbb{Q}^{>0}$.
Let $y = (r/2, 1)$. Then for all $\varepsilon \in \mathbb{Q}^{>0}$, $B_{(r/2,1,\varepsilon)} \cap B_b$ is inhabited, so $y \in cl(B_b) \subseteq B_a$. But $(r/2, 1) \notin B_{(0,1,1)}$, so X cannot be R. □

The next example can be used to show that the implications $T_i \Rightarrow T_i^+ \Rightarrow T_i^\sharp$ cannot be proved in CZF.

Example 11.10 'Discrete' topology[2] on X.

Given a set X, let $S = X$ and $B_a = \{a\}$ for each $a \in S$.

Lemma 11.11 *Let X be a set with the discrete topology, as defined above. Then for $i = 0, 1, 2, 3$:*

(i) *X is T_i*
(ii) *(X is T_i^+) $\Leftrightarrow$ (X is R^+) $\Leftrightarrow$* $(\forall x, y \in X)[\neg\neg(x = y) \to (x = y)]$
(iii) *(X is $T_i^\sharp$) $\Leftrightarrow$ (X is $R^\sharp$) $\Leftrightarrow$* $(\forall x, y \in X)[(x = y) \vee \neg(x = y)]$.

[2]Although this is defined in the same way as the classical discrete topology, the word 'discrete' seems slightly inappropriate as the set X is not necessarily discrete.

Proof. It suffices to prove the proposition for T_0, T_0^+, $T_0^\sharp$ and for R, R^+, $R^\sharp$. These cases follow immediately from the definitions, observing that

- $\alpha_x = \{x\}$
- $x \neq_0 y$ iff $\neg(x = y)$
- $cl(\{x\}) = \{x\}$. □

Corollary 11.12

(i) *(Every T_i space is T_i^+) $\Rightarrow$ REM for $i = 0, 1, 2, 3$*
(ii) *(Every R space is R^+) $\Rightarrow$ REM*
(iii) *(Every T_i^+ space is $T_i^\sharp$) $\Rightarrow$ $wREM$ for $i = 0, 1, 2, 3$*
(iv) *(Every R^+ space is $R^\sharp$) $\Rightarrow$ $wREM$.*

Proof.

(i) Given a restricted formula ϕ, let $p = \{0 \mid \phi \vee \neg\phi\}$, $1 = \{0\}$ and $X = \{1, p\}$ with the discrete topology. X is T_i. If X is T_i^+ then $\neg\neg(1 = p) \to 1 = p$, and so $\neg\neg(\phi \vee \neg\phi) \to (\phi \vee \neg\phi)$. But $\neg\neg(\phi \vee \neg\phi)$ is true in CZF, so $\phi \vee \neg\phi$.

(ii) Identical to (i), replacing T_i with R and T_i^+ with R^+.

(iii) Given a restricted formula ϕ, let $p = \{0 \mid \neg\phi\}$, $1 = \{0\}$ and $X = \{1, p\}$ with the discrete topology. $\neg\neg\neg\phi \to \neg\phi$, so for all $x, y \in X$ we have we have $\neg\neg(x = y) \to (x = y)$. So X is T_i^+. If X is $T_i^\sharp$ then $(1 = p) \vee \neg(1 = p)$, so $\neg\phi \vee \neg\neg\phi$.

(iv) Identical to (iii), replacing T_i^+ with R^+ and $T_i^\sharp$ with $R^\sharp$. □

11.5 Separation Relative to an Inequality

This section is based on ideas in [3], which assumes that a topological space comes with an inequality relation on it. We have reformulated the ideas to our context. We call a relation $\neq$ on a class X an *inequality* if, for all $x, y \in X$,

$$x = y \Rightarrow \neg(x \neq y)$$

and is *tight* if the reverse implication also holds.

Definition 11.13 *Let $\neq$ be an inequality on a* ct*-space X. For each $a \in S$ let*

$$B_a^{\neq} = \{y \in X \mid \forall z \in B_a \; z \neq y\}.$$

- *X is $T_1^{B\&V}$ wrt $\neq$ if, for all $x, y \in X$,*

$$x \neq y \Rightarrow \exists a \in \alpha_x \; y \in B_a^{\neq}.$$

- *X is $T_2^{B\&V}$ wrt $\neq$ if, for all $x, y \in X$,*

$$x \neq y \Rightarrow \exists a \in \alpha_x \exists b \in \alpha_y B_b \subseteq B_a^{\neq}.$$

- *X is topologically cotransitive wrt $\neq$ if, for all $x, y \in X$,*

$$\forall a \in \alpha_x [x \neq y \vee a \in \alpha_y].$$

- *X is locally decomposable wrt $\neq$ if, for all $x \in X$,*

$$\forall a \in \alpha_x \exists b \in \alpha_x [X \subseteq B_b^{\neq} \cup B_a].$$

Note that, for $i = 0, 1, 2$, the definition of a $T_i^{\sharp}$ ct-space can be formulated as follows. A ct-space is $T_i^{\sharp}$ if it is T_0 and topologically cotransitive wrt $\neq_i$.

Proposition 11.14 *Let X be a* ct*-space.*

1. *For $i = 1, 2$ the* ct*-space X is T_i^{+} iff X is $T_i^{B\&V}$ wrt some tight inequality.*
2. *If X is $R^{\sharp}$ then X is locally decomposable wrt $\neq_2$.*
3. *If X is T_1 and X is topologically cotransitive wrt an inequality $\neq$ then $\neq$ is tight.*

Proof. Let X be a ct-space.

1. For $i = 1, 2$ let X be $T_i^{B\&V}$ wrt a tight inequality $\neq$.
   Claim For $x, y \in X$, $x \neq y \Rightarrow x \neq_i y$.
   Proof. Note that $B_a^{\neq} \subseteq X \backslash B_a$.

$$\begin{aligned} i = 1 : x \neq y &\Rightarrow \exists a \in \alpha_x \; y \in B_a^{\neq} \\ &\Rightarrow \exists a \in \alpha_x \backslash \alpha_y \\ &\Rightarrow x \neq_1 y \\ i = 2 : x \neq y &\Rightarrow \exists a \in \alpha_x \exists b \in \alpha_y B_b \subseteq B_a^{\neq} \\ &\Rightarrow \exists a \in \alpha_x \exists b \in \alpha_y B_a \cap B_b = \emptyset \\ &\Rightarrow x \neq_2 y \end{aligned}$$

   By the claim

$$\begin{aligned} \neg(x \neq_i y) &\Rightarrow \neg(x \neq y) \\ &\Rightarrow x = y \end{aligned}$$

   as $\neq$ is tight. So X is T_i^{+}.
   For the converse let X be T_i^{+}. Then $\neq_i$ is tight and it suffices to show that X is $T_i^{B\&V}$ wrt $\neq_i$.
   $i = 1$: For $x, y \in X$

$$\begin{aligned} (x \neq_1 y) &\Rightarrow \exists a \in \alpha_x \backslash \alpha_y \\ &\Rightarrow \exists a \in \alpha_x \forall z \in B_a a \in \alpha_z \backslash \alpha_y \\ &\Rightarrow \exists a \in \alpha_x \forall z \in B_a z \neq_1 y \\ &\Rightarrow \exists a \in \alpha_x \; y \in B_a^{\neq_1} \end{aligned}$$

   So X is $T_1^{B\&V}$ wrt $\neq_1$.

$i = 2$**:** First observe that if $B_a \cap B_b = \emptyset$ then $x \in B_a$ & $y \in B_b \Rightarrow x \neq_2 y$ so that $B_b \subseteq B_a^{\neq}$. So

$$\begin{aligned} x \neq_2 y &\Rightarrow \exists a \in \alpha_x \exists b \in \alpha_y B_a \cap B_b = \emptyset \\ &\Rightarrow \exists a \in \alpha_x \exists b \in \alpha_y B_b \subseteq B_a^{\neq}; \end{aligned}$$

i.e. X is $T_2^{B\&V}$ wrt $\neq_2$.

2. It is enough to observe that $B_b^* \subseteq B_b^{\neq_2}$.
3. Let X be T_1 and topologically cotransitive wrt $\neq$. Then

$$\begin{aligned} \neg(x \neq y) &\Rightarrow \forall a \in \alpha_x \; a \in \alpha_y \\ &\Rightarrow \alpha_x \subseteq \alpha_y \\ &\Rightarrow x = y \end{aligned}$$

as X is T_1. Thus X is tight. □

11.6 $\mathbb{R}^u$-Metric Spaces

The usual definition of a metric space is a set X equipped with a function from $X \times X$ to the real numbers which satisfies certain axioms. For our purposes, we will allow the metric to take values in the class of *upper reals*, $\mathbb{R}^u$, defined below, to allow a wider class of spaces to be metrized. This version of the real numbers is used by Vickers in [11] in his generalized metric spaces, and it is also used in the constructive theory of metric formal topologies. Some other axioms that can be added to this definition are given in [10], and in particular we will consider the strongly monotone and Dedekind reals.

There are, however, some drawbacks in using $\mathbb{R}^u$ instead of $\mathbb{R}$. Since $\mathbb{R}^u$ is a proper class, it cannot itself be an $\mathbb{R}^u$-metric space by our definition, because we shall require the points to be a set. In fact when $\mathbb{R}^u$ is given the Euclidean 'metric', it does not even have a dense subset.

Definition 11.15 *An upper real is a subset R of $\mathbb{Q}^{>0}$ such that*

$$(\forall r \in \mathbb{Q}^{>0})(r \in R \leftrightarrow (\exists r' \in R)(r' < r)).$$

$\mathbb{R}^u$ denotes the class of upper reals. We say that $R \in \mathbb{R}^u$ is:

- *finite if* $(\exists r)(r \in R)$
- *strongly monotone if* $(r < s \;\&\; \neg\neg(r \in R)) \rightarrow s \in R$ *for all* $r, s \in \mathbb{Q}^{>0}$
- *located or Dedekind if* $r < s \rightarrow (r \notin R \vee s \in R)$ *for all* $r, s \in \mathbb{Q}^{>0}$

Note that every located upper real is strongly monotone.

Given $R, S \in \mathbb{R}^u$, define $R + S = \{r + s \mid r \in R \;\&\; s \in S\}$. If $r \in \mathbb{Q}^{\geq 0}$, then let $r^* = \{s \in \mathbb{Q}^{>0} \mid r < s\}$. Write $R \leq S$ iff $S \subseteq R$ and $R < S$ iff $(\exists r \in \mathbb{Q}^{>0})(R + r^* \leq S)$.

It is easy to show that $R < r^*$ iff $r \in R$. From now on we will write r instead of r^* provided the meaning is clear.

Definition 11.16 *An $\mathbb{R}^u$-metric space is a pair (X, d), consisting of a set X together with a function $d : X \times X \to \mathbb{R}^u$ such that for all $x, y, z \in X$:*

- $d(x, y) = 0 \Leftrightarrow x = y$
- $d(x, y) = d(y, x)$
- $d(x, y) + d(y, z) \leq d(x, z)$.

We say that the space is finitary if $d(x, y)$ is finite for all $x, y \in X$, strongly monotone if $d(x, y)$ is strongly monotone and Dedekind if $d(x, y)$ is Dedekind. A metric space is a finitary Dedekind $\mathbb{R}^u$-metric space.

If X is an $\mathbb{R}^u$-metric space, we can define a topology on X by taking $S = X \times \mathbb{Q}^{>0}$ and $B_{(x,\varepsilon)} = \{y \in X \mid d(x, y) < \varepsilon\}$.

Proposition 11.17 *Let X be an $\mathbb{R}^u$-metric space.*

(i) *X is T_3.*
(ii) *If X is strongly monotone then X is T_3^+.*
(iii) *If X is Dedekind then X is $T_3^\sharp$.*

Proof.

(i) First we have to show that X is T_0. Let $x, y \in X$ and suppose that $\alpha_x = \alpha_y$. It that $y \in B_{(x,\varepsilon)}$ for all $\varepsilon \in \mathbb{Q}^{>0}$. So $d(x, y) = 0$, and hence $x = y$.
Next we show that X is R. Suppose that $x \in X$ and $(y, \varepsilon) \in \alpha_x$. Then $(y, \varepsilon') \in \alpha_x$ for some $\varepsilon' < \varepsilon$. If $z \in cl(B_{(y,\varepsilon')})$ then $B_{(z,(\varepsilon-\varepsilon'))} \between B_{(y,\varepsilon')}$, and so by the triangle inequality $z \in B_{(y,\varepsilon)}$. So $cl(B_{(y,\varepsilon')}) \subseteq B_{(y,\varepsilon)}$.

(ii) X is T_0 by part (i), so it remains to prove that X is R^+. Given $x \in X$ and $(y, \varepsilon) \in \alpha_x$, there exist positive rationals $\varepsilon'' < \varepsilon' < \varepsilon$ such that $(y, \varepsilon'') \in \alpha_x$. Suppose that $z \in wcl(B_{(y,\varepsilon'')})$. Then $B_{(z,(\varepsilon'-\varepsilon''))} \cap B_{(y,\varepsilon'')} \neq \emptyset$, so by the triangle inequality $\neg\neg(d(y, z) < \varepsilon')$. Since X is strongly monotone $d(y, z) < \varepsilon$, so $z \in B_{(x,\varepsilon)}$.

(iii) It remains to prove that X is $R^\sharp$. Given $x \in X$ and $(y, \varepsilon) \in \alpha_x$, choose an $\varepsilon' < \varepsilon$ such that $(y, \varepsilon') \in \alpha_x$. For all $z \in X$, either $d(y, z) < \varepsilon$ or $d(y, z) > \varepsilon'$. In the first case $z \in B_{(y,\varepsilon)}$, and in the second $z \in B^*_{(y,\varepsilon')}$. □

Example 11.18 Discrete $\mathbb{R}^u$-metric on X.

Given a set X, we can define the discrete $\mathbb{R}^u$-metric on X by setting

$$d(x, y) = \{r \in \mathbb{Q}^{>0} \mid (x = y) \vee (r > 1)\}.$$

It is easily verified that $d(x, y)$ is a finite upper real, and that d is an $\mathbb{R}^u$-metric on X.

If $\varepsilon \leq 1$ then $B_{(x,\varepsilon)} = \{x\}$, and if $\varepsilon > 1$ then $B_{(x,\varepsilon)} = X$. So the topology induced by this $\mathbb{R}^u$-metric has the same open sets as the discrete topology on X. □

Proposition 11.19 *Let X be a set with the discrete $\mathbb{R}^u$-metric.*

(i) *X is strongly monotone iff $(\forall x, y \in X)(\neg\neg(x = y) \to x = y)$.*

(ii) *X is Dedekind iff $(\forall x, y \in X)(x = y \vee \neg(x = y))$.*

Proof.

(i) The implication from left to right follows from Lemma 11.11 and Proposition 11.17, so assume that $\neg\neg(x = y) \to x = y$ holds for all $x, y \in X$. Let $r, s \in \mathbb{Q}^{>0}$ with $r < s$, and suppose that $\neg\neg(r \in d(x, y))$. If $s > 1$ then $s \in d(x, y)$, so suppose that $s \leq 1$. Then $r < 1$, so $\neg\neg(x = y)$, hence $x = y$, so $s \in d(s, y)$. So X is strongly monotone.

(ii) Again the implication from left to right follows from Lemma 11.11 and Proposition 11.17, so assume that $=$ is decidable on X. Let $r, s \in \mathbb{Q}^{>0}$ with $r < s$, and let $x, y \in X$. If $s > 1$ then $s \in d(x, y)$, so suppose that $r < s \leq 1$. If $x = y$ then $s \in d(x, y)$, and if $\neg(x = y)$ then $r \notin d(x, y)$. So X is Dedekind. □

11.7 Regular Formal Topologies

Definition 11.20 *A regular formal topology is a formal topology such that $a \lhd W_a$ for all $a \in S$, where*

$$W_a = \{b \in S \mid (\forall c \in S)\ c \lhd \{a\} \cup b^*\}.$$

Here $b^* = \{c \in S \mid \forall d \in S\ [d \leq b, c \Rightarrow d \lhd \emptyset]\}$.

Theorem 11.21 *The* ct*-space of formal points of a regular formal topology is $T_3^\sharp$.*

Proof. Let X be the space of formal points of a formal topology on S. Then X is T_0. To show that X is $R^\sharp$, given $\alpha \in X$ and $a \in \alpha$ we want $b \in \alpha$ such that $X \subseteq B_b^* \cup B_a$. As $a \in \alpha$ and $a \lhd W_a$ there is $b \in \alpha$ such that $b \in W_a$. We show that $X \subseteq B_b^* \cup B_a$.

Let $\beta \in X$. As β is inhabited we may choose some $d \in \beta$. It follows that $d \lhd \{a\} \cup b^*$ so that, as $d \in \beta$, there is some $c \in \beta$ such that $c \in \{a\} \cup b^*$. So either $a = c \in \beta$ so that $\beta \in B_a$ or $c \in b^*$.

It only remains to show that, assuming $c \in b^*$, we have $\beta \in B_b^*$. If $\gamma \in B_c \cap B_b$ then $b, c \in \gamma$ so that there is $d \in \gamma$ such that $d \leq b, c$, and hence, as $c \in b^*$, $d \lhd \emptyset$ which contradicts that γ is a formal point. So $B_c \cap B_b = \emptyset$. As $\beta \in B_c$, $\beta \in B_b^*$.

□

11.8 Sober Spaces

The notion of a sober topological space comes from point-free topology. Classically a sober space is one that is isomorphic to the space of formal points of a locale. The notion is classically a separation property lying somewhere between classical T_0 and classical T_2. The following seems an appropriate constructive formulation.

Definition 11.22 *A* ct*-space is sober if every ideal point of X is equal to α_x for a unique point $x \in X$, where an ideal point of X is a subset α of S such that:*

1. $\exists a \in \alpha$
2. $(\forall a, b \in \alpha)(\exists c \in \alpha)[B_c \subseteq B_a \cap B_b]$
3. $(\forall a \in \alpha)(\forall U \in Pow(S))[B_a \subseteq B_U \to \alpha \between U]$.

Note that for every point x of a ct-space X the set α_x is an ideal point of X and every sober space is T_0.

Proposition 11.23 *The* ct-*space of formal points of a formal topology is sober.*

Proof. Let X be the ct-space of formal points of a formal topology on S. Observe that for any formal point β of $\mathcal{A}$,

$$\alpha_\beta = \{a \in S \mid \beta \in B_a\} = \{a \in S \mid a \in \beta\} = \beta.$$

Thus if α is an ideal point of X, to show that there is a unique $\beta \in X$ such that $\alpha = \alpha_\beta$ we have to show that α is itself a formal point of $\mathcal{A}$. To do this, we must prove the following.

1. $a, b \in \alpha \;\Rightarrow\; \exists c \in \alpha\; c \leq a, b$
2. $a \in \alpha \;\&\; a \lhd U \;\Rightarrow\; \alpha \between U$.

1. Given $a, b \in \alpha$, let $U = \{d \in S \mid d \leq a, b\}$. Since α is an ideal point there is a $c \in \alpha$ such that $B_c \subseteq B_a \cap B_b$. Given any $\beta \in B_c$, $a, b \in \beta$ and since β is a $\leq$-point there is a $d \in \beta$ such that $d \in U$. Thus $B_c \subseteq B_U$, so $\alpha \between U$.
2. Suppose that $a \in \alpha$ and $a \lhd U$. If $\beta \in B_a$ then $a \in \beta$ and so $\beta \between U$ since β is a formal point. So $B_a \subseteq B_U$ and hence $a \between U$. □

Classically it is true that every Hausdorff (T_2) space is sober. Constructively we can show that every complete metric space is sober, and more generally that every complete uniform space is sober. Constructively we cannot expect to do much better than this, as the space of rational numbers with the Euclidean topology cannot be shown to be sober. This was proved in [4] using sheaf models, and we give an alternative proof below.

Example 11.24 The rational numbers.

The usual topology on the rational numbers can be given by taking $X = \mathbb{Q}$, $S = \mathbb{Q} \times \mathbb{Q}^{>0}$ and $B_{(x,\varepsilon)} = \{y \in \mathbb{Q} \mid |x - y| < \varepsilon\}$. □

Proposition 11.25 *If* $\mathbb{Q}$ *is sober then* LPO *holds.*

Proof. Given a function $f : \mathbb{N} \to \{0, 1\}$, we may define a sequence (x_n) of rationals such that

$$x_n = \begin{cases} n^{-1} \text{ if } (\forall k \leq n)(f(k) = 0) \\ k^{-1} \text{ if } (k \leq n) \;\&\; (f(k) = 1) \;\&\; (\forall j < k)(f(j) = 0). \end{cases}$$

Let

$$\alpha = \{\, (x, \varepsilon) \in S \mid (\exists \varepsilon' < \varepsilon)(\exists N \in \mathbb{N})(\forall n \geq N)[x_n \in B_{(x,\varepsilon')}] \,\}.$$

We claim that α is an ideal point.

1. For all n, $0 < x_n \leq 1$, so $(1, 2) \in \alpha$.

2. If $(x, \varepsilon), (y, \delta) \in \alpha$, then there exists an $\eta \in \mathbb{Q}^{>0}$ and an $N \in \mathbb{N}$ such that $(\forall n \geq N)(x_n \in B_{(x,\varepsilon-\eta)} \cap B_{(y,\delta-\eta)})$. In particular $B_{(x,\varepsilon-\eta)} \cap B_{(y,\delta-\eta)}$ is inhabited, so by the standard properties of the rational numbers it is an open interval, and so there exists $(z, \zeta) \in S$ with $B_{(z,\zeta)} = B_{(x,\varepsilon-\eta)} \cap B_{(y,\delta-\eta)}$. Now if we take $a = (z, \zeta + \eta)$, then $B_a \in \alpha$ and $B_a = B_{(x,\varepsilon)} \cap B_{(y,\delta)}$.

3. Suppose that $(x, \varepsilon) \in \alpha$, $U \subseteq S$ and $B_{(x,\varepsilon)} \subseteq B_U$. Then there exists an $\varepsilon' < \varepsilon$ and an $N \in \mathbb{N}$ such that $(\forall n \geq N)(x_n \in B_{(x,\varepsilon')})$.
   Either (i) $(x - \varepsilon' > 0)$ or (ii) $(x - \varepsilon' \leq 0)$, so we can treat the two cases separately.

   (i) If $x - \varepsilon' > 0$, then there is a positive integer $n \geq N$ with $n^{-1} < x - \varepsilon'$. $x_n \in B_{(x,\varepsilon')}$, so $\neg(x_n = n^{-1})$, so we must have $x_n = k^{-1}$ for some $k \leq n$ such that $f(k) = 1$ and $f(j) = 0$ for all $j < k$. $k^{-1} \in B_{(x,\varepsilon)}$, so $k^{-1} \in B_{(y,\delta)}$ for some $(y, \delta) \in U$, and hence $k^{-1} \in B_{(y,\delta')}$ for some $\delta' < \delta$. So $x_m \in B_{(y,\delta')}$ for all $m \geq k$, hence $(y, \delta) \in \alpha$.

   (ii) If $x - \varepsilon' \leq 0$, then $x + \varepsilon' > 0$, because otherwise no x_n would lie in $B_{(x,\varepsilon')}$. So $0 \in B_{(x,\varepsilon)}$, and so there is a neighbourhood $(y, \delta) \in U$ with $0 \in B_{(y,\delta)}$. Now choose $n \in \mathbb{N}$ with $n^{-1} < y + \delta$. Either $x_n \in B_{(y,\delta)}$ or $x_n \notin B_{(y,\delta)}$. In the first case we can find a $\delta' < \delta$ with $0, x_n \in B_{(y,\delta')}$, so $x_m \in B_{(y,\delta')}$ for all $m \geq n$, and so $(y, \delta) \in \alpha$. In the second case $x_n \geq y + \delta > n^{-1}$, so $f(k) = 1$ for some $k \leq n$ and we can proceed as in case (i).

This completes the proof that α is an ideal point. So, by the assumption that X is sober, $\alpha = \alpha_x$ for some $x \in \mathbb{Q}$. Note that $(x = 0) \vee (x < 0) \vee (x > 0)$.

If $x = 0$, then for all $n \in \mathbb{N}$ $B_{(0,n^{-1})} \in \alpha$ so there exists an $m \in \mathbb{N}$ such that $x_m < n^{-1}$, hence $(\forall n \in \mathbb{N})(f(n) = 0)$.

If $x < 0$, we get a contradiction since no x_n lies in $B_{(x,x/2)}$.

If $x > 0$, then $(x, x/2) \in \alpha$. Choose $n \in \mathbb{N}$ with $n^{-1} < \frac{1}{2}x$. Then $n^{-1} \notin (x, x/2)$, so $x_n = k^{-1}$ for some $k \in \mathbb{N}$ with $f(k) = 1$.

So in each case $(\forall n \in \mathbb{N})(f(n) = 0) \vee (\exists n \in \mathbb{N})(f(n) = 1)$. □

11.8.1 *Strong ideal points and weak sobriety*

The usual definition of an ideal point arises from the functors Pt and Ω between the categories of topological spaces and formal topologies. In [7], Sambin defines a *balanced formal topology* to be a formal topology with a *binary positivity predicate* $\ltimes$ added, which must satisfy certain axioms. The intuitive meaning of $a \ltimes U$ for $a \in S$ and $U \in Pow(S)$ is that the formal neighbourhood a touches some 'formal closed set' determined by U.

Given a small ct-space X, we obtain a balanced formal topology on S by taking $a \lhd U$ iff $B_a \subseteq B_U$, and $a \ltimes U$ iff $B_a \between \mathsf{rest}\ U$. Sambin's definition of the points of a balanced formal topology leads us to a new definition of ideal points:

Definition 11.26 *A strong ideal point of a* ct*-space X is a subset α of S such that:*

1. $\exists a \in \alpha$
2. $(\forall a, b \in \alpha)(\exists c \in \alpha)[B_c \subseteq B_a \cap B_b]$
3. $(\forall a \in \alpha)(\exists x \in B_a)[\alpha_x \subseteq \alpha]$.

Observe that every strong ideal point is an ideal point and α_x is a strong ideal point for each $x \in X$. We say that X is *weakly sober* if for every strong ideal point α, there is a unique $x \in X$ such that $\alpha = \alpha_x$.

Proposition 11.27 *Every T_2 space is weakly sober.*

Proof. Let X be T_2, and let α be a strong ideal point of X. By (1) and (3) there is an $x \in X$ such that $\alpha_x \subseteq \alpha$.

Given $y \in X$ such that $\alpha_y \subseteq \alpha$, we will show that $x = y$. Given $a \in \alpha_x$ and $b \in \alpha_y$ we have $a, b \in \alpha$, so by (2) and (3) there is a $z \in B_a \cap B_b$. Hence $x \sim_2 y$, so $x = y$.

It remains to prove that $\alpha \subseteq \alpha_x$. If $a \in \alpha$, then by (3) there is a $y \in B_a$ such that $\alpha_y \subseteq \alpha$, and by the previous paragraph $y = x$, so $a \in \alpha_x$. □

References

1. Aczel, P. and Rathjen, M. (2001). Notes on constructive set theory. Technical Report 40, 2000/2001, Institut Mittag-Leffler, The Royal Swedish Academy of Sciences.
2. Bishop, E. and Bridges, D. (1985). *Constructive Analysis*. Springer-Verlag, Berlin, Heidelberg, New York, Tokyo.
3. Bridges, D. and Vita, L. (2003) Separatedness in constructive topology. *Documenta Math.* **8**, 567–576.
4. Fourman, M. and Scott, D. (1979). Sheaves and logic. *Applications of Sheaves, LNM*, **753**, 302–401.
5. Grayson, R. J. (1981). Concepts of general topology in constructive mathematics and in sheaves. *Annals of Math. Logic*, **20**, 1–41.
6. Grayson, R. J. (1982). Concepts of general topology in constructive mathematics and in sheaves, II. *Annals of Math. Logic*, **23**, 55–98.
7. Sambin, G. (2003). Some points in formal topology. *Theoretical Computer Science*, **305**, 347–408.
8. Steen, L. A. and Seebach, Jr., J. A. (1970). *Counterexamples in Topology*. Holt, Rinehart and Winston, Inc.
9. Troelstra, A. S. Intuitionistic general topology. Thesis, Amsterdam, 1966.
10. Troelstra, A. S. and van Dalen, D. (1998). *Constructivism in Mathematics*. North-Holland.
11. Vickers, S. J. Localic completion of generalized metric spaces I. Draft available at *http://www.cs.bham.ac.uk/~sjv*.

12

SPACES AS COMONOIDS

A. BUCALO AND G. ROSOLINI

Abstract

The constructive approach taken to topology from the point of view of type theory brings to light a presentation of topological spaces which exploits another monoidal structure on the category of complete sup-lattices from that given by Galois connections. In this modified context, we characterize the frames which are topologies.

12.1 Introduction

The constructive approach to topology has produced a wealth of new insights about spaces, almost reaching the point of making these more elementary than sets. In particular, the point of view taken by [2] brings forward the relevance of relations between sets as an elementary constituent to understand topological spaces.

In a sense that we hope to make clear in the chapter, one could consider that approach as looking at topological spaces in order to make sense of the notion of function between sets from a logically constructive point of view.

We shall employ the language and notions of category theory as the most apt to deal with mathematical structures, but we shall always consider a 'category' as a metatheoretical notion, extremely useful in classifying objects and transformations, but not necessarily definable within the (constructive) theory we use.

12.2 A Categorical View of Basic Pairs

The notion of a basic pair, as introduced in [3], gives rise naturally to a category. It consists of:

objects: relations $r \subseteq X \times A$ between sets X and A. The triple (X, A, r) is a *basic pair* which we shall write as $r\colon X \longrightarrow A$.

arrows: diagonal compositions in commutative squares

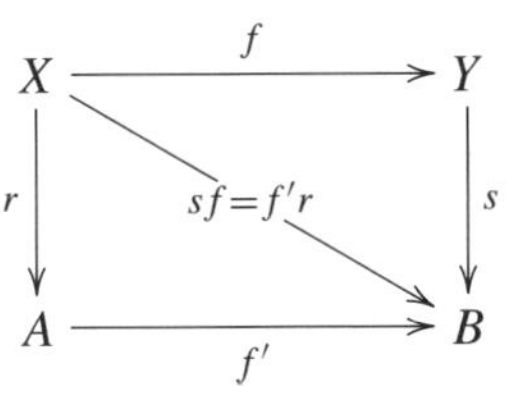

i.e. equivalence classes of pairs of relations (f, f') such that $sf = f'r$ with respect to the relation

$$(f, f') \sim (f_1, f_1') \iff sf = sf_1 \quad [\iff f'r = f_1'r].$$

composition: is given by relational composition pasting commutative squares

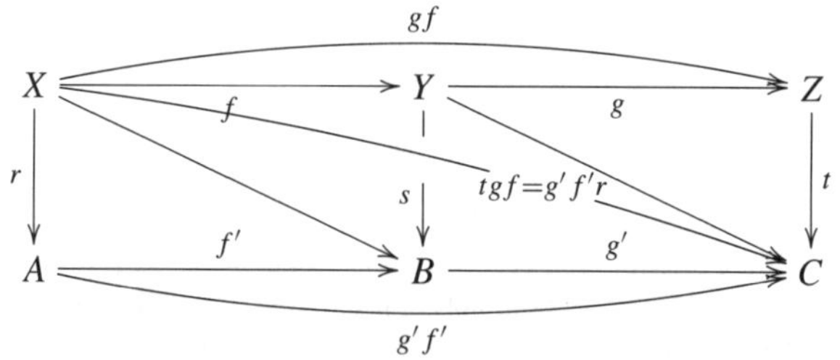

More explicitly, the composite is represented by the pair $(gf, g'f')$.

Remarks The following remark is due to Marco Grandis: The category of basic pairs is the free completion Fr(**Rel**) of the category **Rel** of sets and relations with a (stable) proper factorization system. The general construction goes under the name of 'Freyd completion of a category', see [4, 5].

The following remark is due to Aurelio Carboni: The category of complete sup-lattices and sup-preserving functions is (equivalent to) the free completion $\mathbf{Rel}_{\mathrm{ex}}$ of the category **Rel** of sets and relations to an exact category, see [6].

Since every exact category has a (stable) proper factorization system, by freeness there is a functor

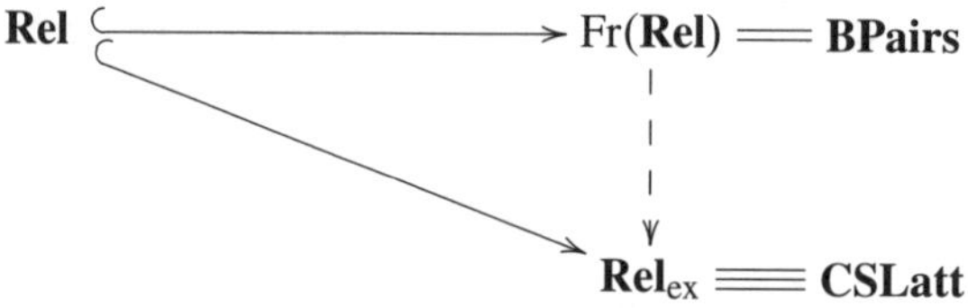

from basic pairs to complete sup-lattices.

Notice that, given an ambient category **T**—say the model of the type theory under consideration—the construction of the category **Rel** of relations requires that **T** is regular. Next, the Freyd completion goes through with no problem. On the other hand, the exact completion requires weak equalizers in **Rel**, which exist, e.g. if **T** is a topos (i.e. **T** has a subobject classifier).

Hence, if the functor Fr(**Rel**) $\longrightarrow$ **CSLatt** is an equivalence, then the Freyd completion of **Rel** is a natural substitute for the category of complete sup-lattices. Alas, assuming the category is strong enough to support both definitions, that functor (in fact, any functor between the two categories above) is an equivalence if and only if the logic of **T** is Boolean.

On the other hand, in the case of an arbitrary topos **T**, the lattices in the image of the functor can be characterized precisely as those L such that there is a sup-preserving embedding into a powerset sup-lattice, see [7].

It is thus clear that basic pairs offer the possibility to present an elementary notion—that of complete sup-lattice—which underlies the notion of a topological space, in a much more general context, possibly at the price of restricting the collection of structures (whatever this may mean from the point of view of the universe **T**).

In any case, that notion by itself is not adequate since, in order to understand spaces, it is crucial to understand transformations between them. Continuous functions are not simply maps between basic pairs, but those maps which further preserve a certain structure. To put this precisely (assuming all data below can be considered within **T**), the elementary functor which takes a topological space to the 'basic pair' which defines the lattice of its open subsets

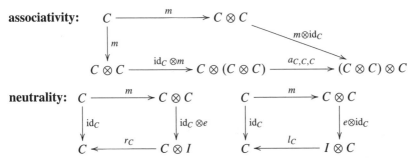

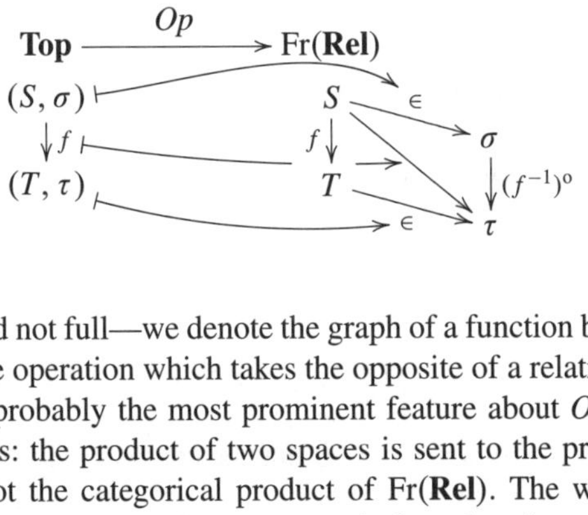

is clearly faithful, and not full—we denote the graph of a function by the function itself, and write $(-)^{\text{o}}$ for the operation which takes the opposite of a relation.

Moreover—and probably the most prominent feature about *Op*—the functor does not preserve products: the product of two spaces is sent to the product of relations in Fr(**Rel**), which is not the categorical product of Fr(**Rel**). The well-known universal solution to making the tensor product a categorical product is to take the category of commutative comonoids and comonoid homomorphisms, see [8].

We recall the main definitions:

Definition 12.1 *In a symmetric monoidal category* $(\mathbf{C}, \otimes, I, a, r, l, c)$*, a commutative comonoid is a triple* (C, m, e) *which consists of an object of* $\mathbf{C}$*, and maps* $m: C \to C \otimes C$ *and* $e: C \to I$ *such that the following diagrams commute*

associativity:

$$\begin{array}{ccccc} C & \xrightarrow{m} & C \otimes C & & \\ \downarrow{\scriptstyle m} & & & \searrow{\scriptstyle m \otimes \mathrm{id}_C} & \\ C \otimes C & \xrightarrow{\mathrm{id}_C \otimes m} & C \otimes (C \otimes C) & \xrightarrow{a_{C,C,C}} & (C \otimes C) \otimes C \end{array}$$

neutrality:

$$\begin{array}{ccc} C & \xrightarrow{m} & C \otimes C \\ \downarrow{\scriptstyle \mathrm{id}_C} & & \downarrow{\scriptstyle \mathrm{id}_C \otimes e} \\ C & \xleftarrow{r_C} & C \otimes I \end{array} \qquad \begin{array}{ccc} C & \xrightarrow{m} & C \otimes C \\ \downarrow{\scriptstyle \mathrm{id}_C} & & \downarrow{\scriptstyle e \otimes \mathrm{id}_C} \\ C & \xleftarrow{l_C} & I \otimes C \end{array}$$

commutativity:

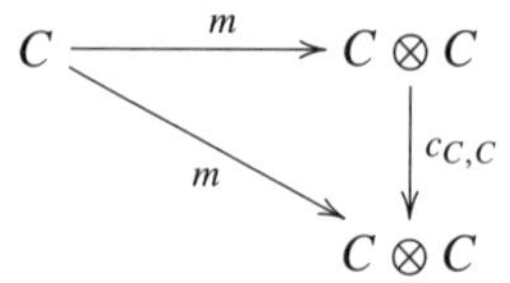

A homomorphism between two comonoids (C, m, e) *and* (D, m', e') *is a map* $f: C \to D$ *such that one has commutative diagrams*

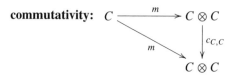

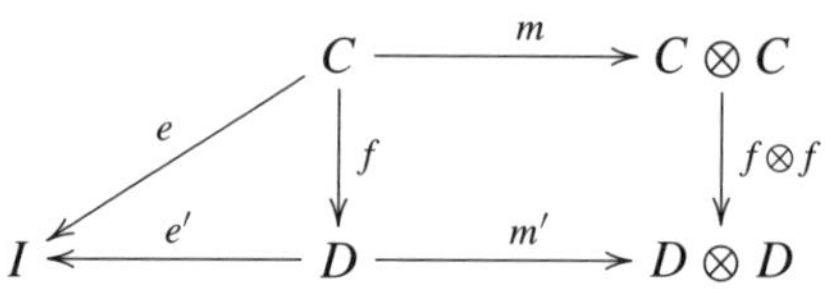

Write CCom$(\mathbf{C}, \otimes, I, a, r, l, c)$ *for the category of commutative comonoids and homomorphisms—we shall always discard those items in the list describing the monoidal structure on* $\mathbf{C}$ *which we consider clear from the context.*

Example 12.2 Leaving all the obvious examples aside, such as commutative monoids, abelian groups, vector spaces, or commutative rings, we recall a single example, which we found extremely useful for developing a correct kind of intuition for comonoids with relations: In the category **Rel** of relations, the cartesian product extends to a tensor: given two relations $r_1: X_1 \longrightarrow Y_1$ and $r_2: X_2 \longrightarrow Y_2$, their tensor product is defined 'componentwise' as follows

$$\langle x_1, x_2 \rangle \; (r_1 \times r_2) \; \langle y_1, y_2 \rangle \Leftrightarrow x_1 \; r_1 \; y_1 \wedge x_2 \; r_2 \; y_2.$$

We pause to note that the notation $r_1 \times r_2$ may be confusing: it is not the set-theoretical product of the two sets (of pairs) r_1 and r_2. We are convinced though that it is very convenient, and worth that (little) risk of confusion.

Consider the following structure on a given set X. As a multiplication we take the diagonal function $\Delta_X: X \longrightarrow X \times X$, seen as a relation, as a neutral element the constant function $t: X \longrightarrow 1$. It is easy to check that the data give a commutative comonoid.

For two such comonoids on the sets X and Y, one can check that a relation $r: X \longrightarrow Y$ is a comonoid homomorphism if and only if r is a function from X to Y. A reader interested in pursuing this example further should first read the fundamental paper [9].

The monoidal structure on Fr(**Rel**) is induced from that of cartesian product on **Rel** after one notices that the product preserves the equivalence relation on arrows. So, for basic pairs $r_1: X_1 \longrightarrow Y_1$ and $r_2: X_2 \longrightarrow Y_2$, the tensor is defined as above. For maps represented by (f_1, f_1') and (f_2, f_2'), one checks immediately that $(f_1 \times f_2, f_1' \times f_2')$ determines an equivalence class which does not depend on the choice of representatives.

Remark Since the functor Op: **Top** $\to$ Fr(**Rel**) takes a product of topological spaces to a product of basic pairs and the terminal topological space to the identity on the one-point set, by the theorem in [8] it determines a product-preserving functor

$$\mathbf{Top} \xrightarrow{Op_c} \mathrm{CCom}(\mathrm{Fr}(\mathbf{Rel}), \times, 1)$$

Explicitly, the comonoid structure on $Op(S, \sigma) = S \xrightarrow{\in} \sigma$ is

$$\begin{array}{ccccc} 1 & \xleftarrow{t} & S & \xrightarrow{\Delta_S} & S \times S \\ {\scriptstyle \mathrm{id}_1}\downarrow & & \downarrow{\scriptstyle \in} & & \downarrow{\scriptstyle \in\times\in} \\ 1 & \xleftarrow{S^{\circ}} & \sigma & \xrightarrow{\cap^{\circ}} & \sigma \times \sigma \end{array}$$

where S is the constant function $1 \to \sigma$ with value the open set S, and $\cap$ is the (binary) operation of intersection.

Moreover, the functor Op_c has a right adjoint, see [7].

It follows that, even when the category Fr(**Rel**) is equivalent to **CSLatt**, the tensor product defined by taking the product of the basic pairs, as above, is not that of Galois connections, since the functor from topological spaces into locales does not preserve products, see [1, 10].

12.3 The Characterization

In any elementary topos **E**, the functor Op_c restricts to an equivalence on sober spaces, see [7]. We now intend to characterize the category of comonoids equivalent to that of sober spaces via Op_c. As in [7], we shall work in a model of intuitionistic impredicative type theory, an arbitrary elementary topos **E**, and construct categories therein.

First we notice a property of basic pairs.

Lemma 12.3 *Given a basic pair* $r\colon X \longrightarrow A$, *there is one* $r'\colon X' \longrightarrow A'$ *isomorphic to it such that*

(i) $\forall x' \in X' \exists a' \in A' \left[x' \; r' \; a' \wedge \left[\forall x'' \in X' \forall a' \in A' \left(x' \; r' \; a' \Leftrightarrow x'' \; r' \; a'\right) \Longrightarrow x' = x''\right]\right]$

(ii) $\forall a' \in A' \left[\exists x' \in X' \; x' \; r' \; a' \wedge \left[\forall a'' \in A' \forall x' \in X' \left(x' \; r' \; a' \Leftrightarrow x' \; r' \; a''\right) \Longrightarrow a' = a''\right]\right]$.

Proof. Easy: impose on the domain of r the equivalence relation given by

$$\forall a \in A(x' \; r \; a \Leftrightarrow x'' \; r \; a)$$

and verify that the quotient map defines an isomorphism. Similarly for A. □

Say that a basic pair is T_0 if it satisfies the two conditions of the lemma. Note that this notion is not categorical. And one also has the following property.

Lemma 12.4 *Suppose the basic pair $r\colon X \longrightarrow A$ is T_0. Then the largest relations f and g producing a commutative diagram*

$$\begin{array}{ccc} X & \xrightarrow{\ f\ } & X \\ {\scriptstyle r}\downarrow & \searrow^{r} & \downarrow{\scriptstyle r} \\ A & \xrightarrow[\ g\ ]{} & A \end{array}$$

are, respectively, the partial order $x_1 \prec_X x_2$ given by

$$\forall a \in A(x_2\ r\ a \Longrightarrow x_1\ r\ a)$$

and the partial order $a_1 \prec_A a_2$ given by

$$\forall x \in X(x\ r\ a_2 \Longrightarrow x\ r\ a_1).$$

Hence the identity map on the basic pair r can be represented by the pair $(\prec_X, \prec_A)$ or any other pair between that and $(\mathrm{id}_X, \mathrm{id}_A)$.

Theorem 12.5 *Suppose* $\mathrm{id}_1 \xleftarrow{\ e\ } c \xrightarrow{\ m\ } c \times c$ *is a comonoid in* Fr(**Rel**). *If it is isomorphic to $Op_c(S, \sigma)$ for some topological space (S, σ), then among the representatives of the unit there is one of the form*

$$\begin{array}{ccc} Z & \xrightarrow{\ t\ } & 1 \\ {\scriptstyle c}\downarrow & & \downarrow \\ D & \xrightarrow[\ e'\ ]{} & 1 \end{array}$$

and among the representatives of the multiplication there is one of the form

$$\begin{array}{ccc} Z & \xrightarrow{\ \Delta_Z\ } & Z \times Z \\ {\scriptstyle c}\downarrow & & \downarrow{\scriptstyle c\times c} \\ D & \xrightarrow[\ m'\ ]{} & D \times D \end{array}$$

In each case, the second component can be chosen as the largest relation making the square commute.

Proof. With no loss of generality, we may assume that the basic pair c is T_0. Let

$$\begin{array}{ccccc} Z & \xrightarrow{\ h\ } & S & \xrightarrow{\ k\ } & Z \\ {\scriptstyle c}\downarrow & & \downarrow{\scriptstyle \in} & & \downarrow{\scriptstyle c} \\ D & \xrightarrow[\ h'\ ]{} & \sigma & \xrightarrow[\ k'\ ]{} & D \end{array}$$

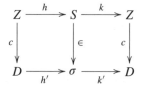

be an isomorphism as in the hypothesis. The first part of the statement follows immediately because the first component $h\colon Z \longrightarrow S$ is total because c is T_0.

The second part is similar, with a diagram chase of relational compositions: consider the diagram

$$\begin{array}{ccccc}
S & \xrightarrow{\Delta_S} & S \times S & & \\
{\scriptstyle k}\downarrow \; \uparrow{\scriptstyle h} & & {\scriptstyle k\times k}\downarrow \; \uparrow{\scriptstyle h\times h} & & \\
Z & \xrightarrow[m]{} & Z \times Z & \xrightarrow[c\times c]{} & D \times D
\end{array}$$

Note that the compositions from Z to $D \times D$ give the same relation. To see that $(c \times c)m \subseteq (c \times c)\Delta_Z$, consider that, applying the general fact $\Delta_S h \subseteq (h \times h)\Delta_Z$ to the identity in the diagram, one obtains

$$(c \times c)m = (c \times c)(k \times k)\Delta_Z h \subseteq ((ckh) \times (ckh))\Delta_Z = (c \times c)\Delta_Z$$

using that $ckh = c$ because (k, k') and (h, h') are inverse to each other.

Note then that m is total because c is T_0, so $(c \times c)\Delta_Z \subseteq (c \times c)m$. □

12.4 Comparison with the Presentations of Formal Topologies

It is clear that the mathematical approach we present takes a suggestion of [11] very seriously: the operation of 'meeting"

$$u \between v \stackrel{\text{def}}{=} \exists x \in X.\, u(x) \wedge v(x)$$

since this is just another notation for relational composition once other variables on which u and v may depend are made explicit—in fact, these are often left implicit in type theory.

It is presented in [11] as 'the dual of inclusion', which is a very particular case of the much more general property that each unary operation $(-)v$ and $u(-)$, respectively pre-composition with v and post-composition with u, has a right adjoint in any category **T** with some reasonable logical power.

A particular remark is that the largest relation $\overline{m}$ producing a commutative diagram

$$\begin{array}{ccc}
Z & \xrightarrow{\Delta_Z} & Z \times Z \\
{\scriptstyle c}\downarrow & & \downarrow{\scriptstyle c\times c} \\
D & \xrightarrow[\overline{m}]{} & D \times D
\end{array}$$

is that defined by $d\ \overline{m}\ (d_1, d_2)$ if

$$\forall z \in Z (z\ c\ d \Longrightarrow z\ c\ d_1 \wedge z\ c\ d_2).$$

This leads to the following property.

Theorem 12.6 *A commutative comonoid in* Fr(**Rel**) *is isomorphic to one of the form* $Op_c(S, \sigma)$ *for some sober space* (S, σ) *if and only if it is a basic pair* $c\colon Z \to D$ *such that*

(i) the total relation $t\colon Z \to 1$ *factors through* $c\colon Z \to D$*, and*

(ii) the composition $(c \times c)\Delta_Z\colon Z \to D \times D$ *factors through* $c\colon Z \to D$*—in which case the relation* $\overline{m}\colon D \to D \times D$ *defined above can be used.*

Proof. For a sober space (S, σ), the comonoid $Op_c(S, \sigma)$ clearly satisfies the two conditions. Conversely, for a basic pair $c\colon Z \to D$ conditions (i) and (ii) state precisely that the family of subsets defined by

$$U_d = \{z \in Z \mid z \; c \; d\}, \qquad \text{as } d \text{ varies in } D$$

is a basis for a topology. Indeed, the basic pairs $c\colon Z \to D$ and $\in\colon Z \to B$ are isomorphic and the isomorphism preserves the comonoid structure. By Theorem 5.11 in [7], the comonoid defined by a topology is isomorphic to that defined by the soberification of the topology. □

One can compare this result with the discussion on basic topologies in [11]: in particular (i) in Theorem 12.6 relates to positivity and (ii) relates to condition B2 of [11].

Acknowledgements

Discussions with Peter Schuster helped in clarifying some aspects of the chapter. The comments of an anonymous referee provided very useful suggestions to improve the presentation of the subject.

References

1. Joyal, A. and Tierney, M. (1984). An extension of the Galois theory of Grothendieck. *Mem. Amer. Math. Soc.*, **51**(309), vii+71.
2. Sambin, G. and Gebellato, S. (1999). A preview of the basic picture: a new perspective on formal topology. In *Types for Proofs and Programs (Irsee, 1998)*, Volume 1657 of *Lecture Notes in Comput. Sci.*, pp. 194–207. Springer, Berlin.
3. Sambin, G. (2000). Formal topology and domains. *Electronic Notes in Theoretical Computer Science*, **35**, 177–190.
4. Freyd, P. J. (1966). Stable homotopy. In *Proceedings of the Conference on Categorical Algebra* (ed. S. Eilenberg, D. Harrison, S. Mac Lane, and H. Röhrl), La Jolla 1965, pp. 121–176. Springer-Verlag.
5. Grandis, M. (2000). Weak subobjects and the epi-monic completion of a category. *J. Pure Appl. Algebra*, **154**(1–3), 193–212. Category theory and its applications (Montreal, QC, 1997).
6. Carboni, A. and Vitale, E. M. (1998). Regular and exact completions. *J. Pure Appl. Alg.*, **125**, 79–117.

7. Bucalo, A. and Rosolini, G. (2005). Completions, comonoids, and topological spaces. In: *Second Workshop on Formal Topology* (eds B. Banaschewski, Th. Coquand, and G. Sambin), Venice 2002. *Ann. Pure Appl. Logic* (special issue), to appear
8. Fox, Th. (1976). Coalgebras and Cartesian categories. *Comm. Algebra*, **4**(7), 665–667.
9. Carboni, A. and Walters, R. F. C. (1987). Cartesian bicategories, I. *J. Pure Appl. Alg.*, **49**, 11–32.
10. Johnstone, P. and Sun, S. H. (1988). Weak products and Hausdorff locales. In: *Categorical Algebra and its Applications (Louvain-La-Neuve, 1987)*, pp. 173–193. Springer, Berlin.
11. Sambin, G. (2003). Some points in formal topology. *Theoret. Comput. Sci.*, **305**(1–3), 347–408. Topology in computer science (Schloß Dagstuhl, 2000).

13

PREDICATIVE EXPONENTIATION OF LOCALLY COMPACT FORMAL TOPOLOGIES OVER INDUCTIVELY GENERATED TOPOLOGIES

MARIA EMILIA MAIETTI

Abstract

We provide an intuitionistic predicative proof of the well-known result proved by M. Hyland that locally compact locales are the exponentiable objects in the category of locales.

More precisely, by adapting Hyland's proofs to the context of formal topology, we prove that locally compact formal covers are the exponentiable objects in the category of inductively generated formal covers. Then, we deduce the analogous result for inductively generated formal topologies.

13.1 Introduction

It is well known that the category of topological spaces is not closed and that the exponentiable objects are exactly those spaces whose frame is continuous (see for example the overview [1]).

M. Hyland in [2] proved the analogous result for locales, namely that locally compact locales are the exponentiable objects in the category of locales. Later, I. Sigstam in [3] adapted his proof to show that locally compact locales are exponentiable in the category of (effectively given) formal spaces, but still working within a (classical) impredicative setting.[1] Moreover, Vickers reproduced most of Hyland's results using geometric reasoning (see [4]).

Here, we want to give an intuitionistic predicative proof of the whole characterization in [2].

To this purpose we consider the category of inductively generated formal covers [5] and continuous relations [6] which constitutes a predicative presentation of the category of locales (see [7]), and also of its opposite, namely that of frames, by taking the opposite relations. Then, we also consider the full subcategory of inductively generated formal topologies which provides a predicative presentation of open locales (see [8]). In both categories, we address the question of characterizing the exponentiable objects predicatively.

[1] Actually, she proves exponentiation of a locally compact formal space only over formal spaces admitting generators made of directed subsets.

A predicative study of exponentiation in the context of formal topology has been started in [9] where it is proved predicatively that unary formal topologies are exponentiable in the category of inductively generated formal topologies. This result was deduced after proving the corresponding result for inductively generated formal covers and the existence of a coreflection of inductively generated formal covers into inductively generated formal topologies thanks to which exponentiation within the subcategory of formal topologies is inherited from the category of inductively generated formal covers.

Here, we show predicatively that locally compact formal covers are the exponentiable objects in the category of inductively generated formal covers by adapting Hyland's proofs in [2]. In particular, the construction of the exponential topology of a locally compact formal cover over an inductively generated one follows the original idea in [2] of generating the exponential topology by means of the 'way-below' component of a locale morphism, since this construction fits perfectly with the use of the predicative method for the inductive generation of a formal cover in [5], provided that we can define predicatively the 'way-below' relation between open subsets, which was shown to be possible for inductively generated formal covers in [10]. Hence, inspired by Hyland's construction and thanks to the analysis in [9] of the conditions defining a continuous relation, we characterize a continuous relation via its way-below component in terms of conditions that can be put as basic axioms from which to generate the exponential topology candidate as in [2, 3, 7]. Then, the proof of exponentiation mostly consists of inductions over the inductive generation of the covers involved.

Afterwards, we prove that any exponentiable inductively generated formal cover is locally compact following very closely the original proof in [2] that, being almost completely based on categorical properties related to exponentiation and on properties of the Sierpinski locale, is already essentially predicative.

Finally, thanks to the coreflection proved in [9], we deduce the analogous characterization for inductively generated formal topologies, namely that locally compact formal topologies are the exponentiable objects in the category of inductively generated formal topologies.[2]

13.2 Preliminaries on Formal Covers

There are at least two different equivalent presentations of formal topology: one based on a monoid [6] or on a preorder [5]. Here, we deal with formal covers and formal topologies based on a preorder, called $\leq$-formal covers and $\leq$-formal topologies, that are inductively generated [5].

Note that when we speak of 'set' we mean a set in Martin-Löf constructive set theory [12] and we adopt the set-theoretic notation in [13].

We start with recalling the notion of formal cover based on a preorder [5] corresponding predicatively to that of locale [7]:

[2] More detailed proofs of the main results in this chapter can be found in [11].

Definition 13.1 ($\leq$-Formal cover) *A $\leq$-formal cover is a structure*

$$\mathcal{A} \equiv (A, \leq_{\mathcal{A}}, \lhd_{\mathcal{A}})$$

where A is a set of basic opens (or base), $\leq_{\mathcal{A}}$ is a preorder between elements of A and $\lhd_{\mathcal{A}}$ is an infinitary relation, called a cover relation, between elements and subsets of A satisfying the following conditions:

$$(reflexivity)\ \frac{a \varepsilon U}{a \lhd U} \qquad (transitivity)\ \frac{a \lhd U \quad U \lhd V}{a \lhd V}$$

$$(\leq\text{-}left)\ \frac{a \leq b \qquad b \lhd U}{a \lhd U} \qquad (\leq\text{-}right)\ \frac{a \lhd U \qquad a \lhd V}{a \lhd U \downarrow_{\leq} V}$$

where $U \lhd V$ is an abbreviation for $(\forall u \varepsilon U)\ u \lhd V$ and $U \downarrow_{\leq} V$ for $\{x \in A \mid \exists u \varepsilon U\ x \leq u \ \&\ \exists v \varepsilon V\ x \leq v\}$.

Then, we give the associated notion of formal topology based on a preorder [5] corresponding predicatively to that of open locale [8]:

Definition 13.2 ($\leq$-Formal topology) *A $\leq$-formal topology is a structure*

$$\mathcal{A} \equiv (A, \leq_{\mathcal{A}}, \lhd_{\mathcal{A}}, \mathsf{Pos}_{\mathcal{A}})$$

where $(A, \leq_{\mathcal{A}}, \lhd_{\mathcal{A}})$ is a formal cover and $\mathsf{Pos}_{\mathcal{A}}$ is a predicate over A, called a positivity predicate, satisfying the following conditions:

$$(monotonicity)\ \frac{\mathsf{Pos}_{\mathcal{A}}(a) \qquad a \lhd U}{(\exists u \varepsilon U)\ \mathsf{Pos}_{\mathcal{A}}(u)} \qquad (positivity\ axiom)\ a \lhd a^{+}$$

where $a^{+} \equiv \{x \in A \mid x = a \ \&\ \mathsf{Pos}_{\mathcal{A}}(x)\ \}$.

Now, we recall the notion of inductively generated formal cover [5], closely related to Aczel's notion of the set-based closure operation in [14] and that of inductive definition in [15]:

Definition 13.3 (I.g. formal cover) *A $\leq$-formal cover $\mathcal{A} \equiv (A, \leq_{\mathcal{A}}, \lhd_{\mathcal{A}})$ is inductively generated (abbreviated to i.g.) if and only if there exists a localized axiom-set $I(-)$ and $C(-,-)$ (as defined in [5]) such that $\lhd_{\mathcal{A}}$ is inductively generated by using reflexivity, $\leq$-left and*

$$(infinity)\ \frac{i \in I(a) \qquad C(a, i) \lhd U}{a \lhd U}.$$

We recall from [5] that in general the relation generated by an axiom-set by means of *reflexivity*, *$\leq$-left* and *infinity* does not necessarily satisfy *$\leq$-right* unless the axiom-set is localized. However, given any axiom-set we can always localize it as shown in [5].

Then, we give the definition of inductively generated formal topology, namely an inductively generated formal cover with a positivity predicate (further readings on this are [16–18]):

Definition 13.4 (I.g. formal topology) $\mathcal{A} \equiv (A, \leq_{\mathcal{A}}, \lhd_{\mathcal{A}}, Pos_{\mathcal{A}})$ *is an inductively generated (abbreviated to i.g.)* $\leq$*-formal topology if and only if the formal cover* $(A, \leq_{\mathcal{A}}, \lhd_{\mathcal{A}})$ *is inductively generated.*

Then, we recall the definition of continuous relation between $\leq$-formal covers (as in [9] we take the opposite relation of that defining a frame morphism in [6]):

Definition 13.5 (Continuous relation) *Suppose that* $\mathcal{A} = (A, \leq_{\mathcal{A}}, \lhd_{\mathcal{A}})$ *and* $\mathcal{B} = (B, \leq_{\mathcal{B}}, \lhd_{\mathcal{B}})$ *are two formal covers. Then a continuous relation from* $\mathcal{A}$ *to* $\mathcal{B}$ *is a binary proposition* $a\,F\,b$*, for* $a \in A$ *and* $b \in B$*, satisfying the following conditions:*

(totality) $\quad A \lhd_{\mathcal{A}} F^{-}(B)$ $\qquad$ *($\leq$-convergence)* $\quad \dfrac{aFb \qquad aFd}{a \lhd_{\mathcal{A}} F^{-}(b \downarrow_{\leq} d)}$

(saturation) $\quad \dfrac{a \lhd_{\mathcal{A}} W \qquad (\forall w \varepsilon W)\, wFb}{aFb}$ $\qquad$ *(continuity)* $\quad \dfrac{aFb \qquad b \lhd_{\mathcal{B}} V}{a \lhd_{\mathcal{A}} F^{-}(V)}.$

The definition of continuous relation simplifies when we consider continuous relations between i.g. formal covers as proved in [9]:

Lemma 13.6 *Let* $\mathcal{A}$ *and* $\mathcal{B}$ *be i.g. formal covers having* $I_{\mathcal{A}}(-)$*,*$C_{\mathcal{A}}(-,-)$ *and* $I_{\mathcal{B}}(-)$*,* $C_{\mathcal{B}}(-,-)$ *as axiom-sets respectively. Then, a relation* F *from* A *to* B *is continuous if and only if it satisfies the following conditions:*

(totality) $\quad A \lhd_{\mathcal{A}} F^{-}(B)$ $\qquad$ *($\leq$-convergence)* $\quad \dfrac{aFb \qquad aFd}{a \lhd_{\mathcal{A}} F^{-}(b \downarrow_{\leq} d)}$

($\leq$-saturation) $\quad \dfrac{a \leq c \quad c\,F\,b}{a\,F\,b}$ $\qquad$ *(axiom-saturation)* $\quad \dfrac{i \in I_{\mathcal{A}}(a) \quad (\forall x \varepsilon C_{\mathcal{A}}(a,i))\; x\,F\,b}{a\,F\,b}$

($\leq$-continuity) $\quad \dfrac{a\,F\,b \quad b \leq d}{a\,F\,d}$ $\qquad$ *(axiom-continuity)* $\quad \dfrac{a\,F\,b \quad j \in I_{\mathcal{B}}(b)}{a \lhd F^{-}(C_{\mathcal{B}}(b,j))}.$

Note that to get a continuous relation it is sufficient to saturate a relation satisfying all the above conditions except *axiom saturation*:

Proposition 13.7 *Let* $\mathcal{A}$ *and* $\mathcal{B}$ *be two i.g. formal covers and suppose that* F *is a relation satisfying totality,* $\leq$*-convergence,* $\leq$*-saturation,* $\leq$*-continuity and axiom-continuity. Then*

$$a\,F^{\lhd}\,b \equiv a \lhd_{\mathcal{A}} \{c \in A \mid cFb\}$$

is the minimal continuous relation extending F*, called the saturation of* F*.*

Moreover, it is sufficient to prove *axiom-continuity* of a relation just with respect to an axiom-set whose localization gives the proper axiom-set generating the cover:

Lemma 13.8 *Let $\mathcal{A}$ be an i.g. formal cover and $\mathcal{B}$ be an i.g. formal cover having $I'_{\mathcal{B}}(-)$, $C'_{\mathcal{B}}(-,-)$ as the axiom-set localizing $I_{\mathcal{B}}(-)$, $C_{\mathcal{B}}(-,-)$ according to [5]. Let F be a relation from $\mathcal{A}$ to $\mathcal{B}$ satisfying totality, $\leq$-continuity, $\leq$-saturation, $\leq$-convergence and*

$$\text{(non-loc-axiom-continuity)}\ \frac{a F b \quad j \in I_{\mathcal{B}}(b)}{a \lhd_{\mathcal{A}} F^{-}(C_{\mathcal{B}}(b, j))}.$$

Then F satisfies axiom-continuity with respect to $I'_{\mathcal{B}}(-)$, $C'_{\mathcal{B}}(-,-)$, and, hence its saturation $F^{\lhd}$ is a continuous relation from $\mathcal{A}$ to $\mathcal{B}$.

We denote by $\mathsf{FTop}_\mathsf{i}^{-}$ the category of inductively generated $\leq$-formal covers and continuous relations. We recall that the composition $G * F$ of a continuous relation G from $\mathcal{B}$ to $\mathcal{C}$ with a continuous relation F from $\mathcal{A}$ to $\mathcal{B}$ is defined as $G * F \equiv (G \circ F)^{\lhd}$ that is

$$a\, G * F\, b \equiv a \lhd_{\mathcal{A}} \{x \in A \mid \exists y \in B\ x\, F\, y\ \&\ y\, G\, b\}$$

with A the base of $\mathcal{A}$, and two continuous relations are equal if they are equivalent relations. This category provides a constructive presentation of the category of locales [19]. Moreover, we denote by FTop_i the full subcategory of $\mathsf{FTop}_\mathsf{i}^{-}$ with inductively generated formal topologies. This category provides a constructive presentation of the category of open locales [20].

Here, we recall a lemma proved in [9] that is useful when we need to compose a continuous relation with one obtained by saturation:

Lemma 13.9 *Let $\mathcal{A}$, $\mathcal{B}$ and $\mathcal{C}$ be formal covers, F be a relation from $\mathcal{A}$ to $\mathcal{B}$ which satisfies continuity and G be a relation from $\mathcal{B}$ to $\mathcal{C}$. Then $(G \circ F^{\lhd})^{\lhd} = (G \circ F)^{\lhd}$ and $(G^{\lhd} \circ F)^{\lhd} = (G \circ F)^{\lhd}$.*

In [9], the reader may also find the construction of a binary product of two formal covers:

Lemma 13.10 *Let $\mathcal{A}$ and $\mathcal{B}$ be two i.g. formal covers whose axiom-sets are respectively $I_A(-)$, $C_A(-,-)$ and $I_B(-)$, $C_B(-,-)$. Then, a binary product of $\mathcal{A}$ and $\mathcal{B}$ in $\mathsf{FTop}_\mathsf{i}^{-}$ is the formal cover $\mathcal{A} \times \mathcal{B}$ over the set $A \times B$, with order relation*

$$(a_1, b_1) \leq_{\mathcal{A}\times\mathcal{B}} (a_2, b_2) \equiv (a_1 \leq_{\mathcal{A}} a_2)\ \&\ (b_1 \leq_{\mathcal{B}} b_2)$$

whose cover relation is inductively generated from the axiom-set

$$\begin{aligned} I((a,b)) &\equiv I_A(a) + I_B(b) \\ C((a,b),i) &\equiv \begin{cases} C_A(a, i_a) \times \{b\} & \text{if } i \equiv \mathsf{inl}(i_a) \\ \{a\} \times C_B(b, i_b) & \text{if } i \equiv \mathsf{inr}(i_b) \end{cases} \end{aligned}$$

and whose first and second projections are defined as follows: $\Pi_1 \equiv p_1^{\lhd}$ where $(a,b)\, p_1\, a' \equiv a \lhd_A a'$ and $\Pi_2 \equiv p_2^{\lhd}$ where $(a,b)\, p_2\, b' \equiv b \lhd_B b'$. Moreover, we *indicate with $\langle F, G\rangle$ the relation from $\mathcal{C}$ to $\mathcal{A} \times \mathcal{B}$ that is the pairing of the relation F from $\mathcal{C}$ to $\mathcal{A}$ with the relation G from $\mathcal{C}$ to $\mathcal{B}$.*

Then, from [9] we report the following useful lemma about the binary product:

Lemma 13.11 *Let $\mathcal{A}$ and $\mathcal{B}$ be i.g. formal covers and let $a \in A$, $b \in B$, $U \subseteq A$ and $V \subseteq B$.*

Then the following conditions are valid:

$$(1)\ \frac{a \lhd_{\mathcal{A}} U}{(a, b) \lhd_{A\times B} U \times \{b\}} \qquad (2)\ \frac{b \lhd_{\mathcal{B}} V}{(a, b) \lhd_{A\times B} \{a\} \times V} \qquad (3)\ \frac{a \lhd_{\mathcal{A}} U \qquad b \lhd_{\mathcal{B}} V}{(a, b) \lhd_{A\times B} U \times V}.$$

About binary products, we recall that one reason to work with inductively generated formal covers is that we know how to build a binary product of two formal covers predicatively only if the formal covers are inductively generated.

13.2.1 *Join formal covers*

Observe that any inductively generated formal cover is isomorphic to an i.g. join-formal cover, that is a formal cover whose base is closed under a binary join operation and has a bottom element.

Definition 13.12 *We call $\mathcal{A} \equiv (A, \leq_{\mathcal{A}}, \vee_{\mathcal{A}}, \perp_{\mathcal{A}}, \lhd_{\mathcal{A}})$ a join-formal cover (abbreviated $\vee$-formal cover) if $(A, \leq_{\mathcal{A}}, \lhd_{\mathcal{A}})$ is a formal cover whose base A is closed under the binary join operation $\vee_{\mathcal{A}}$ and has a bottom element $\perp_{\mathcal{A}}$ satisfying*

$$(\perp\text{-left})\quad \perp_{\mathcal{A}} \lhd_{\mathcal{A}} U \qquad\qquad (\vee\text{-left})\quad \frac{a \lhd_{\mathcal{A}} U \qquad b \lhd_{\mathcal{A}} U}{a \vee_{\mathcal{A}} b \lhd_{\mathcal{A}} U}$$

$$(\vee\text{-right})\quad \frac{a \lhd_{\mathcal{A}} b}{a \lhd_{\mathcal{A}} \{b \vee_{\mathcal{A}} c\}} \qquad \frac{a \lhd_{\mathcal{A}} b}{a \lhd_{\mathcal{A}} \{c \vee_{\mathcal{A}} b\}}.$$

We call $\mathsf{FTop}_{\mathsf{i}_\vee}^-$ the full subcategory of $\mathsf{FTop}_{\mathsf{i}}^-$ with $\vee$-formal covers. Actually, we can prove that these two categories are equivalent since the following holds:

Lemma 13.13 *Any i.g. formal cover $\mathcal{A} \equiv (A, \leq_{\mathcal{A}}, \lhd_{\mathcal{A}})$ with axiom-set $I_{\mathcal{A}}(-)$, $C_{\mathcal{A}}(-,-)$, is isomorphic to a $\vee$-formal cover $\mathcal{A}^\vee \equiv (A^\vee, \leq_{\mathcal{A}^\vee}, \vee_{\mathcal{A}^\vee}, \perp_{\mathcal{A}^\vee}, \lhd_{\mathcal{A}^\vee})$ inductively generated with an axiom-set $I^\vee(-)$, $C^\vee(-,-)$ generating the cover $\lhd_{\mathcal{A}^\vee}$ by means of the rules of* reflexivity, $\leq$-left *and* infinity.

We call $\phi_\vee : \mathcal{A} \to \mathcal{A}^\vee$ the isomorphism and $\phi_\vee^{-1} : \mathcal{A}^\vee \to \mathcal{A}$ its inverse.

Proof. We can build $\mathcal{A}^\vee$ by taking the set of lists of elements of A as the basic set, ordered by inclusion of elements. Then, we can define the join operation as the append operation of lists and the bottom element as the empty list. □

Note that elements covered by the bottom element behave like non-positive elements:

Lemma 13.14 *Let $\mathcal{A} \equiv (A, \leq_{\mathcal{A}}, \perp_{\mathcal{A}}, \lhd_{\mathcal{A}})$ and $\mathcal{B} \equiv (B, \leq_{\mathcal{B}}, \perp_{\mathcal{B}}, \lhd_{\mathcal{B}})$ be $\vee$-formal covers and let F be a continuous relation from $\mathcal{A}$ to $\mathcal{B}$. If $c \lhd_{\mathcal{A}} \perp_{\mathcal{A}}$, or $c \lhd_{\mathcal{A}} \emptyset$, then $c\ F\ b$ follows for every $b \in B$.*

13.3 Locally Compact Formal Covers

Here, we give a definition of locally compact formal cover suggested in [10]. This definition is equivalent to the original one in [20] and it is based on the fact that the way-below relation between subsets is definable in an inductively generated formal cover as proved by [10]:

Proposition 13.15 *Let $\mathcal{A} \equiv (A, \leq_{\mathcal{A}}, \lhd_{\mathcal{A}})$ be an i.g. formal cover. We can predicatively define the way-below relation $U \ll V$ between subsets $U, V \subseteq A$, meaning that for all $W \subseteq A$, if $V \lhd W$ then there is a finite (possibly empty) list $w_1, \ldots, w_n$ of elements in W such that $U \lhd \{w_1, \ldots, w_n\}$.*

Proof. We can assume from [5] that the axiom-set $I_A(-), C_A(-,-)$ is set-based according to [5], that is

$$a \lhd W \ \equiv\ \exists j \in I(a) \quad C(a, j) \subseteq W$$

Then, following page 61 in [10], we can define the way-below relation between subsets $U \ll V$ as follows:

$$\forall f \in \Pi_{v \varepsilon V}\, I(v) \quad \exists c_1 \in A \ldots \exists c_n \in A\, \{c_1, \ldots, c_n\} \subseteq \bigcup_{v \varepsilon V} C(v, f(v))\ \&\ U \lhd \{c_1, \ldots, c_n\}$$

where $\Pi_{v \varepsilon V}\, I(v)$ is an abbreviation for $\Pi_{z \in \Sigma_{x \in A} V(x)} I(\pi_1(z))$. □

Note that the predicative definability of the way-below relation between subsets is a crucial point to build the exponential formal cover according to the idea in [2, 3].

Then, following [10] we give the definition of locally compact formal cover.

Definition 13.16 *A formal cover $\mathcal{A}$ is locally compact if it is inductively generated and $a \lhd \{b : b \ll a\}$, for all $a \in S$, where $b \ll a$ stands for the predicative definition mentioned in Proposition 13.15.*

The notion of locally compact formal cover is preserved by isomorphisms as shown in [10]:

Lemma 13.17 *Let $\mathcal{A}$ and $\mathcal{A}'$ be two formal covers isomorphic via continuous relations. If $\mathcal{A}$ is locally compact, then $\mathcal{A}'$ is locally compact, too.*

Moreover, the following properties about the way-below relation can be proved predicatively:

Lemma 13.18 *Let $\mathcal{A} \equiv (A, \leq_{\mathcal{A}}, \vee_{\mathcal{A}}, \perp_{\mathcal{A}}, \lhd_{\mathcal{A}})$ be a locally compact $\vee$-formal cover.*

(1) If $U \ll V$ then $U \lhd_{\mathcal{A}} V$.
(2) If $U \ll V$ and $V \lhd_{\mathcal{A}} W$ then $U \ll W$.

(3) If $U \lhd_{\mathcal{A}} V$ and $V \ll W$ then $U \ll W$.
(4) (Interpolation property) If $a \ll V$ there exists $a' \in A$ such that $a \ll a' \ll V$.

Remark 13.19 Note that we can express the above *interpolation property* because of the presence of joins in the base.

In the absence of joins in the base, we need to express the interpolation property by saying that: if $a \ll V$ then there exists a finite list $a_1, \ldots, a_n$, possibly empty, such that $a \ll \{a_1, \ldots, a_n\}$ and $\{a_1, \ldots, a_n\} \ll V$.

Now, we state a lemma which is auxiliary to prove the next lemma on binary products.

Lemma 13.20 *Let $\mathcal{A} \equiv (A, \leq_{\mathcal{A}}, \vee_{\mathcal{A}}, \perp_{\mathcal{A}}, \lhd_{\mathcal{A}})$ be a locally compact $\vee$-formal cover. If $d \ll a$ and $a \lhd_{\mathcal{A}} U$, then there exist a list $u_1 \ll u'_1, \ldots, u_n \ll u'_n$ such that $d \lhd_{\mathcal{A}} \{u_1, \ldots, u_n\}$ and $u'_i \,\varepsilon\, U$ for $i = 1, \ldots, n$, or $d \lhd_{\mathcal{A}} \perp_{\mathcal{A}}$.*

Then, we prove a lemma on the product of an i.g. formal cover with a locally compact $\vee$-formal cover that will be crucial to prove that the abstraction relation relative to the exponential formal cover is continuous. It expresses that, in some cases, if $(c, a) \lhd_{\mathcal{C} \times \mathcal{A}} W$ then it is possible to decompose W in an open U covering c and in a suitable finite subset covering any given d way-below a.

Lemma 13.21 *Consider the product of an i.g. formal cover $\mathcal{C} \equiv (C, \leq_{\mathcal{C}}, \lhd_{\mathcal{C}})$ having $I_{\mathcal{C}}(-)$, $C_{\mathcal{C}}(-,-)$ as axiom-set and a locally compact $\vee$-formal cover $\mathcal{A} \equiv (A, \leq_{\mathcal{A}}, \vee_{\mathcal{A}}, \perp_{\mathcal{A}}, \lhd_{\mathcal{A}})$. Suppose also that $c \in C$, $a \in A$, $(c, a) \lhd_{\mathcal{C} \times \mathcal{A}} W$ where $W \subseteq C \times A$ satisfies the following properties:*

(1) if $(c', a') \,\varepsilon\, W$, then $(c'', a') \,\varepsilon\, W$ for every c'' such that $c'' \leq c'$;
(2) $(c, \perp_{\mathcal{A}}) \,\varepsilon\, W$.

Then, for any $d \ll a$ there exists a subset U of C such that $c \lhd_{\mathcal{C}} U$ and, for every $u \,\varepsilon\, U$ there exists $e_1 \ll e'_1, \ldots, e_n \ll e'_n \in A$ with $d \lhd_{\mathcal{A}} \{e_1, \ldots, e_n\}$ and $(u, e'_j) \,\varepsilon\, W$ for $j = 1, \ldots, n$.

Proof. Suppose $d \ll a$. We prove the statement by induction on the derivation of $(c, a) \lhd_{\mathcal{C} \times \mathcal{A}} W$. If $(c, a) \lhd_{\mathcal{C} \times \mathcal{A}} W$ follows by *reflexivity* then $(c, a) \,\varepsilon\, W$. Hence, take $U \equiv \{c\}$ and for $u \equiv c$ take $e_1 \equiv d$ and $e'_1 \equiv a$. Instead, if $(c, a) \lhd_{\mathcal{C} \times \mathcal{A}} W$ follows by *infinity* then we have to consider two cases, namely that it follows from $\{c\} \times C_{\mathcal{A}}(a, j) \lhd_{\mathcal{C} \times \mathcal{A}} W$ or from $C_{\mathcal{C}}(c, j) \times \{a\} \lhd_{\mathcal{C} \times \mathcal{A}} W$.

If it follows from $\{c\} \times C_{\mathcal{A}}(a, j) \lhd_{\mathcal{C} \times \mathcal{A}} W$, we apply Lemma 13.20 to $d \ll a$ and $a \lhd_{\mathcal{A}} C_{\mathcal{A}}(a, j)$. If $d \lhd_{\mathcal{A}} \perp_{\mathcal{A}}$ we put $U \equiv \{c\}$ and $e_1 = e'_1 \equiv \perp_{\mathcal{A}}$, since $\perp_{\mathcal{A}} \ll \perp_{\mathcal{A}}$ and $(c, \perp_{\mathcal{A}}) \in W$. Instead, if we have $h_1 \ll h'_1, \ldots, h_m \ll h'_m$ with $h'_i \,\varepsilon\, C_{\mathcal{A}}(a, j)$ for $i = 1, \ldots, m$ such that $d \lhd_{\mathcal{A}} \{h_1, \ldots, h_m\}$, then by the inductive hypothesis for every $h'_i \,\varepsilon\, C_{\mathcal{A}}(a, j)$ with $i = 1, \ldots, m$ there exists a subset $U_{h'_i}$ of C such that $c \lhd_{\mathcal{C}} U_{h'_i}$ and for every $u_{h'_i} \,\varepsilon\, U_{h'_i}$ there exist $e_1^{h'_i} \ll (e^{h'_i})'_1, \ldots, e_n^{h'_i} \ll (e^{h'_i})'_{n_i}$ such that $h_i \lhd_{\mathcal{A}} \{e_1^{h'_i}, \ldots, e_{n_i}^{h'_i}\}$ and $(u_{h'_i}, (e^{h'_i})'_j) \,\varepsilon\, W$ for $j = 1, \ldots, n_i$. Therefore, take

$U \equiv U_{h'_1} \downarrow_{\leq} \ldots \downarrow_{\leq} U_{h'_m}$. Indeed, for every $u \,\varepsilon\, U$ there exist $u_{h'_i} \ldots, u_{h'_m}$ such that $u \leq u_{h'_1} \ldots u \leq u_{h'_m}$. So, we take all the $e_1^{h'_i} \ll (e^{h'_i})'_1, \ldots e_{n_i}^{h'_i} \ll (e^{h'_i})'_{n_i}$ associated to each $u_{h'_i}$ for $i = 1, \ldots, m$ since $d \lhd_{\mathcal{A}} \cup_{i=1,\ldots,m}\{e_1^{h'_i}, \ldots, e_{n_i}^{h'_i}\}$. Moreover, observe also that for any $j = 1, \ldots, n$ and $i = 1, \ldots, m$ then $(u, (e^{h'_i})'_j) \,\varepsilon\, W$ follows from the hypothesis on W since $u \leq u_{h'_j}$ and $(u_{h'_i}, (e^{h'_i})'_j) \,\varepsilon\, W$.

Instead, if $(c, a) \lhd_{C\times A} W$ follows by using *infinity* on $C_{\mathcal{C}}(c, j) \times \{a\} \lhd_{C\times A} W$, then take $U \equiv \bigcup_{v \varepsilon C_{\mathcal{C}}(c,j)} U_v$ with U_v obtained by the inductive hypothesis on $(v, a) \lhd_{C\times A} W$ for $v \varepsilon C_{\mathcal{C}}(c, j)$. Finally, if $(c, a) \lhd_{C\times A} W$ follows from $\leq$*-left* on $(c, a) \leq (c', a')$ and $(c', a') \lhd_{C\times A} W$, take U obtained by the inductive hypothesis on $(c', a') \lhd_{C\times A} W$. □

13.4 Exponentiation

In this section we prove that any locally compact $\vee$-formal cover is exponentiable in $\mathsf{FTop}_{\mathsf{i}_\vee}^{-}$. This suffices to prove that any locally compact formal cover is exponentiable in $\mathsf{FTop}_{\mathsf{i}}^{-}$ since by Lemma 13.17 and Lemma 13.13 we know that any locally compact formal cover is isomorphic to a locally compact $\vee$-formal cover and that $\mathsf{FTop}_{\mathsf{i}_\vee}^{-}$ and $\mathsf{FTop}_{\mathsf{i}}^{-}$ are equivalent.

To exponentiate a locally compact $\vee$-formal cover $\mathcal{A}$ over any $\vee$-formal cover $\mathcal{B}$ in $\mathsf{FTop}_{\mathsf{i}_\vee}^{-}$, we need to define the exponential cover of $\mathcal{A}$ over $\mathcal{B}$ in such a way that its formal points are in bijection with the continuous relations from $\mathcal{A}$ to $\mathcal{B}$.

As seen in [9] we are able to perform such a correspondence by encoding a continuous relation via lists of pairs of basic opens (a, b) for $a \in A$ and $b \in B$ (where A is the set of basic opens for $\mathcal{A}$ and B the set of basic opens for $\mathcal{B}$) if the conditions defining a continuous relation can be formulated according to one of the forms

$$\frac{a\ R\ b \qquad P(a, b, a', b')}{a'\ R\ b'} \qquad \frac{a\ R\ b \qquad Q(a, b, V)}{(\exists y \,\varepsilon\, V)\, a\ R\ y}.$$

But, note that the *axiom-saturation* condition of a continuous relation from a locally compact formal cover to any other i.g. formal cover is not of the above forms, and it does not seem to be reducible to one of the above forms, contrary to what happens with continuous relations from a unary topology (as we noticed in [9]). However, inspired by the fact that the exponential topology built in [2], [7], [3] is generated from axioms expressing the properties of the relation $a \ll f^*(b)$ for $a \in A$ and $b \in B$ and f locale morphism, observe that to any continuous relation R, from a locally compact $\vee$-formal cover to any i.g. $\vee$-formal cover, we can associate its way-below component, defined as $a \ll R^-(b)$, whose properties can be expressed by conditions of the above forms and hence can be used to generate an exponential cover. Therefore, we give the definition of *wb-relation* in terms of the conditions satisfied by $a \ll R^-(b)$. Then, after proving that wb-relations from a locally compact $\vee$-formal cover to any i.g. $\vee$-formal cover are in bijection with the corresponding continuous relations, we build the exponential topology according to the semantics given in [9] with reference to wb-relations instead of referring to continuous relations.

Definition 13.22 *Let $\mathcal{A}$ be a locally compact $\vee$-formal cover and $\mathcal{B}$ be an i.g. $\vee$-formal cover with axiom-set $I_{\mathcal{B}}(-)$, $C_{\mathcal{B}}(-,-)$. Then, F^{wb} is a* wb-relation *from $\mathcal{A}$ to $\mathcal{B}$ if it is a relation from A to B and it satisfies the following conditions:*

(wb-bottom continuity)
$$\frac{c\ F^{wb}\ \perp_{\mathcal{B}}}{c \lhd_{\mathcal{A}} \perp_{\mathcal{A}}}$$

(wb-bottom saturation)
$$\frac{b \in B}{\perp_{\mathcal{A}}\ F^{wb}\ b}$$

(wb-totality)
$$\frac{d \ll a}{\exists\, y_1,\ldots,y_n \in B \quad \exists\, x_1,\ldots,x_n \in A \quad x_i\ F^{wb}\ y_i \quad i=1,\ldots,n \quad \& \quad d \lhd_{\mathcal{A}} \{x_1,\ldots,x_n\}}$$

(wb-$\leq$-convergence)
$$\frac{a\ F^{wb}\ b \qquad a\ F^{wb}\ c \qquad d \ll a}{\exists\, y_1,\ldots,y_n \in (b \downarrow_{\leq} c) \cup \{\perp_{\mathcal{B}}\}\ \exists\, x_1,\ldots,x_n \in A\ x_i\ F^{wb}\ y_i\ i=1,\ldots,n\ \&\ d \lhd_{\mathcal{A}} \{x_1,\ldots,x_n\}}$$

(wb-finite-saturation)
$$\frac{a \lhd_{\mathcal{A}} \{c_1,\ldots,c_n\} \qquad \forall x \,\varepsilon\, \{c_1,\ldots,c_n\} \qquad x\ F^{wb}\ b}{a\ F^{wb}\ b}$$

(wb-axiom-continuity)
$$\frac{a\ F^{wb}\ b \qquad j \in J(b) \qquad d \ll a}{\exists y_1,\ldots,y_n \,\varepsilon\, C(b,j) \cup \{\perp_{\mathcal{A}}\}\ \exists\, x_1,\ldots,x_n \in A\ x_i\ F^{wb}\ y_i\ i=1,\ldots,n\ \&\ d \lhd_{\mathcal{A}} \{x_1,\ldots,x_n\}}$$

(wb-$\leq$-continuity)
$$\frac{a\ F^{wb}\ b \qquad b \leq d}{a\ F^{wb}\ d}$$

(interpolation)
$$\frac{a\ F^{wb}\ b}{\exists x \in A \quad x\ F^{wb}\ b \quad \& \quad a \ll x}.$$

Lemma 13.23 *Let $\mathcal{A}$ be a locally compact $\vee$-formal cover and $\mathcal{B}$ any i.g. $\vee$-formal cover with axiom-set $I_{\mathcal{B}}(-)$, $C_{\mathcal{B}}(-,-)$. Then, the collection of continuous relations from $\mathcal{A}$ to $\mathcal{B}$ is in bijection with the collection of wb-relations from $\mathcal{A}$ to $\mathcal{B}$. Indeed,*

(1) Given any continuous relation F from $\mathcal{A}$ to $\mathcal{B}$, then $a\ F^{wb}\ b \equiv a \ll F^{-}(b)$, called the wb-component of F, is a wb-relation from $\mathcal{A}$ to $\mathcal{B}$.

(2) Given any wb-relation R from $\mathcal{A}$ to $\mathcal{B}$ then $a\ R^{\lhd}\ b \equiv a \lhd_{\mathcal{A}} \{x \in S \mid x\ R\ b\}$ is a continuous relation from $\mathcal{A}$ to $\mathcal{B}$.

(3) Given any continuous relation F from $\mathcal{A}$ to $\mathcal{B}$ then $F = (F^{wb})^{\lhd}$.

(4) Given any wb-relation R from $\mathcal{A}$ to $\mathcal{B}$ then $R = (R^{\lhd})^{wb}$.

13.4.1 *The exponential formal cover*

For any locally compact $\vee$-formal cover $\mathcal{A} \equiv (A, \leq_{\mathcal{A}}, \vee_{\mathcal{A}}, \perp_{\mathcal{A}}, \lhd_{\mathcal{A}})$ and any i.g. $\vee$-formal cover $\mathcal{B} \equiv (B, \leq_{\mathcal{B}}, \vee_{\mathcal{B}}, \perp_{\mathcal{B}}, \lhd_{\mathcal{B}})$ with axiom-set $I_{\mathcal{B}}(-)$, $C_{\mathcal{B}}(-,-)$, the exponential cover candidate $\mathcal{B}^{\mathcal{A}}$ of $\mathcal{A}$ over $\mathcal{B}$ is defined as the $\vee$-formal cover isomorphic to a formal cover $[\mathcal{A} \to \mathcal{B}]$ whose base does not have natural joins.

As for the exponential formal cover in [9], the set of basic neighbourhoods of $[\mathcal{A} \to \mathcal{B}]$ is *List*$(A \times B)$, that is lists of pairs of elements in the cartesian product $A \times B$, where we indicate the empty list with [ ] and the list of elements $(a_1, b_1), \ldots, (a_n, b_n)$ with $[(a_1, b_1), \ldots, (a_n, b_n)]$.

The intended meaning of a list $l \in List(A \times B)$ is to give partial information on a wb-relation from $\mathcal{A}$ to $\mathcal{B}$. The order on $List(A \times B)$ is

$$l \leq_{[\mathcal{A}\to\mathcal{B}]} m \equiv (\forall (a, b) \in A \times B) \quad (a, b)\, \varepsilon\, m \to (a, b)\, \varepsilon\, l$$

stating that the list l is more precise, that is, it approximates fewer wb-relations than the list m. This order relation is a refinement of the reverse sublist relation, which states that m is a sublist of l, because it does not consider the order among the elements in a list and their repetitions.

We define the cover relation $\lhd_{[\mathcal{A}\to\mathcal{B}]}$ according to the usual semantics of formal topology (see [21]) as follows: $l \lhd_{[\mathcal{A}\to\mathcal{B}]} U$ *means that every formal point containing* l *also contains a basic neighbourhood of* U. Recalling that formal points are expected to be wb-relations, this means that we have to prove that *for any wb-relation* R, *if* $a\,R b$ *holds for each pair* $(a, b)\, \varepsilon\, l$, *then there exists* $m\, \varepsilon\, U$ *such that* $a\, R\, b$ *holds for each pair* $(a, b)\, \varepsilon\, m$.

Since we want to obtain an i.g. formal cover, we apply the method in [5] and we define the axiom-set $I_{[\mathcal{A}\to\mathcal{B}]}(-)$, $C_{[\mathcal{A}\to\mathcal{B}]}(-,-)$ as the localization of another set of axioms $I'_{[\mathcal{A}\to\mathcal{B}]}(-)$, $C'_{[\mathcal{A}\to\mathcal{B}]}(-,-)$ that we describe in the following: we express each axiom in the form $l \lhd_{[\mathcal{A}\to\mathcal{B}]} U$ by intending that for some index $j \in I'_{[\mathcal{A}\to\mathcal{B}]}(l)$ then $C'_{[\mathcal{A}\to\mathcal{B}]}(l, j) \equiv U$. We hope that it will be clear how such a formalization can actually be performed.

The general recipe is that each axiom is in correspondence with a condition defining a wb-relation in Definition 13.22.

The set of axioms $I'_{[\mathcal{A}\to\mathcal{B}]}(-)$ and $C'_{[\mathcal{A}\to\mathcal{B}]}(-,-)$ include only:

(wb-bottom continuity)
$[(a, \bot_{\mathcal{B}})] \lhd_{[\mathcal{A}\to\mathcal{B}]} \{[(\bot_{\mathcal{A}}, \bot_{\mathcal{B}})] \mid a \lhd_{\mathcal{A}} \bot_{\mathcal{A}}\}$

(wb-bottom saturation)
$[\,] \lhd_{[\mathcal{A}\to\mathcal{B}]} [(\bot_{\mathcal{A}}, b)]$
provided that $b \in B$

(wb-totality)
$[\,] \lhd_{[\mathcal{A}\to\mathcal{B}]} \{[(x_1, y_1), \ldots, (x_n, y_n)] \mid y_1, \ldots, y_n \in B \;\;\&\;\; d \lhd_{\mathcal{A}} \{x_1, \ldots, x_n\}\}$
provided that $d \ll a$ for some $a \in A$

(wb-$\leq$-convergence)
$[(a, b), (a, b')] \lhd_{[\mathcal{A}\to\mathcal{B}]} \{[(x_1, y_1), \ldots, (x_n, y_n)] \mid y_i\, \varepsilon\, (b \downarrow_{\leq} b') \cup \{\bot_{\mathcal{B}}\}$ for $i = 1, \ldots, n$
$\& \; d \lhd_{\mathcal{A}} \{x_1, \ldots, x_n\}\}$
provided that $d \ll a$

(wb-finite-saturation)
$[(d_1, b), \ldots, (d_n, b)] \lhd_{[\mathcal{A}\to\mathcal{B}]} [(a, b)]$
provided that $a \lhd_{\mathcal{A}} \{d_1, \ldots, d_n\}$ for some $a \in A$

(*wb-axiom-continuity*)
$[(a, b)] \lhd_{[\mathcal{A}\to\mathcal{B}]} \{[(x_1, y_1), \dots, (x_n, y_n)] \mid y_i \,\varepsilon\, C(b, j) \cup \{\bot_{\mathcal{B}}\} \text{ for } i = 1, \dots, n$
$\& \; d \lhd_{\mathcal{A}} \{x_1, \dots, x_n\}\}$

provided that $d \ll a$ and $j \in J(b)$

(*wb-$\leq$-continuity*)
$[(a, b)] \lhd_{[\mathcal{A}\to\mathcal{B}]} [(a, b')]$
provided that $b \leq b'$

(*interpolation*)
$[(a, b)] \lhd_{[\mathcal{A}\to\mathcal{B}]} \{[(x, b)] \mid a \ll x\}.$

Then, as said, we localize such an axiom-set $I'_{[\mathcal{A}\to\mathcal{B}]}(-)$ and $C'_{[\mathcal{A}\to\mathcal{B}]}(-, -)$ according to [5] and we obtain the axiom-set $I_{[\mathcal{A}\to\mathcal{B}]}(-)$, $C_{[\mathcal{A}\to\mathcal{B}]}(-, -)$ that generates the formal cover $[\mathcal{A} \to \mathcal{B}]$.

Note that the formal cover $[\mathcal{A} \to \mathcal{B}]$ lives in $\mathsf{FTop}_{\mathsf{i}}^{-}$ but not in $\mathsf{FTop}_{\mathsf{i}_\vee}^{-}$. Hence, we take its isomorphic copy whose base is closed under joins $[\mathcal{A} \to \mathcal{B}]_\vee$ and we define

$$\mathcal{B}^{\mathcal{A}} \equiv [\mathcal{A} \to \mathcal{B}]_\vee.$$

Remark 13.24 Note that since the formal cover $[\mathcal{A} \to \mathcal{B}]$ lives in $\mathsf{FTop}_{\mathsf{i}}^{-}$ one could suggest working within $\mathsf{FTop}_{\mathsf{i}}^{-}$. But then we would need to express the exponential cover axioms without the use of a symbol for a bottom element and for joins both for $\mathcal{A}$ and for $\mathcal{B}$. Now, it is immediate to observe that to express the exponential axioms we do not need joins in $\mathcal{B}$ but just a bottom element and this is why in [11] we worked in the full subcategory $\mathsf{FTop}_{\mathsf{i}_\bot}^{-}$ of $\mathsf{FTop}_{\mathsf{i}}^{-}$ with formal covers having a bottom element. The bottom element in $\mathcal{B}$ seems necessary to express *wb-bottom continuity* and *wb-$\leq$-convergence* by including the case in which an element is covered by the empty set, that is to circumvent the fact that we work with formal covers that do not necessarily have a positivity predicate.

In principle, we could work in $\mathsf{FTop}_{\mathsf{i}_\bot}^{-}$ with a locally compact cover having a base without joins, but then to express the *interpolation* axiom, as remarked in 13.19 about the interpolation property, we would need to use finite lists in the first component of the pairs, namely we would need to change the base of the exponential to $List(List(A) \times B)$ making the various proofs more complicated.

Let us recall now useful lemmas on the exponential topology proved in [9]. Given $l_1, l_2 \in List(A \times B)$, we indicate with $l_1 \cdot l_2$ the append operation of the list l_1 with l_2.

Lemma 13.25 *Let $l \in List(A \times B)$ and $U_1, U_2 \subseteq List(A \times B)$. Then the following rule holds:*

$$(\cdot\text{-}right) \quad \frac{l \lhd_{[\mathcal{A}\to\mathcal{B}]} U_1 \qquad l \lhd_{[\mathcal{A}\to\mathcal{B}]} U_2}{l \lhd_{[\mathcal{A}\to\mathcal{B}]} U_1 \cdot U_2}$$

where $U_1 \cdot U_2 \equiv \{m_1 \cdot m_2 \mid m_1 \,\varepsilon\, U_1 \,\&\, m_2 \,\varepsilon\, U_2\}$.

Lemma 13.26 *Let F be a continuous relation from $\mathcal{C}$ to $[\mathcal{A} \to \mathcal{B}]$. Then, for any $c \in C$ and any $l_1, l_2 \in List(A \times B)$, the following rule holds*

$$\frac{c\,F\,l_1 \qquad c\,F\,l_2}{c\,F\,l_1 \cdot l_2}.$$

As proved in [9], the cover of $[\mathcal{A} \to \mathcal{B}]$ does not depend on the particular axiom-set generating the cover of $\mathcal{B}$ since we can show:

Lemma 13.27 *Let $\mathcal{A}$ be a locally compact $\vee$-formal cover and $\mathcal{B}$ be an i.g. $\vee$-formal cover. Then, for any list $l \in List(A \times B)$ and any $V \subseteq B$, if $(a, b) \,\varepsilon\, l$ and $b \lhd_{\mathcal{B}} V$ then $l \lhd_{[\mathcal{A}\to\mathcal{B}]} \{(a, y) \mid y \,\varepsilon\, V\}$.*

13.4.2 *Application and abstraction*

Now we are going to show that the formal cover $\mathcal{B}^{\mathcal{A}} \equiv [\mathcal{A} \to \mathcal{B}]_{\vee}$ is the exponential of the locally compact $\vee$-formal cover $\mathcal{A}$ over the i.g. $\vee$-formal cover $\mathcal{B}$.

From a categorical point of view, we prove that for any locally compact $\vee$-formal cover $\mathcal{A}$ the functor $- \times \mathcal{A} \; : \; \mathsf{FTop}_{\mathsf{i}_{\vee}}^{-} \Rightarrow \mathsf{FTop}_{\mathsf{i}_{\vee}}^{-}$ has got a right adjoint $(-)^{\mathcal{A}}$: $\mathsf{FTop}_{\mathsf{i}_{\vee}}^{-} \Rightarrow \mathsf{FTop}_{\mathsf{i}_{\vee}}^{-}$. Equivalently, this amounts to defining, for any locally compact $\vee$-formal cover $\mathcal{A}$ and any i.g. $\vee$-formal cover $\mathcal{B}$, an *application relation* Ap from $(\mathcal{B}^{\mathcal{A}}) \times \mathcal{A}$ to $\mathcal{B}$ such that for any continuous relation F from $\mathcal{C} \times \mathcal{A}$ to $\mathcal{B}$ there exists a continuous relation $\Lambda(F)$ from $\mathcal{C}$ to $\mathcal{B}^{\mathcal{A}}$, called *abstraction* of F, such that $\mathsf{Ap} * \langle \Lambda(F) * \Pi_1, \Pi_2 \rangle = F$ holds, and for any continuous relation G from $\mathcal{C}$ to $\mathcal{B}^{\mathcal{A}}$ then also $\Lambda(\mathsf{Ap} * \langle G * \Pi_1, \Pi_2 \rangle) = G$ holds.

Since we prefer to work with $[\mathcal{A} \to \mathcal{B}]$ than with $[\mathcal{A} \to \mathcal{B}]_{\vee}$, we give the definitions of application and abstraction with respect to $[\mathcal{A} \to \mathcal{B}]$ and we use such definitions to give those of application and abstraction for $[\mathcal{A} \to \mathcal{B}]_{\vee}$ by the isomorphisms $\phi_{\vee}$: $[\mathcal{A} \to \mathcal{B}] \longrightarrow [\mathcal{A} \to \mathcal{B}]_{\vee}$ and $\phi_{\vee}^{-} : [\mathcal{A} \to \mathcal{B}]_{\vee} \longrightarrow [\mathcal{A} \to \mathcal{B}]$.

Indeed, we define the application $\mathsf{Ap} : [\mathcal{A} \to \mathcal{B}]_{\vee} \times \mathcal{A} \longrightarrow \mathcal{B}$ as the composition of the application $\widetilde{\mathsf{Ap}}^{\lhd} : [\mathcal{A} \to \mathcal{B}] \times \mathcal{A} \longrightarrow \mathcal{B}$ relative to $[\mathcal{A} \to \mathcal{B}]$ with the morphism $\phi_{\vee}^{-} \times \mathsf{id} : [\mathcal{A} \to \mathcal{B}]_{\vee} \times \mathcal{A} \longrightarrow [\mathcal{A} \to \mathcal{B}] \times \mathcal{A}$, that is

$$\mathsf{Ap} \equiv \widetilde{\mathsf{Ap}}^{\lhd} * (\phi_{\vee}^{-} \times \mathsf{id})$$

where

$$(l, a) \, \widetilde{\mathsf{Ap}} \, b \equiv l \lhd_{[\mathcal{A}\to\mathcal{B}]} [(a, b)]$$

for any $l \in List(A \times B)$, $a \in A$ and $b \in B$.

Then, for any given relation $F : \mathcal{C} \times \mathcal{A} \longrightarrow \mathcal{B}$, we define the abstraction $\Lambda(F)$: $\mathcal{C} \longrightarrow [\mathcal{A} \to \mathcal{B}]_{\vee}$ as the composition of $\phi_{\vee} : [\mathcal{A} \to \mathcal{B}] \longrightarrow [\mathcal{A} \to \mathcal{B}]_{\vee}$ with the abstraction $\widetilde{\Lambda}(F)^{\lhd} : \mathcal{C} \longrightarrow [\mathcal{A} \to \mathcal{B}]$ relative to $[\mathcal{A} \to \mathcal{B}]$, that is

$$\Lambda(F) \equiv \phi_{\vee} * (\widetilde{\Lambda}(F)^{\lhd})$$

where

$$c \, \widetilde{\Lambda}(F) \, l \equiv (\forall (a, b) \,\varepsilon\, l) \; a \ll \{x \in A \mid (c, x) \, F \, b\}$$

for any $l \in List(A \times B)$ and $c \in C$.

To prove that Ap is a continuous relation, we just need to show that $\widetilde{\mathsf{Ap}}^{\lhd}$ is a continuous relation. For this purpose it is sufficient to show that $\widetilde{\mathsf{Ap}}$ satisfies all the conditions in Proposition 13.7. Before doing this we prove the following useful lemma.

Lemma 13.28 *Let $\mathcal{A}$ be a locally compact $\vee$-formal cover and $\mathcal{B}$ be an i.g. $\vee$-formal cover. Then, for any subset U of B*

$$\widetilde{\mathsf{Ap}}^{-}(U \cup \{\bot_{\mathcal{B}}\}) \lhd_{[\mathcal{A}\to\mathcal{B}]\times\mathcal{A}} \widetilde{\mathsf{Ap}}^{-}(U).$$

Proof. Suppose $(l, a) \,\varepsilon\, \widetilde{\mathsf{Ap}}^{-}(U \cup \{\bot_{\mathcal{B}}\})$. Then, either there exists $b \in U$ such that $(l, a)\, \widetilde{\mathsf{Ap}}\, b$, and hence $(l, a) \lhd_{(A\to B)_{\vee}\times A} \widetilde{\mathsf{Ap}}^{-}(U)$, or $(l, a)\, \widetilde{\mathsf{Ap}}\, \bot_{\mathcal{B}}$. In this latter case, by definition we get $l \lhd_{[\mathcal{A}\to\mathcal{B}]} [(a, \bot_{\mathcal{B}})]$ and hence by *transitivity* with the *wb-bottom continuity* axiom we obtain $l \lhd_{[\mathcal{A}\to\mathcal{B}]} \{[(\bot_{\mathcal{A}}, \bot_{\mathcal{B}})] \mid a \lhd_{\mathcal{A}} \bot_{\mathcal{A}}\}$, from which by Lemma 13.11 $(l, a) \lhd_{[\mathcal{A}\to\mathcal{B}]\times\mathcal{A}} \{[(\bot_{\mathcal{A}}, \bot_{\mathcal{B}})] \mid a \lhd_{\mathcal{A}} \bot_{\mathcal{A}}\} \times \{a\}$ follows. Then, we conclude $(l, a) \lhd_{[\mathcal{A}\to\mathcal{B}]\times\mathcal{A}} \widetilde{\mathsf{Ap}}^{-}(U)$ since $\{[(\bot_{\mathcal{A}}, \bot_{\mathcal{B}})] \mid a \lhd_{\mathcal{A}} \bot_{\mathcal{A}}\} \times \{a\} \lhd_{[\mathcal{A}\to\mathcal{B}]\times\mathcal{A}} \emptyset$. □

Lemma 13.29 *Let $\mathcal{A}$ be a locally compact $\vee$-formal cover and $\mathcal{B}$ be an i.g. $\vee$-formal cover. Then, $\widetilde{\mathsf{Ap}}$ is a relation from $List(A \times B) \times A$ to B which satisfies* totality, $\leq$-convergence, axiom continuity, $\leq$-continuity *and* $\leq$-saturation.

Proof. We just prove (*$\leq$-convergence*). Suppose $(l, a)\, \widetilde{\mathsf{Ap}}\, b$ and $(l, a)\, \widetilde{\mathsf{Ap}}\, b'$, that is $l \lhd_{[\mathcal{A}\to\mathcal{B}]} [(a, b)]$ and $l \lhd_{[\mathcal{A}\to\mathcal{B}]} [(a, b')]$. Then, $l \lhd_{[\mathcal{A}\to\mathcal{B}]} [(a, b), (a, b')]$ follows by Lemma 13.25, and by the *wb-$\leq$-convergence* axiom and *transitivity* we obtain for $d \ll a$

$$l \lhd_{[\mathcal{A}\to\mathcal{B}]} \{[(x_1, y_1), \ldots, (x_n, y_n)] \mid y_i \,\varepsilon\, (b \downarrow_{\leq} b') \cup \{\bot_{\mathcal{B}}\} \text{ for } i = 1, \ldots, n \;\&\; d \lhd_{\mathcal{A}} \{x_1, \ldots, x_n\}\}$$

and by Lemma 13.11

$$(l, d) \lhd_{[\mathcal{A}\to\mathcal{B}]\times\mathcal{A}} \{[(x_1, y_1), \ldots, (x_n, y_n)] \mid y_i \,\varepsilon\, (b \downarrow_{\leq} b') \cup \{\bot_{\mathcal{B}}\} \text{ for } i = 1, \ldots, n \;\&\; d \lhd_{\mathcal{A}} \{x_1, \ldots, x_n\}\} \times \{d\}.$$

Then, for every $[(x_1, y_1), \ldots, (x_n, y_n)]$ in the above subset covering l we obtain

$$([(x_1, y_1), \ldots, (x_n, y_n)], d) \lhd_{[\mathcal{A}\to\mathcal{B}]\times\mathcal{A}} \{[(x_1, y_1), \ldots, (x_n, y_n)]\} \times \{x_1, \ldots, x_n\}.$$

Since $[(x_1, y_1), \ldots, (x_n, y_n)] \leq [(x_i, y_i)]$, we get $([(x_1, y_1), \ldots, (x_n, y_n)], x_i) \lhd_{[\mathcal{A}\to\mathcal{B}]} ([(x_i, y_i)], x_i)$ and $([(x_i, y_i)], x_i) \lhd_{[\mathcal{A}\to\mathcal{B}]\times\mathcal{A}} \widetilde{\mathsf{Ap}}^{-}((b \downarrow b') \cup \{\bot_{\mathcal{B}}\})$ for $i = 1, \ldots, n$, since $([(x_i, y_i)], x_i)\, \widetilde{\mathsf{Ap}}\, y_i$ holds with $y_i \,\varepsilon\, (b \downarrow_{\leq} b') \cup \{\bot_{\mathcal{B}}\}$. Then, we conclude $([(x_i, y_i)], x_i) \lhd_{[\mathcal{A}\to\mathcal{B}]\times\mathcal{A}} \widetilde{\mathsf{Ap}}^{-}((b \downarrow b') \cup \{\bot_{\mathcal{B}}\})$ and by *transitivity* and Lemma 13.28 also that $([(x_1, y_1), \ldots, (x_n, y_n)], d) \lhd_{[\mathcal{A}\to\mathcal{B}]\times\mathcal{A}} \widetilde{\mathsf{Ap}}^{-}(b \downarrow b')$.

Since this holds for any list $[(x_1, y_1), \ldots, (x_n, y_n)]$ under the above hypothesis, we conclude $(l, d) \lhd_{[\mathcal{A}\to\mathcal{B}]\times\mathcal{A}} \widetilde{\mathsf{Ap}}^{-}(b \downarrow b')$. Lastly, since this holds for any $d \ll a$, from $(l, a) \lhd_{[\mathcal{A}\to\mathcal{B}]\times\mathcal{A}} \{(l, d) \mid d \ll a\}$, obtained by Lemma 13.11 from $a \lhd_{\mathcal{A}} wb(a)$, we conclude $(l, a) \lhd_{[\mathcal{A}\to\mathcal{B}]\times\mathcal{A}} \widetilde{\mathsf{Ap}}^{-}(b \downarrow b')$. □

Finally, by Lemma 13.29 and Lemma 13.7 we conclude:

Corollary 13.30 *Let $\mathcal{A}$ be a locally compact $\vee$-formal cover and $\mathcal{B}$ be an i.g. $\vee$-formal cover. Then,* Ap *is a continuous relation from $[\mathcal{A} \to \mathcal{B}]_\vee \times \mathcal{A}$ to $\mathcal{B}$.*

To prove that the *abstraction* $\Lambda(F)$ of a continuous relation F is a continuous relation, too, we just show that $\widetilde{\Lambda}(F)^{\lhd}$ is a continuous relation. For this purpose it is sufficient to show that $\widetilde{\Lambda}(F)$ satisfies all the conditions in Lemma 13.8. Before showing this, note this useful lemma.

Lemma 13.31 *Let $\mathcal{A}$ be a locally compact $\vee$-formal cover and $\mathcal{C}$ and $\mathcal{B}$ be i.g. $\vee$-formal covers. Suppose that F is any continuous relation from $\mathcal{C} \times \mathcal{A}$ to $\mathcal{B}$. Then $a \ll \{x \mid (c, x)\, F\, b\}$ if and only if $\exists x\ a \ll x$ and $(c, x)\, F\, b$.*

Note that in the following lemma to prove *non-loc-axiom-continuity* for $\widetilde{\Lambda}(F)$ it is crucial to apply Lemma 13.21 to decompose some $(c, a') \lhd_{\mathcal{C}\times\mathcal{A}} F^-(U)$ for some subset U into a component covering c and some other component covering any given way-below element of a'.

Lemma 13.32 *Let $\mathcal{A}$ be a locally compact $\vee$-formal cover and $\mathcal{C}$ and $\mathcal{B}$ be two i.g. $\vee$-formal covers. Suppose that F is any continuous relation from $\mathcal{C} \times \mathcal{A}$ to $\mathcal{B}$. Then $\widetilde{\Lambda}(F)$ is a relation from C to $List(A \times B)$ satisfying* totality, $\leq$-convergence, non-loc-axiom-continuity, $\leq$-continuity *and* $\leq$-saturation.

Proof. We just prove some conditions:

(*totality*) Let $c \in C$. Then, $c \lhd_{\mathcal{C}} \widetilde{\Lambda}(F)^-(List(A \times B))$ follows by *reflexivity* since $c\, \widetilde{\Lambda(F)}\, [\,]$ holds by intuitionistic logic.

($\leq$-*convergence*) If $c\, \widetilde{\Lambda}(F)\, l_1$ and $c\, \widetilde{\Lambda}(F)\, l_2$ then by definition we immediately get $c\, \widetilde{\Lambda}(F)\, l_1 \cdot l_2$. and hence by *reflexivity* $c \lhd_{\mathcal{C}} \widetilde{\Lambda}(F)^-(l_1 \downarrow l_2)$ since $l_1 \cdot l_2 \in l_1 \downarrow l_2$.

(*non-loc-axiom-continuity*) We must check that if $c\, \widetilde{\Lambda}(F)\, l$ and $j \in J_{[\mathcal{A}\to\mathcal{B}]}(l)$ then we obtain $c \lhd_{\mathcal{C}} \widetilde{\Lambda}(F)^-(C_{[\mathcal{A}\to\mathcal{B}]}(l, j))$. The proof depends on the particular shape of the axiom indexed by j. We just check the most relevant ones:

(*wb-bottom continuity*) Suppose $c\, \widetilde{\Lambda}(F)\, [(a, \bot_{\mathcal{B}})]$, that is $a \ll \{x \in A \mid (c, x)\, F\, \bot_{\mathcal{B}}\}$. Since, by *continuity* of F, $F^-(\bot_{\mathcal{B}}) \lhd_{\mathcal{C}\times\mathcal{A}} F^-(\emptyset)$, then for any $a' \in A$ such that $(c, a')\, F\ \bot_{\mathcal{B}}$ we get $(c, a') \lhd_{\mathcal{C}\times\mathcal{A}} \emptyset$, from which by *transitivity* $(c, a') \lhd_{\mathcal{C}\times\mathcal{A}} \{(x, \bot_{\mathcal{A}}) \mid x \leq c\}$ and on this we apply Lemma 13.21 since its hypothesis are satisfied. Therefore, for any $d \ll a'$ we get a subset U of C such that $c \lhd_{\mathcal{C}} U$ and, for every $u\ \varepsilon\ U$ there exists $e_1 \ll e'_1, \ldots, e_n \ll e'_n \in A$ such that $d \lhd_{\mathcal{A}} \{e_1, \ldots, e_n\}$ and $(u, e'_j)\ \varepsilon\ \{(x, \bot_{\mathcal{A}}) \mid x \leq c\}$ for $j = 1, \ldots, n$. Now, note that $e'_j = \bot_{\mathcal{A}}$ for every $j = 1, \ldots, n$ and hence $d \lhd_{\mathcal{A}} \emptyset$. Since $a' \lhd_{\mathcal{A}} wb(a')$ and for any $d \ll a'$ we have proved $d \lhd_{\mathcal{A}} \emptyset$, we conclude $a' \lhd_{\mathcal{A}} \emptyset$. Since this holds for any $a' \varepsilon \{x \varepsilon A \mid (c, x)\, F\ \bot_{\mathcal{B}}\}$ we get $\{x \in A \mid (c, x)\, F\ \bot_{\mathcal{B}}\} \lhd_{\mathcal{A}} \emptyset$ and by Lemma 13.18 we obtain $a \ll \emptyset$ and hence $a \lhd_{\mathcal{A}} \emptyset$. Therefore, from this, since $c\, \widetilde{\Lambda}(F)\, (\bot_{\mathcal{A}}, \bot_{\mathcal{B}})$ trivially holds, we conclude $c \lhd_{\mathcal{C}} \widetilde{\Lambda}(F)^-(\{[(\bot_{\mathcal{A}}, \bot_{\mathcal{B}})] \mid a \lhd_{\mathcal{A}} \bot_{\mathcal{A}}\})$.

(*wb-continuity axiom*) Suppose $c\, \widetilde{\Lambda}(F)\, [(a, b)]$ and $j \in J(b)$ and $d \ll a$. Then, by Lemma 13.31 there exists a' such that $(c, a')\, F\, b$ and $a \ll a'$.

So, $(c, a') \lhd_{\mathcal{C}\times\mathcal{A}} F^-(C(b, j))$ follows by *axiom-continuity* of F from which we obtain $(c, a') \lhd_{\mathcal{C}\times\mathcal{A}} F^-(C(b, j) \cup \{\bot_\mathcal{B}\})$ by *continuity* of F considering that $C(b, j) \lhd_\mathcal{B} C(b, j) \cup \{\bot_\mathcal{B}\}$.

Now, note that $F^-(C(b, j)\cup\{\bot_\mathcal{B}\})$ satisfies the hypothesis of Lemma 13.21: indeed the first condition is satisfied by $\leq$ *saturation*, while the second is satisfied since $(c, \bot_\mathcal{A}) \lhd_{\mathcal{C}\times\mathcal{A}} \emptyset$ follows by Lemma 13.11 from $\bot_\mathcal{A} \lhd_\mathcal{A} \emptyset$ and by Lemma 13.14 we get $(c, \bot_\mathcal{A})\ F\ \bot_\mathcal{B}$, that is $(c, \bot_\mathcal{A})\ \varepsilon\ F^-(C(b, j) \cup \{\bot_\mathcal{B}\})$.

Then, we apply Lemma 13.21 on $d \ll a'$, which follows from $d \ll a$, and $(c, a') \lhd_{\mathcal{C}\times\mathcal{A}} F^-(C(b, j) \cup \{\bot_\mathcal{B}\})$ by getting a subset V of C satisfying $c \lhd_\mathcal{C} V$ and such that for every $v\ \varepsilon\ V$ there exists a finite list $e_1 \ll e'_1, \ldots, e_n \ll e'_n \in A$ with $d \lhd_\mathcal{A} \{e_1, \ldots, e_n\}$ and $(v, e'_i)\ \varepsilon\ F^-(C(b, j) \cup \{\bot_\mathcal{B}\})$ for $i = 1, \ldots, n$. This means that there exist $y_1, \ldots, y_n\ \varepsilon\ C(b, j) \cup \{\bot_\mathcal{B}\}$ such that $(v, e'_i)\ F\ y_i$ for $i = 1, \ldots, n$, and hence $v\ \widetilde{\Lambda}(F)\ [(e_i, y_i)]$ follows by Lemma 13.31 for $i = 1, \ldots, n$. By Lemma 13.26 we get $v\ \widetilde{\Lambda}(F)\ [(e_1, y_1), \ldots, (e_n, y_n)]$ and hence we obtain that each $v\ \varepsilon\ V$ is in $\widetilde{\Lambda}(F)^-$ $\{[(x_1, v_1), \ldots, (x_n, v_n)] \mid v_i\ \varepsilon\ C(b, j) \cup \{\bot_\mathcal{B}\}$ for $i = 1, \ldots, n\ \&\ d \lhd_\mathcal{A} \{x_1, \ldots, x_n\}\}$ and hence we conclude $c \lhd_\mathcal{C} \widetilde{\Lambda}(F)^-\{[(x_1, v_1), \ldots, (x_n, v_n)] \mid v_i\ \varepsilon\ C(b, j) \cup \{\bot_\mathcal{B}\}$ for $i = 1, \ldots, n\ \&\ d \lhd_\mathcal{A} \{x_1, \ldots, x_n\}\}$ by *transitivity*. □

Then, by Lemma 13.32 and Lemma 13.8 we conclude:

Corollary 13.33 *Let $\mathcal{A}$ be a locally compact $\vee$-formal cover and $\mathcal{C}$ and $\mathcal{B}$ be i.g. $\vee$-formal covers. Suppose that F is any continuous relation from $\mathcal{C} \times \mathcal{A}$ to $\mathcal{B}$. Then $\Lambda(F)$ is a continuous relation from $\mathcal{C}$ to $[\mathcal{A} \to \mathcal{B}]_\vee$.*

Note that $\widetilde{\Lambda}(F)^\lhd$ enjoys this useful property:

Lemma 13.34 *Let $\mathcal{A}$ be a locally compact $\vee$-formal cover and $\mathcal{C}$ and $\mathcal{B}$ be i.g. $\vee$-formal covers. Suppose that F is any continuous relation from $\mathcal{C} \times \mathcal{A}$ to $\mathcal{B}$. If $c\ \widetilde{\Lambda}(F)^\lhd\ [(a, b)]$ holds, then $(c, a)\ F\ b$ holds, too.*

To finish the proof that the formal cover $\mathcal{B}^\mathcal{A} \equiv [\mathcal{A} \to \mathcal{B}]_\vee$ is the exponential cover of $\mathcal{B}$ over $\mathcal{A}$ we need to show the validity of the adjunction equations about abstraction and application.

Note that the equations are reducible to the corresponding equations about abstraction and application relative to $[\mathcal{A} \to \mathcal{B}]$, because they are the same up to the isomorphisms $\phi_\vee$ and $\phi_\vee^-$.

Lemma 13.35 *Let $\mathcal{A}$ be a locally compact $\vee$-formal cover and $\mathcal{C}$ and $\mathcal{B}$ be i.g. $\vee$-formal covers. The following holds:*

1. *For every continuous relation F from $\mathcal{C} \times \mathcal{A}$ to $\mathcal{B}$ then*

$$\mathsf{Ap} * \langle \Lambda(F) * \Pi_1, \Pi_2\rangle = F \qquad \textit{iff} \qquad (\widetilde{\mathsf{Ap}}^{\lhd}) * \langle \widetilde{\Lambda}(F)^\lhd * \Pi_1, \Pi_2\rangle = F.$$

2. *For every continuous relation G' from $\mathcal{C}$ to $\mathcal{B}^\mathcal{A} \equiv [\mathcal{A} \to \mathcal{B}]_\vee$ we have*

$$\Lambda(\mathsf{Ap} * \langle G' * \Pi_1, \Pi_2\rangle) = G'$$

if and only if for every continuous relation G from $\mathcal{C}$ to $[\mathcal{A} \to \mathcal{B}]$ we have

$$\widetilde{\Lambda}(\widetilde{\mathsf{Ap}}^{\lhd} * \langle G * \Pi_1, \Pi_2\rangle)^{\lhd} = G.$$

Now, in the following we prove the equations for abstraction and application relative to $[\mathcal{A} \to \mathcal{B}]$. But, first, we state a lemma useful to compute the compositions involved in the equations and that can be proved by means of Lemma 13.9:

Lemma 13.36 *Let $\mathcal{A}$ be a locally compact $\vee$-formal cover and $\mathcal{C}$ and $\mathcal{B}$ be i.g. $\vee$-formal covers. For any continuous relation G from $\mathcal{C}$ to $[\mathcal{A} \to \mathcal{B}]$ we can prove:*

(1) $(\widetilde{\mathsf{Ap}}^{\lhd}) * \langle G * \Pi_1, \Pi_2\rangle = (\widetilde{\mathsf{Ap}} \cdot \langle G \cdot p_1, p_2\rangle)^{\lhd}$;

(2) if $(c, a)\ (\widetilde{\mathsf{Ap}}^{\lhd}) * \langle G * \Pi_1, \Pi_2\rangle\ b$ *then* $(c, a) \lhd_{\mathcal{C}\times\mathcal{A}} \{\, (x, y) \in C \times A \mid x\, G\, [(y, b)]\,\}$.

When checking the adjunction equations relative to $[\mathcal{A} \to \mathcal{B}]$, we make a crucial use of Lemma 13.21 to unravel the saturation of a relation with domain $\mathcal{C} \times \mathcal{A}$:

Proposition 13.37 *Let $\mathcal{A}$ be a locally compact $\vee$-formal cover and $\mathcal{C}$ and $\mathcal{B}$ be i.g. $\vee$-formal covers. Then,*

(1) for every continuous relation F from $\mathcal{C} \times \mathcal{A}$ to $\mathcal{B}$ we have

$$(\widetilde{\mathsf{Ap}}^{\lhd}) * \langle(\widetilde{\Lambda}(F)^{\lhd}) * \Pi_1, \Pi_2\rangle = F;$$

(2) for every continuous relation G from $\mathcal{C}$ to $[\mathcal{A} \to \mathcal{B}]$ we have

$$\widetilde{\Lambda}((\widetilde{\mathsf{Ap}}^{\lhd}) * \langle G * \Pi_1, \Pi_2\rangle)^{\lhd} = G.$$

Proof. We prove the two implications of the considered equations one after the other.

(1. Right to left) Suppose $(c, a)\ F\ b$. Then, for any $d \ll a$ by Lemma 13.31 we get $c\ \widetilde{\Lambda}(F)^{\lhd}\ [(d, b)]$ and hence $(c, d)\ (\widetilde{\mathsf{Ap}}^{\lhd}) * \langle(\widetilde{\Lambda}(F)^{\lhd}) * \Pi_1, \Pi_2\rangle\ b$ since $([(d, b)], d)\ (\widetilde{\mathsf{Ap}}^{\lhd})\ b$ holds. By *saturation* on $(c, a) \lhd_{C\times A} \{c\} \times wb(a)$, we conclude $(c, a)\ (\widetilde{\mathsf{Ap}}^{\lhd}) * \langle(\widetilde{\Lambda}(F))^{\lhd} * \Pi_1, \Pi_2\rangle\ b$.

(1. Left to right) Suppose $(c, a)\ (\widetilde{\mathsf{Ap}}^{\lhd}) * \langle(\widetilde{\Lambda}(F))^{\lhd} * \Pi_1, \Pi_2\rangle\ b$. Then, by Lemma 13.36 we get

$$(c, a) \lhd_{\mathcal{C}\times\mathcal{A}} \{\, (x, y) \in C \times A \mid x\, (\widetilde{\Lambda}(F))^{\lhd}\, [(y, b)]\,\}$$

To this we apply Lemma 13.21. Indeed, note that the first condition of the lemma on W is satisfied by $\leq$-*saturation* of $(\widetilde{\Lambda}(F))^{\lhd}$, while the second condition is satisfied since for any $b \in B$ then $c\, \widetilde{\Lambda}(F)\, [(\bot_{\mathcal{A}}, b)]$, that is $\bot_{\mathcal{A}} \ll \{\, x \in A \mid (c, x)\, F\, b\,\}$, trivially holds.

Therefore, by Lemma 13.21 for any $d \ll a$ we find a subset U such that $c \lhd_{\mathcal{C}} U$ and for every $u \varepsilon U$ there exists $e_1 \ll e'_1, \ldots, e_n \ll e'_n \in A$ such that $d \lhd_{\mathcal{A}} \{e_1, \ldots, e_n\}$ and $(u, e'_j) \varepsilon \{\, (x, y) \in C \times A \mid x\, (\widetilde{\Lambda}(F))^{\lhd}\, [(y, b)]\,\}$, that is $u\, (\widetilde{\Lambda}(F))^{\lhd}\, [(e'_j, b)]$, for $j = 1, \ldots, n$. Hence, by Lemma 13.26 we deduce $u\, (\widetilde{\Lambda}(F))^{\lhd}\, [(e'_1, b), \ldots, (e'_n, b)]$.

Then, by *continuity* and *saturation* of $(\widetilde{\Lambda}(F))^{\lhd}$ on *wb-finite-saturation* axiom considering that $d \lhd_{\mathcal{A}} \{e'_1, \ldots, e'_n\}$, which follows from $d \lhd_{\mathcal{A}} \{e_1, \ldots, e_n\}$ and $e_j \ll e'_j$ for $j = 1, \ldots, n$ by Lemma 13.18 and *transitivity*, we get $u\,(\widetilde{\Lambda}(F))^{\lhd}\,[(d, b)]$.

Since this holds for each $u \,\varepsilon\, U$ we conclude $c\,(\widetilde{\Lambda}(F))^{\lhd}\,[(d, b)]$ by *saturation* of $(\widetilde{\Lambda}(F))^{\lhd}$. Then, we obtain $(c, d)\, F\, b$ by Lemma 13.34. Finally, by *saturation* of F on $(c, a) \lhd_{C \times A} \{c\} \times wb(a)$, obtained by Lemma 13.11 from $a \lhd_{\mathcal{A}} wb(a)$, we conclude $(c, a)\, F\, b$.

(2. Right to left) Suppose $c\, G\, l$. If $l \equiv [\,]$ then we conclude immediately $c\ \widetilde{\Lambda}((\widetilde{\mathsf{Ap}}^{\lhd}) * \langle G * \Pi_1, \Pi_2\rangle)^{\lhd}\ l$ by definition of abstraction.

Otherwise, if l is not empty, then for every $(a, b)\epsilon l$ we get $c\ G\ [(a, b)]$ by *continuity* and *saturation* of G on $l \lhd_{[\mathcal{A}\to\mathcal{B}]} [(a, b)]$ which follows by *≤-left* and *reflexivity* from $l \leq [(a, b)]$. By *continuity* of G on the *interpolation* axiom we get $c \lhd_{\mathcal{C}} G^{-}(\{\,[(x, b)] \mid a \ll x\})$.

Now, for any $c' \,\varepsilon\, G^{-}(\{\,[(x, b)] \mid a \ll x\})$, that is $c'\, G\, [(a', b)]$ for some a' with $a \ll a'$, since $([(a', b)], a')\ (\widetilde{\mathsf{Ap}}^{\lhd})\ b$ holds, we get $(c, a')\ (\widetilde{\mathsf{Ap}}^{\lhd}) * \langle G * \Pi_1, \Pi_2\rangle\ b$. Therefore, by Lemma 13.31 we get $c\ \widetilde{\Lambda}((\widetilde{\mathsf{Ap}}^{\lhd}) * \langle G * \Pi_1, \Pi_2\rangle)^{\lhd}\ [(a, b)]$ and by successive application of Lemma 13.26 on each $(a, b)\epsilon l$ we conclude $c\ \widetilde{\Lambda}((\widetilde{\mathsf{Ap}}^{\lhd}) * \langle G * \Pi_1, \Pi_2\rangle)^{\lhd}\ l$.

(2. Left to right) Suppose $c\ \widetilde{\Lambda}((\widetilde{\mathsf{Ap}}^{\lhd}) * \langle G * \Pi_1, \Pi_2\rangle)^{\lhd}\ l$. If l is the empty list then we conclude $c\ G\ [\,]$ by *continuity* and *saturation* of G since by *totality* $c \lhd_{\mathcal{C}} G^{-}([A \to B])$ holds and $[A \to B] \lhd_{[\mathcal{A}\to\mathcal{B}]} [\,]$ also holds.

Instead, if l is not the empty list, the assumption $c\ \widetilde{\Lambda}((\widetilde{\mathsf{Ap}}^{\lhd}) * \langle G * \Pi_1, \Pi_2\rangle)^{\lhd}\ l$ is rewritten as $c \lhd_{\mathcal{C}} \{\, y \in C \mid y\ \widetilde{\Lambda}((\widetilde{\mathsf{Ap}}^{\lhd}) * \langle G * \Pi_1, \Pi_2\rangle)\ l\,\}$.

Now, for any c' such that $c'\ \widetilde{\Lambda}((\widetilde{\mathsf{Ap}}^{\lhd}) * \langle G * \Pi_1, \Pi_2\rangle)\ l$, by definition and Lemma 13.31 we get that for every $(a, b)\epsilon l$, there exists a' such that $a \ll a'$ and $(c', a')\ (\widetilde{\mathsf{Ap}}^{\lhd}) * \langle G * \Pi_1, \Pi_2\rangle\ b$. Now, by Lemma 13.36 we get

$$(c', a') \lhd_{C \times A} \{(x, y) \in C \times A \mid x\, G\, [(y, b)]\}$$

and here we proceed analogously to the proof of *(1. Left to right)* since we can apply Lemma 13.21. Indeed, the first condition on W of the lemma is satisfied by *≤-saturation* of G, while the second condition is satisfied since $c'\, G\, [(\bot_{\mathcal{A}}, b)]$ follows by *continuity* and *saturation* of G on the axiom $[\,] \lhd_{[\mathcal{A}\to\mathcal{B}]} [(\bot_{\mathcal{A}}, b)]$ being $c'\, G\, [\,]$ valid.

Then, analogously to the proof of *(1. Left to right)* we obtain $c'\, G\, [(a, b)]$ and, since this holds for every $(a, b)\epsilon l$, by successive applications of Lemma 13.26 we conclude $c'\, G\, l$.

Since c' was arbitrary, by *saturation* for G we conclude also $c\ G\ l$. □

So, we are arrived at the main theorem of this section.

Theorem 13.38 *Locally compact ∨-formal covers are exponentiable in the category of i.g. ∨-formal covers* $\mathsf{FTop}^{-}_{\mathsf{i}_\vee}$.

Since $\mathsf{FTop}^{-}_{\mathsf{i}_\vee}$ is equivalent to $\mathsf{FTop}^{-}_{\mathsf{i}}$, by Lemmas 13.13 and 13.17, we deduce the following corollary:

Corollary 13.39 *Locally compact formal covers are exponentiable in the category of i.g. formal covers* $\mathsf{FTop}_\mathsf{i}^-$.

Thanks to the predicative coreflection of inductively generated formal covers into formal topologies in [9] we conclude:

Corollary 13.40 *Locally compact formal topologies are exponentiable in the category of inductively generated formal topologies.*

Remark 13.41 After seeing such exponentiation results, a natural question to ask is whether the exponential formal cover between locally compact formal covers happens to be locally compact, too. Even classically this is not true and a counterexample is the formal cover of the Baire space (see [2, 3, 22]).

13.5 Any Exponentiable Formal Cover is Locally Compact

Here, we show that any formal cover that is exponentiable in the category of inductively generated formal covers is locally compact, and we deduce the same result for inductively generated formal topologies. The proof is obtained by following closely the corresponding proof in [2], which is essentially based on categorical properties related to exponentiation and on properties of the Sierpinski locale and hence it is essentially predicative. We recall that $\top \equiv (\{1\}, \leq_1, \lhd_1)$ indicates the terminal formal cover in $\mathsf{FTop}_\mathsf{i}^-$.

Then, we recall the definition of the Sierpinski formal cover:

Definition 13.42 *Let* $\mathcal{S} \equiv (\{0, 1\}, \leq, \lhd_{\mathcal{S}})$ *be the Sierpinski formal cover where* $0 \leq 1$ *and* $a \lhd_{\mathcal{S}} U \equiv \exists u \varepsilon U \; a \leq u$.

Clearly, the Sierpinski formal cover is inductively generated being a unary formal topology [5].

Then, we observe that the morphisms from any given formal cover $\mathcal{A}$ to the Sierpinski cover are in bijection with the opens of $\mathcal{A}$, where the opens of a formal cover are its saturated subsets, that is subsets $U \subseteq A$ such that $U = \mathcal{F}(U)$ where $\mathcal{F}(U) \equiv \{x \in A \mid x \lhd_{\mathcal{A}} U\}$:

Proposition 13.43 *The saturated subsets of a formal cover* $\mathcal{A} \equiv (A, \leq_{\mathcal{A}}, \lhd_{\mathcal{A}})$ *are in bijection with the continuous relations from* $\mathcal{A}$ *to* $\mathcal{S}$, *that is for any saturated subset* U *there exists a continuous relation* $R_U : \mathcal{A} \to \mathcal{S}$ *and conversely, given a relation* $R : \mathcal{A} \to \mathcal{S}$ *then* $R^-(0)$ *is a saturated subset of* $\mathcal{A}$.

Moreover, the bijection preserves the order associated to each collection as follows: if $R \leq H$, *that is* $\forall a \in A \quad \forall s \in \{0, 1\} \quad a\,R\,s \;\to\; a\,H\,s$, *then* $R^-(0) \subseteq H^-(0)$, *and if* $U \subseteq V$ *then* $R_U \leq R_V$.

Proof. We just give the definition of R_U as $a\,R_U\,s \equiv s =_{\{0,1\}} 0 \to a \varepsilon U$. □

The main idea of the proof in [2] that any exponentiable locale is locally compact is that we can use the exponential of a locally compact formal cover over the Sierpinski one to characterize the way-below relation between basic elements (see also [4]):

Lemma 13.44 *Let* $\mathcal{A} \equiv (A, \leq_{\mathcal{A}}, \vee_{\mathcal{A}}, \perp_{\mathcal{A}}, \lhd_{\mathcal{A}})$ *be an i.g.* $\vee$*-formal cover which is exponentiable in the category of i.g formal covers. Then, for any* $a \in A$*:*

$$\text{if} \quad a' \,\varepsilon\, wb(a) \equiv \{x \in A \mid \exists l \quad (l, x)\,\mathsf{Ap}\,0 \quad \& \quad 1\, P_{\mathcal{F}(a)}\, l\,\} \quad \text{then} \quad a' \ll a.$$

Moreover,

$$a \lhd wb(a)$$

where $P_{\mathcal{F}(a)} \equiv \Lambda(R_{\mathcal{F}(a)} * \Pi_{\mathcal{A}})$ *and* $\Pi_{\mathcal{A}} : \top \times \mathcal{A} \to \mathcal{A}$ *is the second projection of the cartesian product* $\top \times \mathcal{A}$*.*

The proof of this lemma makes essential use of the fact that the order between saturated subsets of an exponentiable formal cover $\mathcal{A}$ corresponds to the order between the corresponding global elements of $\mathcal{S}^{\mathcal{A}}$, that is, for any saturated subsets $U, V \subseteq A$, then $U \subseteq V$ if and only if $P_U \leq P_V$ where $P_U \equiv \Lambda(R_U * \Pi_{\mathcal{A}})$.

Then, from Lemma 13.44 (whose detailed proof, following closely that in [2], can be found in [11]) we conclude:

Theorem 13.45 *Let* $\mathcal{A}$ *be an inductively generated formal cover that is exponentiable in the category of inductively generated formal covers. Then, it is locally compact.*

Considering that the Sierpinski cover and the other formal covers used in the proof are formal topologies (with $Pos(a)$ always true for every element of the corresponding base) we also conclude:

Corollary 13.46 *An inductively generated formal topology that is exponentiable in the category of inductively generated formal topologies is locally compact.*

Acknowledgements

I am very grateful to Giovanni Curi for various discussions on this topic and for pointing out to me the results in his PhD thesis [10] stating that we can represent predicatively the way-below relation on subsets thanks to which morphisms from a locally compact join-formal cover to any inductively generated join-formal cover could be put in bijection with formal points of an exponential topology candidate following [3]. I also thank Giovanni Sambin for general discussions on formal topology and Silvio Valentini for his collaboration on exponentiation in formal topology.

References

1. Escardo, M. and Heckmann, R. (2001–2002). Topologies on spaces of continuous functions. *Topology Proceedings*, **26**, 545–564.
2. Hyland, J. M. E. (1981). Function spaces in the category of locales. *Lecture Notes in Mathematics*, **871**, 264–281.
3. Sigstam, I. (1995). Formal spaces and their effective presentations. *Arch Math Logic*, **34**, 211–246.
4. Vickers, S. J. (2004). The double powerlocale and exponentiation: A case study in geometric reasoning. *Theory and Applications of Categories*, **12**, 372–422.

5. Coquand, T., Sambin, G., Smith, J., and Valentini, S. (2003). Inductively generated formal topologies. *Annals of Pure and Applied Logic*, **124(1-3)**, 71–106.
6. Sambin, G. (1987). Intuitionistic formal spaces – a first communication. *Mathematical Logic and its Applications*. pp. 187–204.
7. Johnstone, P. T. (1982). *Stone Spaces*. Cambridge University Press.
8. Joyal, A. and Tierney, M. (1984). An extension of the Galois theory of Grothendieck. *Mem Amer Math Soc*, **51**, no. 309.
9. Maietti, M. E. and Valentini, S. (2004). A structural investigation on formal topology: coreflection of formal covers and exponentiability. *Journal of Symbolic Logic*. **69**, 967–1005.
10. Curi, G. (2004). Geometry of observations: some contributions to (constructive) point-free topology. Tesi di dottorato, University of Siena.
11. Maietti, M. E. (2004). A predicative proof of exponentiation locally compact formal topologies. Technical Report, University of Padova, also in *http:\\www.math.unipd.it\~maietti\pubb.html*.
12. Nordström, B., Peterson, K., and Smith, J. (1990). *Programming in Martin Löf's Type Theory*. Clarendon Press, Oxford.
13. Sambin, G. and Valentini, S. (1998). Building up a tool-box for Martin-Löf's type theory. In: G Sambin J Smith, editor. *Twenty Five years of Constructive Type Theory, Venice 1995*. Oxford Science Publications. pp. 221–244.
14. Aczel, P. and Rathjen, M. (2000/2001). Notes on constructive set theory. Mittag-Leffler Technical Report No. 40.
15. Aczel, P. (1997). An introduction to inductive definitions. In: *Handbook of mathematical logic*. vol. 90 of *Studies in Logic and the Foundations of Mathematics*. (eds J Barwise With the cooperation of H J Keisler, Y N Moschovakis, K Kunen, A S Troelstra), pp. 739–782. North-Holland, Amsterdam.
16. Martin-Löf, P. and Sambin, G. (2003). Generating positivity by coinduction. In *The Basic Picture*, Preprint n. 08, Dipartimento di Matematica, Università di Padova.
17. Coquand, T. (1996). Formal topologies with posets. See *http:\\www.cs.chalmers.se\~coquand*.
18. Berardi, S. and Valentini, S. (2004). Between formal topology and game semantics: an explicit solution of the conditions for an inductive generation of formal topologies. Draft.
19. Battilotti, G. and Sambin, G. (2001). Pretopologies and a uniform presentation of sup-lattice, quantales and frames. Draft.
20. Negri, S. (2002). Continuous domains as formal spaces. *Mathematical Structures in Computer Science*, **12**, 19–52.
21. Sambin, G. (2003). Some points in formal topology. *Theoretical Computer Science*, **305(1-3)**, 347–408.
22. Sigstam, I. (1990). On formal spaces and their effective presentations. PhD thesis. Department of Mathematics, University of Uppsala.

14

SOME CONSTRUCTIVE ROADS TO TYCHONOFF

STEVEN VICKERS

Abstract

The Tychonoff theorem is discussed with respect to point-free topology, from the point of view of both topos-valid and predicative mathematics.

A new proof is given of the infinitary Tychonoff theorem using predicative, choice-free methods for a possibly undecidable index set. It yields a complete description of the finite basic covers of the product.

14.1 Introduction

The Tychonoff theorem says that a product of compact topological spaces is still compact. (I do not assume Hausdorff separation here. By 'compact' I mean having the finite subcover property, as in the Heine–Borel theorem; Bourbaki calls this 'quasicompact'.) For finitary products this is fairly elementary. Surprisingly, the result extends to infinitary products, but there is a price—the axiom of choice has to be assumed. In fact, as is well known, the infinitary Tychonoff theorem is equivalent to the axiom of choice [13].

This is sometimes offered as a reason for using the axiom of choice: with it, one can prove many genuinely useful results such as infinitary Tychonoff. There are more that don't rely on the full power of the axiom of choice, but still need classical principles. These include the Heine–Borel theorem (that any bounded closed interval in the real line is compact) [5] and the Hofmann–Mislove theorem (that for a sober space there is a bijection between compact saturated subspaces and Scott open filters in the topology) [7, 19]. The suggestion is that the theory of topology is unavoidably impoverished if it cannot call on all the classical reasoning principles.

It therefore comes as a surprise to discover that in *point-free* topology, such theorems can often be proved constructively, at least if stated correctly. In this chapter we illustrate this with the Tychonoff theorem. Its validity is quite unequivocal, with no special assumptions. For instance, we do not have to assume excluded middle either, and the indexing set does not have to have decidable equality. Neither do we have to use impredicative constructions. Tychonoff is actually a robust part of constructive topology.

The problem lies not in forsaking choice, but in insisting on a point-set formulation of topology. Three things are jointly incompatible: Tychonoff, constructivity, and point-set topology. If we wish to keep Tychonoff then we must drop one of the others. However, it does not have to be the constructivity. The aim of this chapter is to describe how it works if we decide to drop the point-set formulation.

14.1.1 *Point-free topology*

In the usual point-*set* topology, a topological space is a *set* equipped with additional topological structure that can be axiomatized in various ways, classically equivalent to each other. Our preferred form here is via the open sets, defining a *topology* to be a family of subsets of the set of points, closed under finite intersection and arbitrary union.

By contrast I shall use the phrase 'point-free topology' as a generic label for approaches that do not start by assuming a *set* of points of the space. In these approaches, a topological space cannot in general be described as a set of points equipped with extra structure.

In practice, points are not excluded from the 'point-free' discussion. Normally they are at least helpful for keeping in touch with topological intuitions, and there are tricks of categorical logic by which rigorous arguments can be conducted in terms of points. However, we cannot assume that the totality of all points can be collected together as a set. A space is something more general than a set. What is more, when a set of points can be extracted, we cannot assume that it adequately represents the collection of *all* points.

There are two main versions of point-free topology that I shall consider, and they are genuinely different. They amount to the elaboration of point-free ideas in two radically different foundational settings, namely topos theory and predicative type theory. (I know more about the topos theory side, so you must excuse me if sometimes my knowledge of type theory is deficient.)

The first version, used in topos theory, is *locales* [8]. I shall explain those first because in many ways it helps to clarify what the second version is achieving. The second version, used in predicative type theory, is *formal topology* [17].

14.1.2 *Locales*

Locale theory is based on the most direct interpretation of the phrase 'point-free topology'. In point-set topology, the topology is the collection of open sets. The idea is to use this in a point-free way by treating it as an abstract lattice, forgetting that it was ever a set of subsets of some set of points.

The standard introduction is [8] (or see also [19]), and we give just a brief overview.

Definition 14.1 *A frame is a complete lattice in which binary meet distributes over arbitrary join.*

A frame homomorphism is a function between frames that preserves arbitrary joins and finite meets.

(Naturally, the arbitrary joins and finite meets here correspond to the arbitrary unions and finite intersections of open sets.)

We write **Fr** for the category of frames and frame homomorphisms, and **Loc** for its opposite—the category of *locales* and *(continuous) maps*. By this definition a locale 'is' just a frame, but it is best to keep them notationally distinct since the language for locales is quite different from that for frames. For instance, products of locales are coproducts of frames; sublocales are quotient frames. Locales are frames 'pretending to be' topological spaces, and it is best to keep up the pretence. If X is a locale we shall write ΩX for the corresponding frame; and if $f : X \to Y$ then we write $f^* : \Omega Y \to \Omega X$ for the corresponding frame homomorphism.

Locales, then, are the spaces in this version of point-free topology.

The points of a space X should be the maps from the one-point space 1 to X, and we can implement this with locales. The (discrete) topology on 1 is the powerset $\mathcal{P}1$. In the internal language of toposes $\mathcal{P}1$ is just the subobject classifier Ω, and we shall write it as such. The best way to think of Ω is as the 'set of truthvalues', which classically is $\{\mathbf{true}, \mathbf{false}\}$. A point x of a locale X is then a frame homomorphism $x^* : \Omega X \to \Omega$. Now any function to Ω is equivalently described by its true kernel, the inverse image of $\{\mathbf{true}\}$, and the function x^* is a frame homomorphism iff its true kernel is a completely prime filter, an upper set that is closed under finite meets and inaccessible by arbitrary joins. (Note that some of the other common characterizations of point [8, 19] are not constructively equivalent. Some of this is because in the absence of excluded middle the homomorphism x^* is not determined by its *false* kernel $(x^*)^{-1}\{\mathbf{false}\}$.)

In topos-valid mathematics, we can construct the set $\mathrm{pt}(X)$ of points of X. However, as we shall see it may be defective. Even classically there may fail to be enough points to distinguish between the opens, so that ΩX does not embed in $\mathcal{P}\,\mathrm{pt}(X)$.

There are two mismatches between locales and topological spaces.

The first is that locales are intrinsically *sober*. This just means that the space contains *all* the points that can be reconstructed from the topology, and so must be a feature of any point-free approach. A non-sober topological space may have distinct elements of the point-set that cannot be distinguished topologically because they are in the same opens. In other words, the space may fail to be T_0. A non-sober space may also lack points such as the directed joins (with respect to the specialization order) present in every sober space. For example, a poset with its Alexandrov topology (the opens are the upper sets) is not sober in general, and to make it so you have to add directed joins by going to the ideal completion.

That first mismatch is perhaps not so serious. One can argue that all decent spaces should be sober. Alternatively, you can express the duplications and omissions of points in a non-sober space by a map from a discrete locale X ($\Omega X = \mathcal{P}X$) to another locale. (These are the *topological systems* of [19].)

The second mismatch is more fundamental, and that is that locales do not always have enough points—they may fail to be *spatial*. Indeed, some non-trivial locales fail to have any points at all. The technical manifestation of this is that the frame homomorphism $\Omega X \to \mathcal{P}\,\mathrm{pt}(X)$ fails to be 1-1. In classical mathematics the non-spatial locales are generally pathological, since the axiom of choice can be used to show the existence of enough points for wide classes of useful locales. Constructively, however, even necessary locales need not be spatial. A good example is the real line. The localic real line for which good mathematics holds—for instance, the Heine–Borel theorem—is the one presented as $L(\mathbb{R})$ in [8, IV.1.1]. Constructively this can easily be non-spatial [6].

Such non-spatiality may seem pathological wherever it occurs. After all, what kind of topological structure can it be that is not supported by the points? However, the well-known topological theorems work better with the non-spatial locales—the purpose of this chapter is to illustrate this for Tychonoff, and Heine–Borel has also been mentioned. Here's one way to imagine how the points might not be the whole story. Often we are interested in other pieces within the space, for example line elements (maps from [0, 1]). Say a *generalized point* of X is a map from some domain Y (the 'stage of definition') to X. Even if there are insufficient 'global' points (stage of definition is 1), there are still plenty of generalized points. In fact, this comes rather cheaply, since the generic point (the identity map from X to itself) is enough for most purposes.

It was understood quite early (see, e.g. [12]) that locale theory is constructive in the topos-valid sense. The notion of internal frame can be defined in any topos, and the constructions one needs (for example, coproduct of frames for product of locales) can be carried through. Moreover, there is an extremely important relativization principle. Suppose X is a locale. There is a topos associated with it, namely its topos of sheaves which I shall write $\mathcal{S}X$. What are the internal frames in $\mathcal{S}X$? It turns out they are equivalent to continuous maps (of locales) with codomain X. Hence a constructive result about locales, interpreted in $\mathcal{S}X$, can be turned into a result about maps into X—in other words, generalized points of X. Thus the topos-valid constructivist discipline delivers a *payoff*. It is not merely a claim to moral superiority.

One benefit is that constructive arguments about points can be applied also to the generalized points (as 'points at another stage of definition'). The sufficiency of these can therefore validate spatial reasoning about point-free topologies. This is exploited in [22, 24], which also explain why the more stringent *geometric* constructivism is needed to ensure that the arguments can be transferred from one stage of definition to another.

Topos-valid constructivism is completely choice-free. In general it is not even possible to choose one element out of two. Consider for example the topos of sheaves over the circle O. If the circle is represented as the complex numbers of unit modulus, then the squaring function $z \mapsto z^2$, the Möbius double cover of the circle, is a local homeomorphism and hence equivalent to a sheaf. In the internal language of the topos $\mathcal{S}O$, it is a set X, finite with decidable equality, satisfying

$$\exists x, y \in X. (x \neq y \wedge \forall z \in X. (z = x \vee z = y))$$

but with no element $1 \to X$.

14.1.3 *Formal topologies*

Despite the success of locales in topos-valid mathematics, its use of impredicative constructions troubles some constructivists. These are constructions that presuppose a collection that already includes what one is trying to construct. The question often arises in connection with powersets $\mathcal{P}X$, since if one is trying to construct some subset of X it would be impredicative to presuppose that $\mathcal{P}X$—the set of *all* subsets including the one being constructed—is already to hand. In general predicative mathematics would not admit $\mathcal{P}X$ as a set.

Unfortunately, many of the constructions of locale theory are impredicative. This includes the construction of pt(X), though I have already argued that we may be able to do without it. More seriously, however, the frames themselves are impredicative.

This is most obvious for discrete locales, whose frames are powersets. Then other frames such as $\Omega L(\mathbb{R})$ mentioned above, require impredicative constructions. And the very definition of frame, as complete lattice, describes joins in A by a function from $\mathcal{P}A$ to A.

One way to understand formal topology is that it is obliged to do 'locale theory without the frames'. In fact, techniques for this are already present in the ordinary practice of locale theory.

The impredicative step is normally in ensuring that *all* joins (of subsets) are present. But it is often enough to work with a *base* of the topology, so that every open is a join of basic opens, and the base can often be constructed predicatively. More generally one might use a subbase, so that every open is a join of finite meets of subbasics. However, this makes no difference to the predicativity, since the *finite* powerset construction is inductive. (As elsewhere in this paper, 'finite' means *Kuratowski* finite. The finite powerset $\mathcal{F}X$ can also be represented algebraically as the free semilattice over X.)

In locale theory this use of bases or subbases appears in algebraic form, as presenting a frame by generators and relations. In [8] it underlies the construction of $\Omega L(\mathbb{R})$. This is generated by basics (p, q) $(p \in \mathbb{Q} \cup \{-\infty\}, q \in \mathbb{Q} \cup \{\infty\})$ subject to relations

$$\begin{aligned}
1 &= (-\infty, \infty)\\
(p,q) \wedge (p',q') &= (\max(p,p'), \min(q,q'))\\
(p,q) &\leq 0 \text{ if } p \geq q\\
(p,s) &\leq (p,r) \vee (q,s) \text{ if } p \leq q < r \leq s\\
(p,q) &\leq \bigvee\{(p',q') \mid p < p' < q' < q\} \text{ if } p < q.
\end{aligned}$$

The presentation itself is predicative: the set of generators, the set of relations and the sets of disjuncts in infinitary joins are all constructed predicatively. (This idea is explored in great detail in [24].) Hence, within predicative mathematics, the presentation can be used as a surrogate for the frame. This, roughly speaking, is what a formal topology is. More precisely, this is an *inductively generated* formal topology.

One sees many different definitions in formal topology. To give some shape to the issues involved, I mention three different modes of variation.

1. There are different kinds of structure that can be interpreted as generators and relations. The different forms of structure tend to come out as different definitions of formal topology. For example, the *site* as described in [8, II.2.11] provides one particular form of generators and relations. The generators are required to form a meet semilattice, and there are implied relations to say that the semilattice meet is preserved in the frame.

2. It used to be customary in formal topology to require spaces to be *open* in the sense of [12], namely that the unique map to 1 should be an open map. (Classically, all locales are open. But constructively it becomes an important issue.) For this a positivity predicate is needed on the basics in order to say in a positive way which are non-empty [9, 15]. A formal topology without positivity predicate is often called a *formal cover*.

3. The original definitions of formal topology required a specification of the full cover relation $\lhd$, i.e. to say for each set U of generators which generators a were to be less than $\bigvee U$. Of course, $\lhd$ is not itself a predicative set. But the information amounts to describing how proofs of $a \lhd U$ may be constructed. More recently [3] showed how to use an axiom set, effectively a set of relations, to generate the full cover relation $\lhd$. Such a structure is called an inductively generated formal cover. Not all formal covers can be inductively generated.

In what follows, we shall use the following definition of inductively generated formal cover.

Definition 14.2 *A flat site is a structure $(P, \leq, \lhd_0)$ where $(P, \leq)$ is a preorder (i.e. transitive and reflexive), and $\lhd_0 \subseteq P \times \mathcal{P}P$ has the following flat stability property: if $a \lhd_0 U$ and $b \leq a$, then there is some $V \subseteq b \downarrow U$ such that $b \lhd_0 V$.*

(For subsets or elements U and V, we write $\downarrow U$ for the down closure of U with respect to $\leq$ and $U \downarrow V$ for $\downarrow U \cap \downarrow V$.)

The reason for calling this 'flat' is as follows. In category theory there is a notion of *flat functor* from C to **Set** such that if C is cartesian (has all finite limits) then flatness is equivalent to the functor being cartesian (preserves finite limits). See, e.g. [14]. (This is also related to the notion of flat module in ring theory, using the idea from enriched category theory that a functor from C to **Set** can be considered a kind of module over C.) The notion of ordinary site [8] is essentially a special case of our flat site in which P is a meet semilattice, i.e. a cartesian poset. In categorical logic, points of the corresponding locale can be understood as certain cartesian functors from P to **Set**, and in the flat site these generalize to the flat functors from P to **Set**.

Definition 14.2 is just a rephrasing of the *localized axiom sets* of [3]. Their axiom set is an indexed family $I(a)$ set $[a : P]$ together with a family of subsets $C(a, i) \subseteq P$ $[a : P, i : I(a)]$. 'Localized' means that for any $a \leq c$ and $i \in I(c)$, there exists $j \in I(a)$ such that $C(a, j) \subseteq a \downarrow C(c, i)$. Then our $\lhd_0$ comprises the instances of $a \lhd_0 C(a, i)$.

The full formal cover $\lhd$ is generated from this by rules

- $\dfrac{a \in U}{a \lhd U}$ (*reflexivity*)
- $\dfrac{a \leq b \quad b \lhd U}{a \lhd U}$ ($\leq$-*left*)
- $\dfrac{a \lhd_0 V \quad V \lhd U}{a \lhd U}$ (*infinity*).

The flat site gives rise to a frame presentation in which the generators are the elements of P, and the relations are:

$$1 \leq \bigvee P$$
$$a \wedge b = \bigvee \{c \mid c \leq a, c \leq b\}$$
$$a \leq \bigvee U \ (a \lhd_0 U).$$

14.2 Compactness

The notion of compactness translates easily from spaces to locales. A locale X is compact if the top open $1 \in \Omega X$ has the property that every cover has a finite subcover. Alternatively, if a *directed* subset S has its join equal to 1, then 1 must already be in S, i.e. $\{1\}$ is Scott open.

This is straightforward, but notice that compactness of the locale and its space of points $\mathrm{pt}(X)$ become two unrelated properties. Let $\Omega\,\mathrm{pt}(X)$ be the topology induced on $\mathrm{pt}(X)$, the image of the frame homomorphism $\Omega X \to \mathcal{P}\,\mathrm{pt}(X)$, and let us write F for the filter of ΩX comprising those opens that map to top in $\Omega\,\mathrm{pt}(X)$. If X is not spatial, so $\Omega X \to \Omega\,\mathrm{pt}(X)$ is not 1-1, then F may be different from $\{1\}$. Compactness for $\mathrm{pt}(X)$ is equivalent to saying that F is Scott open—if a directed join $\bigvee S$ is in F, then S already has an element in F. Scott openness of F neither implies nor is implied by Scott openness of $\{1\}$.

In fact, this explains something of the gap between spatial Tychonoff (requiring choice in general) and localic Tychonoff (no choice needed). Even for spatial locales, the product need not be spatial. Hence compactness of the product locale does not imply compactness of the product space—the two questions are separate.

Despite the simplicity of the definition of localic compactness, in practice it is a non-trivial question. This is because it is rare for the frame structure to be given explicitly in a concrete form. For instance, if the frame is presented by basic generators and relations it is not in general clear when one open is covered by a family of basics. The impredicative definition of the cover relation—in effect 'the least frame congruence containing the relations'—is of little help. In particular this is a problem with a product $\prod_i X_i$ of locales, whose coproduct frame is most easily presented by a 'disjoint union' of presentations for the frames ΩX_i. This coproduct frame may also be described as a tensor product of complete join semilattices, but that is no real help here because—just as with linear tensor products—the elements cannot be expressed in any canonical form.

A direct predicative approach requires some knowledge of the full cover relation $\lhd$. Let us outline some sharper approaches to the question.

14.2.1 *Preframes*

A *preframe* is a poset with finite meets and directed joins, with meet distributing over directed joins. A preframe homomorphism preserves finite meets and directed joins.

The importance of preframes lies in the fact that for a subset F of a frame ΩX, F is a Scott open filter iff its characteristic function to Ω is a preframe homomorphism—filteredness and Scott openness correspond to preservation of finite meets and directed joins respectively. A simple proof of Tychonoff using preframe techniques was given in [11].

This can be expressed neatly within locale theory by the *upper powerlocale* $P_U X$ (see [21], and also [24]). By definition its frame $\Omega P_U(X)$ is generated as frame by the elements of ΩX, respecting the preframe structure of ΩX. Hence maps from Y to $P_U(X)$ are equivalent to preframe homomorphisms from ΩX to ΩY and the points of $P_U X$ are the Scott open filters of ΩX.

Johnstone's localic version of the Hofmann–Mislove theorem—see [21], deriving from [10]—says that the Scott open filters of ΩX correspond to compact fitted sublocales of X, where a sublocale is *fitted* if it is a meet of open sublocales. (Classically this corresponds to subspaces that are *saturated*, i.e. upper closed under the specialization order.) Hence $P_U X$ is indeed a *power*locale, its points being certain sublocales of X. The correspondence is order reversing, and a bottom point of $P_U X$ corresponds to the greatest possible compact sublocale of X, namely X itself.

[20] shows that proving compactness of X is equivalent to finding a bottom point $\bot$ of $P_U(X)$, in the strong sense that the composite $!; \bot : P_U X \to 1 \to P_U X$ is less than the identity map in the specialization order. This condition says that $\bot$ is not just least amongst the global points $1 \to P_U X$. It is also less than the generic point $\mathrm{id} : P_U X \to P_U X$, and this makes it least amongst all generalized points $Y \to P_U X$.

All this is impredicative, but it can be made predicative. The 'preframe coverage theorem' of [11] shows how to convert presentations of frames by generators and relations into preframe presentations of the same frames, and so shows how to convert frame presentations of ΩX into frame presentations of $\Omega P_U X$. This can be made into a predicative construction within formal topology.

Proposition 14.3 *Let $(P, \leq, \lhd_0)$ be a flat site presenting locale X. Then $P_U X$ is presented by the flat site $(\mathcal{F}P, \sqsubseteq_L, \lhd_0)$ where $\mathcal{F}P$ is the (Kuratowski) finite powerset of P, $\sqsubseteq_L$ is the lower order on $\mathcal{F}P$, defined by $S \sqsubseteq_L T$ iff for every $x \in S$ there is some $y \in T$ with $x \leq y$, and $\lhd_0$ is given by the following.*

Suppose $a_i \lhd_0 U_i$ $(1 \leq i \leq n)$. Let $A = \{a_i \mid 1 \leq i \leq n\}$. Then for every $S \in \mathcal{F}P$,

$$A \cup S \lhd_0 \left\{ T \cup S \,\middle|\, T \sqsubseteq_L \bigcup_{i=1}^{n} U_i \right\}.$$

Proof. We merely sketch the proof here. ΩX is presented as frame by generators P and relations as given after Definition 14.2. We can write it as

$$\begin{aligned} \mathbf{Fr}\langle P \text{ (qua preorder) } | 1 &\leq \bigvee P \\ a \wedge b &\leq \bigvee (a \downarrow b) \\ a &\leq \bigvee U \quad (\text{if } a \lhd_0 U)\rangle \end{aligned}$$

where 'qua preorder' indicates extra implicit relations to say the preorder structure of P is preserved in the frame. The free join semilattice over the preorder P is $\mathcal{F}P/\sqsubseteq_L$, with join represented by union. We can therefore transform the presentation into an equivalent one,

$$\mathbf{Fr}\langle \mathcal{F}P/\sqsubseteq_L \text{ (qua } \vee\text{-semilattice)} \mid 1 \leq \bigvee\nolimits^{\uparrow} \mathcal{F}P$$
$$(\{a\}\cup S) \wedge (\{b\}\cup S) \leq \bigvee\nolimits^{\uparrow}\{(T\cup S) \mid T \sqsubseteq_L \{a\}, T \sqsubseteq_L \{b\}\}$$
$$(\{a\}\cup S) \leq \bigvee\nolimits^{\uparrow}\{(T\cup S) \mid T \sqsubseteq_L U\} \quad (\text{if } a \lhd_0 U, S \in \mathcal{F}P)\rangle.$$

($\bigvee^{\uparrow}$ indicates a join that is known to be directed. We have not distinguished between on the one hand the elements of $\mathcal{F}P/\sqsubseteq_L$, equivalence classes with respect to the equivalence relation corresponding to $\sqsubseteq_L$, and on the other hand the elements of $\mathcal{F}P$ that represent them.)

Here the relations have been put in the join stable form required for the preframe coverage theorem [11], and from that we find that $\Omega P_U X$ can be given exactly the same presentation, except that 'qua $\vee$-semilattice' is replaced by 'qua preorder under $\sqsubseteq_L$'.

Using some induction on the finite sets, we find that the middle relation scheme is equivalent to

$$(A) \wedge (B) \leq \bigvee\nolimits^{\uparrow}\{(T) \mid T \sqsubseteq_L A, T \sqsubseteq_L B\}$$

for $A, B \in \mathcal{F}P$. Thus those first two relation schemes are equivalent to the implicit relations in a flat site on $(\mathcal{F}P, \sqsubseteq_L)$. The final relation scheme does not satisfy the flat stability condition, but it is equivalent to the relation scheme given in the statement of the theorem, which does. □

It follows that the upper powerlocale can also be accessed in predicative theories. The same compactness criterion—existence of a suitable point of the powerlocale—can then be expressed. The argument given so far for its correctness has used the impredicative results about preframes, but [23] gives a direct predicative proof. After a little simplification, it appears there as

Theorem 14.4 *Let $(P, \leq, \lhd_0)$ be a flat site presenting a locale X. Then X is compact iff there is a subset F of $\mathcal{F}P$ such that*

1. *F is upper closed with respect to $\sqsubseteq_L$.*
2. *F is inhabited.*
3. *If $a \lhd_0 U$ and $\{a\}\cup T \in F$, then $U_0 \cup T \in F$ for some $U_0 \in \mathcal{F}U$.*
4. *If $S \in F$ then $P \lhd S$ (i.e. $\forall g \in P.\, g \lhd S$).*

In that case, F is necessarily the set of all finite covers of X by basics:

We note briefly that it is not only preframe homomorphisms that can be captured predicatively in this way. So too can arbitrary dcpo morphisms (Scott continuous functions) between frames. (A dcpo is a directed complete poset, i.e. a poset with all directed joins.) This is done using the *double powerlocale* $\mathbb{P}X$ [11, 24], for which the frame $\Omega\mathbb{P}X$ is the free frame generated by ΩX and preserving its dcpo structure. The maps from Y to $\mathbb{P}X$ are equivalent to Scott continuous functions from ΩX to ΩY, and $\mathbb{P}X$ can be constructed by predicative constructions on presentations [24, 26]. Thus the double powerlocale can also be defined on inductively generated formal topologies.

14.3 Tychonoff

We can now illustrate the techniques with a proof of Tychonoff that is valid both in topos theory and in predicative mathematics. It assumes neither finiteness nor decidability of equality for the indexing set for the locales of which the product is taken.

In topos-valid locale theory this result appears to be due to Vermeulen [18]. In formal topology, following an earlier treatment of [1], the infinitary Tychonoff was proved in [16] without choice but under the assumption that the indexing set had decidable equality. This arose from the way that basic opens for the product $\prod_i X_i$—finite meets of opens taken from the components—were normalized into elements of $\prod_i \Omega X_i$ in which all but finitely many components are 1. This normalization can only be done effectively if there is decidable equality for indexes.

Subsequently, [2] gave a simple choice-free predicative proof without decidable equality. His argument rests on the fact that for any spectral locale X, there is a least compact sublocale Y whose fitted hull (= saturation) is the whole of X. It follows that every sublocale between Y and X is compact. Coquand shows how to describe a product locale in this way. (Coquand has remarked separately that the underlying construction is a localic version of the 'maximal spectrum' described spatially in [8, II.3.5].)

Though elegant, Coquand's proof requires some preparation before it can be put into effect. It relies on having each locale presented using a distributive lattice of generators for which the order coincides with the order in the presented frame, and getting that is non-trivial. We now give a proof that shows how from general flat sites, the finite covers of the product can be calculated.

Proposition 14.5 *Let $(P_i, \leq, \lhd_0)$ be a flat site for each $i \in I$. Then the product of the corresponding locales is presented by a flat site $(P, \leq, \lhd_0)$ defined as follows.*

First, let $(P', \leq) = \sum_{i\in I} P_i$ be the poset coproduct. As a set it is the disjoint union, $\{(i, x) \mid i \in I, x \in P_i\}$, with $(i, x) \leq (j, y)$ iff $i = j$ and $x \leq y$ in P_i. Its elements are subbasics.

Now define $P = \mathcal{F}P'$ ordered by $\sqsubseteq_U$, i.e. $A \leq B$ iff $\forall b \in B.\ \exists a \in A.\ a \leq b$. Its elements represent finite meets of subbasics. $P/\sqsubseteq_U$ is in fact the free meet semilattice over the poset P', meet being represented by union.

Covers are defined as follows:

1. *If $i \in I$ and $B \in P$, then*

$$B \lhd_0 \{\{(i, a)\} \cup B \mid a \in P_i\}.$$

2. *If $i \in I$, $a, a' \in P_i$ and $B \in P$, then*

$$\{(i, a), (i, a')\} \cup B \lhd_0 \{\{(i, c)\} \cup B \mid c \leq a, c \leq a'\}.$$

3. *If $i \in I$, $a \lhd_0 U$ in P_i, and $B \in P$, then*

$$\{(i, a)\} \cup B \lhd_0 \{\{(i, u)\} \cup B \mid u \in U\}.$$

Proof. First note that this is indeed a flat site; in fact it is an ordinary site. (P is a meet semilattice and the coverage has meet stability.)

The frame for the product is presented by putting together the presentations for the original frames. For clarity, let us write α_i for the injections of generators. Then the frame is presented as

$$\begin{aligned}
\mathbf{Fr}\langle \alpha_i(a)\ (i \in I, a \in P_i) \mid \\
\alpha_i(a) \le \alpha_i(a') \quad (i \in I, a \le a' \text{ in } P_i) \\
1 \le \textstyle\bigvee_{a \in P_i} \alpha_i(a) \quad (i \in I) \\
\alpha_i(a) \wedge \alpha_i(a') \le \textstyle\bigvee \{\alpha_i(c) \mid c \le a, c \le a'\} \quad (i \in I, a, a' \in P_i) \\
\alpha_i(a) \le \textstyle\bigvee_{u \in U} \alpha_i(u) \quad (i \in I, a \lhd_0 U \text{ in } P_i)\rangle.
\end{aligned}$$

This is isomorphic to

$$\begin{aligned}
\mathbf{Fr}\langle P \text{ (qua } \wedge = \cup \text{ semilattice)} \mid \\
B \le \textstyle\bigvee_{a \in P_i} (\{(i, a)\} \cup B) \quad (i \in I, B \in P) \\
\{(i, a), (i, a')\} \cup B \le \textstyle\bigvee \{\{(i, c)\} \cup B \mid c \le a, c \le a'\} \quad (i \in I, a, a' \in P_i, B \in P) \\
\{(i, a)\} \cup B \le \textstyle\bigvee_{u \in U} \{(i, u)\} \cup B \quad (i \in I, a \lhd_0 U \text{ in } P_i, B \in P)\rangle.
\end{aligned}$$

The 'qua' notation denotes additional relations to preserve the $\wedge$-semillatice structure (concretely $\cup$) of P.

In one direction, the isomorphism takes $\alpha_i(a) \mapsto (\{(i, a)\})$, while in the other it takes $A \longmapsto \bigwedge \{\alpha_i(a) \mid (i, a) \in A\}$.

To say this predicatively, we are describing two mutually inverse continuous maps between the corresponding formal topologies.

This second presentation corresponds to the product site described in the statement. □

To prove Tychonoff, we use Theorem 14.4. Suppose we have flat sites $(P_i, \le, \lhd_0)$ and are given sets F_i describing compactness for the P_is. We show how to construct a corresponding set F for the product. The main point of interest is that F itself can be defined without reference to the full coverage $\lhd$. The full coverage and its inductive generation need to be considered only when showing that every set in F covers the product space; but this is hardly surprising, because the corresponding facts for the F_is were described in terms of $\lhd$.

We must find a way to characterize the finite covers by basics. We give an informal argument as motivation; applying the theorem will confirm its correctness.

Each subbasic (i, a) in P' is of the form (spatially) $\{(x_j)_{j \in I} \mid x_i \in a\}$: think of this as a product of a (at i) $\times P_j$ (everywhere else). A basic in P is a set of these representing a meet, and that can be thought of as a product of specified a's at finitely many specified i's, times P_j everywhere else. (However, we must also allow for the fact that some i may occur more than once.)

We want to know when a join of these meets covers the entire product, and the trick is to use distributivity to change it to a meet of joins. Then every one of the joins must be the whole product.

By distributivity,

$$\bigvee_{A\in\mathcal{A}}\bigwedge A = \bigwedge_{\gamma\in\mathrm{Ch}(\mathcal{A})}\bigvee \mathrm{Im}\,\gamma$$

where $\mathrm{Ch}(\mathcal{A})$ is the set of *choices* of $\mathcal{A}$, i.e. [25] the finite total relations γ from $\mathcal{A}$ to $\bigcup\mathcal{A}$ such that if $(A, a) \in \gamma$ then $a \in A$, and $\mathrm{Im}\,\gamma$ is the image of γ (under the second projection from $\mathcal{A} \times \bigcup\mathcal{A}$ to $\bigcup\mathcal{A}$).

Now consider a finite join of subbasics $\bigvee B$. This is (it will turn out) a cover of the entire product iff at some i its components cover P_i. Classically one sees this as follows. Suppose at every i we have some point x_i that is not in any subbasic b in B. Then the point $(x_i)_{i\in I}$ is not in $\bigvee B$. Hence (classically) if $\bigvee B$ does cover the product, then there is some i and some finite cover S of P_i (so $S \in F_i$) such that $\{i\} \times S \subseteq B$.

This idea lies behind our definition of F in the theorem.

Theorem 14.6 (Infinitary Tychonoff) *Let $(P_i, \leq, \lhd_0)$ $(i \in I)$ be flat sites for compact spaces, equipped with sets $F_i \subseteq \mathcal{F}P_i$ satisfying the conditions of Theorem 14.4. Let the product site P be defined as above.*

Let $F \subseteq \mathcal{F}P$ be defined such that $\mathcal{A} \in F$ iff for every $\gamma \in \mathrm{Ch}(\mathcal{A})$ there is some i and some $S \in F_i$ such that for every $a \in S$ we have $(i, a) \in \mathrm{Im}(\gamma)$.

Then F satisfies the conditions of Theorem 14.4 for P, and hence shows that P is compact.

Proof. In the definition of F, we should like to say that for some i, $\mathrm{Im}\,\gamma$ covers P_i: or $\{a \mid (i, a) \in \mathrm{Im}\,\gamma\} \in F_i$. But we have to be somewhat careful, since if I does not have decidable equality then $\{a \mid (i, a) \in \mathrm{Im}\,\gamma\}$ need not be finite. Nonetheless, let us abuse language and say '$\mathrm{Im}\,\gamma$ covers P_i'. Note also that if some F_i contains $\emptyset$, so that P_i gives an empty locale and so does the whole product, then every $\mathcal{A}$ is in F.

We verify the four conditions in Theorem 14.4.

Condition 1, F is upper closed with respect to $\sqsubseteq_L$. Suppose $\mathcal{A} \in F$ and $\mathcal{A} \sqsubseteq_L \mathcal{B}$. Let $\delta \in \mathrm{Ch}(\mathcal{B})$. If $A \in \mathcal{A}$ then $A \sqsubseteq_U B$ for some $B \in \mathcal{B}$. There is some $b \in B \cap \mathrm{Im}\,\delta$, and $a \leq b$ for some $a \in A$. In short, $\forall A \in \mathcal{A}.\ \exists a \in A.\ \exists b \in \mathrm{Im}\,\delta.\ a \leq b$. It follows that there is some $\gamma \in \mathrm{Ch}(\mathcal{A})$ such that $\mathrm{Im}\,\gamma \sqsubseteq_L \mathrm{Im}\,\delta$. Now because $\mathcal{A} \in F$ we deduce that $\mathrm{Im}\,\gamma$ covers some P_i, and it follows that $\mathrm{Im}\,\delta$ covers the same P_i so $\mathcal{B} \in F$.

Condition 2, F is inhabited. $\{\emptyset\}$ is vacuously in F, because it has no choices.

Condition 3. There are three parts to check, corresponding to the three axiom schemes in Proposition 14.5. We use Lemma 14.7, which is proved separately.

For scheme 1, we need that if $i \in I$, $B \in P$ and $\{B\} \cup \mathcal{C} \in F$, then there is some $S' \in \mathcal{F}P_i$ such that $\{\{(i, a)\} \cup B \mid a \in S'\} \cup \mathcal{C} \in F$. In the lemma, take $S = \emptyset$ and ϕ the whole of P_i. Given T, there is some S' such that $S' \cup T$ covers P_i; just choose S' to be any element of F_i (which is inhabited by hypothesis).

For scheme 2, if $i \in I$, $a_1, a_2 \in P_i$, $B \in P$ and $\{\{(i, a_1), (i, a_2)\} \cup B\} \cup \mathcal{C} \in F$, then there is some $S' \in \mathcal{F}(a_1 \downarrow a_2)$ such that $\{\{(i, a)\} \cup B \mid a \in S'\} \cup \mathcal{C} \in F$. Here $S = \{a_1, a_2\}$ and $\phi = a_1 \downarrow a_2$. If $\{a_1\} \cup T$ and $\{a_2\} \cup T$ both cover P_i then so does $(\{a_1\} \cup T) \downarrow (\{a_2\} \cup T)$ and hence so does some finite subset. This enables us to find S'.

For scheme 3, if $i \in I$, $a \lhd_0 U$ in P_i, $B \in P$ and $\{\{(i,a)\} \cup B\} \cup \mathcal{C} \in F$, then there is some $U_0 \in \mathcal{F}U$ such that $\{\{(i,u)\} \cup B \mid u \in U_0\} \cup \mathcal{C} \in F$. Here $S = \{a\}$ and $\phi = U$. If $\{a\} \cup T \in F_i$ then by hypothesis $U_0 \cup T \in F_i$ for some $U_0 \in \mathcal{F}U$.

Condition 4, if $\mathcal{A} \in F$ then $P \lhd \mathcal{A}$. Let us write

$$\mathcal{B}' = \{\mathrm{Im}\,\gamma \mid \gamma \in \mathrm{Ch}(\mathcal{A})\}.$$

For every $\gamma \in \mathrm{Ch}(\mathcal{A})$ we can find S in some F_i such that $\{i\} \times S \subseteq \mathrm{Im}\,\gamma$, and it follows that we can find $\mathcal{B} \in \mathcal{F}P$ such that

- every B in $\mathcal{B}$ is $\{i\} \times S$ for some i and $S \in F_i$;
- every B in $\mathcal{B}$ is included in some $\mathrm{Im}\,\gamma$ in $\mathcal{B}'$;
- every $\mathrm{Im}\,\gamma$ in $\mathcal{B}'$ includes some B in $\mathcal{B}$.

The last two imply that $\mathcal{B}' \sqsubseteq_U \mathcal{B}$ and $\mathcal{B}' \sqsubseteq_L \mathcal{B}$ (recalling that the order used on $P = \mathcal{F}P'$ is $\sqsubseteq_U$, which includes $\supseteq$).

Now let

$$\mathcal{C} = \{\mathrm{Im}\,\delta \mid \delta \in \mathrm{Ch}(\mathcal{B})\}.$$

We show (i) $\mathcal{C} \sqsubseteq_L \mathcal{A}$, and (ii) $(\emptyset) \lhd \mathcal{C}$, and these together imply that $(\emptyset) \lhd \mathcal{A}$.

For the first, take $\delta \in \mathrm{Ch}(\mathcal{B})$. For every $\gamma \in \mathrm{Ch}(\mathcal{A})$ we have that $\mathrm{Im}\,\gamma$ includes some $B \in \mathcal{B}$ and so meets $\mathrm{Im}\,\delta$. By the diagonalization lemma of [25] it follows that $A \subseteq \mathrm{Im}\,\delta$ for some $A \in \mathcal{A}$. (Classically, if no $A \in \mathcal{A}$ is included in $\mathrm{Im}\,\delta$ then there is a choice that avoids $\mathrm{Im}\,\delta$. But with these finite sets there is a constructive proof.)

For the second, we use induction on $\mathcal{B}$. If $\mathcal{B} = \emptyset$, then it has only one choice, which is empty, and so $\mathcal{C} = \{\emptyset\}$. Now suppose it holds for $\mathcal{B}_0$; we prove it for $\mathcal{B} = \mathcal{B}_0 \cup \{\{i\} \times S\}$ where $S \in F_i$. We have

$$\mathcal{C} \supseteq \{\mathrm{Im}\,\delta \cup \{(i,b)\} \mid \delta \in \mathrm{Ch}(\mathcal{B}_0), b \in S\}.$$

By induction,

$$(\emptyset) \lhd \{\mathrm{Im}\,\delta \mid \delta \in \mathrm{Ch}(\mathcal{B}_0)\}.$$

By definition of $\lhd_0$ for P,

$$\mathrm{Im}\,\delta \lhd_0 \{\mathrm{Im}\,\delta \cup \{(i,a)\} \mid a \in P_i\}.$$

Since $S \in F_i$, we have $a \lhd S$ for each $a \in P_i$, and a straightforward induction on the proof of $a \lhd S$ then shows that

$$\mathrm{Im}\,\delta \cup \{(i,a)\} \lhd \{\mathrm{Im}\,\delta \cup \{(i,b)\} \mid b \in S\}.$$

We can now use transitivity of $\lhd$. □

Now here is the lemma that was promised in proving condition 3.

Lemma 14.7 *Under the hypotheses of Theorem 14.6, suppose we have $i \in I$, $S \in \mathcal{F}P_i$ and $\phi \subseteq P_i$ with the property that, for every $T \in \mathcal{F}P_i$, if $\forall a \in S.\ \{a\} \cup T \in F_i$ then there is some $S' \in \mathcal{F}P_i$ with $S' \subseteq \phi$ and $S' \cup T \in F_i$.*

Then if $B \in P$, $\mathcal{C} \in \mathcal{F}P$ and $\{(\{i\} \times S) \cup B\} \cup \mathcal{C} \in F$, there is some $S' \in \mathcal{F}P_i$ with $S' \subseteq \phi$ and $\{\{(i, a)\} \cup B \mid a \in S'\} \cup \mathcal{C} \in F$.

Proof. Suppose B and $\mathcal{C}$ are given. Now suppose $\gamma \in \mathrm{Ch}(\mathcal{C})$. For every $a \in S$ we have $\{(i, a)\} \cup \mathrm{Im}\,\gamma$ covers some P_j, so there is some $S \in F_j$ such that $\{j\} \times S \subseteq \{(i, a)\} \cup \mathrm{Im}\,\gamma$. We can deduce that either $\mathrm{Im}\,\gamma$ covers some P_j, or $\{(i, a)\} \cup \mathrm{Im}\,\gamma$ covers P_i. We can therefore decompose $\mathrm{Ch}(\mathcal{C})$ as a union of finite sets, $D \cup D'$, such that if $\gamma \in D$ then $\mathrm{Im}\,\gamma$ covers some P_j, and if $\gamma \in D'$ then $\{(i, a)\} \cup \mathrm{Im}\,\gamma$ covers P_i for every $a \in S$. If $\gamma \in D'$ then for each $a \in S$ we can find $T \in P_i$ such that $\{a\} \cup T \in F_i$ and $\{i\} \times T \subseteq \mathrm{Im}\,\gamma$, and by taking their union we can assume that a single T does for all the a's. Then we can find S' with $S' \subseteq \phi$ and $S' \cup T \in F_i$. By taking the union of the S's we can assume a single S' such that $(\{i\} \times S') \cup \mathrm{Im}\,\gamma$ covers P_i for all $\gamma \in D'$.

We now show that $\{\{(i, a)\} \cup B \mid a \in S'\} \cup \mathcal{C} \in F$. For any choice of $\{\{(i, a)\} \cup B \mid a \in S'\} \cup \mathcal{C}$, its image contains a set of the form $\mathrm{Im}\,\gamma \cup \mathrm{Im}\,\delta$, where $\gamma \in \mathrm{Ch}(\mathcal{C})$ and $\delta \in \mathrm{Ch}(\{\{(i, a)\} \cup B \mid a \in S'\})$. For each $a \in S'$, we have either $\mathrm{Im}\,\delta$ meets B or $(i, a) \in \mathrm{Im}\,\delta$. If the former holds for some a, then there is some choice of $\{(\{i\} \times S) \cup B\} \cup \mathcal{C}$ whose image is a subset of $\mathrm{Im}\,\gamma \cup \mathrm{Im}\,\delta$, and from $\{(\{i\} \times S) \cup B\} \cup \mathcal{C} \in F$ we deduce that $\mathrm{Im}\gamma \cup \mathrm{Im}\,\delta$ covers some P_j. Alternatively, suppose $\{i\} \times S' \subseteq \mathrm{Im}\,\delta$. It suffices then to know that $(\{i\} \times S') \cup \mathrm{Im}\,\gamma$ covers some P_j. If $\gamma \in D$ then $\mathrm{Im}\,\gamma$ covers some P_j, while if $\gamma \in D'$ then by construction of S' we have $(\{i\} \times S') \cup \mathrm{Im}\,\gamma$ covers P_i. In either case we are done. □

14.4 Synthetic Locale Theory

We conclude with some remarks on an approach that promises to lay bare many issues of topology, both point-set and point-free. This is the 'synthetic topology' of Escardó [4]. It uses the lambda calculus to express maps, and the Tychonoff theorem (at least, binary Tychonoff) is a good illustration.

Recall that X is compact iff there is a Scott continuous map $\forall_! : \Omega X \to \Omega$ that is right adjoint to the unique frame homomorphism. For sober spaces, *all* continuous maps are Scott continuous (with respect to the specialization order). Also, the opens of X are equivalent to continuous maps from X to the Sierpinski space $\mathbb{S}$, so we can identify ΩX with the function space $\mathbb{S}^X$. Hence, compactness of X can be expressed by a map $\forall_X : \mathbb{S}^X \to \mathbb{S}$ right adjoint to the map $\mathbb{S}^! : \mathbb{S} \to \mathbb{S}^X$. We can think of the points of $\mathbb{S}$ as being truth values, the top (open) point $\top$ being **true**, and then $\forall(a)$ is the truth value of '$a = X$'.

If X and Y are both compact, then the corresponding map for $X \times Y$ can be expressed very easily as

$$\forall_{X\times Y}(u) = \forall_Y(\lambda y.\ \forall_X(\lambda x.\ u(x, y))).$$

To put it another way, $\forall_{X\times Y} : \mathbb{S}^{X\times Y} \to \mathbb{S}$ is the composite $\cong; (\forall_X)^Y; \forall_Y : \mathbb{S}^{X\times Y} \cong (\mathbb{S}^X)^Y \to \mathbb{S}^Y \to \mathbb{S}$. This is the required right adjoint, and if everything preserves continuity then we get the required Scott continuity.

But there's an obvious flaw in the argument! The function space $\mathbb{S}^X$ only exists if X is locally compact (this holds for locales as well as spaces). Apparently, it proves Tychonoff only for locally compact (and sober) spaces.

However, it is possible to get round this by embedding one's category of spaces in a larger category in which the exponentials exist. Escardó refers to 'real' spaces (in the original category) and 'complex' spaces (in the supercategory). It then remains only to show that morphisms between the complex function spaces do indeed give the required Scott continuous functions between frames.

For locales, the requisite results have been proved in [26]. There the category **Loc** of locales is embedded (by the Yoneda embedding) in the category $\mathbf{Set}^{\mathbf{Loc}^{op}}$ of presheaves over **Loc**. The fundamental lemma then is that presheaf morphisms (natural transformations) from $\mathbb{S}^X$ to $\mathbb{S}^Y$ correspond to Scott continuous functions from ΩX to ΩY. This allows us to use the above construction of $\forall_{X \times Y}$ as a proof of binary Tychonoff.

Despite the set-theoretic difficulties, it is to be hoped that a predicative argument can also be found to justify such synthetic methods in formal topology.

The infinitary Tychonoff theorem is less well understood from this point of view, but it seems to play the role of a termination principle for recursive algorithms.

References

1. Coquand, T. (1992). An intuitionistic proof of Tychonoff's theorem. *Journal of Symbolic Logic*, **57**, 28–32.
2. Coquand, T. (2003). Compact spaces and distributive lattices. *Journal of Pure and Applied Algebra*, **184**, 1–6.
3. Coquand, T., Sambin, G., Smith, J., and Valentini, S. (2003). Inductively generated formal topologies. *Annals of Pure and Applied Logic*, **124**, 71–106.
4. Escardó, Martín (2004). Synthetic topology of data types and classical spaces. In: *Bellairs Workshop: Domain Theoretic Methods in Probabilistic Processes,* (ed P. Panangaden and J. Desharnais), Volume 87 of *Electronic Notes in Theoretical Computer Science*. Elsevier.
5. Fourman, M. P. and Grayson, R. J. (1982). Formal spaces. In: *The L.E.J. Brouwer Centenary Symposium,* (ed Troelstra and van Dalen), pp. 107–122. North Holland.
6. Fourman, M. P. and Hyland, J. M. E. (1979). Sheaf models for analysis. In *Applications of Sheaves,* (ed M. Fourman, C. Mulvey, and D. Scott), Number 753 in Lecture Notes in Mathematics, pp. 280–301. Springer.
7. Hofmann, Karl H. and Mislove, Michael W. (1981). Local compactness and continuous lattices. In *Continuous Lattices: Proceedings, Bremen, 1979* (ed B. Banaschewski and R.-E. Hoffmann), Number 871 in Lecture Notes in Mathematics, pp. 209–248. Springer.
8. Johnstone, P. T. (1982). *Stone Spaces*. Number 3 in Cambridge Studies in Advanced Mathematics. Cambridge University Press.
9. Johnstone, P. T. (1984). Open locales and exponentiation. *Contemporary Mathematics*, **30**, 84–116.

10. Johnstone, P. T. (1985). Vietoris locales and localic semi-lattices. In: *Continuous Lattices and their Applications* (ed R.-E. Hoffmann), Number 101 in Pure and Applied Mathematics, pp. 155–18. Marcel Dekker.
11. Johnstone, P. T. and Vickers, S. J. (1991). Preframe presentations present. In: *Category Theory – Proceedings, Como 1990* (ed Carboni, A. Pedicchio, M. and Rosolini, G.), Number 1488 in Lecture Notes in Mathematics, pp. 193–212. Springer.
12. Joyal, A. and Tierney, M. (1984). An extension of the Galois theory of Grothendieck. *Memoirs of the American Mathematical Society*, **309**.
13. Kelley, J. L. (1950). The Tychonoff product theorem implies the axiom of choice. *Fundamenta Mathematicae*, **37**, 75–76.
14. Mac Lane, S. and Moerdijk, I. (1992). *Sheaves in Geometry and Logic*. Springer.
15. Negri, S. (2002). Continuous domains as formal spaces. *Mathematical Structures in Computer Science*, **12**, 19–52.
16. Negri, S. and Valentini, S. (1997). Tychonoff's theorem in the framework of formal topologies. *Journal of Symbolic Logic*, **62(4)**, 1315–1332.
17. Sambin, G. (1987). Intuitionistic formal spaces – a first communication. In: *Mathematical Logic and its Applications* (ed Skordev), pp. 187–204. Plenum.
18. Vermeulen, J. J. C. (1986). Constructive Techniques in Functional Analysis. PhD thesis, University of Sussex.
19. Vickers, S. J. (1989). *Topology via Logic*. Cambridge University Press.
20. Vickers, S. J. (1995). Locales are not pointless. In: *Theory and Formal Methods of Computing 1994* (ed C. Hankin, I. Mackie, and R. Nagarajan), London, pp. 199–216. Imperial College Press.
21. Vickers, S. J. (1997). Constructive points of powerlocales. *Math. Proc. Cam. Phil. Soc.*, **122**, 207–222.
22. Vickers, S. J. (1999). Topical categories of domains. *Mathematical Structures in Computer Science*, **9**, 569–616.
23. Vickers, S. J. (2002). Compactness in locales and formal topology. *Ann. Pure Appl. Logic*, to appear.
24. Vickers, S. J. (2004*a*). The double powerlocale and exponentiation: A case study in geometric reasoning. *Theory and Applications of Categories*, **12**, 372–422.
25. Vickers, S. J. (2004*b*). Entailment systems for stably locally compact locales. *Theoretical Computer Science*, **316**, 259–296.
26. Vickers, S. J. and Townsend, C. F. (2004). A universal characterization of the double powerlocale. *Theoretical Computer Science*, **316**, 297–321.

15

AN ELEMENTARY CHARACTERIZATION OF KRULL DIMENSION

THIERRY COQUAND, HENRI LOMBARDI AND MARIE-FRANÇOISE ROY

Abstract

We give an elementary characterization of Krull dimension for distributive lattices and commutative rings. This follows the geometrical intuition that an algebraic variety is of dimension $\leq k$ if and only if each subvariety has a boundary of dimension $< k$. Since our results hold for distributive lattices, they hold, by Stone duality, for spectral spaces.

15.1 Introduction

The notion of dimension is one of the most fundamental in geometry and topology. It was an idea of the ancient Greek mathematicians that a curve segment was something bounded by points and that a surface was something bounded by curves. A precise topological definition of dimension was given by Brouwer (1913), working from ideas of Poincaré. Later, in 1922, Menger and Urysohn independently found a similar definition. This definition is inductive, and is reminiscent of the ancient notion of dimension, but more complicated. A space has dimension $\leq k$ if every point can be surrounded arbitrarily closely by neighbourhoods with boundaries of dimension $< k$. In algebraic geometry there appears another definition of dimension, due to Krull (1937), which a priori does not seem to be connected in any way to the Menger–Urysohn definition [1]. This definition is also a priori non-effective, since it refers to the notion of prime ideals, which may fail to exist constructively. The Krull dimension of a ring is the supremum of the lengths of chains of distinct prime ideals in this ring. We present here a new inductive definition of Krull dimension, which follows the geometrical intuition that an algebraic variety is of dimension $\leq k$ if and only if each subvariety has a boundary of dimension $< k$. This definition makes sense constructively. The same idea applies to distributive lattices, and by Stone duality [2], to spectral spaces. In this case, the Krull dimension appears as a simplified version of the Menger–Urysohn notion of dimension.

15.2 Boundaries of an Element in a Distributive Lattice

By *distributive lattice* we mean a lattice with a minimum and a maximum (so that all finite parts have a supremum and an infimum) which is distributive.

Let L be a distributive lattice. An *ideal* of L is a subset $I \subseteq L$ such that

$$\begin{aligned} 0 &\in I \\ x, y \in I &\Longrightarrow x \vee y \in I \\ x \in I,\ z \in L &\Longrightarrow x \wedge z \in I. \end{aligned}$$

The last property can be written as $(x \in I,\ y \leq x) \Rightarrow y \in I$.

The dual notion is the notion of *filter*. A filter F is a subset of L such that

$$\begin{aligned} 1 &\in F \\ x, y \in F &\Longrightarrow x \wedge y \in F \\ x \in F,\ z \in L &\Longrightarrow x \vee z \in F. \end{aligned}$$

A *prime ideal* is an ideal I such that $1 \notin I$ and

$$x \wedge y \in I \Rightarrow [x \in I \ or \ y \in I]$$

and dually a *prime filter* is a filter F such that $0 \notin F$ and

$$x \vee y \in F \Rightarrow [x \in F \ or \ y \in F].$$

Notice that an ideal (resp. a filter) is prime if and only if its complement is a filter (resp. an ideal).

If $x \in L$ we denote by $D(x)$ the set of prime ideals I such that $x \notin I$. We have $D(0) = \emptyset$ and $D(x) \cap D(y) = D(x \wedge y)$. The set of all prime ideals of a distributive lattice L has a natural structure of a topological space, called the *spectrum* $Sp(L)$ of L. We take for basic open sets the sets $D(x),\ x \in L$. It can be shown that each $D(x)$ is compact and that the compact open sets of $Sp(L)$ are exactly the subsets of the form $D(x),\ x \in L$ [2].

The spaces (homeomorphic to spaces) of the form $Sp(L)$ are called *spectral spaces* and it is possible to characterize directly these spaces by topological properties [2, 3]. Most topological spaces used in commutative algebra (Zariski spectrum of a ring, spaces of valuations of a field, etc.) are spectral spaces.

The set $Sp(L)$ is ordered by inclusion, and the *Krull dimension* of L is defined as the upper bound of the length of chains of prime ideals (or equivalently chains of prime filters).

If $x \in L$ we define the *boundary ideal* of x as the ideal generated by x and the elements $y \in L$ such that $x \wedge y = 0$. Dually, we define the *boundary filter* of x as being the filter generated by x and the elements $y \in L$ such that $x \vee y = 1$.

Definition 15.1 *The upper boundary of $x \in L$ in the distributive lattice L is the distributive lattice $L^{\{x\}}$ quotient of L by the boundary ideal of x. Thus it is the lattice $L, \wedge, \vee$ with the order*

$$a \leq^x b \qquad \Longleftrightarrow \qquad \exists y \in L \ \ (x \wedge y = 0 \ \ \& \ \ a \leq x \vee y \vee b).$$

When L is implicative the definition becomes $a \leq x \vee \neg x \vee b$.

By considering the dual lattice, one defines the lower boundary $L_{\{x\}}$*, which is the distributive lattice quotient of* L *by the boundary filter of* x*. Thus it is the lattice* $L, \wedge, \vee$ *with the order*

$$a \leq_x b \qquad \Longleftrightarrow \qquad \exists y \in L \ \ (x \vee y = 1 \ \ \& \ \ a \wedge x \wedge y \leq b).$$

It can be checked that the boundary of the open $D(x)$, viewed as a subspace of $Sp(L)$, is a spectral space (as a closed set in a spectral space) and corresponds by Stone duality to the distributive lattice $L^{\{x\}}$.

15.3 Krull Dimension of a Distributive Lattice

The duality between distributive lattices and spectral spaces relies on classical logic and the axiom of choice. From a constructive point of view, this duality is seen as a way to develop the theory of spectral spaces, using distributive lattices as a point-free presentation of these spaces [4]. One is thus led to look for direct definitions of topological notions in terms of distributive lattices, and for instance, a direct definition of the Krull dimension.

A first constructive definition of Krull dimension was sketched in [5]. This definition was analysed in the work [6]. The author gave an elementary characterization of the Krull dimension of a lattice L in terms of the Boolean algebra generated by L. In [7], following the idea in [5], the two first authors proved the following result, which gives yet another concrete characterization of Krull dimension.

Theorem 15.2 *Let* L *be a distributive lattice generated by a subset* S *and* ℓ *a non-negative integer. The following are equivalent*

(1) *L has Krull dimension* $\leq \ell$*.*

(2) *For all* $x_0, \ldots, x_\ell \in S$ *there exist* $a_0, \ldots, a_\ell \in L$ *such that*

$$a_0 \wedge x_0 \leq 0, \quad a_1 \wedge x_1 \leq a_0 \vee x_0\,, \ldots, \quad a_\ell \wedge x_\ell \leq a_{\ell-1} \vee x_{\ell-1}, \quad 1 \leq a_\ell \vee x_\ell.$$

In particular a distributive lattice L is of dimension ≤ 0 if and only if L is a boolean algebra (any element has a complement).

The goal of this chapter is to present a simpler inductive characterization of Krull dimension, which provides also a simple proof of the equivalence between (1) and (2) in the previous theorem. This inductive characterization corresponds to the following geometrical intuition: a variety is of dimension $\leq k$ if and only if any subvariety has a boundary of dimension $< k$ (the induction begins with dimension -1 which defines the trivial lattice).

Theorem 15.3 *Let* L *be a distributive lattice generated by a subset* S *and* ℓ *a non-negative integer. The following are equivalent:*

(1) *L has Krull dimension* $\leq \ell$*.*

(2) *For all* $x \in S$ *the boundary* $L^{\{x\}}$ *is of Krull dimension* $\leq \ell - 1$*.*

(3) *For all* $x \in S$ *the boundary* $L_{\{x\}}$ *is of Krull dimension* $\leq \ell - 1$*.*

Proof. (1) $\Leftrightarrow$ (2): We show first that any maximal filter F of L becomes trivial in $L^{\{x\}}$, i.e. it contains 0. This means that one can find $a \in F$ such that $a \leq^x 0$. If $x \in F$ this holds since $x \leq^x 0$. If $x \notin F$ there exists $z \in F$ such that $x \wedge z = 0$ (since the filter generated by F and x is trivial) and we have then $z \leq^x 0$. This shows that the Krull dimension of $L^{\{x\}}$ becomes one less than that of L (if it is finite).

Next, we show that if $F' \subset F$, F maximal and $x \in F \setminus F'$, then F' does not become trivial in $L^{\{x\}}$ (which shows that $dim\ L^{\{x\}}$ is $dim\ L - 1$ with a good choice of x). Indeed, we would get otherwise $z \in F'$ such that $z \wedge x = 0$, which is impossible since both z and x are in F.

We finally notice that if $F' \subset F$ are distinct prime filters and S generates L one can find $x \in S$ such that $x \in F \setminus F'$.
(1) $\Leftrightarrow$ (3) is a consequence of (1) $\Leftrightarrow$ (2) by duality. □

By Stone duality [2], we get the following result.

Theorem 15.4 *A spectral space X is of Krull dimension $\leq k$ if and only if any open compact of X has a boundary of dimension $< k$.*

15.4 The Two Boundaries of an Element in a Commutative Ring

Let R be a commutative ring. We write $\langle J \rangle$ for the ideal of R generated by the subset $J \subseteq R$. We write $\mathcal{M}(U)$ for the monoid (a monoid will always be multiplicative) generated by the subset $U \subseteq R$. Given a commutative ring R the *Zariski lattice* $\mathrm{Zar}(R)$ has for elements the radicals of finitely generated ideals. The order relation is the inclusion and we get

$$\sqrt{I} \wedge \sqrt{J} = \sqrt{IJ}, \quad \sqrt{I} \vee \sqrt{J} = \sqrt{I+J}.$$

We shall write $\widetilde{a}$ for $\sqrt{\langle a \rangle}$. We have

$$\widetilde{a_1} \vee \cdots \vee \widetilde{a_m} = \sqrt{\langle a_1, \ldots, a_m \rangle} \quad \text{and} \quad \widetilde{a_1} \wedge \cdots \wedge \widetilde{a_m} = \widetilde{a_1 \cdots a_m}.$$

Let U and J be two finite subsets of R, we have

$$\bigwedge_{u \in U} \widetilde{u} \leq_{\mathrm{Zar}(R)} \bigvee_{a \in J} \widetilde{a} \iff \prod_{u \in U} u \in \sqrt{\langle J \rangle} \iff \mathcal{M}(U) \cap \langle J \rangle \neq \emptyset.$$

This describes completely the lattice $\mathrm{Zar}(R)$. More precisely [7] we have:

Proposition 15.5 *The lattice* $\mathrm{Zar}(R)$ *of a commutative ring R is (up to isomorphism) the lattice generated by symbols $D(x)$, $x \in R$ with the relations*

$$D(0) = 0, \quad D(1) = 1, \quad D(fg) = D(f) \wedge D(g), \quad D(f+g) \leq D(f) \vee D(g).$$

The spectrum of the distributive lattice $\mathrm{Zar}(R)$ is naturally isomorphic to the Zariski spectrum of the ring R. So the Krull dimension of a commutative ring R is the same as the Krull dimension of its Zariski lattice $\mathrm{Zar}(R)$.

Definition 15.6 *Let R be a commutative ring and $x \in R$.*

(1) *The boundary $R^{\{x\}}$ of x in R is the quotient ring $R/I^{\{x\}}$ where $I^{\{x\}} = xR + (\sqrt{0} : x)$.*

(2) *The boundary $R_{\{x\}}$ of x in R is the localized ring $R_{S_{\{x\}}}$ where $S_{\{x\}} = x^{\mathbb{N}}(1 + xR)$.*

The next proposition is easy.

Proposition 15.7 *Let $L = \mathrm{Zar}(R)$ and $x \in R$. Then $L^{\{\widetilde{x}\}}$ is naturally isomorphic to $\mathrm{Zar}(R^{\{x\}})$ and $L_{\{\widetilde{x}\}}$ is naturally isomorphic to $\mathrm{Zar}(R_{\{x\}})$.*

We get an elementary inductive characterization of Krull dimension of commutative rings. Recall that a ring R has Krull dimension -1 if and only if it is trivial (i.e. $1_R = 0_R$).

Theorem 15.8 *Let R be a commutative ring and $\ell \geq 0$ an integer. The following are equivalent*

(1) *The Krull dimension of R is $\leq \ell$.*

(2) *For all $x \in R$ the Krull dimension of $R^{\{x\}}$ is $\leq \ell - 1$.*

(3) *For all $x \in R$ the Krull dimension of $R_{\{x\}}$ is $\leq \ell - 1$.*

These equivalences are immediate consequences of Theorem 15.3 and Proposition 15.7.

Corollary 15.9 (cf. [7,8]) *Let ℓ be a non-negative integer. The Krull dimension of R is $\leq \ell$ if and only if for all $x_0, \ldots, x_\ell$ in R there exist $a_0, \ldots, a_\ell \in R$ and $m_0, \ldots, m_\ell \in \mathbb{N}$ such that*

$$x_0^{m_0}(\cdots(x_\ell^{m_\ell}(1 + a_\ell x_\ell) + \cdots) + a_0 x_0) = 0. \tag{15.1}$$

Proof. Since dimension -1 corresponds to the trivial ring, the equivalence for the case $\ell = 0$ is clear.

Assume the equivalence has been established for all integers $< \ell$ and all R. We deduce that the dimension of a localization $S^{-1}R$ is $< \ell$ if and only if for all $x_0, \ldots, x_{\ell-1} \in R$ there exist $a_0, \ldots, a_{\ell-1} \in R$, $s \in S$ and $m_0, \ldots, m_{\ell-1} \in \mathbb{N}$ such that

$$x_0^{m_0}(x_1^{m_1} \cdots (x_{\ell-1}^{m_{\ell-1}}(s + a_{\ell-1}x_{\ell-1}) + \cdots + a_1 x_1) + a_0 x_0) = 0\,. \tag{15.2}$$

Notice that s replaces 1 in the similar equality (15.1) with R instead of $S^{-1}R$. It remains only to replace s by an arbitrary element in $S_{\{x_\ell\}}$, i.e. an element $x_\ell^{m_\ell}(1 + a_\ell x_\ell)$. □

The advantage of this definition is, besides its elementary character, to allow simple proofs by induction on the dimension. In this way we can, for instance, prove directly the following non-Noetherian version of Bass' stable range theorem.

Theorem 15.10 *If the dimension of R is $< n$ and $1 = D(a, b_1, \ldots, b_n)$, then there exist $x_1, \ldots, x_n$ such that $1 = D(b_1 + ax_1, \ldots, b_n + ax_n)$.*

Examples

If $A = \mathbb{Z}$ and $n \neq 0, 1, -1$, then $\mathbb{Z}^{\{n\}} = \mathbb{Z}/n\mathbb{Z}$ and $\mathbb{Z}_{\{n\}} = \mathbb{Q}$. These are two 0-dimensional rings. For $n = 0$, 1 or -1 the two boundaries are trivial. Thus the Krull dimension of $\mathbb{Z}$ is 1.

Let $\mathbb{K}$ be a field contained in an algebraically closed field $\mathbb{L}$, and J be a finitely generated ideal of $\mathbb{K}[X_1, \dots, X_n]$ and $A = \mathbb{K}[X_1, \dots, X_n]/J$. If V is the algebraic variety corresponding to J in $\mathbb{L}^n$, if $f \in A$ defines the subvariety W of V and if B is the boundary of W in V, defined as the intersection of W with the Zariski closure of its complement in V, then the affine variety B corresponds to the ring $A^{\{f\}}$.

Acknowledgements

We thank the referee for the careful rereading and valuable comments.

References

1. Eisenbud, D. (1995). *Commutative Algebra with a View Towards Algebraic Geometry*. Springer-Verlag.
2. Stone, M. H. (1937). Topological representations of distributive lattices and Brouwerian logics. *Cas. Mat. Fys.*, **67**, 1–25.
3. Hochster, M. (1969). Prime ideal structure in commutative rings. *Trans. Amer. Math. Soc.*, **142**, 43–60.
4. Johnstone, P. (1986). *Stone Spaces.* Cambridge Studies in Advanced Mathematics **3**. Cambridge University Press, Cambridge.
5. Boileau, A. and Joyal, A. (1981). La logique des topos, *J. Symbolic Logic*, **46**, no. 1, 6–16.
6. Español, L. (1982). Constructive Krull dimension of lattices. *Rev. Acad. Cienc. Zaragoza*, **37**(2), 5–9.
7. Coquand, T. and Lombardi, H. (2002). Hidden constructions in abstract algebra (3). Krull dimension of distributive lattices and commutative rings. In: *Commutative ring theory and applications*, (eds Fontana M., Kabbaj S.-E., and S. Wiegand.), pp. 477–499. Lecture notes in pure and applied mathematics vol 131. M. Dekker.
8. Lombardi, H. (2002). Dimension de Krull, Nullstellensätze et évaluation dynamique. *Math. Zeitschrift,* **242**, 23–46.
9. Cederquist, J. and Coquand, T. (2000). *Entailment relations and distributive lattices.* Logic Colloquium '98 (Prague), pp. 127–139, Lect. Notes Log. **13**. Assoc. Symbol. Logic, Urbana.
10. Coquand, T. (2004). Sur un théorème de Kronecker concernant les variétés algébriques. *C. R. Acad. Sci. Paris, Ser. I* **338**, 291–294.
11. Curry, H. B. (1963). *Foundations of Mathematical Logic*. McGraw-Hill Book Co.
12. Joyal, A. (1975). Le théorème de Chevalley-Tarski. *Cahiers de Topologie et Géometrie Différentielle*, **16**, 256–258.

16

CONSTRUCTIVE REVERSE MATHEMATICS: COMPACTNESS PROPERTIES

HAJIME ISHIHARA

Abstract

We first propose a base formal system for constructive reverse mathematics aiming at classifying various theorems in intuitionistic, constructive recursive and classical mathematics by logical principles, function existence axioms and their combinations. The system is weak enough to compare results, in the system, with the results in classical reverse mathematics and to prove theorems in Bishop's constructive mathematics. Then we formalize some results on compactness properties, such as the Heine–Borel theorem, the Cantor intersection theorem, the Bolzano–Weierstraß theorem, and sequential compactness, in the base formal system as test cases of its adequacy and faithfulness for the purpose of constructive reverse mathematics. We also investigate the computability of function existence axioms and their combination with logical principles, identifying them with closure conditions on a class of functions.

16.1 Introduction

The purpose of *constructive reverse mathematics* is to classify various theorems in intuitionistic, constructive recursive and classical mathematics by logical principles, function existence axioms and their combinations. Classifying mathematical theorems means finding logical principles and/or function existence axioms which are not only sufficient but also necessary to prove the theorems in a fairly weak formal system.

Bishop's constructive mathematics [1–5] is an informal mathematics using intuitionistic logic and assuming some function existence axioms—the axiom of countable choice, the axiom of dependent choice, and the axiom of unique choice. It is a core of the varieties of mathematics in the sense that it can be extended not only to intuitionistic mathematics (by adding the principle of continuous choice and the fan theorem) [3, 5–7, 9] and constructive recursive mathematics (by adding Markov's principle and the extended Church's thesis) [3, 5, 10], but also to classical mathematics practised by most mathematicians today (by adding the principle of the excluded middle). More on philosophy and practice of Bishop's and his followers' constructive mathematics can be found in [11–13].

The Friedman–Simpson program, called (classical) reverse mathematics [14, 15], is a formal mathematics using classical logic and assuming, in its base system, a very weak set existence axiom—the Δ^0_1 comprehension scheme. Its main question is 'Which set existence axioms are needed to prove the theorems of ordinary mathematics?', and many theorems have been classified by set existence axioms of various strengths. Since classical reverse mathematics is formalized with classical logic, we cannot

- classify theorems in intuitionistic mathematics or in constructive recursive mathematics which are inconsistent with classical mathematics (for example, continuity of mappings from $\mathbb{N}^{\mathbb{N}}$ to $\mathbb{N}$ [5, 4.6]);
- distinguish theorems from their contrapositions (for example, the fan theorem from the weak König lemma, see Theorem 16.17 and Theorem 16.21 in the following, and [14, IV.1]).

Although non-constructive logical principles such as the limited principle of omniscience (LPO) have been rejected within any constructive framework, some recent proofs in Bishop's (forward) mathematics [16–19] have made use of such non-constructive principles, and many theorems in classical, intuitionistic and constructive recursive mathematics have been classified using such principles within the framework of Bishop's constructive mathematics [20]. For example, Mandelkern [21], Ishihara [22], Bridges, Ishihara and Schuster [23], and Ishihara and Schuster [27] dealt with LPO, the lesser limited principle of omniscience (LLPO), the fan theorem (FAN) and various compactness principles; Ishihara [25, 26], Bridges, Ishihara, Schuster and Vîţă [18], and Bridges, Ishihara and Schuster [23] treated the weak limited principle of omniscience (WLPO), weak Markov's principle (WMP), a boundedness principle (BD-N), FAN and various continuity principles. Since Bishop's constructive mathematics is informal, and assumes the function existence axioms, we cannot

- compare those results with the results in classical reverse mathematics (for example, the equivalence in Bishop's constructive mathematics between a function existence axiom, the weak König lemma (WKL), and a logical principle, LLPO, was proved in [22]);
- prove neither underivability nor separability of those principles (for example, underivability of LPO and separability of WLPO from LPO, see [28, chapter 6] for such results in **HA** and $\mathbf{HA}^{\omega}$).

In this chapter, we first propose a base formal system for constructive reverse mathematics which is weak enough to compare results in the system with the results in classical reverse mathematics and to prove theorems in Bishop's constructive mathematics. Then we formalize some of the above results on compactness properties, such as the Heine–Borel theorem, the Cantor intersection theorem (which was called the pseudo-Heine–Borel theorem in [22]), the Bolzano–Weierstraß theorem, and sequential compactness in the base formal system as test cases of its adequacy and faithfulness for the purpose of constructive reverse mathematics. We also investigate the computability of function existence axioms and their combination with logical principles, identifying them with closure conditions on a class of functions as in [8].

Although current research of computability in mathematics follows the lines of [29, 30], uses classical logic, and is mainly concerned with sufficiency aspects only, constructive reverse mathematics deals not only with the sufficiency of logical principles and function existence axioms, but also with the necessity of them, so as to exactly classify mathematical theorems.

16.2 A Formal System

Definition 16.1 *The class of Kalmár's elementary functions,* ELEM, *is generated by the following clauses*

1. Z, S, prd, $+$, $\dot{-}$, $\cdot$, $\boldsymbol{p}_i^n$ *for* $i < n$, $1 \leq n$ *belong to* ELEM. *Here* Z *is the zero-function satisfying* $Z(x) = 0$, S *the successor function,* prd *the predecessor function,* $+$ *the addition,* $\dot{-}$ *the cut-off subtraction,* $\cdot$ *the multiplication, and* $\boldsymbol{p}_i^n$ *is a projection determined by* $\boldsymbol{p}_i^n(x_0, \ldots, x_{n-1}) = x_i$.
2. ELEM *is closed under composition: if* $f, g_i \in$ ELEM, *with* $f \in \mathbb{N}^n \to \mathbb{N}$, $g_i \in \mathbb{N}^k \to \mathbb{N}$ $(1 \leq i \leq n)$, *then there is an* $h \in$ ELEM *satisfying*
$$h(\vec{x}) = f(g_1(\vec{x}), \ldots, g_n(\vec{x})).$$
3. ELEM *is closed under bounded sum and bounded product: if* $f \in$ ELEM *with* $f \in \mathbb{N}^{n+1} \to \mathbb{N}$, *then there are* $g, h \in$ ELEM *such that*
$$\begin{aligned} g(0, \vec{x}) &= 0, \quad & g(Sy, \vec{x}) &= f(y, \vec{x}) + g(y, \vec{x}) \\ h(0, \vec{x}) &= 1, \quad & h(Sy, \vec{x}) &= f(y, \vec{x}) \cdot h(y, \vec{x}). \end{aligned}$$
We usually write $\sum_{z<y} f(z, \vec{x})$ *and* $\prod_{z<y} f(z, \vec{x})$ *for* g *and* h, *respectively.*

It is well known that ELEM is closed under the *bounded minimum operator*: if $f \in$ ELEM with $f \in \mathbb{N}^{n+1} \to \mathbb{N}$, then

$$\min_{z \leq y}(g(z, \vec{x}) = 0) := \textstyle\sum_{u<Sy} \operatorname{sg}\left(\prod_{z<Su} g(z, \vec{x})\right),$$

where $\operatorname{sg}(x) := 1 \dot{-} (1 \dot{-} x)$ is the *signum function*. Thus $\min_{z \leq y}(g(z, \vec{x}) = 0)$ is the least $z \leq y$ with $f(z, \vec{x}) = 0$, if it exists, $y + 1$ otherwise. The elementary relations, that is relations whose characteristic functions belong to ELEM, are closed under intersection, union, complementation, bounded quantification, and substitution [32, Chapter 1], [5, 3.1.1]. Within ELEM, we can code n-tuples of natural numbers and finite sequence of natural numbers. For the coding of pairs we have a mapping $j \in \mathbb{N}^2 \to \mathbb{N}$ with inverses j_1, j_2 satisfying

$$j_1 j(x, y) = x, \quad j_2 j(x, y) = y, \quad j(j_1 z, j_2 z) = z.$$

There are many possible choices of j, a convenient one is

$$\begin{aligned} j(x, y) &:= 2^x \cdot (2y + 1) \dot{-} 1, \\ j_1(z) &:= \min_{x \leq z}[\exists y \leq z(2^x \cdot (2y + 1) = Sz], \\ j_2(z) &:= \min_{y \leq z}[\exists x \leq z(2^x \cdot (2y + 1) = Sz]. \end{aligned}$$

From j, j_1, j_2 we can construct codings ν^n for n-tuples with inverse j_i^n such that $j_i^n \nu^n (x_1, \dots, x_n) = x_i$ $(1 \le i \le n)$. We can code finite sequences $x_0, \dots, x_{n-1}$ with $x_i < 2^b$ for $i < n$ by a number

$$\langle x_0, \dots, x_{n-1} \rangle := j\left(n, j\left(b, \sum_{i<n} x_i \cdot 2^{b \cdot i}\right)\right)$$

with the *length* function $\mathrm{lh}(a) := j_1(a)$ and the *decoding* function

$$\pi(a, i) := r(q(j_2 j_2(a), 2^{i \cdot j_1 j_2(a)}), 2^{j_1 j_2(a)}),$$

where $q(x, y) := \min_{z \le x}(x < Sz \cdot y)$ and $r(x, y) := x \dot{-} q(x, y) \cdot y$ are quotient and remainder functions, respectively. We usually write $(u)_i$ for $\pi(u, i)$. Note that letting $\mathrm{pow}(x) := \min_{y \le x}(x < 2^y)$ and $\sigma(M, n) := 2^{(\mathrm{pow}(M)+1)\cdot(n+1)+1}$ we have

$$\langle x_0, \dots, x_{n-1} \rangle < \sigma(M, n)$$

whenever $x_i \le M$ for all $i < n$.

Our formal system $\mathbf{EL}_{\mathrm{ELEM}}$ is a subsystem of the *elementary analysis* **EL** [33, 34], [5, 3.6 and 4.4.6] taking as our basis Kalmár's elementary functions. Systems based on subrecursive functions can be also found in [35, 36].

Definition 16.2 *The language of* $\mathbf{EL}_{\mathrm{ELEM}}$ *contains numerical variables, numerical constant* 0*, unary function variables, a unary function constant S, a function symbol for each Kalmár's elementary function, the application operator* Ap*, the abstraction operator* λ*, the bounded minimum operator* $\boldsymbol{\mu}$*, and the binary predicate* $=$*. As usual we write* $\varphi(t)$*, or even* φt*, for* $\mathrm{Ap}(\varphi, t)$*.*

Functors and terms are defined simultaneously by

1. *numerical variables and the numerical constant are terms;*
2. *if* f *is an* n*-ary function constant and* $t_1, \dots, t_n$ *terms, then* $f(t_1, \dots, t_n)$ *is a term;*
3. *function variables and unary function constants are functors;*
4. *if* φ *is a functor and* t *a term, then* $\varphi(t)$ *is a term;*
5. *if* t *is a term, then* $\lambda x.t$ *is a functor;*
6. *if* t *and* t' *are terms, and* φ *is a functor, then* $\boldsymbol{\mu}(t, \varphi, t')$ *is a term.*

Formulas on the basis of $\wedge, \vee, \to, \perp, \forall, \exists$ *as primitives, are defined as usual.* Δ_0 *denotes the class of quantifier-free formulas.*

The logic of $\mathbf{EL}_{\mathrm{ELEM}}$ *is two-sorted intuitionistic predicate logic. As non-logical axioms we have the equality axioms*

REFL $\quad x = x$

REPL $\quad A(x) \wedge x = y \to A(y)$,

the axiom for the successor

$\neg(S0 = 0)$,

defining equations for all Kalmár's elementary functions: for examples

$$\begin{aligned} \mathrm{prd}(0) &= 0, & \mathrm{prd}(Sx) &= x, \\ x \dot{-} 0 &= x, & x \dot{-} Sy &= \mathrm{prd}(x \dot{-} y), \\ x + 0 &= x, & x + Sy &= S(x+y), \\ x \cdot 0 &= 0, & x \cdot Sy &= x \cdot y + x, \end{aligned}$$

etc., the induction axiom schema for quantifier-free formulas

Δ_0-IND $\quad A(0) \wedge \forall x(A(x) \rightarrow A(Sx)) \rightarrow \forall x A(x) \quad (A \in \Delta_0)$,

the axiom for λ*-conversion*

CON $\quad (\lambda x.t)s = t[x/s]$,

the axiom for the bounded minimum operator (φ *unary,* $\varphi(x, y) = \varphi j(x, y)$*)*

$$\text{BMIN} \quad \begin{cases} \boldsymbol{\mu}(t, \varphi, 0) = \mathrm{sg}(\varphi(0, t)), \\ \boldsymbol{\mu}(t, \varphi, St') = \boldsymbol{\mu}(t, \varphi, t') + \mathrm{sg}(\boldsymbol{\mu}(t, \varphi, t') \dot{-} t') \cdot \mathrm{sg}(\varphi(St', t))). \end{cases}$$

We write $\min_{x \leq t'}(\varphi(x, t) = 0)$ *for* $\boldsymbol{\mu}(t, \varphi, t')$.

Equality between functors is defined by $\varphi = \psi := \forall x(\varphi x = \psi x)$. Coding of n-tuples, finite sequences, etc. can be lifted from numbers to functions via λ-abstraction:

$$\begin{aligned} \nu^n(\alpha_1, \ldots, \alpha_n) &:= \lambda x.\nu^n(\alpha_1 x, \ldots, \alpha_n x), \\ j_i^n(\alpha) &:= \lambda x.j_i^n(\alpha x) \quad (1 \leq i \leq n), \\ \langle \alpha_0, \ldots, \alpha_m \rangle &:= \lambda x.\langle \alpha_0 x, \ldots, \alpha_m x \rangle. \end{aligned}$$

It is straightforward to show that for every Δ_0 formula A there exists a term t such that $\mathbf{EL}_{\mathrm{ELEM}} \vdash \forall \vec{x}[A(\vec{x}) \leftrightarrow t(\vec{x}) = 0]$, and the following lemma using the bounded minimum operator.

Lemma 16.3 *The class* Δ_0 *of all quantifier free formulas is closed under bounded quantification in the sense that for every formula* A *in* Δ_0 *there exist formulas* B *and* C *in* Δ_0 *such that* $\mathbf{EL}_{\mathrm{ELEM}} \vdash \forall x \vec{y}[B(x, \vec{y}) \leftrightarrow \exists z(z \leq x \wedge A(z, \vec{y}))]$ *and* $\mathbf{EL}_{\mathrm{ELEM}} \vdash \forall x \vec{y}[C(x, \vec{y}) \leftrightarrow \forall z(z \leq x \rightarrow A(z, \vec{y}))]$.

We write $\exists z \leq t A(z, \vec{y})$, and $\forall z \leq t A(z, \vec{y})$ for $\exists z(z \leq t \wedge A(z, \vec{y}))$, and $\forall z(z \leq t \rightarrow A(z, \vec{y}))$, respectively.

Definition 16.4 *The* Σ_1^0*-closure* $\Sigma_1^0(\mathcal{C})$ *and the* Π_1^0*-closure* $\Pi_1^0(\mathcal{C})$ *of a class* $\mathcal{C}$ *of formulas are simultaneously inductively defined by the clauses*

1. *every formula in* $\Delta_0 \cup \mathcal{C}$ *is in* $\Sigma_1^0(\mathcal{C})$ *and in* $\Pi_1^0(\mathcal{C})$,
2. *if* A *and* B *are in* $\Sigma_1^0(\mathcal{C})$, *then so are* $A \wedge B$ *and* $A \vee B$,
3. *if* A *and* B *are in* $\Pi_1^0(\mathcal{C})$, *then so is* $A \wedge B$,
4. *if* A *is in* Δ_0 *and* B *is in* $\Pi_1^0(\mathcal{C})$, *then* $A \vee B$ *and* $B \vee A$ *are in* $\Pi_1^0(\mathcal{C})$,

5. *if A is in Δ_0 and B is in $\Sigma_1^0(\mathcal{C})$, then $A \to B$ is in $\Sigma_1^0(\mathcal{C})$,*
6. *if A is in $\Sigma_1^0(\mathcal{C})$ and B is in $\Pi_1^0(\mathcal{C})$, then $A \to B$ is in $\Pi_1^0(\mathcal{C})$,*
7. *if A is in $\Sigma_1^0(\mathcal{C})$, then so is $\exists x A$,*
8. *if A is in $\Pi_1^0(\mathcal{C})$, then so is $\forall x A$.*

We put $\Sigma_1^0 := \Sigma_1^0(\Delta_0)$ and $\Pi_1^0 := \Pi_1^0(\Delta_0)$. The class $\mathcal{A}$ of arithmetical formulas is defined as the class of those formulas which have no quantifiers over function variables.

Lemma 16.5 *Let $\mathcal{C}$ be a class of formulas containing Δ_0. Then*

1. *for every formula A in $\Sigma_1^0(\mathcal{C})$ there exists a formula B in $\mathcal{C}$ such that $\mathbf{EL}_{\mathrm{ELEM}} \vdash \forall \vec{x}[A(\vec{x}) \leftrightarrow \exists y B(\vec{x}, y)]$,*
2. *for every formula A in $\Pi_1^0(\mathcal{C})$ there exists a formula B in $\mathcal{C}$ such that $\mathbf{EL}_{\mathrm{ELEM}} \vdash \forall \vec{x}[A(\vec{x}) \leftrightarrow \forall y B(\vec{x}, y)]$.*

Proof. By simultaneous induction on the complexity of A. □

The introduction of the integers $\mathbb{Z}$ and the rationals $\mathbb{Q}$, and the development of the elementary theory of arithmetical operations on $\mathbb{Z}$ and $\mathbb{Q}$ are unproblematic using the coding of pairs. The introduction of the reals $\mathbb{R}$ via Cauchy sequences of rationals (fundamental sequences), and the development of the theory of the reals in [5, 5.2–5.4] can be carried out within $\mathbf{EL}_{\mathrm{ELEM}}$ without problems. We adopt fundamental sequences with a fixed rate of convergence: a *real number* is a sequence $\langle p_n \rangle_n$ of rationals such that

$$\forall mn(|p_m - p_n| < 2^{-m} + 2^{-n}).$$

For a real number $x := \langle p_n \rangle_n$, we write $(x)_n$ for p_n.

The *universal tree* T_ω consists of all finite sequences of natural numbers; so $T_\omega = \mathbb{N}$ with the equality

$$a = b := \mathrm{lh}(a) = \mathrm{lh}(b) \wedge \forall i < \mathrm{lh}(a)((a)_i = (b)_i).$$

The *concatenation* $*$ is defined by

$$a * b := \min_{c < \sigma(\max\{M(a), M(b)\}, \mathrm{lh}(a)+\mathrm{lh}(b))} [\mathrm{lh}(c) = \mathrm{lh}(a) + \mathrm{lh}(b) \\ \wedge \forall i < \mathrm{lh}(a)((c)_i = (a)_i) \wedge \forall i < \mathrm{lh}(b)((c)_{\mathrm{lh}(a)+i} = (b)_i)],$$

where $M(a) := (a)_{\min_{i<\mathrm{lh}(a)}[\forall j < \mathrm{lh}(a)((a)_j \le (a)_i)]}$, and the *initial segment* function by

$$\overline{a}(k) := \min_{c < \sigma(M(a), \min\{\mathrm{lh}(a), k\})} [\mathrm{lh}(c) = \min\{\mathrm{lh}(a), k\} \\ \wedge \forall i < \min\{\mathrm{lh}(a), k\}((c)_i = (a)_i)].$$

We can also define the initial segment of a function $\alpha \in \mathbb{N} \to \mathbb{N}$ by

$$\overline{\alpha}(k) := \min_{c < \sigma(M(\alpha, k), k)} [\mathrm{lh}(c) = k \wedge \forall i < k((c)_i = \alpha i)],$$

where $M(\alpha, k) := \alpha(\min_{i<k}[\forall j < k(\alpha j \le \alpha i)])$.

We shall adopt a *virtual* theory of sets [37], and for a class $\mathcal{C}$ of formulas, a $\mathcal{C}$ *set* S is a formula A in $\mathcal{C}$ such that $\forall x[x \in S \leftrightarrow A(x)]$.

16.3 CSM-Spaces and CTB-Spaces

Definition 16.6 *A complete separable metric space, CSM-space for short, is a pseudometric $d \in \mathbb{N}^2 \to \mathbb{R}$, that is for all i, j, k*

1. $d(i, i) = 0$,
2. $d(i, j) = d(j, i)$,
3. $d(i, j) \leq d(i, k) + d(k, j)$.

A point of the CSM-space is a function $\alpha \in \mathbb{N} \to \mathbb{N}$ such that

$$\alpha \in X_d := \forall mn(d(\alpha m, \alpha n) < 2^{-m} + 2^{-n}).$$

The pseudometric on the set X_d of points is defined by

$$d(\alpha, \beta) := \lim_{k \to \infty} d(\alpha k, \beta k).$$

We put $\alpha = \beta := d(\alpha, \beta) = 0$ and $\alpha \# \beta := d(\alpha, \beta) > 0$, and regard a natural number i as a point $\lambda x.i$. Note that $d(\alpha, \alpha k) := \lim_{m \to \infty} d(\alpha m, \alpha k) \leq 2^{-k}$ for all k. A complete totally bounded metric space, CTB-space for short, is a CSM-space d with a function $\nu \in \mathbb{N} \to \mathbb{N}$ such that

$$\forall k i \exists j \leq \nu k(d(i, j) < 2^{-k}).$$

We can show completeness of CSM-spaces by a straightforward modification of the proof of [5, 4.4.2].

Definition 16.7 *A sequence of points of a CSM-space d is a function $\alpha \in \mathbb{N}^2 \to \mathbb{N}$ such that $\alpha_n := \lambda x.\alpha(n, x) \in X_d$. A sequence $\langle \alpha_n \rangle_n$ is said to be a Cauchy sequence with modulus $\gamma \in \mathbb{N} \to \mathbb{N}$ if $\forall k \forall mn \geq \gamma k(d(\alpha_m, \alpha_n) < 2^{-k})$;it converges to a limit $\beta \in X_d$ with modulus $\gamma \in \mathbb{N} \to \mathbb{N}$ if $\forall k \forall n \geq \gamma k(d(\beta, \alpha_n) < 2^{-k})$.*

Proposition 16.8 *Every Cauchy sequence of a CSM-space converges to a limit.*

Definition 16.9 *If d is a CSM-space, $\alpha \in X_d$, and $r \in \mathbb{R}$, then the open ball $U(\alpha, r)$ of radius r with centre α is*

$$\beta \in U(\alpha, r) := d(\alpha, \beta) < r.$$

Lemma 16.10 *Let d be a CSM-space. Then*

$$\forall \alpha \in X_d \forall r \in \mathbb{Q} \forall i[\alpha \in U(i, r) \leftrightarrow \exists k((d(\alpha k, i))_k < r - 2^{-k+1})].$$

Proof. Let $\alpha \in U(i, r)$. Then $d(\alpha, i) < r$, and hence there exists k such that $d(\alpha k, i) + 2^{-k+2} < r$. Therefore $(d(\alpha k, i))_k \leq d(\alpha k, i) + 2^{-k} < r - 2^{-k+1}$. Conversely suppose that $(d(\alpha k, i))_k < r - 2^{-k+1}$ for some k. Then

$$d(\alpha, i) \leq d(\alpha, \alpha k) + d(\alpha k, i) \leq (d(\alpha k, i))_k + 2^{-k+1} < r. \qquad \square$$

Definition 16.11 *A detachable subset T of T_ω is a function $\chi_T \in \mathbb{N} \to \mathbb{N}$ such that $\forall ab(\chi_T(a) = 0 \wedge a = b \to \chi_T(b) = 0)$. We write $a \in T$ for $\chi_T(a) = 0$. A tree is a detachable subset T of T_ω such that $\langle\rangle \in T$ and $\forall ak(a \in T \wedge k \leq \mathrm{lh}(a) \to \overline{a}(k) \in T)$; T is infinite if $\forall m \exists a \in T(\mathrm{lh}(a) = m)$; T is finitely branching if there exists a bounding function $\nu \in \mathbb{N} \to \mathbb{N}$ such that*

$$\forall kax(a * \langle x\rangle \in T \wedge \mathrm{lh}(a) \leq k \to x \leq \nu k).$$

(Note that we may assume that ν is non-decreasing.) A sequence α is a branch of the tree T if $\alpha \in T := \forall m(\overline{\alpha}m \in T)$. A spread is a tree T with a function $\Gamma_T \in \mathbb{N}^2 \to \mathbb{N}$ such that

$$\forall x(\Gamma_T(x) \in T) \text{ and } \forall a \in T(\Gamma_T(a) \in a),$$

where $\Gamma_T(x) := \lambda y.\Gamma_T(x, y)$ and $\alpha \in a := (\overline{\alpha}(\mathrm{lh}(a)) = a)$. (In a strong system, this definition of a spread is equivalent to the usual definition of spread, that is $\forall a \in T\exists x(a * \langle x\rangle \in T)$; in fact, $\forall a \in T \exists \alpha \in T(\alpha \in a)$ by primitive recursion, $\forall x \exists \alpha \in T(x \in T \to \alpha \in x)$ as T is detachable, and then apply countable choice.) A fan is a finitely branching spread. The n-ary tree consisting of all finite sequences of $\{0, \ldots, n-1\}$ will be denoted by T_n.

Proposition 16.12 *Let T be a spread (respectively, fan), and let*

$$d_T(a, b) := \inf\left\{2^{-k} \mid \overline{\Gamma_T a}k = \overline{\Gamma_T b}k\right\}.$$

Then d_T is a CSM-space (respectively, CTB-space) such that

1. *$\forall \alpha \in X_{d_T} \forall k(\overline{\overline{\Gamma_T \alpha}k}k = \overline{\lambda x.((\Gamma_T \alpha(x+1))xk} \in T)$,*
2. *$\forall \alpha \in X_{d_T} \forall k(d_T(\alpha, \overline{\Gamma_T \alpha k}k) < 2^{-k+1})$,*
3. *$\forall \alpha \in T(\lambda x.\overline{\alpha}x \in X_{d_T})$,*
4. *$\forall \alpha \in T \forall a \in T(d_T(\lambda x.\overline{\alpha}x, a) < 2^{-\mathrm{lh}(a)+1} \to \alpha \in a)$.*

Proof. It is trivial that d_T is a pseudometric on $\mathbb{N}$. Noting that $\overline{\Gamma_T \alpha k}k = \overline{\Gamma_T \alpha(k+1)}k$ as $d(\alpha k, \alpha(k+1)) < 2^{-k+1}$, it is straightforward by Δ_0-IND to see that $\forall k(\overline{\Gamma_T k}k = \overline{\lambda x.((\Gamma_T(x+1))xk}$.

(1): Since $\Gamma_T \alpha k \in T$, we have $\overline{\Gamma_T \alpha k}k \in T$.

(2): Since $\Gamma_T(\overline{\Gamma_T \alpha k}k) \in \overline{\Gamma_T \alpha k}k$, we have $\overline{\Gamma_T \alpha k}k = \overline{\Gamma_T(\overline{\Gamma_T \alpha k}k)}k$, and hence $d_T(\alpha k, \overline{\Gamma_T \alpha k}k) \leq 2^{-k}$. Therefore, since $\alpha \in X_{d_T}$, we have

$$d_T(\alpha, \overline{\Gamma_T \alpha k}k) \leq d_T(\alpha, \alpha k) + d_T(\alpha k, \overline{\Gamma_T \alpha k}k) \leq 2^{-k} + 2^{-k} = 2^{-k+1}.$$

(3): Since $\alpha \in T$, we have $\Gamma_T \overline{\alpha}m \in \overline{\alpha}m$ and $\Gamma_T \overline{\alpha}n \in \overline{\alpha}n$, and hence $d_T(\overline{\alpha}m, \overline{\alpha}n) \leq 2^{-\min\{m,n\}} < 2^{-m} + 2^{-n}$.

(4): For given $\alpha \in T$ and $a \in T$, if $d_T(\lambda x.\overline{\alpha}x, a) < 2^{-\mathrm{lh}(a)+1}$, then there exists $k \geq \mathrm{lh}(a)$ such that $d_T(\overline{\alpha}k, a) < 2^{-\mathrm{lh}(a)+1}$, and hence $\overline{\alpha}(\mathrm{lh}(a)) = \overline{\Gamma_T \overline{\alpha}k}(\mathrm{lh}(a)) = \overline{\Gamma_T a}(\mathrm{lh}(a)) = a$. □

For CSM-space, and CTB-space d, we define a spread, and fan T_d by specifying its branches:

$$\alpha \in T_d := \forall k((d(\alpha k, \alpha(k+1)))_{k+1} < 2^{-k+1}),$$
$$\alpha \in T_d := \forall k(\alpha k \leq \nu(k+2) \wedge (d(\alpha k, \alpha(k+1)))_{k+1} < 2^{-k+1}),$$

respectively. Γ_{T_d} is defined by $\Gamma_{T_d} x := \lambda y.0$ if $x = \langle\rangle$ or $x \notin T_d$, and $\Gamma_{T_d} x := x * (\lambda z.(x)_{\mathrm{lh}(x) \dot{-} 1})$ otherwise. Then we can show properties of T_d similar to [5, 7.2.4 and 7.4.3].

Proposition 16.13 *Let d be a CSM-space, and let $\hat{\alpha} := \lambda x.\alpha(x+3)$ and $\hat{V}_a := \{\hat{\alpha} \mid \hat{\alpha} = \hat{\beta} \wedge \beta \in T_d \wedge \beta \in a\}$. Then*

1. $\forall \alpha \in T_d(\hat{\alpha} \in X_d)$,
2. $\forall \alpha \in T_d \forall k(\hat{V}_{\overline{\alpha}(k+5)} \subset U(\hat{\alpha}, 2^{-k}))$,
3. $\forall \alpha \in X_d(\alpha \in T_d \wedge \alpha = \hat{\alpha})$,
4. $\forall \alpha \in X_d \exists \beta \in T_d(\alpha = \hat{\beta} \wedge \forall k(U(\alpha, 2^{-k}) \subset \hat{V}_{\overline{\beta}k}))$.

Proof. (1), (2), and (3): Straightforward.

(4): Note that $\alpha \in T_d$ implies $\lambda x.\alpha(x+1) \in T_d$; in fact,

$$\begin{aligned}(d(\alpha(k+1), \alpha(k+2)))_{k+1} &\leq (d(\alpha(k+1), \alpha(k+2)))_{k+2} + 2^{-k-1} + 2^{-k-2} \\ &< 2^{-k+1}.\end{aligned}$$

Let $\alpha \in X_d$, and let $\beta := \lambda x.\alpha(x+1)$. Then $\beta \in T_d$ and $\alpha = \hat{\beta}$. For $\gamma \in U(\alpha, 2^{-k})$, letting $\delta := \overline{\beta}k * (\lambda x.\gamma(x+k))$, since $\gamma \in T_d$ and

$$\begin{aligned}(d(\delta(k-1), \delta k))_k &= (d(\beta(k-1), \gamma k))_k \leq d(\beta(k-1), \gamma k) + 2^{-k} \\ &\leq d(\alpha k, \gamma k) + 2^{-k} \leq d(\alpha, \gamma) + 2^{-k+1} + 2^{-k} < 2^{-k+2},\end{aligned}$$

we have $\delta \in T_d$ and $\gamma = \hat{\delta}$. □

Proposition 16.14 *Let (d, ν) be a CTB-space, and let $\tilde{\alpha}k := \min_{j \leq \nu(k+2)}((d(\alpha(k+2), j))_{k+2} < 2^{-k-1})$. Then*

1. $\forall \alpha \in T_d(\hat{\alpha} \in X_d)$,
2. $\forall \alpha \in T_d \forall k(\hat{V}_{\overline{\alpha}(k+5)} \subset U(\hat{\alpha}, 2^{-k}))$,
3. $\forall \alpha \in X_d(\tilde{\alpha} \in T_d \wedge \alpha = \hat{\tilde{\alpha}})$,
4. $\forall \alpha \in X_d \exists \beta \in T_d(\alpha = \hat{\beta} \wedge \forall k(U(\alpha, 2^{-k}) \subset \hat{V}_{\overline{\beta}k}))$.

Proof. (1), (2) and (4): Similar to the proofs of Proposition 16.13 (1), (2) and (4), respectively.

(3): Let $\alpha \in X_d$. Then for each k there exists $j \leq \nu(k+2)$ such that $d(\alpha(k+2), j) < 2^{-k-2}$, and hence $(d(\alpha(k+2), j))_{k+2} < d(\alpha(k+2), j) + 2^{-k-2} < 2^{-k-1}$. Therefore

$(d(\alpha(k+2), \tilde{\alpha}k))_{k+2} < 2^{-k-1}$, and so

$$\begin{aligned}
&(d(\tilde{\alpha}k, \tilde{\alpha}(k+1)))_{k+1} \le d(\tilde{\alpha}k, \tilde{\alpha}(k+1)) + 2^{-k-1} \\
&\le d(\tilde{\alpha}k, \alpha(k+2)) + d(\alpha(k+2), \alpha(k+3)) + d(\alpha(k+3), \tilde{\alpha}(k+1)) + 2^{-k-1} \\
&< (d(\tilde{\alpha}k, \alpha(k+2)))_{k+2} + (d(\alpha(k+3), \tilde{\alpha}(k+1)))_{k+3} + 2^{-k} + 2^{-k-2} \\
&< 2^{-k+1}.
\end{aligned}$$

Thus $\tilde{\alpha} \in T_d$ and $\alpha = \hat{\tilde{\alpha}}$. □

16.4 The Heine–Borel Theorem

In this section, we deal with the *Heine–Borel theorem* and the *fan theorem* (FAN) [5, 4.7]: for a given fan T,

$$\forall \alpha \in T \exists m A(\overline{\alpha}m) \to \exists m \forall \alpha \in T \exists n \le m A(\overline{\alpha}n).$$

The fan theorem restricting the formula A in FAN to formulas in a class $\mathcal{C}$ is called $\mathcal{C}$-FAN. Here we assume that the A respects the equality on finite sequences.

Proposition 16.15 *Let $\mathcal{C}$ be a class of formulas containing Δ_0 and closed under term substitution. Then the following are equivalent.*

1. $\mathcal{C}$-FAN.
2. $\Sigma^0_1(\mathcal{C})$-FAN.

Proof. (1) $\implies$ (2): Let T be a fan, A a $\Sigma^0_1(\mathcal{C})$ formula, and suppose that $\forall \alpha \in T \exists m A(\overline{\alpha}m)$. Then by Lemma 16.5, there exists a $\mathcal{C}$ formula B such that

$$\forall a (A(a) \leftrightarrow \exists k B(a, k)),$$

and hence $\forall \alpha \in T \exists mk B(\overline{\alpha}m, k)$. Therefore since $m \le j(m, k)$, we have $\forall \alpha \in T \exists mk B\left(\overline{\overline{\alpha} j(m,k)}(m), k\right)$, and defining a $\mathcal{C}$ formula C by

$$C(a) := B(\overline{a}(j_1(\mathrm{lh}(a))), j_2(\mathrm{lh}(a))),$$

we have $C(a) \to \exists n' \le \mathrm{lh}(a) A(\overline{a}(n'))$ and $\forall \alpha \in T \exists m C(\overline{\alpha}m)$. Applying $\mathcal{C}$-FAN, we have $\exists m \forall \alpha \in T \exists n \le m C(\overline{\alpha}n)$ or $\exists m \forall \alpha \in T \exists n \le m \exists n' \le n A(\overline{\alpha}n')$, and hence $\exists m \forall \alpha \in T \exists n \le m A(\overline{\alpha}n)$.

(2) $\implies$ (1): Trivial. □

Definition 16.16 *Let d be a CSM-space and $\mathcal{C}$ be a class of formulas. A $\mathcal{C}$ open cover of X_d is a $\mathcal{C}$ set I such that*

$$\forall \alpha \in X_d \exists m \in I (\alpha \in U(j_1 m, 2^{-j_2 m})).$$

A set J is a finite subcover of I if J is a $\mathcal{C}$ open cover of X_d, and

$$\exists k \forall m (m \in J \to m \le k \wedge m \in I).$$

In classical reverse mathematics, the Heine–Borel theorem for Δ_0 open covers of [0, 1] is equivalent to the weak König lemma [14, IV.1.2], and the weak König lemma is equivalent to its contraposition Δ_0-FAN. The following theorem shows that, in constructive reverse mathematics, we can classify the Heine–Borel theorem into FAN, and especially, the Heine–Borel theorem for Δ_0 open covers of [0, 1] into Δ_0-FAN. Similar results can be found in [38].

Theorem 16.17 *Let $\mathcal{C}$ be a class of formulas containing Δ_0 and closed under conjunction and term substitution. Then the following are equivalent.*

1. $\mathcal{C}$-FAN.
2. *A $\mathcal{C}$ open cover of a CTB-space has a finite subcover.*

Proof. (1) $\Longrightarrow$ (2): Let d be a CTB-space, and I a $\mathcal{C}$ open cover of X_d. Then $\forall \alpha \in X_d \exists m \in I(\alpha \in U(j_1 m, 2^{-j_2 m}))$, and hence $\forall \alpha \in T_d \exists m \in I(\hat{\alpha} \in U(j_1 m, 2^{-j_2 m}))$. Therefore

$$\forall \alpha \in T_d \exists m \in I \exists k((d(\alpha(k+3), j_1 m))_k < 2^{-j_2 m} - 2^{-k+1}).$$

Define a $\mathcal{C}$ formula A by

$$A(a) := j_1(\mathrm{lh}(a)) \in I \wedge \exists k \leq \mathrm{lh}(a)[k+3 < \mathrm{lh}(a) \\ \wedge (d((a)_{k+3}, j_1 j_1(\mathrm{lh}(a))))_k < 2^{-j_2 j_1(\mathrm{lh}(a))} - 2^{-k+1}].$$

Then for each $\alpha \in T_d$ there exists m and k such that

$$m \in I \wedge (d((\overline{\alpha} j(m, k+4))_{k+3}, j_1 m))_k < 2^{-j_2 m} - 2^{-k+1},$$

and hence letting $m' := j(m, k+4)$, we have $A(\overline{\alpha} m')$. Thus $\forall \alpha \in T_d \exists m A(\overline{\alpha} m)$, and hence $\exists m \forall \alpha \in T_d \exists n \leq m A(\overline{\alpha} n)$ by $\mathcal{C}$-FAN. Therefore there exists m such that

$$\forall \alpha \in T_d \exists n \leq m[j_1 n \in I \wedge \exists k((d(\alpha(k+3), j_1 j_1 n))_k < 2^{-j_2 j_1 n} - 2^{-k+1})],$$

and hence, letting $n' := j_1 n$, we have

$$\forall \alpha \in T_d \exists n' \leq m(n' \in I \wedge \hat{\alpha} \in U(j_1 n', 2^{-j_2 n'})).$$

Thus $\forall \alpha \in X_d \exists n' \leq m(n' \in I \wedge \alpha = \hat{\hat{\alpha}} \in U(j_1 n', 2^{-j_2 n'}))$, and so a $\mathcal{C}$ set J defined by $n \in J := n \in I \wedge n \leq m$ is a finite subcover of I.

(2) $\Longrightarrow$ (1): Let T be a fan, A a $\mathcal{C}$ formula, and suppose that $\forall \alpha \in T \exists m A(\overline{\alpha} m)$. Then d_T is a CTB-space and $\forall \alpha \in X_{d_T} \exists m A(\overline{\lambda x.(\Gamma_T \alpha(x+1))} m)$ by Proposition 16.12, and hence, letting

$$m \in I := j_1 m \in T \wedge j_2 m + 1 = \mathrm{lh}(\gamma_T j_1 m) \wedge A(\overline{\gamma_T j_1 m}(j_2 m)),$$

we have

$$\forall \alpha \in X_{d_T} \exists m[j(\overline{\Gamma_T \alpha(m+1)}(m+1), m) \in I \wedge \alpha \in U(\overline{\Gamma_T \alpha(m+1)}(m+1), 2^{-m})].$$

Therefore I is a $\mathcal{C}$ cover of X_{d_T}, and so it has a finite subcover. There exists m such that $\forall \alpha \in T \exists n \leq m[n \in I \wedge \lambda x.\overline{\alpha} x \in U(j_1 n, 2^{-j_2 n})]$, and hence

$$\forall \alpha \in T \exists n \leq m[j_1 n \in T \wedge j_2 n + 1 = \mathrm{lh}(j_1 n) \wedge A(\overline{j_1 n}(j_2 n)) \wedge \alpha \in j_1 n].$$

Thus $\forall \alpha \in T \exists n \leq m A(\overline{\alpha} n)$. □

16.5 The Cantor Intersection Theorem

In this section, we discuss the *Cantor intersection theorem*, the *weak König lemma* (WKL):

every finitely branching infinite tree has an infinite branch,

and a closure condition:

$$\text{(Sep)} \quad \forall\alpha\beta[\forall m\neg(\exists n(\alpha(m,n) \neq 0) \wedge \exists n(\beta(m,n) \neq 0)) \\ \rightarrow \exists\gamma\forall m[(\gamma m = 0 \rightarrow \forall n(\alpha(m,n) = 0)) \wedge (\gamma m \neq 0 \rightarrow \forall n(\beta(m,n) = 0))].$$

We first show that (Sep) is equivalent to a logical axiom, *De Morgan's law*

DML $\quad \neg(A \wedge B) \rightarrow \neg A \vee \neg B$

for Σ^0_1 formulas, and a function existence axiom, the *axiom of countable choice for disjunctions*

$$\mathrm{AC}^{\vee}_0 \quad \forall m(A(m) \vee B(m)) \\ \rightarrow \exists\alpha \in T_2\forall m[(\alpha m = 0 \rightarrow A(m)) \wedge (\alpha m = 1 \rightarrow B(m))]$$

for Π^0_1 formulas.

Proposition 16.18 *The following are equivalent.*

1. Σ^0_1-DML + Π^0_1-$\mathrm{AC}^{\vee}_0$.
2. (Sep).

Proof. Straightforward. □

Proposition 16.19 *The axiom of dependent choice for disjunctions*

$$\mathrm{DC}^{\vee} \quad \forall u \in T_2(A(u) \vee B(u)) \\ \rightarrow \exists\alpha \in T_2\forall m[(\alpha m = 0 \rightarrow A(\overline{\alpha}m)) \wedge (\alpha m = 1 \rightarrow B(\overline{\alpha}m))]$$

for Δ_0 formulas holds in $\mathbf{EL}_{\mathrm{ELEM}}$.

Proof. Let A and B be Δ_0 formulas such that $\forall u \in T_2(A(u) \vee B(u))$. We may assume without loss of generality that $\forall u \in T_2\neg(A(u) \wedge B(u))$. Let $\gamma \in \mathbb{N}^2 \rightarrow \mathbb{N}$ be such that

$$\gamma(u,k) = 0 \leftrightarrow u \in T_2 \wedge \mathrm{lh}(u) = k \\ \wedge\forall i < \mathrm{lh}(u)[((u)_i = 0 \rightarrow A(\overline{u}i)) \wedge ((u)_i = 1 \rightarrow B(\overline{u}i))].$$

Then letting $\alpha k := (\min_{u \leq \sigma(2,k+1)}(\gamma(u,k+1) = 0))_k$, we have $\alpha \in T_2$ and

$$\forall m[(\alpha m = 0 \rightarrow A(\overline{\alpha}m)) \wedge (\alpha m = 1 \rightarrow B(\overline{\alpha}m))]. \qquad \square$$

We define two kinds of closed sets of a CSM-space which are analogous to r.e and co-r.e. closed sets in [30, 5.1] and [31].

Definition 16.20 *An enumerably closed set of a CSM-space d is a function $\varphi \in \mathbb{N} \to \mathbb{N}$, and for $\alpha \in X_d$*

$$\alpha \in_{e.c.} \varphi := \forall m(0 < j_2\varphi m \to \alpha \notin U(j_1\varphi m, 2^{-j_2\varphi m+1}).$$

A co-enumerably closed set of a CSM-space d is a function $\varphi \in \mathbb{N} \to \mathbb{N}$, and for $\alpha \in X_d$

$$\alpha \in_{co\text{-}e.c.} \varphi := \forall k \exists m(k \leq j_2\varphi m \wedge \alpha \in U(j_1\varphi m, 2^{-k})).$$

(Intuitively, a sequence φ, as an enumerably closed set, enumerates open balls whose union is the complement of the closed set; whereas, as a co-enumerably closed set, it enumerates open balls which intersect with the closed set.) A sequence of enumerably closed and co-enumerably closed sets of d is a function $\varphi \in \mathbb{N}^2 \to \mathbb{N}$ such that $\varphi_n := \lambda x.\varphi(n, x) \in \mathbb{N} \to \mathbb{N}$ is enumerably closed and co-enumerably closed, respectively. A sequence φ of enumerably and co-enumerably closed sets has the finite intersection property if

$$\forall m \exists \alpha \in X_d \forall n \leq m(\alpha \in_{e.c.} \varphi_n) \text{ and } \forall m \exists \alpha \in X_d \forall n \leq m(\alpha \in_{co\text{-}e.c.} \varphi_n),$$

respectively.

The following theorem shows that, in constructive reverse mathematics, the Cantor intersection theorem for sequences of enumerably closed subsets, which is classically equivalent to the Heine–Borel theorem for Δ_0 open covers, is classified into WKL.

Theorem 16.21 *The following are equivalent.*

1. (Sep).
2. WKL.
3. *Every sequence of enumerably closed subsets of a CTB-space with the finite intersection property has inhabited intersection.*

Proof. (1) $\Longrightarrow$ (2): It is enough to show WKL for binary trees [14, Lemma IV.1.4]. The proof is similar to the proof of [14, Lemma IV.4.4] using Δ_0-DC$^\vee$.

(2) $\Longrightarrow$ (3): Let d be a CTB-space, and φ a sequence of enumerably closed sets of d with the finite intersection property. Then $\forall m \exists \alpha \in X_d \forall n \leq m(\alpha \in_{\text{e.c.}} \varphi_n)$, and hence

$$\forall m \exists \alpha \in T_d \forall n \leq m \forall m'[0 < j_2\varphi_n m' \to \hat{\alpha} \notin U(j_1\varphi_n m', 2^{-j_2\varphi_n m'+1})].$$

Define the detachable set T by

$$a \in T := a \in T_d \wedge \forall n \leq \text{lh}(a) \forall m' \leq \text{lh}(a)(0 < j_2\varphi_n m' \to \\ \forall k \leq \text{lh}(a) \neg(k + 3 < \text{lh}(a) \wedge (d((a)_{k+3}, j_1\varphi_n m'))_k < 2^{-j_2\varphi_n m'+1} - 2^{-k+1})).$$

Then clearly T is a finitely branching tree, and since for given n' there exists $\alpha \in T_d$ such that

$$\forall n \leq n' \forall m' \leq n'(0 < j_2\varphi_n m' \to \\ \forall k \leq n' \neg(k + 3 < n' \wedge (d((\overline{\alpha}n')_{k+3}, j_1\varphi_n m'))_k < 2^{-j_2\varphi_n m'+1} - 2^{-k+1})),$$

we have $\overline{\alpha}n' \in T$ for all n', and hence T is infinite. Thus there exists $\alpha \in T$ by WKL. Suppose that $0 < j_2\varphi_n m'$ and $(d(\hat{\alpha}k, j_1\varphi_n m'))_k < 2^{-j_2\varphi_n m'+1} - 2^{-k+1}$ for some n, m' and k. Then letting $n' := n + m' + k + 4$, we have

$$\exists n \leq n' \exists m' \leq n' \exists k \leq n'(k+3 < n' \\ \wedge (d((\overline{\alpha}n')_{k+3}, j_1\varphi_n m'))_k < 2^{-j_2\varphi_n m'+1} - 2^{-k+1}),$$

and hence $\overline{\alpha}n' \notin T$, a contradiction. Thus

$$\forall nm'[0 < j_2\varphi_n m' \rightarrow \hat{\alpha} \notin U(j_1\varphi_n m', 2^{-j_2\varphi_n m'+1})].$$

(3) $\implies$ (1): Let $\alpha, \beta \in \mathbb{N}^2 \rightarrow \mathbb{N}$ be such that $\forall m \neg(\exists n(\alpha(m,n) \neq 0) \wedge \exists n(\beta(m,n) \neq 0))$. Define a sequence φ of enumerably closed sets of the CTB-space d_{T_2} by

$$\varphi(n,m) := \begin{cases} j(m * \langle 0 \rangle, \mathrm{lh}(m)+1) & \text{if } m \in T_2 \wedge \alpha(\mathrm{lh}(m), n) \neq 0), \\ j(m * \langle 1 \rangle, \mathrm{lh}(m)+1) & \text{if } m \in T_2 \wedge \beta(\mathrm{lh}(m), n) \neq 0), \\ j(\langle\rangle, 0) & \text{otherwise.} \end{cases}$$

To see that φ has the finite intersection property, for each n let $\tau_n k := 0$ if $\exists n' \leq n(\beta(k,n') \neq 0)$ and $\tau_n k := 1$ otherwise. If $0 < j_2\varphi_{n'} m$ and $\tau_n \in U(j_1\varphi_{n'} m, 2^{-j_2\varphi_{n'} m+1})$ for $n' \leq n$ and m, then either $m \in T_2 \wedge \alpha(\mathrm{lh}(m), n') \neq 0$ or $m \in T_2 \wedge \beta(\mathrm{lh}(m), n') \neq 0$; in the former case, we have $0 = \tau_n(\mathrm{lh}(m)) = 1$, a contradiction; in the latter case, similarly we have a contradiction. Hence $\forall n' \leq n(\tau_n \in_{\text{e.c.}} \varphi_{n'})$. Therefore there exists $\gamma \in T_2$ such that $\forall n(\gamma \in_{\text{e.c.}} \varphi_n)$. If $\gamma m = 0$ and $\alpha(m,n) \neq 0$ for some n, then $\varphi(n, \overline{\gamma}m) = j(\overline{\gamma}(m+1), m+1)$, and hence $\gamma \notin U(\overline{\gamma}(m+1), 2^{-m})$, a contradiction. Similarly, we have a contradiction if $\gamma m = 1$ and $\beta(m,n) \neq 0$ for some n. □

The following theorem which connects WKL and FAN was proved directly in [24], and indirectly in [22, Corollary].

Theorem 16.22 WKL *implies* Δ_0-FAN.

16.6 The Bolzano–Weierstraß Theorem

In this section, we deal with the *Bolzano–Weierstraß theorem*, and a closure condition:

$$(\mathrm{E}_0) \quad \forall\alpha\exists\beta\forall m[(\beta m \neq 0 \rightarrow \exists n(\alpha(m,n) \neq 0) \wedge (\beta m = 0 \rightarrow \forall n(\alpha(m,n) = 0))].$$

Lemma 16.23 *Assume* (E_0). *Then for every arithmetical formula A*

$$\exists\alpha\forall\vec{x}(\alpha\vec{x} \neq 0 \leftrightarrow A(\vec{x})).$$

Proof. By induction on the complexity of A. □

Corollary 16.24 (E_0) *implies* $\mathcal{A}$-PEM *and* $\mathcal{A}$-IND.

Proof. Straightforward by Lemma 16.23. □

We show that (E_0) is equivalent to a logical axiom, the *principle of the excluded middle*

PEM $A \vee \neg A$

for Σ_1^0 formulas, and $\mathrm{AC}_0^\vee$ for arithmetical formulas.

Proposition 16.25 *The following are equivalent.*

1. Σ_1^0-PEM + $\mathcal{A}$-$\mathrm{AC}_0^\vee$.
2. $\mathcal{A}$-PEM + $\mathcal{A}$-$\mathrm{AC}_0^\vee$.
3. (E_0).

Proof. Straightforward. □

Definition 16.26 *Let d be a CSM-space. Then $\alpha \in X_d$ is an accumulation point of a subset S of X_d if $\forall k \exists \beta \in S(\alpha \# \beta \wedge d(\alpha, \beta) < 2^{-k})$. A subset S of X_d is infinite if there exists a strongly one-one function from $\mathbb{N}$ onto S ($f \in \mathbb{N} \to S$ is strongly one-one if $m \neq n \to f(m) \# f(n)$).*

The Bolzano–Weierstraß theorem is equivalent to the arithmetical comprehension axiom in classical reverse mathematic [14, III.2.9], and the following theorem shows that it is equivalent to the closure condition (E_0) in constructive reverse mathematics.

Theorem 16.27 *The following are equivalent.*

1. (E_0).
2. *Every infinite subset of a CTB-space has an accumulation point.*
3. *Every sequence of co-enumerably closed subsets of a CTB-space with the finite intersection property has inhabited intersection.*

Proof. (1) $\Longrightarrow$ (3): Let φ be a sequence of co-enumerably closed subsets of a CTB-space d with the finite intersection property. Then by Lemma 16.23 there exists a function $\beta \in \mathbb{N}^2 \to \mathbb{N}$ such that

$$\beta(n, m) \neq 0 \leftrightarrow \exists k \neg \exists m'(k \leq j_2\varphi_n m' \wedge d(j_1 m, j_1\varphi_n m') < 2^{-j_2 m} + 2^{-k}).$$

Let $\varphi' \in \mathbb{N}^2 \to \mathbb{N}$ be a sequence of enumerable closed subsets of d defined by

$$\varphi'(n, m) := \begin{cases} j(j_1 m, j_2 m + 1) & \text{if } \beta(n, m) \neq 0, \\ j(j_1 m, 0) & \text{otherwise.} \end{cases}$$

We first show that $\forall n \forall \alpha(\alpha \in_{\text{co-e.c.}} \varphi_n \leftrightarrow \alpha \in_{\text{e.c.}} \varphi'_n)$. To this end, suppose that $\alpha \in \varphi_n$ and $0 < j_2\varphi'_n m \wedge \alpha \in U(j_1\varphi'_n m, 2^{-j_2\varphi'_n m})$ for some m. Then $\beta(n, m) \neq 0$, and hence there exists k such that

$$\neg \exists m'(k \leq j_2\varphi_n m' \wedge d(j_1 m, j_1\varphi_n m') < 2^{-j_2 m} + 2^{-k}).$$

Since $\alpha \in_{\text{co-e.c.}} \varphi_n$, there exists m' such that $k \leq j_2\varphi_n m' \wedge \alpha \in U(j_1\varphi_n m', 2^{-k})$, and hence

$$\begin{aligned} d(j_1 m, j_1\varphi_n m') &\leq d(j_1 m, \alpha) + d(\alpha, j_1\varphi_n m') < d(j_1\varphi'_n m, \alpha) + 2^{-k} \\ &< 2^{-j_2\varphi'_n m} + 2^{-k} = 2^{-j_2 m - 1} + 2^{-k} < 2^{-j_2 m} + 2^{-k}, \end{aligned}$$

a contradiction. Therefore $\forall m(0 < j_2\varphi'_n m \to \alpha \notin U(j_1\varphi'_n m, 2^{-j_2\varphi'_n m}))$ or $\alpha \in_{\text{e.c.}} \varphi'_n$. Conversely suppose that $\alpha \in_{\text{e.c.}} \varphi'_n$. For given k, assume that

$$\neg\exists m'(k \leq j_2\varphi_n m' \wedge \alpha \in U(j_1\varphi_n m', 2^{-k})),$$

and let $k' := k+2$ and $m := j(\alpha(k'+1), k')$. If $k' \leq j_2\varphi_n m'$ and $d(j_1 m, j_1\varphi_n m') < 2^{-j_2 m} + 2^{-k'}$ for some m', then $k \leq j_2\varphi_n m'$ and

$$\begin{aligned} d(\alpha, j_1\varphi_n m') &\leq d(\alpha, j_1 m) + d(j_1 m, j_1\varphi_n m') \\ &= d(\alpha, \alpha(k'+1)) + d(j_1 m, j_1\varphi_n m') \\ &< 2^{-k'} + 2^{-j_2 m} + 2^{-k'} < 2^{-k} \end{aligned}$$

or $\alpha \in U(j_1\varphi_n m', 2^{-k})$, a contradiction. Hence

$$\neg\exists m'(k' \leq j_2\varphi_n m' \wedge d(j_1 m, j_1\varphi_n m') < 2^{-j_2 m} + 2^{-k'})$$

or $\beta(n, m) \neq 0$, and therefore $0 < j_2\varphi'_n m$. Thus $\alpha \notin U(j_1\varphi'_n m, 2^{-j_2\varphi'_n m+1})$ or $\neg(d(\alpha, \alpha(k'+1)) < 2^{-k'})$, a contradiction, and so

$$\exists m'(k \leq j_2\varphi_n m' \wedge \alpha \in U(j_1\varphi_n m', 2^{-k})),$$

by Corollary 16.24. Therefore $\alpha \in_{\text{co-e.c.}} \varphi_n$.

Since (E_0) implies (Sep), we have φ', and hence φ, has inhabited intersection by Theorem 16.21.

(3) $\implies$ (2): Let S be an infinite subset of a CTB-space d. Then there exists a strongly one-one function $n \mapsto \alpha_n$ from $\mathbb{N}$ onto S. Define a sequence φ of co-enumerably closed subsets of T by

$$\varphi_n m := j(\alpha_{n+j_1 m}(j_2 m + 1), j_2 m).$$

Then for given n, and for each $n' \leq n$ and k, letting $m := j(n \dot{-} n', k)$, we have $\varphi_{n'}(m) = j(\alpha_n(k+1), k)$, and hence $k \leq j_2\varphi_{n'}m \wedge \alpha_n \in U(j_1\varphi_{n'}m, 2^{-k})$. Therefore φ has the finite intersection property, and so there exists $\beta \in T$ such that $\beta \in \varphi_n$ for all n. For each k, since $\beta \in \varphi_0$, there exists m such that

$$k+1 \leq j_2 m \wedge \beta \in U(\alpha_{j_1 m}(j_2 m + 1), 2^{-k-1}),$$

and since $\beta \in \varphi_{j_1 m+1}$, there exists m' such that

$$k+1 \leq j_2 m' \wedge \beta \in U(\alpha_{j_1 m + j_1 m' + 1}(j_2 m' + 1), 2^{-k-1}).$$

Noting that $\alpha_{j_1 m} \# \alpha_{j_1 m + j_1 m' + 1}$, either $\alpha_{j_1 m} \# \beta$ or $\alpha_{j_1 m + j_1 m' + 1} \# \beta$: in the first case, we have

$$\begin{aligned} d(\alpha_{j_1 m}, \beta) &\leq d(\alpha_{j_1 m}, \alpha_{j_1 m}(j_2 m + 1)) + d(\alpha_{j_1 m}(j_2 m + 1), \beta) \\ &< 2^{-j_2 m} + 2^{-k-1} \leq 2^{-k}; \end{aligned}$$

in the second case, similarly we have $d(\alpha_{j_1 m + j_1 m' + 1}, \beta) < 2^{-k}$. Thus β is an accumulation point of S.

(2) $\Longrightarrow$ (1): Let $\alpha \in \mathbb{N}^2 \to \mathbb{N}$ and let $\tau \in \mathbb{N}^2 \to \mathbb{N}$ be such that

$$\forall mn[(\tau(m,n) = 1 \to \exists k \leq n(\alpha(m,k) \neq 0)) \\ \wedge(\tau(m,n) = 0 \to \forall k \leq n(\alpha(m,k) = 0))].$$

Then letting $\sigma_n := (\overline{\lambda x.\tau(x,n)}(n)) * (\lambda x.2)$, the set $S := \{\lambda x.\overline{\sigma_n}x : n \in \mathbb{N}\}$ is an infinite set of a CTB-space d_{T_3}. Hence S has an accumulation point β. If $(\beta(m+1))_m \neq 0$, then there exists $n > m$ such that $d_{T_3}(\lambda x.\overline{\sigma_n}x, \beta) < 2^{-m}$ and hence $d_{T_3}(\lambda x.\overline{\sigma_n}x, \overline{\beta(m+1)}(m+1)) < 2^{-m+1}$; therefore $\lambda x.\overline{\sigma_n}x \in \overline{\beta(m+1)}(m+1)$, and so $\tau(m,n) = \sigma_n m = (\overline{\sigma_n}(m+1))_m = (\overline{\lambda x.\overline{\sigma_n}x}(m+1))_m = (\overline{\beta(m+1)}(m+1))_m = (\beta(m+1))_m \neq 0$. If $(\beta(m+1))_m = 0$ and $\exists n(\alpha(m,n) \neq 0)$, then there exists $n > m$ such that $\exists k \leq n(\alpha(m,k) \neq 0)$ and $d_{T_3}(\lambda x.\overline{\sigma_n}x, \beta) < 2^{-m}$, and hence $0 = (\beta(m+1))_m = \sigma_n m = \tau(m,n) \neq 0$, a contradiction; thus $\forall n(\alpha(m,n) = 0)$. Therefore ($E_0$) holds. □

16.7 Sequential Compactness

The present section is devoted to *sequential compactness* and a closure condition:

(μ) $\forall\alpha\exists\beta\forall m[\exists n(\alpha(m,n) \neq 0) \to \alpha(m, \beta m) \neq 0]$.

We first show that (μ) is equivalent to a logical principle, PEM for Σ_1^0 formulas, and a function existence axiom, the *axiom of countable choice*

AC_{00} $\forall m\exists n A(m,n) \to \exists\alpha\forall m A(m, \alpha m)$

for Π_1^0 formulas.

Lemma 16.28 *The following are equivalent.*

1. Σ_1^0-PEM.
2. $\forall\alpha\forall m\exists n[\exists n(\alpha(m,n) \neq 0) \to \exists k \leq n(\alpha(m,k) \neq 0)]$.

Proof. (1) $\Longrightarrow$ (2): Let $\alpha \in \mathbb{N}^2 \to \mathbb{N}$. Then by Σ_1^0-PEM, we have either $\exists n(\alpha(m,n) \neq 0)$ or $\neg\exists n(\alpha(m,n) \neq 0)$. In the former case, suppose that $\alpha(m,n) \neq 0$. Then $n \leq n \wedge \alpha(m,n) \neq 0$, and hence $\exists k \leq n(\alpha(m,k) \neq 0)$. Therefore $\exists n(\alpha(m,n) \neq 0) \to \exists k \leq n(\alpha(m,k) \neq 0)$, and so $\exists n[\exists n(\alpha(m,n) \neq 0) \to \exists k \leq n(\alpha(m,k) \neq 0)]$. In the latter case, assuming $\exists n(\alpha(m,n) \neq 0)$, we have a contradiction, and hence $\exists k \leq n(\alpha(m,k) \neq 0)$. Therefore $\exists n(\alpha(m,n) \neq 0) \to \exists k \leq n(\alpha(m,k) \neq 0)$, and so

$$\exists n[\exists n(\alpha(m,n) \neq 0) \to \exists k \leq n(\alpha(m,k) \neq 0)].$$

Thus $\forall m\exists n[\exists n(\alpha(m,n) \neq 0) \to \exists k \leq n(\alpha(m,k) \neq 0)]$.

(2) $\Longrightarrow$ (1): Let $A(m)$ be a Σ_1^0 formula. Then there exists a functor φ such that $A(m) \leftrightarrow \exists n(\varphi(m,n) \neq 0)$, and hence

$$\forall m\exists n[\exists n(\varphi(m,n) \neq 0) \to \exists k \leq n(\varphi(m,k) \neq 0)].$$

Either $\exists k \leq n(\varphi(m,k) \neq 0)$ or $\forall k \leq n(\varphi(m,k) = 0)$: in the former case, we have $A(m)$; in the latter case, $\neg A(m)$. Thus Σ_1^0-PEM holds. □

Proposition 16.29 *The following are equivalent.*

1. Σ^0_1-PEM + Π^0_1-AC$_{00}$.
2. $\mathcal{A}$-PEM + $\mathcal{A}$-AC$_{00}$.
3. (E$_0$) + Δ_0-AC$_{00}$.
4. (μ).

Proof. (1) $\Longrightarrow$ (4): Let $\alpha \in \mathbb{N}^2 \to \mathbb{N}$. Then by Lemma 16.28, we have $\forall m \exists n[\exists n (\alpha(m,n) \neq 0) \to \exists k \leq n(\alpha(m,k) \neq 0)]$ or

$$\forall m \exists n \forall k[\alpha(m,k) \neq 0 \to \exists k \leq n(\alpha(m,k) \neq 0)].$$

Hence by Π^0_1-AC$_{00}$, there exists $\beta' \in \mathbb{N} \to \mathbb{N}$ such that $\forall m \forall k[\alpha(m,k) \neq 0 \to \exists k \leq \beta' m(\alpha(m,k) \neq 0)]$ therefore, letting $\beta m := \min_{k \leq \beta' m}(\alpha(m,k) \neq 0)$, we have $\forall m[\exists n (\alpha(m,n) \neq 0) \to \alpha(m, \beta m) \neq 0]$.

(4) $\Longrightarrow$ (3): Let $\alpha \in \mathbb{N}^2 \to \mathbb{N}$. Then there exists $\beta \in \mathbb{N} \to \mathbb{N}$ such that

$$\forall m[\exists n(\alpha(m,n) \neq 0) \to \alpha(m, \beta m) \neq 0],$$

and hence letting $\gamma m := \alpha(m, \beta m)$, we have

$$\forall m[(\gamma m \neq 0 \to \exists n(\alpha(m,n) \neq 0)) \wedge (\gamma m = 0 \to \forall n(\alpha(m,n) = 0))].$$

Suppose that $\forall m \exists n A(m,n)$ for a Δ_0 formula $A(m,n)$. Then there exists a functor φ such that $\forall mn(\varphi(m,n) \neq 0 \leftrightarrow A(m,n))$, and hence there exists β such that $\forall m[\exists n (\varphi(m,n) \neq 0) \to \varphi(m, \beta m) \neq 0]$. Therefore $\forall m A(m, \beta m)$.

(3) $\Longrightarrow$ (2): Straightforward by Corollary 16.24.

(2) $\Longrightarrow$ (1): Trivial. □

The following proposition shows that, assuming (μ), the closure condition, *primitive recursion*:

(Prim) $\forall \alpha \beta \exists \gamma \forall m[\gamma(0,m) = \alpha m \wedge \forall k(\gamma(Sk,m) = \beta(k,m,\gamma(k,m)))]$

and the *axiom of dependent choice*

DC$_0$ $\forall m \exists n A(m,n) \to \forall n_0 \exists \alpha[\alpha 0 = n_0 \wedge \forall m A(\alpha m, \alpha(m+1))]$

for arithmetical formulas hold.

Proposition 16.30 *The following are equivalent.*

1. Σ^0_1-IND + Δ_0-AC$_{00}$.
2. (Prim) + Δ_0-AC$_{00}$.
3. Δ_0-DC$_0$.

Proof. (1) $\Longrightarrow$ (2): For given α and β, define a Δ_0 formula A by

$$A(k,m,a) := \mathrm{lh}(a) = k+1 \wedge (a)_0 = \alpha m \\ \wedge \forall i < \mathrm{lh}(a) \dot{-} 1((a)_{i+1} = \beta(i,m,(a)_i)).$$

Then by Σ^0_1-IND, we can show that $\forall km\exists a A(k, m, a)$, and hence by Δ_0-AC$_{00}$, there exists γ such that $\forall km A(k, m, \gamma(k, m))$. Letting $\gamma'(k, m) := (\gamma(k, m))_k$, we have $\forall k \forall i \leq k(\gamma'(i, m) = (\gamma(k, m))_i)$ by Δ_0-IND, and therefore

$$\forall m[\gamma'(0, m) = \alpha m \wedge \forall k(\gamma'(Sk, m) = \beta(k, m, \gamma'(k, m)))].$$

(2) $\Longrightarrow$ (3): Straightforward.

(3) $\Longrightarrow$ (1): Let A be a Δ_0 formula, and suppose that

$$\exists y A(0, y) \wedge \forall x(\exists y A(x, y) \rightarrow \exists y A(Sx, y)).$$

Then $A(0, y_0)$ for some y_0 and $\forall x \forall y \exists z(A(x, y) \rightarrow A(Sx, z))$, and hence

$$\forall m \exists n[Sj_1 m = j_1 n \wedge (A(j_1 m, j_2 m) \rightarrow A(j_1 n, j_2 n))].$$

Applying Δ_0-DC$_0$, there exists α such that

$$\alpha 0 = j(0, y_0) \wedge \forall k[Sj_1\alpha k = j_1\alpha Sk \wedge (A(j_1\alpha k, j_2\alpha k) \rightarrow A(j_1\alpha Sk, j_2\alpha Sk))].$$

By Δ_0-IND, we have $\forall k(j_1\alpha k = k \wedge A(j_1\alpha k, j_2\alpha k))$, and hence $\forall x \exists y A(x, y)$. Therefore Σ^0_1-IND holds. The proof of Δ_0-AC$_{00}$ is similar to the proof of [33, 2.7.2]. □

In classical reverse mathematics, the sequential compactness and the monotone convergence theorem are equivalent to the arithmetical comprehension axiom [14, III.2.2]. They are classified into the closure condition (μ) which is stronger than (E$_0$) in our base system without a weak axiom of countable choice.

Theorem 16.31 *The following are equivalent.*

1. (μ).
2. *Every sequence of a CTB-space has a convergent subsequence.*
3. *Every monotone sequence of* [0, 1] *has a limit in* [0, 1].

Proof. (1) $\Longrightarrow$ (2): Let $\langle \alpha_n \rangle$ be a sequence of a CTB-space d with $\nu \in \mathbb{N} \rightarrow \mathbb{N}$, and let $\theta \in \mathbb{N}^3 \rightarrow \mathbb{N}$ be such that

$$\forall ikn[\theta(i, k, n) \neq 0 \leftrightarrow (d(\alpha_n(k+2), i))_{k+2} < 2^{-k}].$$

Then by (μ), there exists a function $\beta \in \mathbb{N}^3 \rightarrow \mathbb{N}$ such that

$$\forall ikm[\exists n > m(\theta(i, k, n) \neq 0) \rightarrow \beta(i, k, m) > m \wedge \theta(i, k, \beta(i, k, m)) \neq 0],$$

and hence by (μ) again, there exists a function $\gamma \in \mathbb{N}^2 \rightarrow \mathbb{N}$ such that

$$\begin{aligned}\forall ik[&\exists m \neg(\beta(i, k, m) > m \wedge \theta(i, k, \beta(i, k, m)) \neq 0) \\ &\rightarrow \neg(\beta(i, k, \gamma(i, k)) > \gamma(i, k) \wedge \theta(i, k, \beta(i, k, \gamma(i, k))) \neq 0)]\end{aligned}$$

or $\forall ik[\exists m \forall n > m(\theta(i, k, n) = 0) \rightarrow \forall n > \gamma(i, k)(\theta(i, k, n) = 0)]$. Note that

$$\begin{aligned}\forall ik[&\forall m \exists n > m(\theta(i, k, n) \neq 0) \\ &\leftrightarrow \beta(i, k, \gamma(i, k)) > \gamma(i, k) \wedge \theta(i, k, \beta(i, k, \gamma(i, k))) \neq 0],\end{aligned}$$

and

$$\begin{aligned}&\forall ik[\exists m\forall n > m(\theta(i,k,n) = 0)\\&\qquad\leftrightarrow \neg(\beta(i,k,\gamma(i,k)) > \gamma(i,k) \wedge \theta(i,k,\beta(i,k,\gamma(i,k))) \neq 0)].\end{aligned}$$

Let $m_k := \max_{j \leq \nu(k+2)} \gamma(j, k+1)$, and

$$\tau(i,k) := \min_{j\leq\nu(k+2)}((d(\alpha_{\beta(i,k,m_k)}(k+3), j))_{k+3} < 2^{-k-1}).$$

Then since there exists $j \leq \nu(k+2)$ such that $d(\alpha_{\beta(i,k,m_k)}(k+3), j) < 2^{-k-2}$ and hence $(d(\alpha_{\beta(i,k,m_k)}(k+3), j))_{k+3} < 2^{-k-1}$, we have $\tau(i,k) \leq \nu(k+2)$ and $(d(\alpha_{\beta(i,k,m_k)}(k+3), \tau(i,k)))_{k+3} < 2^{-k-1}$, i.e. $\theta(\tau(i,k), k+1, \beta(i,k,m_k)) \neq 0$. We will show that

$$\begin{aligned}&\forall ik[\forall m\exists n > m(\theta(i,k,n) \neq 0)\\&\qquad\rightarrow d(i,\tau(i,k)) < 2^{-k+2} \wedge \forall m\exists n > m(\theta(\tau(i,k), k+1, n) \neq 0)].\end{aligned}$$

To this end, suppose that $\forall m\exists n > m(\theta(i,k,n) \neq 0)$ and $\forall n > \gamma(\tau(i,k), k+1)(\theta(\tau(i,k), k+1, n) = 0)$. Then $\exists n > m_k(\theta(i,k,n) \neq 0)$ or $\beta(i,k,m_k) > m_k \wedge \theta(i,k,\beta(i,k,m_k)) \neq 0$. Therefore since $m_k \geq \gamma(\tau(i,k), k+1)$, we have $\theta(\tau(i,k), k+1, \beta(i,k,m_k)) = 0$, a contradiction, and so $\forall m\exists n > m(\theta(\tau(i,k), k+1, n) \neq 0)$. Moreover we have

$$\begin{aligned}d(i,\tau(i,k)) &\leq d(i,\alpha_{\beta(i,k,m_k)}) + d(\alpha_{\beta(i,k,m_k)}, \tau(i,k))\\&\leq (d(\alpha_{\beta(i,k,m_k)}(k+2), i))_{k+2} + (d(\alpha_{\beta(i,k,m_k)}(k+3), \tau(i,k)))_{k+3}\\&\quad + 3\cdot 2^{-k-2}\\&< 2^{-k+2}.\end{aligned}$$

Let $i_0 := \min_{i\leq\nu 2}((d(\alpha_{m_0+1}2, i))_2 < 2^{-1})$. Then since there exists $i \leq \nu 2$ such that $d(\alpha_{m_0+1}2, i) < 2^{-2}$ and hence $(d(\alpha_{m_0+1}2, i))_2 < 2^{-1} < 1$, we have $i_0 \leq \nu 2$ and $\theta(i_0, 0, m_0+1) \neq 0$. If $\exists m\forall n > m(\theta(i_0, 0, n) = 0)$, then $\forall n > \gamma(i_0, 0)(\theta(i_0, 0, n) = 0)$, and therefore since $m_0 + 1 > m_0 \geq \gamma(i_0, 0)$, we have $\theta(i_0, 0, m_0+1) = 0$, a contradiction. Hence $\forall m\exists n > m(\theta(i_0, 0, n) \neq 0)$.

Define a function $\psi \in \mathbb{N} \to \mathbb{N}$ by primitive recursion (Prim):

$$\psi(0) = i_0, \quad \psi(k+1) = \tau(k, \psi k).$$

Then $\forall k[d(\psi k, \psi(k+1)) < 2^{-k+2} \wedge \forall m\exists n > m(\theta(\psi k, k, n) \neq 0)]$, and hence defining a function $\varphi \in \mathbb{N} \to \mathbb{N}$ by (Prim):

$$\varphi(0) = 0, \quad \varphi(k+1) = \beta(\psi k, k, \varphi k),$$

we have $\forall k[\varphi(k+1) > \varphi k \wedge \theta(\phi k, k, \varphi(k+1)) \neq 0]$. Thus, noting that $\theta(i,k,n)$ implies $d(\alpha_n, i) < 2^{-k+1}$, we have for all m and n

$$\begin{aligned} d(\alpha_{\varphi(k+m+1)}, \alpha_{\varphi(k+n+1)}) &\leq d(\alpha_{\varphi(k+m+1)}, \psi(k+m)) + d(\psi(k+m), \psi(k+n)) \\ &\quad + d(\psi(k+n), \alpha_{\varphi(k+n+1)}) \\ &< 2^{-k-m+1} + \sum_{i=0}^{\infty} d(\psi(k+i), \psi(k+i+1)) + 2^{-k-n+1} \\ &< 2^{-k+1} + 2^{-k+3} + 2^{-k+1} < 2^{-k+4}. \end{aligned}$$

Therefore $\langle \alpha_{\varphi k} \rangle_k$ is a Cauchy sequence with modulus $\lambda x.(x+5)$, and is a convergent subsequence of $\langle \alpha_n \rangle_n$.

(2) $\implies$ (3): Straightforward.

(3) $\implies$ (1): Let $\alpha \in \mathbb{N}^2 \to \mathbb{N}$, and let $\tau \in \mathbb{N}^2 \to \mathbb{N}$ be such that

$$\begin{aligned} \forall mn[(\tau(m,n) = 1 \rightarrow \exists k \leq n(\alpha(m,k) \neq 0)) \\ \wedge(\tau(m,n) = 0 \rightarrow \forall k \leq n(\alpha(m,k) = 0))]. \end{aligned}$$

Define a monotone, non-decreasing sequence of [0, 1] by

$$x_n := \lambda z. \left(\sum_{i=0}^{n} \tau(i,n) \cdot 2^{-i} \right).$$

Then the sequence $\langle x_n \rangle$ converges to a limit x with a modulus γ. We may assume that $\forall m(m \leq \gamma m)$. Hence

$$\begin{aligned} &(\tau(m, \gamma(m+1) + m') - \tau(m, \gamma(m+1))) \cdot 2^{-m} \\ &\leq \sum_{i=0}^{\gamma(m+1)} (\tau(i, \gamma(m+1) + m') - \tau(i, \gamma(m+1))) \cdot 2^{-i} \\ &\leq \sum_{i=0}^{\gamma(m+1)+m'} \tau(i, \gamma(m+1) + m') \cdot 2^{-i} - \sum_{i=0}^{\gamma(m+1)} \tau(i, \gamma(m+1)) \cdot 2^{-i} \\ &\leq x_{\gamma(m+1)+m'} - x_{\gamma(m+1)} \\ &\leq |x_{\gamma(m+1)+m'} - x| + |x - x_{\gamma(m+1)}| \\ &< 2^{-m-1} + 2^{-m-1} = 2^{-m}, \end{aligned}$$

and so $\forall m'(\tau(m, \gamma(m+1)+m') = \tau(m, \gamma(m+1))$. Therefore, with $\beta m := \min_{k \leq \gamma(m+1)} (\alpha(m,k) \neq 0)$, we have $\forall m[\exists n(\alpha(m,n) \neq 0) \rightarrow \alpha(m, \beta m) \neq 0]$. □

Acknowledgments.

The author wishes to express his hearty thanks to the anonymous referee, Jeremy Clark, and Wim Veldman for helpful comments and suggestions on an earlier version of the chapter. He wishes to thank Josef Berger, Douglas Bridges, Cristian Calude, Laura

Crosilla, Dirk van Dalen, Susumu Hayashi, Bakhadyr Khoussainov, Ulrich Kohlenbach, Peter Schuster, Helmut Schwichtenberg, Klaus Thiel, Wim Veldman, and Luminiţa Vîţă for being interested in his work and giving him many positive suggestions. He thanks the Japan Society for the Promotion of Science (Grant-in-Aid for Scientific Research (C) No.15500005) for partly supporting the research. The final version of the chapter was prepared at the Mathematisches Institut of the Universität München, and at the Computer Science Department of the University of Auckland. He also wishes to thank the DFG-hosted Graduiertenkolleg *Logik in der Informatik* for supporting his stay in München, and the Overseas Advance Educational Research Practice Support Program of the Japanese Ministy of Education, Culture, Sports, Science and Technology for supporting his stay in Auckland.

References

1. Bishop, E. (1967). *Foundations of Constructive Mathematics*. McGraw-Hill, New York.
2. Bishop, E. and Bridges, D. (1985). *Constructive Analysis*. Springer, Berlin.
3. Bridges, D. and Richman, F. (1987). *Varieties of Constructive Mathematics*. Cambridge University Press, Cambridge.
4. Mines, R., Richman, F., and Ruitenburg, W. (1988). *A Course in Constructive Algebra*. Springer-Verlag, New York.
5. Troelstra, A. S. and van Dalen, D. (1988). *Constructivism in Mathematics*. two volumes, North-Holland, Amsterdam.
6. Brouwer, L. E. J. (1981). *Brouwer's Cambridge Lectures on Intuitionism*. ed. D. van Dalen, Cambridge University Press, Cambridge.
7. Dummett, M. (2000). *Elements of Intuitionism*. 2nd ed., Clarendon Press, Oxford.
8. Feferman, S. (1977). Theories of finite type related to mathematical practice. In: *Handbook of Mathematical Logic* (eds J. Barwise), pp. 913–971. North-Holland, Amsterdam.
9. Heyting, A. (1971). *Intuitionism, An Introduction*. 3rd rev. ed., North-Holland, Amsterdam.
10. Kushner, B. (1984). *Lectures on Constructive Mathematical Analysis*. Amer. Math. Soc., Providence.
11. Bishop, E. (1970). Mathematics as a numerical language, In: *Intuitionism and Proof Theory* (eds A. Kino, J. Myhill, and R. E. Vesley), pp. 53–71. North-Holland, Amsterdam.
12. Richman, F. (1990). Intuitionism as generalization. *Philos. Math*, **5**, 124–128.
13. Beeson, M. J. (1985). *Foundations of Constructive Mathematics*. Springer-Verlag, Berlin.
14. Simpson, S. G. (1999). *Subsystems of Second Order Arithmetic*. Springer, Berlin.
15. Kazuyuki, T. (1992). Reverse mathematics and subsystems of second-order arithmetic. *Sugaku Expositions*, **5**, 213–234.
16. Ishihara, H. (1994). A constructive version of Banach's inverse mapping theorem, *New Zealand J. Math.*, **23**, 71–75.
17. Bridges, D., Van Dalen D., and Ishihara, H. (2003). Ishihara's proof technique in constructive analysis. *Indag. Math (N.S.)*, **14**, 163–168.

18. Bridges, D., Ishihara, H., Schuster, P. and Vîţă, L. Apartness continuity implies uniformly sequential continuity, *Arch. Math. Logic*.
19. Spitters, B. (2002). Constructive and intuitionistic integration theory and functional analysis. PhD thesis, University of Nijmegen, Nijmegen.
20. Ishihara, H. (2004). *Informal constructive reverse mathematics*, Feasibility of theoretical arguments of mathematical analysis on computer (Kyoto 2003), Sūrikaisekikenkyūsho Kōkyūroku **1381**, 108–117.
21. Mandelkern, M. (1988). Limited omniscience and the Bolzano–Weierstrass principle. *Bull. London Math. Soc.*, **20**, 319-320.
22. Ishihara, H. (1990). An omniscience principle, the König lemma and the Hahn-Banach theorem. *Z. Math. Logik Grundlagen Math.*, **36**, 237–240.
23. Bridges, D., Ishihara, H., and Schuster, P. (2002). Compactness and continuity, constructively revisited. In: *Computer Science Logic*, (eds J. Bradfield) Springer, Berlin, Lecture Notes in Comput. Sci. **2471**, 89–102.
24. Ishihara, H. (2002). Weak König's lemma implies Brouwer's fan theorem: a direct proof. *Notre Dame J. Formal Logic*, to appear
25. Ishihara, H. (1991). Continuity and nondiscontinuity in constructive mathematics. *J. Symbolic Logic*, **56**, 1349–1354.
26. Ishihara, H. (1992). Continuity properties in constructive mathematics. *J. Symbolic Logic*, **57**, 557–565.
27. Ishihara, H. and Schuster, P. (2004). Compactness under constructive scrutiny. *MLQ Math. Log. Q.*, **50**, 540–550.
28. Kohlenbach, U. (2004). *Proof interpretations and the computational content of proofs*. preprint, April.
29. Pour-El, M. B. and Richards, J. I. (1989). *Computability in Analysis and Physics*. Springer, Berlin.
30. Weihranch, K. (2000). *Computable Analysis*. Springer, Berlin.
31. Brattka, V. and Presser, G. (2003). Computability on subsets of metric spaces, *Theoret. Comput. Sci.*, **305**, 43–76.
32. Rose, H. E. (1984). *Subrecursion: Functions and Hierarchies*. Oxford University Press, Oxford.
33. Kreisel, G. and Troelstra, A. S. (1970). Formal systems for some branches of intuitionistic analysis. *Ann. Math. Logic*, **1**, 229–387.
34. Troelstra, A. S. (1973). *Mathematical Investigation of Intuitionistic Arithmetic and Analysis*. Lecture Notes in Math. **344**. Springer, Berlin.
35. Kohlenbach, U. (1996). Mathematically strong subsystems of analysis with low rate of growth of provably recursive functionals. *Arch. Math. Logic*, **36**, 31–71.
36. Kohlenbach, U. (1998). Relative constructivity. *J. Symbolic Logic*, **63**, 1218–1238.
37. Goodman, N. D. and Myhill, J. (1972). The foundation of Bishop's constructive mathematics, In: *Toposes, Algebraic Geometry and Logic*, (eds Lawvere, F. W.), Springer, Berlin, Lecture Notes in Math. **274**. 83–96.
38. Loeb, I. (2005). Equivalents of the (weak) fan theorem. *Ann. Pure Appl. Logic.*, **132**, 51–66.

17

APPROXIMATING INTEGRABLE SETS BY COMPACTS CONSTRUCTIVELY

BAS SPITTERS

Abstract

In locally compact spaces, (Borel-)measurable sets can be approximated by compact sets. Ulam extended this result to complete separable metric spaces. We give a constructive proof of Ulam's theorem. It is first proved intuitionistically and then, using a logical 'trick' due to Ishihara, a proof acceptable in Bishop-style mathematics is obtained. We feel this proof provides some insight into Ishihara's trick. Finally, we show how several intuitionistic measure theoretic theorems can be extended to regular integration spaces, that is integration spaces where integrable sets can be approximated by compacts. These results may help to understand Bishop's choice of definitions.

17.1 Introduction

When developing measure theory constructively, one often wants to know whether measurable sets can be approximated by compact sets. In fact, for Brouwer [1,2] this is part of the definition of a (Lebesgue-)measurable set. In Bishop's more abstract approach to integration theory one can prove that an integrable set, with respect to a positive measure on a locally compact space, can be approximated by a compact set, see Theorem 6.6.7 in [3].

In this article we will extend this result to integrals defined on complete separable metric spaces. We will work constructively, in the sense of Bishop, that is using intuitionistic logic. However, sometimes one of Brouwer's principles is used as an extra axiom. We will always make clear when we use such axioms. This treatment is similar to that of Bridges and Richman [4], who consider Brouwer's intuitionistic mathematics as an extension of Bishop's constructive mathematics. It should be noted that we do not work within a fixed formal system, but there are various proposals for formalizing Bishop-style mathematics. See for instance [5,6].

This chapter is organized as follows. In Section 17.2 we recall some standard results from the constructive literature. Section 17.3 gives an overview of the Bishop/Cheng integration theory. In Section 17.4 we give a constructive proof of Ulam's theorem. Finally, in Section 17.5, we show how several intuitionistic measure theoretic theorems can be extended to regular integration spaces.

17.2 Preliminaries

We will now recall some results and fix some notation; most of them are standard in the literature on constructivism, see for instance [5, 6].

Definition 17.1 *Let* Seq(**N**) *be the set of finite sequences of natural numbers. The concatenation of two finite sequences is denoted by* $*$ *and we write* $a * n$ *for* $a * \{n\}$, *when* $a \in \mathrm{Seq}(\mathbf{N})$ *and* $n \in \mathbf{N}$. *Define the map* $\bar{\cdot} : \mathbf{N}^{\mathbf{N}} \times \mathbf{N} \to \mathrm{Seq}(\mathbf{N})$ *such that for all* $\alpha \in \mathbf{N}^{\mathbf{N}}$ *and* $n \in \mathbf{N}$, $\overline{\alpha}n$ *is the initial segment of* α *which has length* n. *When* a *is an initial segment of a finite sequence* b *we write* $a \subset b$. *When* a *is an initial segment of an infinite sequence* α, *we write* $\alpha \in a$.

Let d be the metric on $\mathbf{N}^{\mathbf{N}}$ defined by

$$d(\alpha, \beta) := \inf\{2^{-n} : \overline{\alpha}n = \overline{\beta}n\}.$$

We will sometimes implicitly consider $\mathbf{N}^{\mathbf{N}}$ as the metric space $(\mathbf{N}^{\mathbf{N}}, d)$ without mentioning it.

Definition 17.2 *A spread* S *is a decidable subtree of* Seq(**N**) *such that*

1. *when there exists* $n \in \mathbf{N}$ *such that* $a * n \in S$, *then* $a \in S$ *and*
2. *if* $a \in S$, *then* $a * n \in S$, *for some* n.

A spread S *can also be seen as a closed subset of* $\mathbf{N}^{\mathbf{N}}$ *by considering the set* $\{\alpha \in \mathbf{N}^{\mathbf{N}} : \forall n[\overline{\alpha}n \in S]\}$. *In this case we also write* $\alpha \in S$, *for* $\alpha \in \mathbf{N}^{\mathbf{N}}$. *A fan is a spread* F *which is finitely branching. Let* S *be a spread. A bar* B *for* S *is a set of finite sequences of natural numbers such that for all infinite sequences* $\sigma \in S$ *there is* $b \in B$ *such that* $\sigma \in b$. *A subbar of a bar* B *is a subset of* B *that is a bar.*

The fan theorem and the continuity principle are two of the most characteristic principles of intuitionistic mathematics. Brouwer considered them to be theorems. Bishop and Bridges [3, p. 76] seem to believe that these principles hold, but they do not consider them to be mathematical theorems. They avoid using these principles by adapting their definitions. The most well-known example of this is their definition of continuity. They define a function on a locally compact space to be continuous precisely when it is uniformly continuous on compacts. For Bishop a subset of a metric space is *compact*, if it is totally bounded and complete. This is also the definition we will use.

We think that the intuitionists have made it clear that it is worthwhile to find out what the consequences of these statements are, so we will treat these principles as axioms. We will always make it clear when we use these axioms.

Axiom 1 *The continuity principle (CP): Let* S *be a spread and let* $A \subset S \times \mathbf{N}$. *If for all* $\alpha \in S$ *there is an* $n \in N$ *such that* $A(\alpha, n)$, *then for each* $\alpha \in S$, *there are* $n, m \in \mathbf{N}$ *such that if* $\beta \in S$ *and* $\overline{\beta}m = \overline{\alpha}m$, *then* $A(\beta, n)$.

The axiom CP is also known as WC-N. This principle also holds when interpreted in Markov's recursive mathematics.

Axiom 2 *The fan theorem (FAN): Let F be a fan and B a decidable bar for F. Then there is a finite subbar $B' \subset B$ which is also a bar for F.*

Note that the fan theorem also holds when read classically.

Axiom 3 *The extended fan theorem (FAN_{ext}): Let F be a fan and let $A \subset F \times \mathbf{N}$. If for all $\alpha \in F$ there is $n \in \mathbf{N}$ such that $A(\alpha, n)$, then there is $N \in \mathbf{N}$ such that for all $\alpha \in F$ there is an $n \leq N$ such that $A(\alpha, n)$.*

The following theorems are useful for later reference. They can be found for instance as Corollary 5.2.4, Corollary 5.3.7 and Theorem 5.3.6 in [4].

Theorem 17.3 [CP] *Every function on a complete separable metric space is (pointwise) continuous.*

Theorem 17.4 [FAN] *Let h be a continuous function on $[0, 1]$ such that for all $x \in [0, 1]$, $h(x) > 0$. Then there is $\varepsilon > 0$ such that $h(x) > \varepsilon$, for all $x \in [0, 1]$.*

Theorem 17.5 [FAN] *Every pointwise continuous function on a compact space is uniformly continuous.*

17.2.1 *Ishihara's trick*

We will now present what we feel is the essence of what has come to be known as Ishihara's trick, see for instance in [7,8]. The trick provides a way to find a proof which is acceptable in Bishop-style mathematics for certain statements for which we have both an intuitionistic and a classical proof. This technique rests on two observations. First, many theorems in classical analysis can be proved in Bishop-style mathematics assuming only LPO, instead of the full form of the law of excluded middle. Here LPO denotes the 'limited principle of omniscience',

$$\forall \alpha \in \mathbf{N}^{\mathbf{N}}[\exists n[\alpha(n) \neq 0] \vee \forall n[\alpha(n) = 0]].$$

Second, given a discontinuous function on a complete metric space we can prove LPO. In view of Theorem 17.3 no such function exists in intuitionistic mathematics.

A precise statement of Ishihara's trick is the following.

Theorem 17.6 *Let (X, d) be a complete metric space. Let f be a strongly extensional map from X to a metric space (Y, ρ), and let $(x_n)_{n \in \mathbf{N}}$ be a sequence in X converging to a point $x \in X$. Then for all $\epsilon > 0$, either LPO or $\rho(f(x_n), f(x)) \leq \epsilon$ for all sufficiently large n.*

A map f between a metric space X and a metric space Y is called *strongly extensional* when $f(x) \neq f(y)$, whenever x, y in X and $x \neq y$. Here we use Bishop's notation where $a \neq b$ denotes that a is apart from b, that is the distance between a and b is positive. Every continuous function is strongly extensional.

It should be noted that Theorem 17.6 is trivial in classical, intuitionistic and recursive mathematics. Indeed, in classical mathematics LPO holds. In intuitionistic and recursive mathematics every function on a complete separable metric space is pointwise continuous, so one can consider a complete separable subspace of X containing the sequence $(x_n)_{n\in\mathbf{N}}$.

We now outline a proof of Ishihara's trick in three lemmas. Let (X, d) be a complete metric space. Let f be a strongly extensional map from X to a metric space (Y, ρ), let $\epsilon > 0$ and let $(x_n)_{n\in\mathbf{N}}$ be a sequence in X converging to a point $x \in X$.

Lemma 17.7 *Either for all n, $|f(x) - f(x_n)| < 2\varepsilon$ or there exists an n such that $|f(x) - f(x_n)| > \varepsilon$.*

Proof. Define an increasing binary sequence such that

$$\lambda(n) = 0 \Rightarrow |f(x) - f(x_n)| < 2\varepsilon;$$
$$\lambda(n) = 1 \Rightarrow |f(x) - f(x_m)| > \epsilon, \text{ for some } m \leq n.$$

Define $y_n = x_n$ when $\lambda(n) = 0$ and $y_n = x_{n_0}$ when $n_0 \leq n$ is the first m such that $\lambda(m) = 1$. The sequence y_n converges to a limit y. If $|f(y) - f(x)| < \varepsilon$, then for all n, $|f(x) - f(x_n)| < 2\varepsilon$. If on the other hand $|f(y) - f(x)| > \epsilon/2$, then, by the strong extensionality of f, there exists an n such that $|f(x) - f(x_n)| > \varepsilon$. □

Lemma 17.8 *Either for all sufficiently large n, $|f(x) - f(x_n)| < 2\varepsilon$ or there are infinitely many n such that $|f(x) - f(x_n)| > \varepsilon$.*

Proof. Using the previous lemma we define an increasing binary sequence such that

$$\lambda(n) = 0 \Rightarrow \text{there exists } m > n,\ |f(x) - f(x_m)| > \varepsilon;$$
$$\lambda(n) = 1 \Rightarrow \text{for all } m > n,\ |f(x) - f(x_m)| < 2\epsilon.$$

Define $y_n = x_n$ when $\lambda(n) = 0$ and $y_n = x_{n_0}$ when $n_0 \leq n$ is the first m such that $\lambda(m) = 1$. The sequence y_n converges to a limit y. If $|f(y) - f(x)| < \varepsilon$, then there are infinitely many n such that $|f(x) - f(x_n)| > \varepsilon$. If $|f(x) - f(y)| > \epsilon/2$ then, by the strong extensionality of f, for all sufficiently large n, $|f(x) - f(x_n)| < 2\varepsilon$. □

Corollary 17.9 *Either LPO or for all sufficiently large n, $|f(x) - f(x_n)| < 2\varepsilon$.*

Proof. Suppose that x'_n is a subsequence of x_n such that for all n, $|f(x) - f(x'_n)| > \varepsilon$. We will prove LPO. Let $\alpha \in \mathbf{N}^{\mathbf{N}}$. Define $y_n := x$ as long as $\alpha(n) = 0$ and $y_n := x'_{m_0}$, where m_0 is the smallest $m \leq n$ such that $\alpha(m) \neq 0$, when such an m has been found. Then y_n converges to a limit y. If $|f(x) - f(y)| > \epsilon/2$, then by the strong extensionality of f, x is apart from y. So we can find m such that $\alpha(m) \neq 0$. Otherwise, if $|f(x) - f(y)| < \epsilon$, then $\alpha(m) = 0$ for all m. □

17.3 Integration Theory

In this section we recall some results from Bishop and Cheng's constructive version of the Daniell's integration theory [3,9].

A triple (X, L, I) is an *integration space* if X is an inhabited set with an apartness relation $\neq$, L a set of strongly extensional partial functions, and I is a mapping from L into $\mathbf{R}$, called the *integral*, such that the following properties hold.

1. If $f, g \in L$ and $\alpha, \beta \in \mathbf{R}$, then $\alpha f + \beta g$, $|f|$ and $f \wedge 1$ belong to L, and $I(\alpha f + \beta g) = \alpha I(f) + \beta I(g)$.
2. If $f \in L$ and $(f_n)_{n \in \mathbf{N}}$ is a sequence of non-negative functions in L such that $\sum_{n=1}^{\infty} I(f_n)$ converges and $\sum_{n=1}^{\infty} I(f_n) < I(f)$, then there exists x in X such that $\sum_{n=1}^{\infty} f_n(x)$ converges and $\sum_{n=1}^{\infty} f_n(x) < f(x)$.
3. There exists a function p in L with $I(p) = 1$.
4. For each f in L, $\lim_{n \to \infty} I(f \wedge n) = I(f)$ and $\lim_{n \to \infty} I(|f| \wedge n^{-1}) = 0$.

When R denotes the Riemann integral, then $([0, 1], C[0, 1], R)$ is an example of an integration space. More generally, let X be a locally compact space and I any non-zero positive linear functional on the space of test functions $C(X)$; then $(X, C(X), I)$ is an integration space. A linear functional I on $C(X)$ is called *positive* if $I(f) \geq 0$ whenever $f \geq 0$. Such a positive linear functional is also called a *positive measure*.

There is a general completion construction for integration spaces, which allows us to carry out all the usual measure theoretic constructions. The elements of the completion are, almost everywhere defined, partial functions. When this completion is applied to the Riemann integral one obtains the Lebesgue integral.

Intuitively, an integrable set is identified with its characteristic function χ_A. To make this work nicely, one has to consider what is called a complemented set, that is a pair (A^1, A^0) such that for all $a_1 \in A^1$ and $a_0 \in A^0$, $a_1 \neq a_0$. When μ is an integral and A a μ-integrable (complemented) set we will usually write $\mu(A)$ instead of $\mu(\chi_A)$. Similarly, we will sometimes use the notation $\int f d\mu$ instead $\mu(f)$.

The following theorem is a fundamental result in integration theory. Let f be an integrable function and let $\alpha > 0$. Define the complemented sets

$$[f \geq \alpha] := (\{x : f(x) \geq \alpha\}, \{x : f(x) < \alpha\}) \text{ and}$$
$$[f > \alpha] := (\{x : f(x) > \alpha\}, \{x : f(x) \leq \alpha\}).$$

In general, we will not be able to compute the measure of the set $[f > \alpha]$ for all α. However, the next theorem states that $I([f > \alpha])$ can be computed for many α.

Let T be a countable subset of a metric space X and write $X \sim T$ for the set $\{x \in X \,;\, \forall t \in T\, x \neq t\}$. When a predicate P holds for all $x \in X \sim T$, we will say that *P holds for all but countably many x* and we will call the elements of $X \sim T$ admissible (for P).

Theorem 17.10 *Let I be an integral and let f be an integrable function; then for all but countably many $\alpha > 0$, the sets $[f \geq \alpha]$ and $[f > \alpha]$ are integrable and have the same measure. Moreover, for each such admissible $\alpha > 0$ and each $\varepsilon > 0$, there is $\delta > 0$ with*

$$|I[f \geq \alpha] - I[f \geq \alpha']| < \varepsilon,$$

whenever $\alpha' > 0$ is admissible and $|\alpha - \alpha'| < \delta$.

This theorem can be read as a constructive way of stating that the function $\alpha \mapsto I[f \geq \alpha]$ being non-decreasing, is continuous at all but countably many points.

It is important to define measurable functions together with an integral. So the following definition may look different from the ones the reader is used to.

An integration space X is *finite* when X is integrable and (therefore) $\mu(X) < \infty$. We will not recall the general definition of a full set, but in the special case when the integration space X is finite, a set A is *full* when $\mu(A) = \mu(X)$.

Definition 17.11 *A function defined on a full set is measurable if for each integrable set A and each $\varepsilon > 0$, there exist an integrable set $B \subset A$ and an integrable function g such that $\mu(A - B) < \varepsilon$ and $|f - g| < \varepsilon$ on B.*

17.4 Regular Measures and Ulam's Theorem

In this section we study regular measures and prove Ulam's theorem. Regularity is very useful when combined with intuitionistic axioms, as we will show in Section 17.5.

We want to find a good substitute for the classical notion of a Borel measure on a complete separable metric space. Classically, a Borel measure is a measure such that all open sets and hence all Borel sets are measurable. Constructively, even for Lebesgue measure on [0,1] not all open sets are measurable. For measures on locally compact spaces a good substitute is to demand that all test functions, and hence all uniformly continuous functions are measurable. We define a *Borel measure* as a measure on a separable metric space such that all uniformly continuous functions are measurable.

There is an example in recursive mathematics of a pointwise continuous, but not uniformly continuous, function on [0,1] that is not Lebesgue measurable, see for instance [10, Corollary 1, p. 272].

Recall that a total function f on a metric space (X, ρ) is called *Lipschitz* if there is a constant $L > 0$ such that for all $x, y \in X$, $|f(x) - f(y)| \leq L\rho(x, y)$. Define for all $x \in X$, the Lipschitz function ρ_x by $\rho_x(y) := \rho(x, y)$. Then for all $x \in X$ and $\varepsilon > 0$, $B(x, \varepsilon) = [\rho_x < \varepsilon]$. Using classical logic, it follows that all basic open sets, and hence all open sets, are measurable. Consequently, the present definition of Borel measure is equivalent to the usual one in classical mathematics.

Definition 17.12 *Let X be a metric space. A measure μ on X is regular if for every measurable set $A \subset X$ and $\varepsilon > 0$ there is a compact integrable set $C \subset A$ such that $\mu(A - C) < \varepsilon$. A finite measure μ on X is tight if for each $\varepsilon > 0$ there is a compact integrable set $C \subset X$ such that $\mu(X - C) < \varepsilon$.*

What we call 'regular' is called 'inner regular for compacts' by some authors.

Positive measures on locally compact sets are regular [3, Theorem 6.6.7]. Theorem 17.14 states that many more measures are regular.

Lemma 17.13 *A tight finite Borel measure is regular.*

Proof. Let μ be a tight Borel measure on a metric space X and $\varepsilon > 0$. We may assume that $\mu(X) = 1$. Choose a compact integrable set $A \subset X$ such that $\mu(A) > \mu(X) - \varepsilon/2$ and apply Theorem 6.6.7 [3, p. 257] to the restriction of μ to A. □

Theorem 17.14 [Ulam] *Let X be a complete separable metric space and let μ be a finite Borel measure on X. Then μ is regular.*

We will first give a classical proof of this theorem due to Ulam (see [11, footnote 3] and [12, p683]) then give an intuitionistic proof using CP, and finally we will use Ishihara's trick to remove this hypothesis and obtain a proof which is acceptable in Bishop-style mathematics.

We assume that $\mu(X) = 1$.

Proof. [Classical proof] Let $Q = \{q_1, q_2, \ldots\}$ be a countable dense set in X. Let $\varepsilon > 0$ and $m \in \mathbf{N}$. Because $X = \bigcup_{n=1}^{\infty} B(q_n, 1/m)$, there is N_m such that

$$\mu\left(\bigcup_{n=1}^{N_m} B(q_n, 1/m)\right) > 1 - \varepsilon 2^{-m}.$$

Define $C := \bigcap_m \bigcup_{n=1}^{N_m} \overline{B(q_n, 1/m)}$. Then C is compact and $\mu(C) \geq 1 - \varepsilon$. □

There are two main problems when we try to interpret the previous proof constructively. First, we cannot conclude in general that if $(X_n)_{n\in\mathbf{N}}$ is a sequence of integrable sets and $X = \bigcup_{n=1}^{\infty} X_n$, then there exists a natural number N such that $\mu(\bigcup_{n=1}^{N} X_n) > 1/2$. Secondly, it is not clear constructively that a set like the set C defined is totally bounded; one has to choose an appropriate subset of $\{q_1, \ldots, q_{N_m}\}$ as a $1/m$-net, depending on the choices made for $k < m$.

We elaborate on the first problem. In recursive mathematics there is a sequence $(I_n)_{n\in\mathbf{N}}$ of intervals such that $[0, 1] = \bigcup_{n=1}^{\infty} I_n$ but for all $N \in \mathbf{N}$, $\mu(\bigcup_{n=1}^{N} I_n) < 1/2$, see for instance [10]. Here μ denotes Lebesgue measure, which is regular. In intuitionistic mathematics, when $(X_n)_{n\in\mathbf{N}}$ is a sequence of integrable sets and $X = \bigcup_{n=1}^{\infty} X_n$, then there exists a natural number N such that $\mu(\bigcup_{n=1}^{N} X_n) > 1/2$, for any measure μ which is regular (see Theorem 17.20). However, regularity is exactly what we are trying to prove!

With these problems in mind it may seem quite surprising that it is possible to proof Ulam's theorem constructively. In fact, the only way we know how to do this is by first giving an intuitionistic proof and then use Ishihara's trick to transform this into a constructive proof. We will now proceed towards an intuitionistic proof of Ulam's theorem. Let $\overline{\mathbf{N}}$ be the completion of $\mathbf{N}$ with respect to the metric $d(n, m) := |1/(n + 1) - 1/(m + 1)|$ and let ∞ be the point at infinity. Note that $\overline{\mathbf{N}}$ is a complete separable metric space.

Proof. [Proof using CP] It follows from Lemma 17.13 that we only need to prove that μ is tight. To do so we first prove a lemma which contains the key argument of the proof. □

Lemma 17.15 *Let X be a complete separable metric space, D a finite subset of X and μ be a finite Borel measure on X. Then for each $\varepsilon > 0$, there is a closed separable μ-measurable subset A of X such that there is a finite ε-net for A which contains D and $\mu(X) - \mu(A) < \varepsilon$.*

Proof. [Proof of the lemma] Write $D = \{q_1, \dots, q_d\}$ and choose a countable dense set $Q = \{q_1, q_2, \dots\}$ in X. Let $T := \{\rho(x, y) : x, y \in Q\}$, then T is countable. Note that for all $s \in X \sim T$ and $x, y \in Q$, we can decide whether $x \in B(y, s)$ or not. Define for all $N \in \mathbf{N}$,

$$h_N(x) := \min_{n \le N} \rho(x, q_n) \wedge 1.$$

For all $N \in \mathbf{N}$ and all $\alpha \in (0, 1] \sim T$ which are admissible for h_N,

$$[h_N < \alpha] = \{x : \min_{n \le N} \rho(x, q_n) < \alpha\} = \bigcup_{n=1}^{N} B(q_n, \alpha).$$

Since α is admissible for h_N, $\mu[h_N < \alpha] = \mu[h_N \le \alpha]$ and, because h_N is continuous,

$$[h_N \le \alpha] \supset \overline{\bigcup_{n=1}^{N} B(q_n, \alpha)}.$$

For all $n \in \mathbf{N}$, h_n is Lipschitz with constant 1. Because Q is dense the sequence $(h_n)_{n \in \mathbf{N}}$ converges pointwise to 0. So the map $n \mapsto h_n$ can be extended from $\mathbf{N}$ to $\overline{\mathbf{N}}$ such that $h_\infty = 0$. Then h_β is Lipschitz with constant 1, for all $\beta \in \overline{\mathbf{N}}$.

The function $f : \overline{\mathbf{N}} \to \mathbf{R}$ defined by $f(\beta) := \mu(h_\beta)$ is continuous, by CP. Because $f(\infty) = 0$, there is an N such that $\mu(h_N) < \varepsilon$. For all positive integrable g and every $\alpha > 0$ which is admissible for g, $\alpha\mu[g \ge \alpha] \le \int g$. So for fixed $\alpha > 0$, $\mu[h_N \ge \alpha] \to 0$ as $N \to \infty$.

Choose $N \ge d$ and $\alpha \in (0, \varepsilon/2] \sim T$ such that α is admissible for h_N and $\mu[h_N < \alpha] > \mu(X) - \varepsilon$. Define $A := \overline{[h_N < \alpha]}$. We claim that A satisfies the required properties. We only show that A is separable. Indeed, A is separable, because for all $q \in Q$, we can decide, by our construction of T, whether $q \in \bigcup_{n=1}^{N} B(q_n, \alpha)$ or not. So $\bigcup_{n=1}^{N} B(q_n, \alpha)$ is separable and hence A is separable. □

We now complete the intuitionistic proof of Ulam's theorem by applying the lemma to obtain better and better approximations of the compact set.

Let $\varepsilon > 0$. Let A_1 be a closed separable μ-measurable subset of X and a finite ε-net D_1 for A_1 such that $\mu(X) - \mu(A_1) < 2^{-1}\varepsilon$. Suppose that there are closed separable μ-measurable subsets $A_1, \dots, A_n \subset X$ such that for all $i < n$, $A_{i+1} \subset A_i$ and $\mu(A_i) - \mu(A_{i+1}) < 2^{-i}\varepsilon$. Moreover, suppose that finite sets $D_1, \dots, D_n \subset X$ have been defined such that for all $i < n$, $D_i \subset D_{i+1}$ and for all $i \le n$, D_i is a $2^{-i}\varepsilon$-net for A_i. Note that A_n is a complete separable metric space. So we can apply the lemma to the space A_n, the set D_n and the restriction of μ to A_n to obtain a closed separable μ-measurable set $A_{n+1} \subset A_n$ and a finite set $D_{n+1} \subset A_{n+1}$ such that $\mu(A_n) - \mu(A_{n+1}) < 2^{-n}\varepsilon$, $D_n \subset D_{n+1}$ and D_{n+1} is an $2^{-(n+1)}\varepsilon$-net for A_{n+1}. It follows that $A := \bigcap_{i=1}^{\infty} A_i$ is a compact set such that $\mu(X) - \mu(A) < \varepsilon$.

To remove the assumption of CP from the above proof we can use Ishihara's trick. The function f we defined above is a strongly extensional map on the complete space $\bar{\mathbf{N}}$, by [3, Lemma 6.1.3]. Let $\varepsilon > 0$. By Ishihara's trick, either for all sufficiently large n,

$f(n) < \varepsilon$ or LPO. We assume the latter. In this case the descending sequence $f(n) = \mu(h_n)$ of non-negative real numbers converges. So the sequence $(h_n)_{n\in\mathbf{N}}$ is Cauchy in L_1, and as it converges pointwise to 0, it converges in L_1 to the constant function with value 0. Hence $\lim_{n\to\infty} \mu(h_n) = 0$. We conclude that in both cases we can compute an $n \in \mathbf{N}$ such that $f(n) < \varepsilon$ and continue as in the proof above.

17.5 Intuitionistic Theorems

Every compact set can be represented as the image of a fan (see for instance [6, p. 363]). This fact can be used to obtain the following four striking results. Similar results were first proved by van Rootselaar [1, 13] for Lebesgue measure on [0, 1]. The first theorem was proved for positive measures on locally compact spaces by Bridges and Demuth in [10]. They also used intuitionistic axioms.

Theorem 17.16 [FAN,CP] *Let μ be a regular measure on a metric space X. Every μ-a.e. defined function is measurable.*

Proof. Let $\varepsilon > 0$. Let f be an a.e. defined function and let A be an integrable set. Choose a compact integrable set $C \subset A \cap \operatorname{Dom} f$ such that $\mu(A - C) < \varepsilon$. By FAN and CP, f is uniformly continuous on C. Construct by the Tietze extension theorem [3, Theorem 4.6.16] a bounded uniformly continuous extension g of $f|_C$ to X. Then $g\chi_C$ is integrable and $f = g$ on C. So f is measurable. □

A simple, but interesting consequence is the following. If f is measurable on a regular measure space and g is a measurable function on $\mathbf{R}$, such that $f(x) \in \operatorname{Dom} g$ for almost all x, then $g \circ f$ is measurable. In particular, if $f > 0$, then one may choose $g(x) := x^{-1}$ on $(0, \infty)$. This should be compared with [3, Corollary 6.7.10], where, in the absence of intuitionistic axioms, stronger assumptions are necessary to prove that $1/f$ is measurable.

The hypothesis that $f(x) \in \operatorname{Dom} g$ for almost all x is necessary, because, in order to be measurable, the function $g \circ f$ needs to be defined on a full set.

Definition 17.17 *Let $(f_n)_{n\in\mathbf{N}}$ be a sequence of measurable functions, and f a function defined on a full set. The sequence $(f_n)_{n\in\mathbf{N}}$ converges to f almost uniformly if to each integrable A and $\varepsilon > 0$, there is an integrable $B \subset A$ with $\mu(A - B) < \varepsilon$ and the sequence $(f_n)_{n\in\mathbf{N}}$ converges uniformly on B.*

Theorem 17.18 [Egoroff] *Let μ be a regular finite measure on a metric space. If a sequence $(f_n)_{n\in\mathbf{N}}$ of measurable functions converges on a full set to f, then the sequence $(f_n)_{n\in\mathbf{N}}$ converges almost uniformly.*

Proof. [FAN_{ext}] The sequence $(f_n)_{n\in\mathbf{N}}$ converges to f on a large compact set. By the extended fan theorem it converges uniformly on this set. □

Bishop and Bridges [3, Theorem 6.8.16] proved the previous theorem without the fan theorem, replacing convergence on a full set by the following strong definition of convergence a.e.

Definition 17.19 *Let* $(f_n)_{n\in\mathbf{N}}$ *be a sequence of measurable functions, and* f *a function defined on a full set. The sequence* $(f_n)_{n\in\mathbf{N}}$ *converges to* f *almost everywhere if to each integrable* A *and* $\varepsilon > 0$*, there is* $N \in \mathbf{N}$ *and an integrable* $B \subset A$ *with* $\mu(A - B) < \varepsilon$ *and* $|f - f_n| < \varepsilon$ *on* B *for all* $n \geq N$.

The following result motivates this definition of convergence a.e. Again we see that Bishop avoids the use of intuitionistic axioms by modifying the more familiar definitions.

Lemma 17.20 [$\mathrm{FAN}_{\mathrm{ext}}$] *Let* μ *be a regular measure. If a sequence* $(f_n)_{n\in\mathbf{N}}$ *of measurable functions converges pointwise on a full set, then the sequence* $(f_n)_{n\in\mathbf{N}}$ *converges almost everywhere.*

Proof. Let A be an integrable set, let $\varepsilon \geq 0$, and let C be a compact subset of A such that $\mu(A - C) < \varepsilon$. By the extended fan theorem the sequence $(f_n)_{n\in\mathbf{N}}$ converges uniformly on this set. □

Note that the fan theorem is crucial here. There is an example in recursive mathematics [10, Corollary 1, p. 272] where f is the pointwise limit of uniformly continuous functions on $[0, 1]$, f is pointwise continuous and bounded, but not Lebesgue integrable.

Note that Lemma 17.20 can also be proved classically.

Theorem 17.21 [Lusin] Let $\varepsilon > 0$, μ be a regular measure on a metric space and let A be an integrable set. If f is measurable then there exist an integrable $B \subset A$ such that $\mu(A - B) < \varepsilon$ and f is uniformly continuous on B.

Proof. [FAN, CP] Choose a large compact set $C \subset A$ on which f is defined. By FAN and CP f is uniformly continuous on C. □

A similar result can also be obtained in Bishop-style mathematics, but then the proof is somewhat longer:

Theorem 17.22 [Lusin] Let $\varepsilon > 0$, μ be a regular positive measure on a locally compact space and let A be an integrable set. If f is measurable then there exist an integrable $B \subset A$ such that $\mu(A - B) < \varepsilon$ and f is uniformly continuous on B.

Proof. First assume that f is a characteristic function of an integrable set $B = (B^1, B^0)$. Take compact sets $B_1 \subset B^1$ and $B_0 \subset B^0$ such that $\mu(B^i - B_i) < \varepsilon/2$ for all $i \in \{0, 1\}$ and the distance between B_0 and B_1 is strictly positive, see Theorem 6.6.7 in [3] for this last fact. The function f is constant on each B_i and hence uniformly continuous on $B_0 \cup B_1$. When f is a simple function, we can construct in a similar way a large integrable set on which f is uniformly continuous. Now let f be the limit of a sequence $(f_k)_{k\in\mathbf{N}}$ of simple functions. There is a sequence $(C_k)_{k\in\mathbf{N}}$ of measurable sets such that for all $k \in \mathbf{N}$, $\mu(A - C_k) < \varepsilon 2^{-k-1}$ and f_k is uniformly continuous on C_k. By applying Theorem 17.18 to this sequence of functions we find an integrable set C_0 such that $\mu(A - C_0) < \varepsilon/4$ and the sequence $(f_n)_{n\in\mathbf{N}}$ converges uniformly on C_0. Now there is a measurable set $C \subset C_0 \cap \bigcap_{k=1}^{\infty} C_k$, such that $\mu(A - C) < \varepsilon$. It follows that f is uniformly continuous on C. □

17.6 Conclusions

We have shown how to constructively approximate integrable sets by compact integrable sets. The proof uses an interesting logical detour, the theorem was first proved intuitionistically and only then these intuitionistic assumptions were removed to obtain a result acceptable in Bishop's constructive mathematics.

Finally, it is also possible to develop integration theory algebraically, see for instance [14]. This seems to be more appropriate for the applications to functional analysis. However, the algebraic approach does not include a notion of convergence almost everywhere.

Acknowledgements

Some of the results in this article can also be found in my PhD thesis [15]. I would like to thank Wim Veldman for his advice during the PhD project and for remarks on an earlier version of this chapter. The anonymous referee provided useful suggestions on the presentation of the chapter. This research was partially supported by NWO.

References

1. Heyting, A. (1956). *Intuitionism. An introduction.* Studies in Logic and the Foundations of Mathematics. North-Holland Publishing Company, Amsterdam, (English).
2. Brouwer, L. E. J. (1975). *Collected works.* North-Holland.
3. Bishop, E. and Bridges, D. (1985). *Constructive analysis.* Grundlehren der Mathematischen Wissenschaften, vol. 279. Springer-Verlag.
4. Bridges, D. and Richman, F. (1987).*Varieties of constructive mathematics.* London Mathematical Society Lecture Notes Series, no. 97. Cambridge University Press, (English).
5. Beeson, M. J. (1985). *Foundations of constructive mathematics.* Springer-Verlag, Berlin, Heidelberg, New York.
6. Troelstra, A. S. and van Dalen, D. (1988) Constructivism in mathematics. An introduction. Volume II. *Studies in Logic and the Foundations of Mathematics*, no. 123. North-Holland.
7. Ishihara, H. (1991). Continuity and nondiscontinuity in constructive mathematics. *Journal of Symbolic Logic*, **56**, no. 4, 1349–1354.
8. Bridges, D., van Dalen, D., and Ishihara, H. (2003). Ishihara's proof technique in constructive analysis. *Proc. Koninklijke Nederlandse Akad. Wetenschappen (Indag. Math.) N.S.* **4**, no. 2, 2749–2752.
9. Bishop, E. and Cheng, H. (1972). *Constructive measure theory*, American Mathematical Society, Providence, R.I., Memoirs of the American Mathematical Society, No. 116.

10. Bridges, D. and Demuth, O. (1991). On the Lebesgue measurability of continuous functions in constructive analysis. *Bulletin of the American Mathematical Society, New Series*, **24**, no. 2, 259–276.
11. Oxtoby, J. C. and Ulam, S. M. (1939). On the existence of a measure invariant under a transformation. *Ann. of Math.*, **40**(2), 560–566.
12. Ulam, S. (1974). Sets, numbers, and universes: selected works. *Mathematicians of Our Time* (eds W. A. Beyer, J. Mycielski and G.-C. Rota), Vol. 9. The MIT Press, Cambridge, Mass.
13. van Rootselaar, B. (1954). Generalization of the Brouwer integral. PhD thesis, Universiteit van Amsterdam, Amsterdam.
14. Spitters, B. (2004). Constructive algebraic integration theory, Proceedings of the Second Workshop on Formal Topology, special issue of *Annals of Pure and Applied Logic* (eds B. Banaschewski, T. Coquand, and G. Sambin).
15. ——. (2002). Constructive and intuitionistic integration theory and functional analysis, PhD thesis, University of Nijmegen.

18

AN INTRODUCTION TO THE THEORY OF C*-ALGEBRAS IN CONSTRUCTIVE MATHEMATICS

HIROKI TAKAMURA

Abstract

In this chapter we introduce an elementary theory of C*-algebras in Bishop's constructive mathematics. The main achievement of the chapter is a proof of the Gelfand–Naĭmark–Segal (GNS) construction theorem in Bishop's constructive mathematics. This important theorem in the theory of operator algebras says that for each C*-algebra and every state there exists a cyclic representation on some Hilbert space.

18.1 Introduction

In this chapter we introduce an elementary theory of C*-algebras in Bishop's constructive mathematics. The main result of the chapter is to give a proof of the Gelfand–Naĭmark–Segal (GNS) construction theorem in Bishop's constructive mathematics. This theorem is one of the important theorems in the theory of operator algebras. Historically, there are two main topics in the theory of operator algebras: von Neumann algebras which was introduced by Murray and von Neumann in the 1930s, and C*-algebras, introduced by Gelfand and Naĭmark in the 1940s. The theory of operator algebras is not only of mathematical interest, but is also a useful tool for describing theoretical (quantum) physics. The GNS construction theorem is a representation theorem which says that for any abstract C*-algebra and any state on a C*-algebra there exists a cyclic representation.

Related works are as follows. Bishop [1] described commutative Banach algebras and a special case of commutative C*-algebras which consists of normable operators on a Hilbert space in constructive mathematics. In his book, he proves a constructive version of the Gelfand representation theorem for the special case mentioned above. These results appear in Bishop and Bridges [2], and Bridges [3]. In Bridges [4] and Havea's PhD thesis [5], the authors study constructive non-commutative Banach algebras. In Spitters' PhD thesis [6], he claims that the proof of the Gelfand representation theorem in Bishop [1] also works well in the general setting of a commutative C*-algebra and he gives a constructive proof of the theorem. The area of constructive von Neumann algebras is studied by Vîţă in her PhD thesis [7], and also by Spitters [6].

In this chapter, we assume familiarity with constructive mathematics, as found in [1–3], and with the theory of C*-algebra, as found in [8–13].

18.2 Constructive C*-Algebras

An *involution* on an algebra $\mathfrak{A}$ over a field $\mathbb{K}$ is a map $^* : \mathfrak{A} \to \mathfrak{A}$ such that for all $x, y \in \mathfrak{A}$ and $a, b \in \mathbb{K}$,

$$(xy)^* = y^* x^*,$$
$$(x^*)^* = x,$$
$$(ax + by)^* = \overline{a}x^* + \overline{b}y^*,$$

where $\overline{a}, \overline{b}$ are conjugates of a and b.
A *$*$-algebra* over a field $\mathbb{K}$ is an algebra with an involution. Let $\mathfrak{A}$ be an algebra over a complex field $\mathbb{C}$ with unit $\mathbf{e}$. An algebra $\mathfrak{A}$ is called a *Banach algebra* if $\mathfrak{A}$ is a Banach space with $||\mathbf{e}|| = 1$, and $||xy|| \leq ||x|| ||y||$ for all x, y in $\mathfrak{A}$.

Note that by definition a Banach space in constructive mathematics is separable, hence a Banach algebra is separable.

Definition 18.1 (Constructive C*-algebra) A Banach $*$-algebra $\mathfrak{A}$ is called a *C*-algebra* if $||x^*x|| = ||x||^2$ for all x in $\mathfrak{A}$.

If $\mathcal{H}$ is a Hilbert space, we denote by $\mathfrak{B}(\mathcal{H})$ the set of bounded linear operators on $\mathcal{H}$. An operator $T \in \mathfrak{B}(\mathcal{H})$ is said to be *compact* if $\{T(x) : ||x|| \leq 1\}$ is totally bounded. Every compact operator is normable and has its adjoint. The set of all compact operators on a Hilbert space is an example of a constructive C*-algebra.

Definition 18.2 (Concrete C*-algebra) A self-adjoint $*$-subalgebra $\mathfrak{R}$ of normable elements of $\mathfrak{B}(\mathcal{H})$ is called a *concrete C*-algebra* if it is complete and separable with respect to the norm.

Classically, the set $\mathfrak{B}(\mathcal{H})$ of all bounded linear operators on a Hilbert space $\mathcal{H}$ is an example of a C*-algebra. However, $\mathfrak{B}(\mathcal{H})$ is not an example of a constructive C*-algebra, since, as the following Brouwerian example [14] shows, every bounded linear operator in $\mathfrak{B}(\mathcal{H})$ is not normable.
Let $\{a_n\}$ be a binary sequence and $\{e_n\}$ a orthonormal basis for ℓ^2. Define a linear operator A on ℓ^2 by

$$Ax = \sum_{n=1}^{\infty} a_n x_n e_n,$$

where x_n is the nth component of x. Then it is easy to see that A is self-adjoint. If A is normable, then either $||A|| > 0$ in which case $a_n = 1$ for some n, or else $||A|| < 1$ and $a_n = 0$ for all n. Thus, the statement

Every bounded operator on a Hilbert space has a norm

entails the Limit Principle of Omniscience (LPO).

18.3 Positive Elements

In this section, we study positive elements in a constructive C*-algebra. An element x in a C*-algebra $\mathfrak{A}$ is *normal* if $x^*x = xx^*$ and is *self-adjoint* if $x = x^*$ holds. Trivially, any self-adjoint element is a normal element: for any x in $\mathfrak{A}$ there exist self-adjoint elements x_1, x_2 such that $x = x_1 + \imath x_2$, where $x_1 = (x + x^*)/2, x_2 = (x^* - x)/2\imath$. Let $\mathfrak{A}$ be a C*-algebra. For each $x \in \mathfrak{A}$, the set

$$Sp(x) = \{\lambda \in \mathbb{C} : (x - \lambda\mathbf{e})^{-1} \text{ does not exist}\},$$

is called the *spectrum of* x.

The following classical result can be found on page 6 of [10].

Theorem 18.3 *(1) x is normal and $Sp(x) \subset [0, \infty)$.*
(2) There is a self-adjoint element y in $\mathfrak{A}$ such that $x = y^2$.
(3) There is an element y in $\mathfrak{A}$ such that $x = y^*y$.
(4) x is self-adjoint and $||a\mathbf{e} - x|| \le a$ for any $a \ge ||x||$.
(5) x is self-adjoint and $||a\mathbf{e} - x|| \le a$ for some $a \ge ||x||$.

Classically, an element x in a C*-algebra $\mathfrak{A}$ is *positive* if it satisfies any of the five conditions (1)–(5).

In the classical theory, condition (1) plays an essential role. However, the spectrum of an element is not easy to treat constructively. In [15], there are Brouwerian examples that we cannot establish elementary properties of spectra which are obtained in classical mathematics. Hence, we need to modify the condition (1) of a positive element to show the equivalence in constructive C*-algebra.

The spectrum Σ of a commutative Banach algebra $\mathfrak{A}$ consists of all non-zero bounded multiplicative linear functionals. Each element in the spectrum Σ of $\mathfrak{A}$ is called a *character*. With regard to commutative constructive C*-algebras, the Gelfand representation theorem states that every commutative C*-algebra is isomorphic to the space of continuous functions on its spectrum.

Note that the spectrum of a Banach algebra is compact classically. Constructively, this is not true for Banach algebras [4, 5], but it is true for C*-algebras.

Theorem 18.4 (Gelfand representation theorem [6]) Let $\mathfrak{A}$ be a commutative C*-algebra and let $\mathsf{C}(\Sigma)$ be the set of all complex valued continuous functions on the spectrum Σ of $\mathfrak{A}$. Then there exists a norm preserving $*$-isomorphism from $\mathfrak{A}$ onto $\mathsf{C}(\Sigma)$.

Let $\mathfrak{A}$ be a C*-algebra and x a normal element in $\mathfrak{A}$. Then $[\mathbf{e}, x]$ denotes the commutative C*-subalgebra of $\mathfrak{A}$ generated by $\mathbf{e}$ and x and $\mathsf{C}(\Sigma)$ denotes the set of all complex valued continuous functions on the spectrum Σ of $[\mathbf{e}, x]$. For a norm preserving $*$-isomorphism ϕ from $[\mathbf{e}, x]$ onto $\mathsf{C}(\Sigma)$ (the existence of such ϕ is assured by Theorem 18.4), we say that $\phi(x)$ is a *name* of x. A name $\phi(x)$ of x is non-negative if it satisfies $\phi(x)u \ge 0$ for all $u \in \Sigma$.

In the rest of the chapter, unless otherwise specified we use the symbol ϕ as a particular $*$-isomorphism from $[\mathbf{e}, x]$ onto $\mathsf{C}(\Sigma)$.

Definition 18.5 (Positive element) An element x in a C*-algebra $\mathfrak{A}$ is called *positive* if $\phi(x)$ is non-negative.

Since $[\mathbf{e}, x]$ is defined for a normal element x, and the name $\phi(x)$ of x is non-negative, it is easy to see that a positive element is self-adjoint.

$\mathfrak{A}_{sa}$ and $\mathfrak{A}_+$ denote the set of all self-adjoint elements in a C*-algebra $\mathfrak{A}$ and the set of all positive elements in $\mathfrak{A}$ respectively. We also denote $-\mathfrak{A}_+ = \{x \in \mathfrak{A} : -x \in \mathfrak{A}_+\}$ and $\mathfrak{A}_{+1} = \{x \in \mathfrak{A}_+ : ||x|| \leq 1\}$. Let x, y in $\mathfrak{A}_{sa}$, then we write $x \geq y$ for $x - y$ is in $\mathfrak{A}_+$.

Lemma 18.6 *Let $\mathfrak{A}$ be a C*-algebra and x in $\mathfrak{A}$. The following conditions are equivalent constructively.*

(1) *x is a positive element in $\mathfrak{A}$.*
(2) *x is a self-adjoint element and $||a\mathbf{e} - x|| \leq a$ for some $||x|| \leq a$.*

Proof. (1) $\Rightarrow$ (2) Suppose $x \in \mathfrak{A}_+$. By Definition 18.5 the name $\phi(x)$ of x is non-negative. Then $||a\mathbf{e} - x|| = \sup\{|\phi(a\mathbf{e} - x)u| : u \in \Sigma\} = \sup\{|a - \phi(x)u| : u \in \Sigma\}$ and $||x|| = \sup\{|\phi(x)u| : u \in \Sigma\}$. Thus $||a\mathbf{e} - x|| \leq a$.

(2) $\Rightarrow$ (1). Suppose $x \in \mathfrak{A}_{sa}$ and $||a\mathbf{e} - x|| \leq a$ for some $||x|| \leq a$. Then $||a\mathbf{e} - x||$ is equal to $\sup\{|\phi(a\mathbf{e} - x)u| : u \in \Sigma\}$. Thus $\sup\{|a - \phi(x)u| : u \in \Sigma\} \leq a$. Hence, $\phi(x)u \geq 0$ for any $u \in \Sigma$. Therefore $\phi(x)$ is non-negative, and so x is positive. □

By Lemma 18.6 we can show that if x, y are positive elements in $\mathfrak{A}$ then $x + y$ is also positive. Indeed, $||(||x|| + ||y||)\mathbf{e} - (x + y)|| \leq ||x|| + ||y||$. So we take $a = ||x|| + ||y|| \geq ||x + y||$.

Note that the following condition (i) is also equivalent to the definition of positivity.

(i) $x = x^*$ and $|||x||\mathbf{e} - x|| \leq ||x||$.

Moreover, the condition (i) is equivalent to the following condition.

(ii) $x = x^*$ and $||a\mathbf{e} - x|| \leq a$ for all $||x|| \leq a$.

Indeed, it is clear that (ii) implies (i). Conversely, if x satisfies (i), then $||a\mathbf{e} - x|| = ||(a - ||x||)\mathbf{e} + ||x||\mathbf{e} - x|| \leq a - ||x|| + |||x||\mathbf{e} - x|| \leq a - ||x|| + ||x|| = a$. So we have (ii).

The following comes from (2.2.11) on page 34 of [8].

Proposition 18.7 *Let $\mathfrak{A}$ be a C*-algebra. Then $\mathfrak{A}_+$ is closed convex cone and $\mathfrak{A}_+ \bigcap (-\mathfrak{A}_+) = \{0\}$.*

Proof. To prove $\mathfrak{A}_+$ is a closed convex cone, it is sufficient to show that $\mathfrak{A}_{+1}$ is a closed convex cone. Trivially, $\mathfrak{A}_{+1}$ is closed under scalar multiplication. By Lemma 18.6 and the following note, $x \in \mathfrak{A}_{+1}$ is equivalent to the condition $||\mathbf{e} - x|| \leq 1$ and x is self-adjoint. Thus $\mathfrak{A}_{+1}$ is closed. For any $x, y \in \mathfrak{A}_{+1}$ and $0 \leq \lambda \leq 1$, $||\mathbf{e} - \lambda x - (1 - \lambda y)|| \leq \lambda||\mathbf{e} - x|| + (1 - \lambda)||\mathbf{e} - y|| \leq 1$. Hence $\lambda x - (1 - \lambda y) \in \mathfrak{A}_+$, and so $\mathfrak{A}_{+1}$ is a convex cone. Finally, suppose $x \in \mathfrak{A}_+ \bigcap(-\mathfrak{A}_+)$. This means that $x, -x \in \mathfrak{A}_+$. Let ϕ be a norm preserving isomorphism from $[\mathbf{e}, x]$ onto $\mathsf{C}(\Sigma)$. Then $\phi(x), -\phi(x)$ are non-negative for any $u \in \Sigma$. Therefore $\phi(x)u = 0$ for any $u \in \Sigma$, and so $x = 0$. □

Lemma 18.8 *Let $\mathfrak{A}$ be a C*-algebra and $\mathsf{C}(\Sigma)$ a norm preserving $*$-isomorphic to a commutative C*-subalgebra $\mathfrak{A}'$ of $\mathfrak{A}$, and ϕ the norm preserving $*$-isomorphism $\mathfrak{A}'$ onto $\mathsf{C}(\Sigma)$. If $\phi(x)$ is non-negative then x is a positive element of $\mathfrak{A}$.*

Proof. By Theorem 18.4, $||x|| = \sup\{|\phi(x)u| : u \in \Sigma\}$. Let a be a real number with $a \geq ||x||$. Then $||a\mathbf{e} - x|| = \sup\{|a - \phi(x)u| : u \in \Sigma\}$. Hence we have $|a\mathbf{e} - x|| \leq a$ and we have also $x = x^*$, since $\phi(x)$ is non-negative. Therefore x is a positive element of $\mathfrak{A}$, by Lemma 18.6 □

The following comes from (1.4.1) on page 7 of [11]. We give a proof without using the spectrum of an element.

Lemma 18.9 *Let x be a positive element in a C*-algebra $\mathfrak{A}$. Then for any positive integer n, there exists an unique positive element y in $\mathfrak{A}$ such that $y^n = x$. This y is denoted by $x^{1/n}$.*

Proof. Since $\phi(x)$ is non-negative in $\mathsf{C}(\Sigma)$, we can define $g(x)$ by $\phi(x)^{1/n}$. Then $g(x)$ is also non-negative and $g(x) = g(x^*)$. Thus $\phi^{-1}(g(x))$ is a positive element of $\mathfrak{A}$. Now we denote $\phi^{-1}(g(x))$ by y. To see the uniqueness, assume that $y'^n = x$ for some other $y' \in \mathfrak{A}_+$. Since $y'x = xy'$ and $[\mathbf{e}, x]$ is isomorphic to $\mathsf{C}(\Sigma)$, y can be expressed by the limit of some polynomial of x. Then we have $yy' = y'y$. Let ϕ' be a norm preserving $*$-isomorphism from $[\mathbf{e}, y, y']$ onto $\mathsf{C}(\Sigma')$. By assumption, $\phi'(y)^n = \phi'(y')^n = \phi'(x)$ and $\phi'(y), \phi'(y')$ is non-negative. Then $(\phi'(y)^n - \phi'(y')^n)u = 0$ for any $u \in \Sigma'$. Thus $(\phi'(y) - \phi'(y'))u = 0$. Hence $\phi'(y) = \phi'(y')$. Since ϕ' is a norm preserving $*$-isomorphism, we conclude that $y = y'$. □

The proof of the following results holds constructively and can be found in [9].

Lemma 18.10 *Let x be a self-adjoint element in a C*-algebra $\mathfrak{A}$. Then there exist unique positive elements x_+ and x_- such that $x = x_+ - x_-$, $x_+x_- = 0$ and $||x|| = \max\{||x_+||, ||x_-||\}$.*

We say that x_+ and x_- are *decompositions* of x.

Let x be a normal element of a C*-algebra $\mathfrak{A}$. Then $||x^{2n}||^2 = ||(x^*x)^{2n}|| = ||x||^{2n+1}$. Thus $||x|| = ||x^{2n}||^{1/2^n}$ for any natural number n. We have also that if x is a normal element of the C*-algebra $\mathfrak{A}$, and $x^k = 0$ for some positive integer k, then $x = 0$. Indeed, for any normal element, x satisfies that $||x|| = ||x^{2n}||^{1/2^n}$ for any natural number n. Thus, $||x|| = ||x^{2k}||^{1/2^k} = 0$.

We are now in position to give the constructive analogue of Theorem 18.3.

Theorem 18.11 *The following conditions are equivalent constructively.*

(1) x is a positive element in a C-algebra $\mathfrak{A}$.*
(2) There is a self-adjoint element y in $\mathfrak{A}$ such that $x = y^2$.
(3) There exists an element y in $\mathfrak{A}$ such that $x = y^*y$.
(4) x is self-adjoint and $||a\mathbf{e} - x|| \leq a$ for any $a \geq ||x||$.
(5) x is self-adjoint and $||a\mathbf{e} - x|| \leq a$ for some $a \geq ||x||$.

Proof. We have the equivalence of (1), (4) and (5), by Lemma 18.6 and by the equivalence of *(i)* and *(ii)*. We have (1) ⇒ (2), by Lemma 18.9. To see (2) ⇒ (1), consider $[y, \mathbf{e}]$. Then the name $\phi(x)$ of x is of the form $\phi(x) = (\phi(y))^2$. It is obvious that $\phi(x)$ is non-negative.

To see (1) $\Rightarrow$ (3), suppose $x \in \mathfrak{A}_+$. By Lemma 18.9 $x^{1/2}$ is also a positive element. Put y for $x^{1/2}$ then $x = y^*y$. (3) $\Rightarrow$ (1). Suppose $x = y^*y$. Since x is a self-adjoint element, by Lemma 18.10 there exists positive elements x_1, x_2 such that $x = x_1 - x_2$, $x_1x_2 = 0$. We will show that $x_2 = 0$. Put $z = yx_2$. Then $z^*z = x_2y^*yx_2 = x_2(x_1 - x_2)x_2 = -x_2{}^3$. Thus $z^*z \in (-\mathfrak{A}_+)$. We also have $zz^* \in (-\mathfrak{A}_+)$. Put $z = z_1 + \imath z_2 (z_1, z_2 \in \mathfrak{A}_{sa})$, then $z^*z + zz^* = 2z_1{}^2 + 2z_2{}^2$. Thus $z^*z = 2z_1{}^2 + 2z_2{}^2 - zz^*$ Therefore $z^*z \in \mathfrak{A}_+$. By Proposition 18.7, $z^*z = 0$. Thus $x_2{}^3 = 0$ and so $x_2 = 0$. Therefore x is positive. □

The following are some properties of elements in a C*-algebra. First, it is easy to show the equivalence $x = 0$ and $x^*x = 0$ (exercise 8.6 on page 54 of [13]).

The proof of the classical counterpart of the next proposition holds constructively and can be found on page 37 of [8].

Proposition 18.12 *For any elements x, y, z in a C*-algebra $\mathfrak{A}$, the following properties hold:*

(1) *If $x \geq y \geq 0$ then $||x|| \geq ||y||$.*
(2) *If $x \geq 0$ then $||x||x \geq x^2$.*
(3) *If $x \geq y$ then $z^*xz \geq z^*yz$.*
(4) *If $x \geq y \geq 0$ then $(x + \lambda\mathbf{e})^{-1} \leq (y + \lambda\mathbf{e})^{-1}$ for any $\lambda > 0$.*

18.4 Positive Linear Functionals and States

Let μ be a linear functional on a C*-algebra $\mathfrak{A}$. μ is called *self-adjoint* if $\mu(x) \in \mathbb{R}$ for all $x \in \mathfrak{A}_{sa}$. A linear functional μ is called *positive* if $\mu(x) \geq 0$ for all x in $\mathfrak{A}_+$. Note that every linear functional on $\mathfrak{A}$ can be represented in the form $\mu = \mu_1 + \imath\mu_2$ where μ_1, μ_2 are both self-adjoint. Since any element x in $\mathfrak{A}$ can be represented as $x = x_1 + \imath x_2$, where x_1, x_2 in $\mathfrak{A}_{sa}$, it follows that $\mu(x^*) = \mu(x_1) - \imath\mu(x_2) = \overline{\mu(x)}$. Hence a self-adjoint linear functional can be also defined by $\mu(x^*) = \overline{\mu(x)}$ for all x in $\mathfrak{A}$.

The proofs of the following results hold constructively and can be found on page 49 of [8] or on page 79 of [13].

Lemma 18.13 *A positive linear functional μ is self-adjoint, and $|\mu(x^*y)|^2 \leq \mu(x^*x)\mu(y^*y)$ for all x, y in $\mathfrak{A}$.*

Proposition 18.14 *The following conditions are equivalent.*
(1) *μ is a positive linear functional.*
(2) *μ is a bounded with $||\mu|| = \mu(\mathbf{e})$.*

Now, we define the state on a C*-algebra $\mathfrak{A}$.

Definition 18.15 (State) A linear functional ρ is a *state* if ρ is positive and $||\rho|| = 1$.

It follows from Proposition 18.14 that ρ is a state if ρ is a linear functional with $\rho(\mathbf{e}) = ||\rho|| = 1$.

Classically the state space $\mathcal{S}(\mathfrak{A})$ is always non-empty and weak $*$-compact. However, the Hahn–Banach theorem does not work well the same way it does in classical mathematics.

The following is a constructive version of the Hahn–Banach theorem (see page 342 of [2]).

Theorem 18.16 *Let Y be a linear subset of a separable normed linear space X, and v a non-zero linear functional on Y whose kernel is located in X. Then for each $\epsilon > 0$ there exists a normable linear functional u on X such that $u(y) = v(y)$ for all y in Y, and $||u|| \leq ||v|| + \epsilon$.*

A consequence of the constructive version of the Hahn–Banach theorem is that we cannot guarantee that the state space of a C*-algebra is non-empty constructively. In [5] on pages 53–57, the author studies a constructive treatment of a state space of Banach algebra.

Here is a sketch of the results in [5]. Let $\mathfrak{A}$ be a Banach algebra and $\mathfrak{A}'$ a dual space of $\mathfrak{A}$. For each $t > 0$, the set

$$\mathcal{S}^t(\mathfrak{A}) = \{\phi \in \mathfrak{A}' : ||\phi|| \leq 1, |1 - \phi(\mathbf{e})| \leq t\}$$

is a t-approximation to the state space $\mathcal{S}(\mathfrak{A})$.

Using (4.9) on page 98 of [2] and (4.5) on page 341 of [2], we have the following.

Proposition 18.17 [5] For all countably many $t > 0$, $\mathcal{S}^t(\mathfrak{A})$ is a non-empty, weak $*$-compact subset of $\mathfrak{A}'$.

We say that $t > 0$ is admissible if $\mathcal{S}^t(\mathfrak{A})$ is weak $*$-compact. Note that $\mathcal{S}(\mathfrak{A}) = \bigcap\{\mathcal{S}^t(\mathfrak{A}) : t > 0 \text{ is admissible}\}$, the intersection of a family of non-empty, weak $*$-compact sets that is descending in the sense that if $0 < t' < t$, then $\mathcal{S}^{t'}(\mathfrak{A}) \subset \mathcal{S}^t(\mathfrak{A})$.

We say the state space $\mathcal{S}(\mathfrak{A})$ is *firm* if it is compact and $\rho_w(\mathcal{S}^{t'}(\mathfrak{A}), \mathcal{S}(\mathfrak{A})) \to 0$ as $t \to 0$, where ρ_w is the Hausdorff metric on the set of weak $*$-compact subsets of $\mathfrak{A}'_1$.

Proposition 18.18 [5] If the state space of $\mathfrak{A}$ is firm, then so is the state space of every separable Banach subalgebra of $\mathfrak{A}$.

A norm of a normed linear space E is said to be *Gâteaux differentiable* at x in E if for each y in E the limit

$$\lim_{t \to 0} \frac{||x + ty|| - ||x||}{t}$$

exists.

Definition 18.19 (Gâteaux differentiable norm) A norm of a normed linear space E is a *Gâteaux differentiable norm* if it is *Gâteaux differentiable* at each point $x \in E$ with $||x|| = 1$.

Ishihara [16] has shown that if the norm of the normed space is Gâteaux differentiable then the approximation ϵ in the constructive version of the Hahn–Banach theorem can be replaced by exact equality. Using this fact, we can show that if a C*-algebra $\mathfrak{A}$ is uniformly convex with the Gâteaux differentiable norm, then the state space $\mathcal{S}(\mathfrak{A})$ of $\mathfrak{A}$ is non-empty, constructively. The following theorem is a modification of a classical result (for example, page 80 of [13]).

Theorem 18.20 *Let $\mathfrak{A}$ be a uniformly convex C*-algebra with the Gâteaux differentiable norm, and x a self-adjoint element in $\mathfrak{A}$. Then for each $u \in \Sigma$ there exists a state μ such that $\mu(x) = \phi(x)u$, where Σ is the spectrum of $[\mathbf{e}, x]$ and ϕ is a norm preserving $*$- isomorphism from $[\mathbf{e}, x]$ onto $\mathsf{C}(\Sigma)$.*

Proof. Fix $x \in \mathfrak{A}_{sa}$ and $\phi(x)u$. Consider $M = \{ax + b\mathbf{e} : a, b \in \mathbb{C}\}$. Then M is a subspace of $\mathfrak{A}$. Define μ_0 by $\mu_0(ax + b\mathbf{e}) = a\phi(x)u + b$, $(a, b \in \mathbb{C})$. Put $\lambda = \phi(x)u$. It is easy to see that μ_0 is a linear functional on M with $\mu_0(x) = \lambda$ and $\mu_0(\mathbf{e}) = 1$. Since $\phi(ax + b\mathbf{e})u = a\lambda + b$ and $|a\lambda + b| \leq \sup\{|\phi(ax + b\mathbf{e})u| : u \in \Sigma\} = ||ax + b\mathbf{e}||$, we have $||\mu_0|| = \sup\{|\mu_0(ax + b\mathbf{e})| : ||ax + b\mathbf{e}|| \leq 1\} = 1$. By the Hahn–Banach theorem, μ_0 extends to a bounded linear functional μ on $\mathfrak{A}$ with $||\mu|| = \mu(\mathbf{e}) = 1$ and $\mu(x) = \lambda = \phi(x)u$. □

The following proposition is the constructive analogue of the classical result on page 81 of [13]. The classical argument applies and we omit the proof.

Proposition 18.21 *Let $\mathfrak{A}$ be a uniformly convex C*-algebra with the Gâteaux differentiable norm and x in $\mathfrak{A}$. The following statements hold:*

(1) *$x = 0$ if and only if $\mu(x) = 0$ for all μ in $\mathcal{S}(\mathfrak{A})$.*
(2) *x is self-adjoint if and only if $\mu(x)$ in $\mathbb{R}$ for all μ in $\mathcal{S}(\mathfrak{A})$.*
(3) *x is positive if and only if $\mu(x) \geq 0$ for all μ in $\mathcal{S}(\mathfrak{A})$.*

18.5 Representations

In this section, we turn our attention to representations. Let π be a homomorphism between $*$-algebras $\mathfrak{A}$ and $\mathfrak{B}$. A homomorphism π is called $*$-homomorphism if $\pi(x^*) = \pi(x)^*$ for all $x \in \mathfrak{A}$. A representation of a $*$-algebra $\mathfrak{A}$ on a Hilbert space $\mathcal{H}$ is a $*$-homomorphism π from $\mathfrak{A}$ into $\mathfrak{B}(\mathcal{H})$. We denote it by $\{\mathcal{H}, \pi\}$.

We can prove that the following lemma, found on page 43 of [8], constructively. Since the proof goes the same way in classical mathematics, we omit the proof.

Lemma 18.22 *Let π be a $*$-homomorphism between C*-algebras $\mathfrak{A}$ and $\mathfrak{B}$. The following statements hold:*

(1) *If x is a positive element in $\mathfrak{A}$ then $\pi(x)$ is also a positive element in $\mathfrak{B}$.*
(2) *$||\pi(x)|| \leq ||x||$ for all x in $\mathfrak{A}$.*

Note that a $*$-homomorphism is defined in the algebraic sense, but Lemma 18.22 tells us it also has topological property. Let π be a $*$-homomorphism between C*-algebras $\mathfrak{A}$ and $\mathfrak{B}$. Classically, we can show that $\pi(\mathfrak{A})$ is a C*-subalgebra of $\mathfrak{B}$ (for example, page 243 of [9]).

Definition 18.23 (Cyclic representation) A representation $\{\mathcal{H}, \pi\}$ is called a *cyclic representation* if there exists a vector ξ in $\mathcal{H}$ such that $\overline{\pi(\mathfrak{A})\xi} = \mathcal{H}$, where $\overline{\pi(\mathfrak{A})\xi}$ is closure of $\pi(\mathfrak{A})\xi$.

We call the vector ξ a *cyclic vector* and denote a cyclic representation by $\{\mathcal{H}, \pi, \xi\}$. Two representations $\{\mathcal{H}, \pi\}$ and $\{\mathcal{H}', \pi'\}$ of a C*-algebra $\mathfrak{A}$ are isometrical (unitarily, respectively) equivalent if there is an isometric (unitary, respectively) operator U of $\mathcal{H}$ onto $\mathcal{H}'$ such that $U\pi(x) = \pi'(x)U$ for all x in $\mathfrak{A}$. It is clear that a unitarily equivalent implies an isometrical equivalent. It is easy to see that if two cyclic representations $\{\mathcal{H}, \pi, \xi_1\}$ and $\{\mathcal{H}', \pi', \xi_2\}$ of a C*-algebra $\mathfrak{A}$ are unitarily equivalent with unitary operator U such that $U\xi_1 = \xi_2$ then $\langle \pi(x)\xi_1, \xi_1\rangle = \langle \pi(x)'\xi_2, \xi_2\rangle$ for all x in $\mathfrak{A}$.

Lemma 18.24 *Let $\{\mathcal{H}, \pi, \xi_1\}$ and $\{\mathcal{H}', \pi', \xi_2\}$ be cyclic representations of a C*-algebra $\mathfrak{A}$. If $\langle \pi(x)\xi_1, \xi_1\rangle = \langle \pi(x)'\xi_2, \xi_2\rangle$ for all x in $\mathfrak{A}$ then two cyclic representations are isometrical equivalent.*

Proof. We define a linear map U from $\pi(\mathfrak{A})\xi_1$ onto $\pi'(\mathfrak{A})\xi_2$ by $U\pi(x)\xi_1 = \pi'(x)\xi_2$. Since

$$||U\pi(x)\xi_1||^2 = \langle \pi'(x)\xi_2, \pi'(x)\xi_2\rangle = \langle \pi'(x^*x)\xi_2, \xi_2\rangle = \langle \pi(x^*x)\xi_2, \xi_2\rangle = ||\pi(x)\xi_1||^2,$$

we can extend U to an isometry of $\overline{\pi(\mathfrak{A})\xi_1}$ to $\overline{\pi(\mathfrak{A})\xi_2}$. However, both ξ_1, ξ_2 are cyclic vectors. Thus U is an isometry $\mathcal{H}$ to $\mathcal{H}'$. Since $U\pi(x)\pi(y)\xi_1 = \pi'(xy)\xi_2 = \pi'(x)U\pi(y)\xi_1$, we have $U\pi(x) = \pi'(x)U$ and so both $\{\mathcal{H}, \pi, \xi_1\}$ and $\{\mathcal{H}', \pi', \xi_2\}$ are isometrical equivalent. □

Classically, we can show that every operator in $\mathfrak{B}(\mathcal{H})$ has its adjoint operator, and we can also prove that a surjective isometric operator U on a Hilbert space is a unitary operator. Hence unitary equivalence and isometric equivalence coincide (on page 48 of [10]). Constructively, it is shown in [14, 17] that the statement

Every operator on a Hilbert space has an adjoint

implies LPO.

Therefore we need to distinguish isometric equivalence and unitary equivalence. It is known that an operator on a separable Hilbert space has its adjoint if and only if it is weakly compact [14].

Let $\{\mathcal{H}, \pi\}$ be a representation of a Banach $*$-algebra $\mathfrak{A}$. A closed subspace $\mathcal{L}$ of $\mathcal{H}$ is an *invariant subspace* if

$$\pi(\mathfrak{A})\mathcal{L} = \{\pi(x)\xi : x \in \mathfrak{A}, \xi \in \mathcal{L}\} \subseteq \mathcal{L}.$$

It is easy to see that if $\mathcal{L}$ is an invariant subspace of $\mathcal{H}$ then

$$\mathcal{L}^{\perp} = \{\xi \in \mathcal{H} : \langle \xi, \eta\rangle = 0, \forall \eta \in \mathcal{L}\}$$

is also an invariant subspace of $\mathcal{H}$.

Let $\mathfrak{R}$ be a subset of $\mathfrak{B}(\mathcal{H})$. Then the commutant $\mathfrak{R}'$ of $\mathfrak{R}$ is a set of elements which commutes for any element of $\mathfrak{R}$. i.e.,

$$\mathfrak{R}' = \{x \in \mathfrak{B}(\mathcal{H}) : xy - yx = 0, \forall y \in \mathfrak{R}\}.$$

The following statement is a well-known fact and also holds constructively.

Proposition 18.25 *Let $\{\mathcal{H}, \pi\}$ be a representation of a Banach $*$-algebra $\mathfrak{A}$ and $\mathcal{L}$ a closed subspace of $\mathcal{H}$. Let P be a projection $\mathcal{H}$ to $\mathcal{L}$. Then $\mathcal{L}$ is an invariant subspace if and only if P is in $\pi(\mathfrak{A})'$.*

Now, we define an irreducible representation.

Definition 18.26 *A representation $\{\mathcal{H}, \pi\}$ of a Banach$*$-algebra $\mathfrak{A}$ is called irreducible if its invariant subspace is trivial, i.e. only $\{0\}$ and $\mathcal{H}$.*

Using Proposition 18.25 it is easy to see that the irreducibility is equivalent to the following condition.

If P in $\pi(\mathfrak{A})'$ is a projection operator, then $P = \mathbf{0}$ or $= \mathbf{1}$.

The following is a standard criterion for irreducibility in classical mathematics (see page 43 of [12].)

Theorem 18.27 *Let $\{\mathcal{H}, \pi\}$ be a representation of a Banach$*$-algebra $\mathfrak{A}$. Then the following conditions are equivalent.*

(1) *$\{\mathcal{H}, \pi\}$ is irreducible.*
(2) *$\pi(\mathfrak{A})'$ is isomorphic to $\mathbb{C}\mathbf{e}$, where $\mathbb{C}$ is a complex field.*

To prove this equivalence constructively, we need to use the constructive spectral resolution theorem. Bishop has proven the spectral resolution of operators of a Banach space [18]. In this paper, he also proves that a bounded self-adjoint operator on a Hilbert space satisfies the spectral resolution theorem (on page 431 of [18]). For more precise information, see page 394 of [2].

18.6 The GNS Construction Theorem

In this section, we prove the Gelfand–Naĭmark–Segal (GNS) construction theorem in Bishop's constructive mathematics (classical reference, for example, on page 84-88 of [13]). First, we define $=_\rho$ on $\mathfrak{A}$ by

$$x =_\rho y \equiv \rho((x-y)^*(x-y)) = 0.$$

It is easy to see that $=_\rho$ is an equivalence relation on $\mathfrak{A}$. We let $\mathfrak{A}_\rho$ denote the structure $\mathfrak{A}$ with new equality $=_\rho$ and also denote $\langle x, y\rangle_\rho \equiv \rho(y^*x)$.

Lemma 18.28 *$\langle , \rangle_\rho$ is an inner product on $\mathfrak{A}_\rho$.*

Proof. First we show that if $x =_\rho x'$, $y =_\rho y'$ then $\langle x, y\rangle_\rho = \langle x', y'\rangle_\rho$. Since $\rho(y^*x) - \rho(y^*x') = \rho(y^*x - y^*x')$ and $|\rho(y^*x - y^*x')|^2 \leq \rho(y^*y)\rho((x-x')^*(x-x')) = 0$, we have $\langle x, y\rangle_\rho = \langle x', y\rangle_\rho$. Similarly, we have $\langle x, y\rangle_\rho = \langle x, y'\rangle_\rho$. Hence $\langle x, y\rangle_\rho = \langle x', y'\rangle_\rho$. Next we show that $\langle , \rangle_\rho$ satisfies the conditions of an inner product. Indeed, $\langle \alpha x, y\rangle_\rho = \rho(\alpha y^* x) = \alpha\rho(y^* x) = \alpha\langle x, y\rangle_\rho$, $\langle x+y, z\rangle_\rho = \rho(z^*(x+y)) = \rho(z^* x + z^* y) = \rho(z^* x) + \rho(z^* y) = \langle x, z\rangle_\rho + \langle y, z\rangle_\rho$, $\langle x, y\rangle_\rho = \rho(y^* x) = \overline{\rho(x^* y)} = \overline{\langle y, x\rangle_\rho}$ and ρ is a state, hence $\rho(x^* x) \geq 0$. If $\rho(x^* x) = 0$ then $x =_\rho 0$ by the definition of $=_\rho$. Thus $\langle , \rangle$ satisfies all of the conditions of an inner product. □

Theorem 18.29 (The GNS construction theorem) If ρ is a state of a C*-algebra $\mathfrak{A}$, there is a cyclic representation $\{\mathcal{H}_\rho, \pi_\rho, \xi_\rho\}$ such that $\rho(x) = \langle \xi_\rho, \pi_\rho(x)\xi_\rho\rangle$ for all x in $\mathfrak{A}$.

Proof. Consider the structure $\mathfrak{A}_\rho$. By Lemma 18.28 $\langle , \rangle_\rho$ is an inner product on $\mathfrak{A}_\rho$. Thus $\mathfrak{A}_\rho$ is a pre-Hilbert space. Let $\mathcal{H}_\rho$ be the Hilbert space which is the completion of $\mathfrak{A}_\rho$. Put $\pi_\rho(x)y =_\rho xy$. Then it is easy to see that for any x in $\mathfrak{A}$, $\pi_\rho(x)$ is a linear operator on $\mathfrak{A}_\rho$. Since $||x||^2\mathbf{e} - x^* x$ is positive, we can write $||x||^2\mathbf{e} - x^* x = z^* z$ for some $z \in \mathfrak{A}$ by Theorem 18.3 and hence $y^*(||x||^2\mathbf{e} - x^* x)y = (zy)^*(zy)$ is also positive. Since ρ is positive, we have $||x|| ||y||_\rho - ||\pi_\rho(x)y||_\rho = \rho(y^*(||x||^2\mathbf{e} - x^* x)y) \geq 0$. Then $||\pi_\rho(x)y||_\rho \leq ||x|| ||y||_\rho$. Hence, $\pi_\rho(x)$ is a bounded linear operator on $\mathfrak{A}_\rho$. Therefore, it can be uniquely extended to a bounded linear operator on $\mathcal{H}_\rho$, also denoted by $\pi_\rho(x)$.

For any $x, y, z \in \mathfrak{A}$, $\pi_\rho(\lambda x + \mu y)z = \lambda xz + \mu yz = \lambda\pi_\rho(x)z + \mu\pi_\rho(y)z$, $\pi_\rho(xy)z = xyz = \pi_\rho(x)\pi_\rho(y)z$, $\langle y, \pi_\rho(x^*)z\rangle = \langle y, x^* z\rangle = \rho(y^* x^* z) = \langle xy, z\rangle = \langle \pi_\rho(x)y, z\rangle = \langle y, \pi_\rho(x)^* z\rangle$, and $\overline{\mathfrak{A}_\rho} = \mathcal{H}_\rho$. It follows that $\pi_\rho(\lambda x + \mu y) = \lambda\pi_\rho(x) + \mu\pi_\rho(y)$, $\pi_\rho(xy) = \pi_\rho(x)\pi_\rho(y)$, $\pi_\rho(x^*) = \pi_\rho(x)^*$. Therefore, π_ρ is a representation of $\mathfrak{A}$ on $\mathcal{H}_\rho$.

Put $\xi_\rho = \mathbf{e}$. Then, $\pi_\rho(x)\xi_\rho = x$. Thus, $\overline{\pi_\rho(\mathfrak{A})\xi_\rho} = \mathcal{H}_\rho$. Therefore, ξ_ρ is a cyclic vector for π_ρ. $\rho(x) = \langle \mathbf{e}, x\rangle = \langle \xi_\rho, \pi_\rho(x)\xi_\rho\rangle$. This completes the theorem. □

Moreover, we can show that the cyclic representation in the main theorem is uniquely determined with respect to its isometrical equivalents. The classical proof (see page 279 of [9]) is constructive.

In our main theorem, there is no guarantee that $\pi_\rho(\mathfrak{A})$ is a C*-subalgebra of $\mathfrak{B}(\mathcal{H}_\rho)$. This is because there is no guarantee that for any x in $\mathfrak{A}$, the norm $||\pi_\rho(x)||$ exists and $\pi_\rho(\mathfrak{A})$ is closed.

Using the same argument discussed in Section 18.4, we can show the following.

Theorem 18.30 *Let $\mathfrak{A}$ be a uniformly convex C*-algebra with the Gâteaux differentiable norm. Then $\pi_\rho(\mathfrak{A})$ is a C*-subalgebras of $\mathfrak{B}(\mathcal{H}_\rho)$ for some state ρ.*

To show this, we prepare two lemmas. For the classical results the reader is referred to pages 9 and 10 of [11].

Lemma 18.31 *Let $\mathfrak{A}$ be a C*-algebra and μ a bounded linear functional on $\mathfrak{A}$ such that $\mu(x) = ||\mu|| ||x||$ for some positive x in $\mathfrak{A}$. Then μ is a positive linear functional.*

Proof. Put $\mu_0 = \mu/||\mu||$, $x_0 = x/||x||$. Then $\mu_0(x_0) = 1$. It is enough to show that $||\mu|| = \mu(\mathbf{e})$ by Proposition 18.14. First we show that the value of $\mu_0(\mathbf{e})$ is in $\mathbb{R}$. Suppose $\mu_0(\mathbf{e}) = a + \imath b$, $(a, b \in \mathbb{R}, |b| > 0)$. Then for any $\lambda \in \mathbb{R}$, $|\mu_0(\mathbf{e} + \imath\lambda x_0)| = |a + \imath(\lambda + b)| \geq |\lambda + b|$. On the other hand $|\mu_0(\mathbf{e} + \imath\lambda x_0)| \leq ||\mathbf{e} + \imath\lambda x_0)|| \leq (1+\lambda^2)^{1/2}$. Hence

$|\lambda + b| \leq (1 + \lambda^2)^{1/2}$. This is a contradiction. Therefore $\mu_0(\mathbf{e})$ is real. Next suppose $\mu_0(\mathbf{e}) < 1$, then $|\mu_0(\mathbf{e} - 2x_0)| = |\mu_0(\mathbf{e}) - 2| \geq 1$ and $|\mu_0(\mathbf{e} - 2x_0)| \leq ||\mathbf{e} - 2x_0|| \leq 1$. Thus $\mu_0(\mathbf{e}) = 1$. Therefore $||\mu|| = \mu(\mathbf{e})$. This completes the lemma. □

Lemma 18.32 *Let $\mathfrak{A}$ be a uniformly convex C*-algebra with the Gâteaux differentiable norm, and x a self-adjoint element in $\mathfrak{A}$. Then $||x|| = \sup_{\rho \in \mathcal{S}(\mathfrak{A})} |\rho(x)|$.*

Proof. By Lemma 18.10 for any self-adjoint element x can be represented in the form $x = x_+ - x_-$, where x_+, x_- are positive elements such that $x_+x_- = x_-x_+ = 0$, $||x|| = \max\{||x_+||, ||x_-||\}$. We need to consider two possibilities, $||x|| = ||x_+||$ and $||x|| = ||x_-||$. Since the proof goes in the same way, we consider only $||x|| = ||x_+||$. By the Hahn–Banach theorem, there exists a bounded linear functional μ on $\mathfrak{A}$ such that $\mu(x_+) = ||x_+||$ and $||\mu|| = 1$. By Proposition 18.14, μ is a state. Now consider $\mu(x_+ + x_-)$. Since $||\langle x\rangle||^2 = ||x^*x|| = ||x||^2 = ||(x_+ - x_-)^2|| = ||(x_+ + x_-)^2|| = ||x_+ + x_-||^2$, thus $||x|| = ||x_+|| = ||x_+ + x_-||$. We have $\mu(x_+) + \mu(x_-) = \mu(x_+ + x_-) \leq ||x_+ + x_-|| = ||x|| = ||x_+||$. Hence $\mu(x_+) + \mu(x_-) = ||x_+|| + \mu(x_-) \leq ||x_1||$. Therefore $\mu(x_-) = 0$. We conclude that $\mu(x) = \mu(x_+) = ||x_+|| = ||x||$. This completes the lemma. □

Proof of Theorem 18.30

Since $\sup_{\rho \in \mathcal{S}(\mathfrak{A})} ||\pi_\rho(x)||^2 \geq \sup_{\rho \in \mathcal{S}(\mathfrak{A})} ||\pi_\rho(x)\mathbf{e}||^2 = \sup_{\rho \in \mathcal{S}(\mathfrak{A})} \rho(x^*x)$, then, by Lemma 18.32 we have $||\pi_\rho(x)|| \geq ||x||$ for some $\rho \in \mathcal{S}(\mathfrak{A})$. On the other hand $||\pi_\rho(x)|| \leq ||x||$. Hence $||\pi_\rho(x)|| = ||x||$ for some $\rho \in \mathcal{S}(\mathfrak{A})$. This means that the representation π_ρ is norm preserving. Thus, for any $x \in \mathfrak{A}$ the norm $||\pi_\rho(x)||$ always exists. Hence, $\pi_\rho(\mathfrak{A})$ is a C*-subalgebra of $\mathfrak{B}(\mathcal{H}_\rho)$.

Finally, we introduce a *universal representation*. Suppose $\{\mathcal{H}_\alpha\}$ is a family of Hilbert spaces. The direct sum $\mathcal{H} = \bigoplus_{\alpha \in A} \mathcal{H}_\alpha$ consists of all $x = \{x_\alpha\}(x_\alpha \in \mathcal{H}_\alpha)$ such that $||x||^2 = \sum_{\alpha \in A} ||x_\alpha||^2 < \infty$. Then $\mathcal{H} = \bigoplus_{\alpha \in A} \mathcal{H}_\alpha$ is a Hilbert space with the inner product $\langle x, y\rangle = \sum_{\alpha \in A} \langle x_\alpha, y_\alpha\rangle$ for all x, y in $\mathcal{H}$. □

Definition 18.33 (Universal representation) Let $\mathcal{S}(\mathfrak{A})$ consist of all states of a C*-algebra $\mathfrak{A}$. Then $\{\bigoplus_{\rho \in \mathcal{S}(\mathfrak{A})} \mathcal{H}_\rho, \bigoplus_{\rho \in \mathcal{S}(\mathfrak{A})} \pi_\rho\}$ is called the *universal representation* of $\mathfrak{A}$.

Theorem 18.34 (Universal representation theorem) Let $\mathfrak{A}$ be a uniformly convex C*-algebra with the *Gâteaux* differentiable norm. Then the universal representation $\{\bigoplus_{\rho \in \mathcal{S}(\mathfrak{A})} \mathcal{H}_\rho, \bigoplus_{\rho \in \mathcal{S}(\mathfrak{A})} \pi_\rho\}$ of $\mathfrak{A}$ is a norm preserving isomorphism.

Therefore, every uniformly convex C*-algebra with the Gâteaux differentiable norm is $*$-isomorphic to a concrete C*-algebra. The proof of Theorem 18.34 is very similar to that of Theorem 18.30 and we will omit it.

Acknowledgements

The author would like to express his sincere gratitude to Dr. Hajime Ishihara for his helpful comments. The author also thanks the anonymous referees for helpful suggestions.

References

1. Bishop, E. (1967). *Foundations of Constructive Analysis*. McGraw-Hill.
2. Bishop, E. and Bridges, D. (1985). *Constructive Analysis*. Springer-Verlag.
3. Bridges, D. (1979). *Constructive Functional Analysis*. Pitman.
4. Bridges, D. (2000). Constructive methods in Banach algebra theory. *Mathematica Japonica*, **52**, 145–161.
5. Havea, R. (2001). Constructive spectral and numerical range. PhD thesis, University of Canterbury,
6. Spitters, Bas. (2003). Constructive and intuitionistic integration theory and functional analysis. PhD thesis, University of Nijmengen.
7. Vîţă, L. (2000). The Constructive theory of operator algebras. PhD thesis, University of Canterbury.
8. Bratteli, O. and Robinson, D. W. (1979). *Operator Algebras and Quantum Statistical Mechanics*. **1**, Springer-Verlag.
9. Kadison, R. V. and Ringrose, J. R. (1983). *Fundamentals of the Theory of Operator Algebras*. vol 1,2, Academic Press.
10. Pedersen, G. K. (1979). *C*-algebras and their Automorphism Groups*. Academic Press.
11. Sakai, S. (1971). *C*-algebras and W*-algebras*. Springer-Verlag.
12. Takesaki, M. (1979). *Theory of Operator Algebras I*. Springer-Verlag.
13. Zhu, K. (1993). *An Introduction to Operator Algebras*. CRC press.
14. Ishihara, H. (1991). Constructive compact operators on a Hilbert space. *Annals of Pure and Applied Logic*, **52**, 31–37.
15. Bridges, D. and Ishihara, H. (1996). Spectral of self-adjoint operators in constructive analysis. *Indagationes Mathematicae*, **7**, 11–35.
16. Ishihara, H. (1989). On the constructive Hahn-Banach theorem. *Bulletin of the London Mathematical Society*, **21**, 79–81.
17. Bridges, D. Richman, F. and Schuster, P. (2000). Adjoint, absolute values and polar decomposition. *Journal of Operator Theory*, **44**, 243–254.
18. Bishop, E. (1957). Spectral theory for operators on a Banach space. *Transactions of the American Mathematical Society*, **86**, 414–445.

19

APPROXIMATIONS TO THE NUMERICAL RANGE OF AN ELEMENT OF A BANACH ALGEBRA

DOUGLAS BRIDGES AND ROBIN HAVEA

Abstract

In constructive mathematics, since the Hahn–Banach theorem does not produce norm-preserving extensions of linear functionals, the numerical range of an element of a Banach algebra has to be described using approximations. Nevertheless, these approximations suffice to produce constructive proofs of theorems such as that of Allan Sinclair on the spectral range of a Hermitian element.

19.1 Introduction

The theory of Banach algebras presents a fine interplay between the topological and the algebraic, and as such is a good challenge for the constructive mathematician.[1] Bishop [2] (Chapter 11) gave the first constructive development of commutative Banach algebra theory, one superseded in [3] (Chapter 9). However, the even more challenging theory of general, non-commutative Banach algebras has only recently been tackled constructively; see [6–9]. In the present chapter we continue our investigation of this lively combination of constructive algebra and topology by examining the spectral radius of a Hermitian element in a Banach algebra.

Throughout this chapter, B will be a complex Banach algebra with identity e and dual space B'. Working constructively and basing our development on the corresponding classical one on pages 51–57 of [10], we use approximations to the state space of B to define *Hermitian* and to prove a number of results that culminate in **Sinclair's theorem**:

Theorem 19.1 *If a is a Hermitian element of a complex unital Banach algebra B, then $\|a^n\|^{1/n} = \|a\|$ for each positive integer n. In particular, the spectral radius*

$$r(a) = \inf\left\{ \|a^n\|^{1/n} : n \geq 1 \right\}$$

of a equals its norm.

---

[1] By *constructively* we mean *with intuitionistic logic and intuitionistic ZF set theory*. Since we use no principles that are incompatible with classical ZF set theory, all our results are classically valid; but they also hold, *mutatis mutandis*, in intuitionistic analysis, recursive analysis, and, to the best of our knowledge, any model for computable analysis (such as Weihrauch's TTE theory [1]). Background information on constructive mathematics can be found in [2–5]. Other constructive aspects of Banach algebra theory are discussed in [6–9].

When we try to develop Banach algebra theory constructively, we immediately run across a major problem: we have no guarantee that a bounded linear functional u on a normed linear space X is **normable**, in the sense that its norm

$$\|u\| = \sup\{|u(x)| : x \in X\}$$

exists. Indeed, u is normable if and only if its kernel K is **located**—that is, the distance

$$\rho(x, K) = \inf\{\|x - y\| : y \in K\}$$

exists for each $x \in X$ ([3], page 303, (1.10)). However, every linear functional on a finite-dimensional normed space X is not only bounded, but normable.

Another major problem occurs with the Hahn–Banach extension theorem, which we state in the general constructive version proved in [3] (page 342, (4.6)).

Theorem 19.2 *Let Y be a linear subset of a separable normed linear space X, and v a non-zero linear functional on Y whose kernel is located in X. Then for each $\varepsilon > 0$ there exists a normable linear functional u on X such that $u(y) = v(y)$ for each $y \in Y$, and $\|u\| < \|v\| + \varepsilon$.*

A recursive example (see [11]) shows that we cannot remove ε from the statement of Theorem 19.2. As a consequence, we may not be able to construct an element of the **state space**

$$V = \{f \in B' : f(e) = 1 = \|f\|\}.$$

of our Banach algebra, let alone prove V weak* compact.

For this reason we introduce, for each $t > 0$, the following approximation to V :

$$V^t = \{f \in B' : \|f\| \leq 1,\ |1 - f(e)| \leq t\}.$$

Applying the Hahn–Banach theorem with $X = B, Y = \mathbf{C}e$, and $v(\lambda e) = \lambda$, we see that V^t is non-empty. In fact, since the mapping $f \rightsquigarrow |1 - f(e)|$ is weak* uniformly continuous on the unit ball B'_1 of B' ([3], page 351, (6.3)), V^t is weak* compact for all but countably many $t > 0$ ([3], page 97, (4.9)). We say that $t > 0$ is **admissible** if V^t is weak* compact. Thus

$$V = \bigcap\{V^t : t > 0 \text{ is admissible}\},$$

which is the intersection of a family of non-empty[2] weak* compact sets that is descending, in the sense that if $0 < t' < t$, then $V^{t'} \subset V^t$. Being the intersection of a family of complete sets, V is weak* complete.

---

[2]When we say that a set S is **non-empty**, we mean that there exists an element of S. This property is constructively stronger than $\neg(S = \emptyset)$.

Since we cannot be certain that V is inhabited, we may be unable to construct an element of the **numerical range**

$$V(x) = \{f(x) : f \in V\}$$

of an element x of B; so it makes sense to consider the approximations to $V(x)$ given by

$$V^t(x) = \left\{f(x) : f \in V^t\right\}$$

where $t > 0$. For each admissible $t > 0$ and each $x \in B$, the weak* uniform continuity of the mapping $f \rightsquigarrow f(x)$ on the weak* compact set V^t ensures that $V^t(x)$ is totally bounded.

19.2 Approximating the Numerical Range

Our aim in this section is to use the sets $V^t(x)$ to present constructive counterparts of a number of classical results on the numerical range. These results will eventually provide the armoury needed for a constructive proof of Sinclair's theorem.

Proposition 19.3 *The following are equivalent conditions on a complex number λ and an element a of B.*

(i) There exists an admissible $t > 0$ such that $\rho\left(\lambda, V^t(a)\right) > 0$.

(ii) There exists $z \in \mathbf{C}$ such that $|\lambda - z| > \|a - ze\|$.

Proof. Assuming that $\rho\left(\lambda, V^t(a)\right) > 0$ for some admissible $t > 0$, first consider the case where $\rho(a, \mathbf{C}e) > 0$. Define a linear functional f_0 on the two-dimensional space $E = \operatorname{span}\{a, e\}$ by

$$f_0(\alpha a + \beta e) = \alpha\lambda + \beta \qquad (\alpha, \beta \in \mathbf{C}).$$

Since E is finite-dimensional, f_0 is normable, so its kernel is located in E; it follows that the kernel is finite-dimensional and hence located in B. Choosing $\delta > 0$ such that

$$\frac{\delta}{1+\delta}(1 + |\lambda|) < \rho\left(\lambda, V^t(a)\right),$$

we show that $\|f_0\| \geq 1 + \delta/2$. To this end, suppose that $\|f_0\| < 1 + \delta/2$. By the Hahn–Banach theorem, there exists a normable linear extension f of f_0 to B such that

$$\begin{aligned} f(e) &= f_0(e) = 1, \\ f(a) &= f_0(a) = \lambda, \end{aligned}$$

and $\|f\| \leq 1 + \delta$. Define $g = \|f\|^{-1} f$; then $\|g\| = 1$ and

$$|1 - g(e)| = \left|1 - \|f\|^{-1} f(e)\right| = \left|1 - \|f\|^{-1}\right| \leq \frac{\delta}{1+\delta} < t,$$

so $g \in V^t$. Moreover,

$$\begin{aligned} |\lambda - g(a)| &= \left|\lambda - \|f\|^{-1} f(a)\right| \\ &= \left|\lambda - \|f\|^{-1} \lambda\right| \\ &= \left|1 - \|f\|^{-1}\right| |\lambda| \leq \frac{\delta}{1+\delta} |\lambda| < \rho\left(\lambda, V^t(a)\right), \end{aligned}$$

a contradiction. Hence, in fact,

$$\|f_0\| \geq 1 + \frac{\delta}{2} > 1,$$

so we can pick $\alpha, \beta \in \mathbf{C}$ such that $\beta \neq 0$ and $|f_0(\alpha e + \beta a)| > \|\alpha e + \beta a\|$. Then

$$\begin{aligned} \left\|a - \left(-\alpha\beta^{-1}\right) e\right\| &= |\beta|^{-1} \|\alpha e + \beta a\| \\ &\leq |\beta|^{-1} |f_0(\alpha e + \beta a)| \\ &= |\beta|^{-1} |\alpha + \beta\lambda| = \left|\lambda - \left(-\alpha\beta^{-1}\right)\right|, \end{aligned}$$

so (ii) holds with $z = -\alpha\beta^{-1}$.

We now remove the restriction that $\rho(a, \mathbf{C}e) > 0$. Let

$$0 < \varepsilon < \min\left\{1, \rho\left(\lambda, V^t(a)\right)\right\}$$

and choose an admissible t' with

$$0 < t' < \min\left\{t, \frac{\varepsilon}{3(1 + \|a\|)}\right\}.$$

Either $\rho(a, \mathbf{C}e) > 0$ and the preceding case applies, or else $\rho(a, \mathbf{C}e) < \varepsilon/3$. In the latter case, choose $z \in \mathbf{C}$ such that $\|a - ze\| < \varepsilon/3$. Then for all $g \in V^{t'}$ we have

$$\begin{aligned} \varepsilon &< |\lambda - g(a)| \\ &\leq |\lambda - z| + |z - g(ze)| + |g(a - ze)| \\ &\leq |\lambda - z| + |z| \, |1 - g(e)| + \|a - ze\| \\ &< |\lambda - z| + (1 + \|a\|) \, |1 - g(e)| + \frac{\varepsilon}{3} \\ &< |\lambda - z| + (1 + \|a\|) \, t' + \frac{\varepsilon}{3} \\ &< |\lambda - z| + \frac{2\varepsilon}{3}. \end{aligned}$$

Hence

$$|\lambda - z| > \frac{\varepsilon}{3} > \|a - ze\|,$$

and again (ii) holds.

Conversely, assume that there exists $z \in \mathbf{C}$ such that $|\lambda - z| > \|a - ze\|$, and let

$$\gamma = \frac{1}{2}\left(|\lambda - z| - \|a - ze\|\right) > 0.$$

Choose an admissible $t > 0$ such that $|z|t < \gamma/2$. Then for all $f \in V^t$ we have

$$\begin{aligned} |f(a) - z| &< |f(a) - zf(e)| + |z - zf(e)| \\ &\leq |f(a - ze)| + |z|\,|1 - f(e)| \\ &\leq \|a - ze\| + |z|t < |\lambda - z| - \gamma + \frac{\gamma}{2} < |\lambda - z| - \frac{\gamma}{2}. \end{aligned}$$

Hence

$$|\lambda - f(a)| \geq |\lambda - z| - |f(a) - z| > |\lambda - z| + \frac{\gamma}{2} - |\lambda - z| > \frac{\gamma}{2}.$$

It follows that $\rho\left(\lambda, V^t(a)\right) \geq \gamma/2$. □

Proposition 19.4 *For each $t > 0$,*

$$V^t(a) \subset \bigcap_{z \in \mathbf{C}} \overline{B}\left(z, \|a - ze\| + t|z|\right).$$

Proof. If $z \in \mathbf{C}$, $t > 0$, and $f \in V^t$, then

$$|f(a) - z| \leq |f(a - ze)| + |z|\,|1 - f(e)| \leq \|a - ze\| + t|z|. \qquad \square$$

Corollary 19.5 $V(a) = \bigcap_{z \in \mathbf{C}} \overline{B}\left(z, \|a - ze\|\right)$.

Proof. Let $\lambda \in V(a)$. By Proposition 19.4, for each $z \in \mathbf{C}$ and each admissible $t > 0$,

$$|\lambda - z| \leq \|a - ze\| + t|z|.$$

Letting $t \to 0$, we obtain $|\lambda - z| \leq \|a - ze\|$. Hence

$$V(a) \subset \bigcap_{z \in \mathbf{C}} \overline{B}\left(z, \|a - ze\|\right).$$

Conversely, if $\lambda \in \bigcap_{z \in \mathbf{C}} \overline{B}\left(z, \|a - ze\|\right)$, then by Proposition 19.3, for each admissible $t > 0$ we have $\lambda \in \overline{V^t(a)} = V^t(a)$; whence $\lambda \in V(a)$. □

We define the **resolvent** and the **spectrum** of the element a of B to be the sets

$$R(a) = \left\{\lambda \in \mathbf{C} : (a - \lambda e)^{-1} \text{ exists}\right\}$$

and

$$\sigma(a) = \left\{\lambda \in \mathbf{C} : \lambda' \in R(a) \quad (\lambda \neq \lambda')\right\},$$

respectively.[3] Relatively minor modifications of the proof of Proposition 6 on page 53 of [10] show that $\sigma(a) \subset V(a)$.

For each unit vector x in B and each $t > 0$ we define

$$V^{t,x} = \left\{f \in B' : \|f\| \leq 1, \quad |1 - f(x)| \leq t\right\}.$$

For each $a \in B$, we then write

$$V^{t,x}(a) = \left\{f(ax) : f \in B', \|f\| \leq 1, \quad |1 - f(x)| \leq t\right\}.$$

If $t > 0$ is admissible, then $V^{t,x}(a)$, being the range of the uniformly continuous mapping $f \rightsquigarrow f(ax)$ on the weak* totally bounded set V^t, is totally bounded in $\mathbf{C}$.

Proposition 19.6 *Let $0 < t < 1/\sqrt{2}$ be such that $2t$ is admissible. Let x be a unit vector in B, $a \in B$, and $\lambda \in V^{t,x}(a)$. Then there exists $\lambda' \in V^{2t}(a)$ such that $|\lambda - \lambda'| \leq 3t\|a\|$.*

Proof. Fixing $\lambda \in V^{t,x}(a)$, choose $g \in V^{t,x}$ such that $\lambda = g(ax)$. Suppose, to begin with, that $\rho(a, \mathbf{C}e) > 0$. Then a and e span a two-dimensional subspace B_0 of B. Define a (necessarily) normable linear functional f on B_0 by

$$f_0(y) = g(yx) \quad (y \in B_0).$$

If $y \in B_0$ and $\|y\| \leq 1$, then

$$|f_0(y)| = |g(yx)| \leq \|yx\| \leq \|y\|\|x\| = \|y\| \leq 1.$$

Since also

$$|1 - f_0(e)| = |1 - g(x)| \leq t,$$

it follows that $1 - t \leq \|f_0\| \leq 1$. By the Hahn–Banach theorem, there exists a normable linear functional f extending f_0 to B such that $1 - t \leq \|f\| \leq 1 + t$. Now define

[3] When we write $\lambda \neq \lambda'$, we mean that $|\lambda - \lambda'| > 0$. This is a stronger property than $\neg(\lambda = \lambda')$, unless we adopt Markov's principle; see Chapter 1 of [4].

$\phi \in B'$ by

$$\phi = \|f\|^{-1} f.$$

Then $\|\phi\| = 1$, $\phi(e) = \|f\|^{-1} g(x)$, and $\phi(a) = \|f\|^{-1} \lambda$. Moreover,

$$\begin{aligned} |1 - \phi(e)| &= \left|1 - \|f\|^{-1} g(x)\right| \\ &\leq \left|1 - \|f\|^{-1}\right| + \left|(1 - g(x)) \|f\|^{-1}\right| \\ &\leq \left|1 - \frac{1}{1+t}\right| + |1 - g(x)| \frac{1}{1-t} \\ &\leq \frac{t}{1+t} + \frac{t}{1-t} = \frac{2t}{1-t^2} < 2t, \end{aligned}$$

as $0 < t < 1/\sqrt{2}$. Hence $\phi(a) \in V^{2t}(a)$. Also,

$$|\lambda - \phi(a)| = |\|f\|\phi(a) - \phi(a)| \leq |\|f\| - 1| \, |\phi(a)| \leq t\|a\|,$$

so the proof in the case $\rho\,(a, \mathbf{C}e) > 0$ is complete.

Now consider the general case. Since $2t$ is admissible, $V^{2t}(a)$ is totally bounded and hence located in B. Either $\rho\left(\lambda, V^{2t}(a)\right) \leq 2t\|a\|$ or else, as we may assume, $\rho\left(\lambda, V^{2t}(a)\right) > t\,\|a\|$. Then, by the first part of the proof, $\rho\,(a, \mathbf{C}e) = 0$, so $a = \alpha e$ and $|\alpha| = \|a\|$ for some $\alpha \in \mathbf{C}$. For any $f \in V^{2t}$ we have

$$\begin{aligned} |\lambda - f(a)| &= |\lambda - \alpha f(e)| \\ &\leq |\lambda - \alpha| + |\alpha| \, |1 - f(e)| \\ &\leq |g(\alpha x) - \alpha| + 2t\|a\| \\ &\leq |\alpha| \, |g(x) - 1| + 2t\,\|a\| \leq 3t\|a\|. \end{aligned}$$

□

19.3 Sinclair's Theorem

In this section we make use of our approximations to the numerical range to sketch the arguments leading to Sinclair's theorem. This requires a number of technical lemmas whose proofs are adaptations of classical counterparts in [10].

Lemma 19.7 *For each admissible $t > 0$ and each unit vector $x \in B$,*

$$\inf\left\{\operatorname{Re}\lambda : \lambda \in V^t(a)\right\} \leq \|ax\|.$$

Proof. First observe that the infimum in question exists, since $V^t(a)$ is totally bounded. Choose an admissible ε such that

$$0 < \varepsilon < \min\left\{\frac{1}{\sqrt{2}}, \frac{t}{2}\right\}.$$

By Proposition 19.6, for each $g \in V^{\varepsilon,x}(a)$ there exists $\lambda' \in V^t(a)$ such that $\left|g(ax) - \lambda'\right| \leq 3\varepsilon \|a\|$. Thus

$$\begin{aligned}\inf\left\{\operatorname{Re}\lambda : \lambda \in V^t(a)\right\} &\leq \operatorname{Re}\lambda' \\ &\leq \operatorname{Re} g(ax) + \left|g(ax) - \lambda'\right| \\ &\leq |g(ax)| + 3\varepsilon\|a\| \\ &\leq \|ax\| + 3\varepsilon\|a\|.\end{aligned}$$

Since ε is arbitrary, the required result follows. □

For convenience we define

$$\mu_t = \sup\{\operatorname{Re}\lambda : \lambda \in V^t(a)\}$$

whenever t is admissible.

Lemma 19.8 *If $t > 0$ is admissible, α is positive number such that $\alpha\mu_t < 1$, and x is a unit vector in B, then*

$$1 - \alpha\mu_t \leq \|(e - \alpha a)x\|.$$

Proof. Given $\lambda \in V^t(e - \alpha a)$, choose $f \in V^t$ such that

$$\lambda = f(e - \alpha a) = f(e) - \alpha f(a).$$

Then

$$\operatorname{Re}\left[\lambda - (1 - \alpha f(a))\right] = \operatorname{Re}\left[(\lambda + \alpha f(a)) - 1\right] = \operatorname{Re}\left(f(e) - 1\right) \geq -t,$$

since $|1 - f(e)| \leq t$. Hence

$$\operatorname{Re}\lambda \geq \operatorname{Re}(1 - \alpha f(a)) - t \geq \left(1 - \alpha\mu_t\right) - t.$$

It follows from Lemma 19.7 that

$$1 - \alpha\mu_t - t \leq \inf\left\{\operatorname{Re}\lambda : \lambda \in V^t(e - \alpha a)\right\} \leq \|(e - \alpha a)x\|.$$

For each admissible $\varepsilon \in (0, t)$, since $\mu_\varepsilon \leq \mu_t$, we now have

$$1 - \alpha\mu_t \leq 1 - \alpha\mu_\varepsilon \leq \|(e - \alpha a)x\| + \varepsilon.$$

Since ε is arbitrary, the desired conclusion follows. □

Lemma 19.9 *If $t > 0$ is admissible, then*

$$\frac{1}{\alpha}\log\|\exp(\alpha a)\| \leq \mu_t$$

for each $\alpha > 0$.

Proof. First consider $\alpha > 0$ such that $\alpha\mu_t < 1$. Applying Lemma 19.8, for each $x \in B$ we have

$$\left(1 - \alpha\mu_t\right)\|x\| \leq \|(e - \alpha a)x\|.$$

Therefore, by induction,

$$\left(1 - \alpha\mu_t\right)^n \|x\| \leq \left\|(e - \alpha a)^n x\right\|$$

for each positive integer n. For any $\alpha > 0$ and for all sufficiently large n, we have $\frac{\alpha}{n}\mu_t < 1$; whence

$$\left(1 - \tfrac{\alpha}{n}\mu_t\right)^n \|x\| \leq \left\|\left(e - \tfrac{\alpha}{n}a\right)^n x\right\|.$$

Taking the limit as $n \to \infty$, we obtain

$$\exp\left(-\alpha\mu_t\right)\|x\| \leq \|\exp(-\alpha a)\, x\|.$$

In particular, the choice $x = \|\exp(\alpha a)\|$ yields

$$\|\exp(\alpha a)\| \leq \exp\left(\alpha\mu_t\right),$$

from which the desired inequality follows. □

Proposition 19.10 *If a is Hermitian, then $\|\exp(\pm \mathrm{i}\alpha a)\| = 1$ for all $\alpha \in \mathbf{R}$.*

Proof. For the moment, take $\alpha > 0$. Given $\varepsilon > 0$, choose an admissible $t > 0$ such that $|\mathrm{Im}\, f(a)| < \varepsilon$ for each $f \in V^t$. Then as

$$\left\{\mathrm{Re}\,\lambda : \lambda \in V^t(\mathrm{i}a)\right\} = \left\{\mathrm{Re}\, f(\mathrm{i}a) : f \in V^t\right\} = \left\{-\mathrm{Im}\, f(a) : f \in V^t\right\},$$

we have

$$\sup\left\{\mathrm{Re}\,\lambda : \lambda \in V^t(\mathrm{i}a)\right\} \leq \sup\left\{|\mathrm{Im}\, f(a)| : f \in V^t\right\} \leq \varepsilon.$$

Now replace a by $\mathrm{i}a$ in Lemma 19.9, to obtain

$$\tfrac{1}{\alpha}\log\|\exp(\mathrm{i}\alpha a)\| \leq \varepsilon.$$

Since $\varepsilon > 0$ is arbitrary,

$$\tfrac{1}{\alpha}\log\|\exp(\mathrm{i}\alpha a)\| \leq 0$$

and therefore $\|\exp(\mathrm{i}\alpha a)\| \leq 1$. Moreover, since $-a$ is also Hermitian, it follows that $\|\exp(-\mathrm{i}\alpha a)\| \leq 1$. Thus for each real $\alpha \neq 0$ we have $\|\exp(\mathrm{i}\alpha a)\| \leq 1$; this inequality holds for every $\alpha \in \mathbf{R}$, by the continuity of the exponential function on B. Since

$$1 = \|e\| \leq \|\exp(\mathrm{i}\alpha a)\|\,\|\exp(-\mathrm{i}\alpha a)\| \leq 1,$$

we conclude that $\|\exp(\pm\mathrm{i}\alpha a)\| = 1$. □

Now note that the holomorphic functional calculus can be developed constructively by suitably modifying the corresponding classical theory on pages 205–207 of [12, page 206, Theorem 3.3.5].

Lemma 19.11 *Let α be a positive number such that $\sigma(a) \subset \left(-\frac{\pi}{2}+\alpha, \frac{\pi}{2}-\alpha\right)$. Then $\sigma(\sin a) \subset B(0,1)$.*

Proof. The hypotheses allow us to choose $r \in (0,1)$ such that some neighbourhood of $\sin\sigma(a)$ is contained in the disk $B(0,r)$ in $\mathbf{C}$. Suppose that $\lambda \in \sigma(\sin a)$ and $|\lambda| > r$. Then $\rho(\lambda, \sin\sigma(a)) > 0$ and so the mapping

$$\zeta \rightsquigarrow (\sin\zeta - \lambda)^{-1}$$

is holomorphic on some open set D well containing $\sigma(a)$. By the holomorphic functional calculus, $(\sin a - \lambda e)^{-1}$ exists as a two-sided inverse to $\sin a - \lambda e$. Hence $\lambda \notin \sigma(\sin a)$, a contradiction. Therefore $|\lambda| \leq r$, and so $\lambda \in B(0,r)$. □

With Lemma 19.11 at hand, it is now fairly straightforward to show, as on page 56 of [10], that $\arcsin(\sin a) = a$ for all $a \in B$. In turn, we can easily continue the development on pages 56–57 of [10] to complete the proof of Sinclair's theorem. We omit the details.

Acknowledgements

The authors thank the Arthington–Davy Foundation for supporting Robin Havea's PhD research from 1998 to 2001.

References

1. Weihrauch, K. (1996). A foundation for computable analysis. In: *Combinatorics, Complexity, & Logic*, (Proceedings of Conference in Auckland, December 1996 (eds D. S. Bridges, C. S. Calude, J. Gibbons, S. Reeves, I. H. Witten), pp. 9–13. Springer–Verlag, Singapore.
2. Bishop, E. (1967). *Foundations of Constructive Analysis*. McGraw–Hill, New York.
3. Bishop, E. and Bridges, D. (1985). *Constructive Analysis*. Grundlehren der Mathematischen Wissenschaften **279**. Springer–Verlag, Berlin.
4. Bridges, D., and Richman, F. (1987). *Varieties of Constructive Mathematics*. London Math. Soc. Lecture Notes **97**. Cambridge University Press.
5. Troelstra, A. S., and van Dalen, D. (1988). *Constructivism in Mathematics: An Introduction* (two volumes), North Holland, Amsterdam.
6. Bridges, D. S. (2000). Constructive methods in Banach algebra theory. *Mathematica Japonica*, **52**(1), 145–161.
7. Bridges, D. S., and Havea, R. S. (2001). A constructive version of the spectral mapping theorem, *Math. Logic Quarterly*, **47**(3), 299–304.
8. Bridges, D. S., and Havea, R. S. (2001). Powers of a Hermitian element in a Banach algebra, preprint, University of Canterbury, Christchurch, New Zealand.
9. Havea, R. S. (2001). Constructive Spectral and Numerical Range Theory, PhD thesis, University of Canterbury, Christchurch, New Zealand.

10. Bonsall, F. F. and Duncan, J. (1973). *Complete Normed Algebras*, Ergebnisse der Mathematik und ihrer Grenzgebiete **80**. Springer–Verlag, Berlin.
11. Metakides, G., Nerode, A., and Shore, R. A. (1984). Recursive limits on the Hahn–Banach theorem, In: *Errett Bishop: Reflections on Him and His Research* (ed M. Rosenblatt), Contemporary Mathematics, *Amer. Math. Soc.*, **39**, 85–91, Providence, R.I.
12. Kadison, R. V. and Ringrose, J. R. (1997). *Fundamentals of the Theory of Operator Algebras: Elementary Theory*, Graduate Studies in Mathematics **15**. *Amer. Math. Soc.*, Providence, RI.
13. Rudin, W. (1970). *Real and Complex Analysis*. McGraw–Hill, New York.

20

THE CONSTRUCTIVE UNIQUENESS OF THE LOCALLY CONVEX TOPOLOGY ON $\mathbb{R}^N$

DOUGLAS BRIDGES AND LUMINITA VÎȚĂ

Abstract

It is proved constructively that every n-dimensional real locally convex space is homeomorphic to $\mathbb{R}^n$.

20.1 Introduction

Consider the following expression of the uniqueness of the topology on $\mathbb{R}^n$:

Every n-dimensional topological vector space X over $\mathbb{R}$ is homeomorphic to the Euclidean space $\mathbb{R}^n$.

The typical classical proofs of this result are non-algorithmic. One proof, using induction over the dimension of X, uses the non-algorithmic proposition that a linear mapping between normed spaces is bounded if its kernel is closed, a proposition that is equivalent to *Markov's principle (MP)*,

$$\forall \boldsymbol{\lambda} \in 2^{\mathbb{N}} \left(\neg \forall n \left(\lambda_n = 0\right) \Rightarrow \exists n \left(\lambda_n = 1\right)\right),$$

a form of unbounded search that we regard as non-algorithmic.[1] Another proof applies the non-algorithmic theorem that a real-valued function on a compact—that is, complete and totally bounded—space attains its infimum, a theorem equivalent to the following weak form of the law of excluded middle known as the *lesser limited principle of omniscience (LLPO)*,

$$\forall \boldsymbol{\lambda} \in 2^{\mathbb{N}} \left(\left(\forall m \forall n \left(m \neq n \Rightarrow \lambda_m \lambda_n = 0\right) \Rightarrow \forall n \left(\lambda_{2n} = 0\right) \vee \forall n \left(\lambda_{2n+1} = 0\right)\right)\right).$$

The strength of the classical principles used in these proofs suggests that a constructive/algorithmic proof of (some version of) the theorem will involve more than mere tampering with one or other of the classical ones. In this chapter we prove the theorem constructively when X carries a locally convex topology defined by a family of seminorms. Our proof embodies an algorithm for computing the homeomorphism between the n-dimensional locally convex space and $\mathbb{R}^n$, an algorithm which, in principle, can be extracted and implemented by any of the several systems designed for such a purpose [1–3]. The core of the algorithm is contained in the proof of Proposition 20.1.

[1] Another reason for not wanting to use Markov's principle in constructive mathematics is that it is independent of Heyting arithmetic: Peano arithmetic with intuitionistic logic.

We understand *constructive mathematics* to mean *mathematics carried out with intuitionistic logic*. Of course, such mathematics requires also some non-logical foundation. We adopt an informal set theory which, when developed with intuitionistic logic, does not enable us to derive as theorems the law of excluded middle in either its full form or any of the weaker versions, such as LLPO, whose informal interpretations are manifestly non-algorithmic. Formal counterparts of our informal set theory are Aczel's CZF and Myhill's CST [4, 5].

Our interest in this work was prompted by the second author's investigations [6], in which constructive results about the Minkowski norm are generalized from normed spaces to locally convex spaces whose topology is defined by a family of seminorms. The present paper shows that such generalizations are nugatory when the space is finite-dimensional. Thus, although essentially topological, our main result has non-trivial implications for constructive analysis.

We assume that the reader has access to [7], [8] or [9] for background material on constructive mathematics in general, and to [10, 11] for the fundamental definitions and results in the constructive theory of locally convex spaces.

20.2 Some Convex Geometry

A *locally convex space* is a pair comprising a linear space X over $\mathbb{R}$ and a family $(p_i)_{i \in I}$ of seminorms on X.[2] The corresponding equality and inequality on X are then defined by

$$\begin{aligned} x = y &\Leftrightarrow \forall i \in I \ (p_i(x - y) = 0), \\ x \neq y &\Leftrightarrow \exists i \in I \ (p_i(x - y) > 0). \end{aligned}$$

The *locally convex topology* on $\left(X, (p_i)_{i \in I}\right)$ is the family τ_X of all subsets of X that are unions of sets of the form

$$U(a, F, \varepsilon) = \left\{ x \in X : \sum_{i \in F} p_i(x - a) < \varepsilon \right\},$$

where $a \in X$, F is a finitely enumerable subset[3] of I, and $\varepsilon > 0$. The seminorms p_i $(i \in I)$ are called the *defining seminorms* of τ_X. The members of τ_X are called *open subsets* of X. On the other hand, if S is a subset of X, then its *closure* $\overline{S}$ is the set of all elements x of X such that $S \cap U(x, F, \varepsilon)$ is inhabited[4] for all finitely enumerable $F \subset I$ and all $\varepsilon > 0$. We say that S is *closed* (in the locally convex topology τ_X) if $S = \overline{S}$.

The *convex hull* of a finitely enumerable subset S of a locally convex space is the closure of the set of convex combinations of the elements of S, and is denoted by $\mathrm{co}(S)$.

---

[2] Following normal practice, we call X itself a locally convex space when it is clear which family of seminorms is under consideration.

[3] Recall that a set F is *finitely enumerable* if there exist a natural number n and a mapping f of $\{1, \ldots, n\}$ onto F. If $n = 0$, then F is the empty set; if $n \geqslant 1$, then we write $F = \{f_1, \ldots, f_n\}$.

[4] A set S is *inhabited*, if there exists a point in S; this condition is constructively stronger than the denial that S is empty.

Every subset S of the locally convex space X has

- a *complement*
$$\sim S := \{x \in X : \forall y \in S\};$$
- and a *metric complement,* consisting of all $x \in X$ that are *bounded away from* S in the sense that there exist a finitely enumerable subset F of I and a positive number ε such that $\sum_{i \in F} p_i(x-y) \geqslant \varepsilon$ for all $y \in S$.

Clearly, $-S \subset \sim S$. Our aim in this section is to prove the following result in convex geometry.

Proposition 20.1 *Let X be an n-dimensional locally convex space with topology defined by the family $(p_i)_{i\in I}$ of seminorms, let $m \leqslant n+1$, and let S be the convex hull of the points $x_1, \ldots, x_m$ of X. Then $\sim S = -S$.*

The proof requires a definition and some preliminary lemmas.

Let J be a finitely enumerable subset of I, and let $\varepsilon > 0$. By an *ε-approximation to S relative to J* we mean a subset T of S such that for each $x \in S$ there exists $y \in T$ with $\sum_{j\in J} p_j(x-y) < \varepsilon$. We say that S is *totally bounded* if for each $\varepsilon > 0$ and each finitely enumerable $J \subset I$, there exists a finitely enumerable ε-approximation to S relative to J.

Lemma 20.2 *Let X be a finite-dimensional locally convex space with topology defined by the family $(p_i)_{i\in I}$ of seminorms, and let S be the set of convex combinations of points in an inhabited finitely enumerable subset of X. Then S is totally bounded in X.*

Proof. The set of convex combinations of a single point x_1 is the image of the totally bounded interval $[0,1]$ under the uniformly continuous mapping $\lambda \rightsquigarrow \lambda x_1$, and so is totally bounded. Suppose that for some $m > 1$ the set of convex combinations of any set of $m-1$ points of X is totally bounded, and consider the set S of convex combinations of m points $x_1, \ldots, x_m$. Fix a finitely enumerable subset J of I, and define

$$K = 1 + \max_{1\leqslant k\leqslant m} \sum_{j\in J} p_j(x_k).$$

Given ε in $(0,1)$, let F be a finite ε-approximation to the interval $[0,1]$ that includes 1, and G a finite ε-approximation, relative to J, to the set of convex linear combinations of $x_1, \ldots, x_{m-1}$. Define a finitely enumerable subset of S by

$$A = \{\lambda x_m + (1-\lambda)\, y : \lambda \in F, y \in G\}.$$

Consider any point $x = \sum_{k=1}^{m} \lambda_k x_k$ with $0 \leqslant \lambda_k \leqslant 1$ $(0 \leqslant k \leqslant m)$ and $\sum_{k=1}^{m} \lambda_k = 1$. Either $\lambda_m > 1-\varepsilon$ or else $\lambda_m < 1$. In the first case we have

$$\begin{aligned}\sum_{j\in J} p_j(x-x_m) &\leqslant (1-\lambda_m)\sum_{j\in J} p_j(x_m) + \sum_{k=1}^{m-1}\lambda_k \sum_{j\in J} p_j(x_k)\\ &< \varepsilon K + (1-\lambda_m)\,K\\ &< 2K\varepsilon.\end{aligned}$$

In the case $\lambda_m < 1$,

$$x = \lambda_m x_m + (1 - \lambda_m) \sum_{k=1}^{m-1} \left(\frac{\lambda_k}{1 - \lambda_m} \right) x_k,$$

where $\sum_{k=1}^{m-1} \left(\frac{\lambda_k}{1-\lambda_m} \right) x_k$ is a convex linear combination of $x_1, \ldots, x_{m-1}$. Choosing $\lambda \in F$ and $y \in G$ such that $|\lambda_m - \lambda| < \varepsilon$ and

$$\sum_{j \in J} p_j \left(\sum_{k=1}^{m-1} \left(\frac{\lambda_k}{1 - \lambda_m} \right) x_k - y \right) < \varepsilon,$$

we have

$$\begin{aligned}
\sum_{j \in J} p_j(y) &\leqslant \sum_{j \in J} p_j \left(\sum_{k=1}^{m-1} \left(\frac{\lambda_k}{1 - \lambda_m} \right) x_k - y \right) + \sum_{j \in J} p_j \left(\sum_{k=1}^{m-1} \left(\frac{\lambda_k}{1 - \lambda_m} \right) x_k \right) \\
&\leqslant \varepsilon + \sum_{k=1}^{m-1} \left(\frac{\lambda_k}{1 - \lambda_m} \right) \sum_{j \in J} p_j(x_k) \\
&\leqslant \varepsilon + \sum_{k=1}^{m-1} \left(\frac{\lambda_k}{1 - \lambda_m} \right) K \\
&= \varepsilon + K.
\end{aligned}$$

Hence

$$\begin{aligned}
&\sum_{j \in J} p_j \left(x - (\lambda x_m + (1 - \lambda)\, y) \right) \\
&\leqslant |\lambda_m - \lambda| \sum_{j \in J} p_j(x_m) \\
&\quad + (1 - \lambda_m) \sum_{j \in J} p_j \left(\sum_{k=1}^{m-1} \left(\frac{\lambda_k}{1 - \lambda_m} \right) x_j - y \right) + |\lambda_m - \lambda| \sum_{j \in J} p_j(y) \\
&\leqslant \varepsilon K + \varepsilon + \varepsilon\,(\varepsilon + K) \\
&\leqslant 2\,(K + 1)\,\varepsilon.
\end{aligned}$$

We now see that A is a finitely enumerable $2\,(K+1)\,\varepsilon$-approximation to S relative to J. This completes the induction. □

Lemma 20.3 *Let S_0 be an inhabited simplex in $\mathbb{R}^n$, and let $(S_n)_{n \geqslant 1}$ be a sequence of simplices in $\mathbb{R}^n$ such that $S_0 \supset S_1 \supset S_2 \supset \cdots$ and each S_n is similar to S_0. Let $0 < c < 1$, and let $(\lambda_n)_{n \geqslant 1}$ be an increasing binary sequence such that*

- *if* $\lambda_n = 0$, *then* $\operatorname{diam}(S_n) = c \operatorname{diam}(S_{n-1})$, *and*
- *if* $\lambda_n = 1$, *then* $S_n = S_{n-1}$.

Then $\bigcap_{n \geqslant 1} S_n$ *is inhabited.*

Proof. For each k let b_k be the barycentre of S_k, and consider positive integers m, n with $m > n$. If $\lambda_n = 0$, then the Euclidean distance between b_m and b_n is at most $\operatorname{diam}(S_n)$, which equals $c^n \operatorname{diam}(S)$; if $\lambda_n = 1$, then $b_m = b_n$. Thus the points b_n form a Cauchy sequence, which converges to a limit $b \in \mathbb{R}^n$. Since $b_m \in S_n$ whenever $m \geqslant n$, we see that b belongs to each of the (closed) simplices S_n. □

Let $\{e_1, \ldots, e_n\}$ be a basis for a finite-dimensional locally convex space X, and $f\left(\sum \lambda_j e_j\right) = (\lambda_1, \ldots, \lambda_n)$. Then f is a linear isomorphism of X onto $\mathbb{R}^n$. Moreover, denoting the Euclidean norm on $\mathbb{R}^n$ by $\|\cdot\|_2$, we see that for each seminorm p on X,

$$p\left(\sum_{j=1}^{n} \lambda_j e_j\right) \leqslant \sum_{j=1}^{n} |\lambda_j| \, p(e_j) \leqslant c \, \|(\lambda_1, \ldots, \lambda_n)\|_2 ,$$

where

$$c = n^{1/2} \max \{p(e_1), \ldots, p(e_n)\} .$$

Hence the map $p \circ f^{-1}$ is uniformly continuous on $\mathbb{R}^n$. It follows that if we use the mapping f to identify X algebraically with $\mathbb{R}^n$, then each seminorm on X is uniformly continuous with respect to the Euclidean norm.

We now have the **proof of Proposition 20.1.** Under the hypotheses of that proposition, we first show that if $0 \in \sim S$, then $0 \in -S$. To do this, we must produce a finitely enumerable subset F of I such that $\sum_{i \in F} p_i$ is bounded away from 0 on S. For convenience, we identify X algebraically with $\mathbb{R}^n$. If $m = 1$, then there is nothing to prove. So let us assume that, for some m with $1 < m \leqslant n + 1$, the desired conclusion holds for the convex hull of $m - 1$ points of X. Consider the case of m points $x_1, \ldots, x_m$. For each k let H_k be the convex hull of $\{x_j : j \neq k\}$. By our induction hypothesis, there exist $\alpha > 0$ and a finitely enumerable subset F_k of I such that $\sum_{i \in F_k} p_i(x) > 2\alpha$ for all $x \in H_k$. It follows from the continuity and convexity of the seminorms that if x_k is sufficiently close to H_k, then $\sum_{i \in F_k} p_i(x) > \alpha$ for all x in S, which is the convex hull of $H_k \cup \{x_k\}$. Thus we may assume that S is a *non-degenerate* simplex: one in which for each k the vertex x_k is bounded away from the opposite face H_k.

Again using our induction hypothesis, we can find a finite sum q of seminorms on X such that

$$0 < \gamma := \inf \left\{ q\left(\sum_{j=1}^{m-1} \lambda_j x_j\right) : \forall j \, (0 \leqslant \lambda_j \leqslant 1) \wedge \sum_{j=1}^{m-1} \lambda_j = 1 \right.$$
$$\left. \wedge \left(\exists j \, (\lambda_j = 0) \vee \exists j \left(\lambda_j = \frac{1}{n+1} \right) \right) \right\} .$$

This infimum exists, since the set over which we take the infimum is the image, under the uniformly continuous mapping q, of a totally bounded set [7] (page 94, Corollary 4.3). This set is shown in Figure 20.1, which, like all the figures below, deals with the case $m = 3$ and $n = 2$. Similarly,

$$\delta := \inf \left\{ q\left(\sum_{j=1}^{m} \lambda_j x_j\right) : \forall j \left(0 \leqslant \lambda_j \leqslant 1\right) \wedge \sum_{j=1}^{m} \lambda_j = 1 \right\}$$

exists. Either $\delta > 0$ and there is nothing to prove, or else $\delta < \gamma$. In the latter case, choose $\xi \in S$ such that $q(\xi) < \gamma$. Let S' be the simplex with vertices $\frac{1}{4}\xi + \frac{3}{4}x_k$ $(1 \leqslant k \leqslant m)$. Then S' lies inside S and is bounded away from the boundary of S; see Figure 20.2. For $0 \leqslant t \leqslant 1/4$ and $x \in H_k$ we have

$$q(t\xi + (1-t)x) \geqslant (1-t)q(x) - tq(\xi) \geqslant \frac{3}{4}\gamma - \frac{1}{4}\gamma = \frac{1}{2}\gamma.$$

It follows from this and the continuity of q that $q \geqslant \gamma/2$ on the closure of the region between S and S'.

Now write $\xi = \sum_{k=1}^{m} \alpha_k x_k$ with $0 \leqslant \alpha_k \leqslant 1$ and $\sum_{k=1}^{m} \alpha_k = 1$. Since $q(\xi) < \gamma$, we see from the continuity of q, that $\alpha_k > 0$ and $\alpha_k \neq 1/(n+1)$ for each k $(1 \leqslant k \leqslant m)$. It follows that there exists j such that $\alpha_j > 1/(n+1)$. Without loss of generality we take $\alpha_1 > 1/(n+1)$. Let A_1 be that portion of S that lies between H_1 and the intersection H_1' of S with the affine subspace of $\mathbb{R}^n$ generated by the face of S' opposite the vertex $\frac{1}{4}\xi + \frac{3}{4}x_1$; and let the line joining x_1 to ξ meet H_1' and H_1 at η, ζ respectively (Figure 20.3). Then $\xi = \alpha_1 x_1 + (1-\alpha_1)\zeta$. Given t with $0 \leqslant t \leqslant \frac{1}{4(n+1)}$,

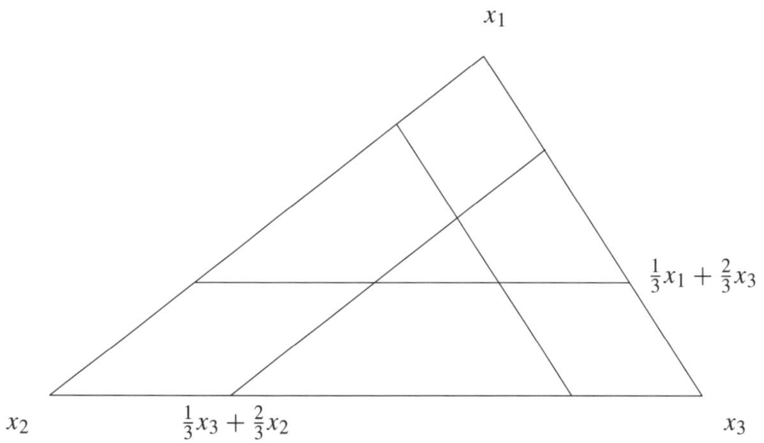

FIG. 20.1. The closure of the union of the lines is the set over which the infimum of q is taken, to give γ

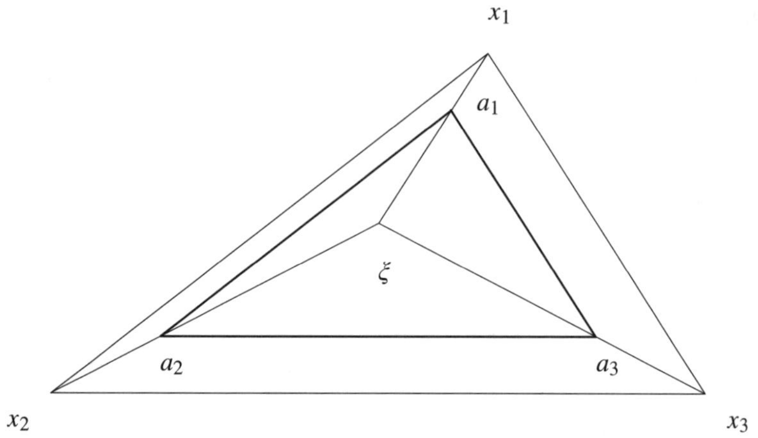

FIG. 20.2. $a_k = \frac{1}{4}\xi + \frac{3}{4}x_k$

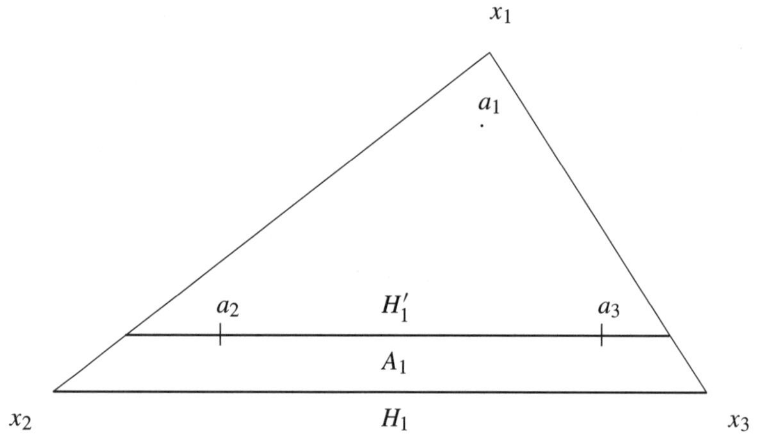

FIG. 20.3. A_1 is the area between H_1 and H_1'

we have

$$\begin{aligned} tx_1 + (1-t)\zeta &= t\left(\frac{1}{\alpha_1}\xi - \frac{1-\alpha_1}{\alpha_1}\zeta\right) + (1-t)\zeta \\ &= \frac{t}{\alpha_1}\xi + \left(1 - t - t\frac{1-\alpha_1}{\alpha_1}\right)\zeta \\ &= \frac{t}{\alpha_1}\xi + \left(1 - \frac{t}{\alpha_1}\right)\zeta, \end{aligned}$$

where

$$0 \leqslant \frac{t}{\alpha_1} < \frac{1}{4}.$$

Thus $tx_1+(1-t)\,\zeta$ belongs to A_1. Elementary Euclidean geometry of similar simplices (see Figure 20.4) now shows that

$$B_1 = \left\{ tx_1 + (1-t)\,y : y \in H_1,\ 0 \leqslant t \leqslant \frac{1}{4\,(n+1)} \right\}$$

is a subset of A_1; whence $q \geqslant \gamma/2$ on B_1. Using the continuity of q with respect to the Euclidean norm on $\mathbb{R}^n$, we now see that S is the union of two overlapping sets: the simplex S_1 obtained by removing the set B_1 from S and taking the closure of what remains, and a slight enlargement C_1 of B_1 on which $q > \gamma/4$. Note that S_1 is similar to S, and that

$$\operatorname{diam}(S_1) = \left(1 - \frac{1}{4\,(n+1)}\right) \operatorname{diam}(S) = \frac{4n+3}{4n+4} \operatorname{diam}(S).$$

Taking $S_0 = S, q_0 = q$, and $\delta_0 = \gamma/4$, and using the foregoing argument inductively, we can now construct

- an increasing binary sequence $(\lambda_k)_{k \geqslant 1}$,
- a descending sequence $(S_k)_{k \geqslant 0}$ of simplices similar to $S_0 = S$,
- a sequence $(C_k)_{k \geqslant 1}$ of subsets of S,
- a sequence $(q_k)_{k \geqslant 0}$ of finite sums of seminorms on $\mathbb{R}^n$, and
- a sequence $(\delta_k)_{k \geqslant 0}$ of positive numbers

such that for each $k \geqslant 1$,

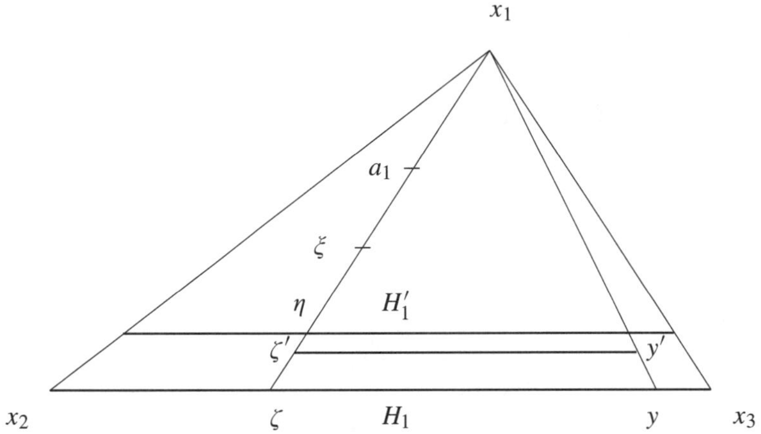

FIG. 20.4. $\zeta' = tx_1 + (1-t)\zeta \quad y' = tx_1 + (1-t)y$

▷ if $\lambda_k = 0$, then

$$\operatorname{diam}(S_k) = \frac{4n+3}{4n+4}\operatorname{diam}(S_{k-1}),$$

$S_{k-1} = S_k \cup C_k$, and $q_k(x) > \delta_k$ for all $x \in C_k$;

▷ if $\lambda_k = 1$, then $S_k = S_{k-1}$, $C_k = C_{k-1}$, $q_k = q_{k-1}$, $\delta_k = \delta_{k-1}$, and q_k is bounded away from 0 on S_k.

Note that if $\lambda_k = 0$ and some seminorm p is bounded away from 0 on S_k, then $p+q_k$ is bounded away from 0 on S_{k-1} $(= S_k \cup C_k)$; so $p+q_k+q_{k-1}$ is bounded away from 0 on S_{k-2} $(= S_{k-1} \cup C_{k-1})$. Arguing inductively, we then see that $p + q_k + \sum_{j=1}^{k-1} q_j$ is bounded away from 0 on S.

Using Lemma 20.3, construct a point a in $\bigcap_{k\geqslant 0} S_k$. Choose a seminorm p such that $p(a) > 0$. Since p is uniformly continuous with respect to the Euclidean metric on $\mathbb{R}^n$, there exists $r > 0$ such that $p(x) > \frac{1}{2}p(a)$ for all x with $\|x-a\| < r$. Choose N such that

$$\left(\frac{4n+3}{4n+4}\right)^N \operatorname{diam}(S) < r.$$

If $\lambda_N = 1$, then, by the remarks at the end of the preceding paragraph, $\sum_{j=1}^{N} q_j$ is bounded away from 0 on S. On the other hand, if $\lambda_N = 0$, then, by our choice of N, $S_N \subset B(a,r)$ and therefore p is bounded away from 0 on S_N; whence $p + q_N + \sum_{j=1}^{N-1} q_j$ is bounded away from 0 on S. This completes the proof that if $0 \in {\sim} S$, then $0 \in -S$.

Now let $x \in {\sim} S$, $0 \leqslant t_j \leqslant 1$ $(1 \leqslant j \leqslant m)$, and $\sum_{j=1}^{m} t_j = 1$. Then

$$\sum_{j=1}^{m} t_j \left(x - x_j\right) = x - \sum_{j=1}^{m} t_j x_j \neq 0,$$

so 0 is in the complement of the convex hull of $\{x-x_1, \ldots, x-x_m\}$. Hence there exist a finitely enumerable set $F \subset I$ and $\varepsilon > 0$ such that

$$\varepsilon \leqslant \sum_{i\in F} p_i \left(\sum_{j=1}^{m} t_j \left(x - x_j\right)\right) = \sum_{i\in F} p_i \left(x - \sum_{j=1}^{m} t_j x_j\right)$$

whenever $0 \leqslant t_j \leqslant 1$ $(1 \leqslant j \leqslant m)$ and $\sum_{j=1}^{m} t_j = 1$. Thus $x \in -S$. This completes the proof of Proposition 20.1. □

20.3 The Locally Convex Topology is Unique

The last lemma before our main result is a special case of a more general result proved in [12] (and given a neater proof in [13]): if a line segment starts in the interior of a located convex subset C of $\mathbb{R}^n$ and ends in the exterior of C, then it crosses the boundary of C (at a unique point). We give our lemma a self-contained, elementary proof.

Lemma 20.4 *Let $\{e_1, \ldots, e_n\}$ be the canonical orthonormal basis of the Euclidean space $\mathbb{R}^n$, and Σ the simplex whose vertices are $\pm e_j$ $(1 \leqslant j \leqslant n)$. Let ξ belong to the metric complement $-\Sigma$ of Σ in $\mathbb{R}^n$. Then the segment $[0, \xi]$ contains points arbitrarily close to the union of the faces of Σ.*

Proof. Let π_i be the hyperplane of $\mathbb{R}^n$ that is orthogonal to the line joining the points $\pm e_i$. Since the located set

$$S := -\bigcup_{i=1}^{n} \pi_i$$

is dense in $\mathbb{R}^n$, we may assume that $\xi \in S$. Elementary Euclidean geometry now enables us to prove that the segment $[0, \xi]$ actually intersects one (and only one) of the faces of Σ; see Figure 20.5. □

At last we come to the result that is the goal of the paper.

Theorem 20.5 *Every n-dimensional locally convex space over $\mathbb{R}$ is homeomorphic to $\mathbb{R}^n$.*

Proof. We may assume that $X = \mathbb{R}^n$. In view of the remarks following the proof of Lemma 20.3, it is enough to prove that the identity linear mapping from $(\mathbb{R}^n, (p_i)_{i \in I})$ to the Euclidean space $\mathbb{R}^n$ is continuous. Let $\{e_1, \ldots, e_n\}$ be the canonical orthonormal basis of $\mathbb{R}^n$, and Σ the simplex whose vertices are $\pm e_j$ $(1 \leqslant j \leqslant n)$. Each face of Σ is an $(n-1)$-dimensional simplex whose vertices form a subset of n points of $\{\pm e_1, \ldots, \pm e_n\}$ that, for any j, does not contain both e_j and $-e_j$; see Figure 20.5. Thus every element of such a face has the form $\sum_{j=1}^{n} \lambda_j e_j$, where $\sum |\lambda_j| = 1$; it follows from the linear independence of the vectors e_j that such a point is non-zero. We now see from Proposition 20.1 that there exist a positive number r and a finite sum q of the seminorms defining the locally convex structure on X, such that $q(x) \geqslant 2r$ for each x in the union of the faces of Σ, and hence for each x in the closure K of that union. Setting

$$V := \{x \in X : q(x) < r\},$$

we obtain a neighbourhood V of 0 in X that is contained in the metric complement of K. Fix a point $\xi \in V$, and suppose that ξ belongs to the metric complement $-\Sigma$ of Σ in the Euclidean space $\mathbb{R}^n$. Then, by Lemma 20.4, the segment $[0, \xi]$ gets arbitrarily close to the union of the faces of Σ. In particular, $[0, \xi]$ contains a point x such that $q(x - y) < r$ for some y in some face of Σ. Since V is convex, $[0, \xi] \subset V$; so $x \in V$,

$$q(y) \leqslant q(x) + q(x - y) < 2r$$

and therefore y is in the metric complement of K in the locally convex space $(\mathbb{R}^n, (p_i)_{i \in I})$. This is absurd, since $y \in K$; so $\xi \notin -\Sigma$ and therefore, as Σ is located in the Euclidean space $\mathbb{R}^n$,

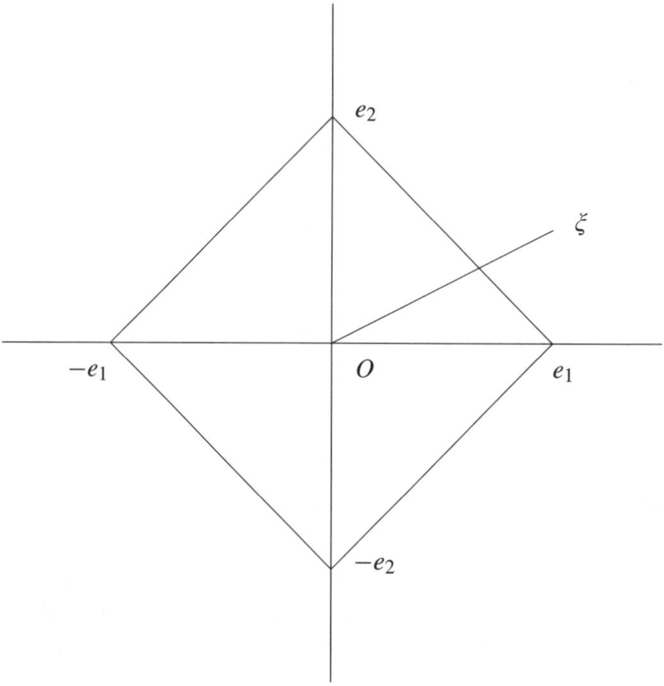

FIG. 5

$$\xi \in \overline{\Sigma} \subset \mathrm{co}\{\pm 2e_1, \ldots, \pm 2e_n\} \subset B(0,2),$$

where $B(0,2)$ denotes the Euclidean ball with centre 0 and radius 2 in $\mathbb{R}^n$. It follows that $V \subset B(0,2)$, and hence that the identity map from $(\mathbb{R}^n, (p_i)_{i \in I})$ to the Euclidean space $\mathbb{R}^n$ is uniformly continuous. □

Acknowledgements

This work was carried out while Luminiţa Vîţă held a New Zealand Science & Technology Postdoctoral Research Fellowship (contract number UOCX0215) at the University of Canterbury. It was completed while she was a visitor, and Douglas Bridges was a DAAD Gastprofessor, at the Mathematisches Institut der Universität München.

References

1. Cruz-Filipe, L. and Spitters, B. (2003). 'Program extraction from large proof developments', In: *Proceedings of TPHOLs 2003* pp. 205–220. Lecture Notes in Computer Science **2758**. Springer-Verlag.

2. Constable, R. L. *et al.*, (1986). *Implementing Mathematics with the Nuprl Proof Development System*. Prentice–Hall, Englewood Cliffs, New Jersey.
3. Hayashi, S. and Nakano, H. (1988). *PX: A Computational Logic*. MIT Press, Cambridge MA.
4. Aczel, P. and Rathjen, M. (20001). *Notes on Constructive Set Theory*, Report No. 40, Institut Mittag–Leffler, Royal Swedish Academy of Sciences.
5. Myhill, J. (1975). 'Constructive set theory', *J. Symbolic Logic*, **40**(3), 347–382.
6. Vîţă, L. (2002). 'Minkowski functionals and located sets', *Bulletin Mathématique de la Société des Sciences Mathématiques de Roumanie*, **45**(93), No. 1–2, 89–95.
7. Bishop, E. and Bridges, D. (1985). *Constructive Analysis*, Grundlagen der Math. Wiss. **279**. Springer–Verlag, Heidelberg.
8. Bridges, D. and Richman, F. (1987). *Varieties of Constructive Mathematics*. London Math. Soc. Lecture Notes **97**, Cambridge Univ. Press.
9. Troelstra A. S. and van Dalen, D. (1988). *Constructivism in Mathematics: An Introduction* (two volumes), North Holland, Amsterdam.
10. Vîţă, L. (2000). The constructive theory of operator algebras. PhD thesis, University of Canterbury.
11. Bridges, D. and Luminiţa Dediu (Vîţă), (1999). 'Constructive notes on uniform and locally convex spaces'. In: Proceedings of 12th International Symposium, FCT'99, Iasi, Romania, pp. 195–203. Lecture Notes in Computer Science **1684**. Springer.
12. Bridges, D. and Popa, G. (2003). 'Exact, continuous boundary crossings out of convex sets in $\mathbb{R}^N$', *Quart. J. Math.*, **54**, 391–398.
13. Bridges, D. and Vîţă, L. *Techniques of Constructive Analysis*. book in preparation.
14. Bishop, E. (1967). *Foundations of Constructive Analysis*. McGraw–Hill, New York.

21

COMPUTABILITY ON NON-SEPARABLE BANACH SPACES AND LANDAU'S THEOREM

VASCO BRATTKA

Abstract

While there is a well-established concept of a computable normed space in the separable case, one can prove that there is no way to represent non-separable normed spaces on Turing machines such that both operations, vector space subtraction and the norm, become computable. However, in certain cases one can at least keep one of these computability properties. We discuss two generalized concepts of computability for normed spaces which allow either the norm or the subtraction to become computable. Using the first concept we can prove a computable version of Landau's theorem for sequence spaces which suggests that priority should be given to the computability of the norm, i.e. to the first concept.

21.1 Introduction

A computable normed space as studied in computable analysis is roughly speaking a normed space with computable norm and limit and computable linear operations (a precise definition follows below). It turns out that any such space is necessarily separable, whereas in classical functional analysis non-separable spaces such as the sequence space ℓ_∞ or the space $\mathcal{B}(X, Y)$ of bounded linear operator $T : X \to Y$ are frequently considered. Hence, the question arises whether the notion of a computable normed space can be generalized to the non-separable case in some meaningful way. We will prove that there are at least two such generalized concepts, a stronger one and a weaker one. The first concept focuses on computability of the norm while the second one preserves computability of the linear operations. One can measure the quality of such a concept by regarding the number of classical results which can be proved computationally by applying the corresponding concept. One reason for the importance of non-separable normed spaces in classical functional analysis is that some of them naturally occur as dual spaces of frequently used separable spaces. One such result is known as Landau's theorem.

Theorem 21.1 *[Landau] Let $p, q > 1$ be real numbers such that $\frac{1}{p} + \frac{1}{q} = 1$ or $p = 1$ and $q = \infty$. Then the dual space ℓ'_p is isometrically isomorphic to ℓ_q.*

Here ℓ_p denotes the sequence space endowed with the well-known norm and ℓ'_p its dual space (precise definitions follow). It can be proved by the Hahn–Banach theorem (and thus by the axiom of choice) that the corresponding result does not hold true in case of $p = \infty$ and $q = 1$. Since there is a canonical choice for the computability structure on ℓ_q for $1 \leq q < \infty$, the computational version of Landau's theorem will force us to choose a specific structure for the dual spaces ℓ'_p. If we accept the same choice in case $p = 1$, then there is only one such choice left for the dual space ℓ_∞. In other words, if we want to have a uniform definition of computability on the sequence spaces ℓ_p and a computable version of the duality result expressed by Landau's theorem, then there is one and only one option for the computability structure on the non-separable space ℓ_∞.

We close the introduction with a short survey of the organization of this chapter. In Section 21.2 we will present some preliminaries from computable analysis. In Section 21.3 we discuss computable metric spaces and ordinary computable normed spaces. In Section 21.4 we introduce general computable normed spaces and we prove a stability theorem and some other results which show that general computable normed spaces are natural generalizations of ordinary computable normed spaces. In Section 21.5 we study some closure properties of general computable normed spaces and in Section 21.6 we prove a computable version of Landau's theorem.

21.2 Preliminaries from Computable Analysis

In this section we briefly summarize some notions from computable analysis. For details the reader is referred to [1]. The basic idea of the representation based approach to computable analysis is to represent infinite objects like real numbers, functions or sets, by infinite strings over some alphabet Σ (which should at least contain the symbols 0 and 1). Thus, a *representation* of a set X is a surjective mapping $\delta :\subseteq \Sigma^\omega \to X$ and in this situation we will call (X, δ) a *represented space*. Here Σ^ω denotes the set of infinite sequences over Σ and the inclusion symbol is used to indicate that the mapping might be partial. If we have two represented spaces, then we can define the notion of a computable function.

Definition 21.2 (Computable function) *Let (X, δ) and (Y, δ') be represented spaces. A function $f :\subseteq X \to Y$ is called (δ, δ')-*computable, *if there exists some computable function $F :\subseteq \Sigma^\omega \to \Sigma^\omega$ such that $\delta' F(p) = f\delta(p)$ for all $p \in \mathrm{dom}(f\delta)$.*

Of course, we have to define computability of functions $F :\subseteq \Sigma^\omega \to \Sigma^\omega$ to make this definition complete, but this can be done via Turing machines: F is computable if there exists some Turing machine, which computes an infinitely long time and transforms each sequence p, written on the input tape, into the corresponding sequence $F(p)$, written on the one-way output tape. If the represented spaces are fixed or clear from the context, then we will simply call a function f *computable*.

For the comparison of representations it will be useful to have the notion of *reducibility* of representations. If δ, δ' are both representations of a set X, then δ is called *reducible* to δ', $\delta \leq \delta'$ in symbols, if there exists a computable function $F :\subseteq \Sigma^\omega \to \Sigma^\omega$ such that $\delta(p) = \delta' F(p)$ for all $p \in \mathrm{dom}(\delta)$. Obviously, $\delta \leq \delta'$ holds if and only if the identity $\mathrm{id} : X \to X$ is (δ, δ')-computable. Moreover, δ and δ' are called *equivalent*, $\delta \equiv \delta'$ in symbols, if $\delta \leq \delta'$ and $\delta' \leq \delta$.

Analogously to the notion of computability we can define the notion of (δ, δ')-*continuity* by substituting a continuous function $F :\subseteq \Sigma^\omega \to \Sigma^\omega$ for the computable function F in the definition above. On Σ^ω we use the *Cantor topology*, which is simply the product topology of the discrete topology on Σ. The corresponding reducibility will be called *continuous reducibility* and we will use the symbols $\leq_t$ and $\equiv_t$ in this case. Again we will simply say that the corresponding function is *continuous*, if the representations are fixed or clear from the context. If not mentioned otherwise, we will always assume that a represented space is endowed with the final topology induced by its representation.

This will lead to no confusion with the ordinary topological notion of continuity, as long as we are dealing with *admissible* representations. A representation δ of a topological space X is called *admissible* if δ is maximal among all continuous representations δ' of X, i.e. if $\delta' \leq_t \delta$ holds for all continuous representations δ' of X. If δ, δ' are admissible representations of topological spaces X, Y, then a function $f :\subseteq X \to Y$ is (δ, δ')-continuous if and only if it is sequentially continuous, cf. [2].

Given a represented space (X, δ), we will use the notions of a *computable sequence* and a *computable point*. A *computable sequence* is a computable function $f : \mathbb{N} \to X$, where we assume that $\mathbb{N} = \{0, 1, 2, ...\}$ is represented by $\delta_{\mathbb{N}}(1^n 0^\omega) := n$, and a point $x \in X$ is called *computable* if there is a constant computable sequence with value x.

Given two represented spaces (X, δ) and (Y, δ'), there is a canonical representation $[\delta, \delta']$ of $X \times Y$ and a representation $[\delta \to \delta']$ of the (δ, δ')-continuous functions $f : X \to Y$. If δ, δ' are *admissible* representations of sequential topological spaces, then $[\delta \to \delta']$ is actually a representation of the set $\mathcal{C}(X, Y)$ of continuous functions $f : X \to Y$. If $Y = \mathbb{R}$, then we write for short $\mathcal{C}(X) := \mathcal{C}(X, \mathbb{R})$. The function space representation can be characterized by the fact that it admits evaluation and type conversion.

Proposition 21.3 (Evaluation and type conversion) Let (X, δ) and (Y, δ') be admissibly represented sequential topological spaces and let (Z, δ'') be a represented space. Then:

1. **(Evaluation)** $\mathrm{ev} : \mathcal{C}(X, Y) \times X \to Y, (f, x) \mapsto f(x)$ is $([[\delta \to \delta'], \delta], \delta')$-computable,
2. **(Type conversion)** $f : Z \times X \to Y$, is $([\delta'', \delta], \delta')$–computable, if and only if the function $\check{f} : Z \to \mathcal{C}(X, Y)$, defined by $\check{f}(z)(x) := f(z, x)$ is $(\delta'', [\delta \to \delta'])$-computable.

The proof of this proposition is based on a version of the smn- and utm-theorem, see [1,2]. If (X, δ), (Y, δ') are admissibly represented sequential topological spaces, then in the following we will always assume that $\mathcal{C}(X, Y)$ is represented by $[\delta \to \delta']$. It follows by evaluation and type conversion that the computable points in $(\mathcal{C}(X, Y), [\delta \to \delta'])$ are just the (δ, δ')-computable functions $f : X \to Y$. Since evaluation and type conversion are even characteristic properties of the function space representation $[\delta \to \delta']$, we can conclude that this representation actually reflects the properties of programs. That is, a name p of a function $f = [\delta \to \delta'](p)$ can be considered as a 'program' of f since it just contains sufficiently much information in order to evaluate f. This corresponds to the well-known fact that the compact-open topology is the appropriate topology for programs and actually, if (X, δ), (Y, δ') are admissibly represented separable normed spaces, one obtains the compact open topology as the final topology of $[\delta \to \delta']$ (see [2]).

If (X, δ) is a represented space, then we will always assume that the set of sequences $X^{\mathbb{N}}$ is represented by $\delta^{\mathbb{N}} := [\delta_{\mathbb{N}} \to \delta]$. The computable points in $(X^{\mathbb{N}}, \delta^{\mathbb{N}})$ are just the computable sequences in (X, δ). Moreover, we assume that X^n is always represented by δ^n, which can be defined inductively by $\delta^1 := \delta$ and $\delta^{n+1} := [\delta^n, \delta]$.

21.3 Computable Metric and Normed Spaces

In this section we will briefly discuss computable metric spaces and computable normed spaces. The notion of a computable normed space will be the central notion for all following results. Computable metric spaces have been used in the literature at least since Lacombe [3]. Pour-El and Richards have introduced a closely related axiomatic characterization of sequential computability structures for Banach spaces [4] which has been extended to metric spaces by Mori, Tsujii, and Yasugi [5].

Definition 21.4 (Computable metric space) *A tuple (X, d, α) is called a* computable metric space *if*

1. $d : X \times X \to \mathbb{R}$ *is a metric on* X,
2. $\alpha : \mathbb{N} \to X$ *is a sequence which is dense in* X,
3. $d \circ (\alpha \times \alpha) : \mathbb{N}^2 \to \mathbb{R}$ *is a computable (double) sequence in* $\mathbb{R}$.

Here, we tacitly assume that the reader is familiar with the notion of a computable sequence of reals, but we will come back to that point below. Occasionally, we will say for short that X is a *computable metric space*. Obviously, a computable metric space is especially separable. Given a computable metric space (X, d, α), its *Cauchy representation* $\delta_X :\subseteq \Sigma^\omega \to X$ can be defined by $\delta_X(01^{n_0+1}01^{n_1+1}01^{n_2+1}\ldots) := \lim_{i\to\infty} \alpha(n_i)$ for all n_i such that $(\alpha(n_i))_{i\in\mathbb{N}}$ converges and $d(\alpha(n_i), \alpha(n_j)) \leq 2^{-i}$ for all $j > i$ (and undefined for all other input sequences). In the following we tacitly assume that computable metric spaces are represented by their Cauchy representations. If X is a computable metric space, then it is easy to see that $d : X \times X \to \mathbb{R}$ is computable [6]. All Cauchy representations are admissible with respect to the corresponding metric topology.

An important computable metric space is $(\mathbb{R}, d_{\mathbb{R}}, \alpha_{\mathbb{R}})$ with the Euclidean metric $d_{\mathbb{R}}(x, y) := |x - y|$ and some standard numbering of the rational numbers $\mathbb{Q}$, as $\alpha_{\mathbb{R}}\langle i, j, k\rangle := (i - j)/(k+1)$. Here, $\langle i, j\rangle := 1/2(i + j)(i + j + 1) + j$ denotes *Cantor pairs* and this definition is extended inductively to finite tuples. Similarly, we can define $\langle p, q\rangle \in \Sigma^\omega$ for sequences $p, q \in \Sigma^\omega$. In the following we assume that $\mathbb{R}$ is endowed with the Cauchy representation $\delta_{\mathbb{R}}$ induced by the computable metric space given above. This representation of $\mathbb{R}$ can also be defined, if $(\mathbb{R}, d_{\mathbb{R}}, \alpha_{\mathbb{R}})$ just fulfils (1) and (2) of the definition above and this leads to a definition of computable real number sequences without circularity. Computationally, we do not have to distinguish the complex numbers $\mathbb{C}$ from $\mathbb{R}^2$. We will use the notation $\mathbb{F}$ for a field which always might be replaced by both $\mathbb{R}$ or $\mathbb{C}$. Correspondingly, we use the notation $(\mathbb{F}, d_{\mathbb{F}}, \alpha_{\mathbb{F}})$ for a computable metric space which might be replaced by both computable metric spaces $(\mathbb{R}, d_{\mathbb{R}}, \alpha_{\mathbb{R}})$ and $(\mathbb{C}, d_{\mathbb{C}}, \alpha_{\mathbb{C}})$ (defined analogously). We will also use the notation $Q_{\mathbb{F}} = \text{range}(\alpha_{\mathbb{F}})$, i.e. $Q_{\mathbb{R}} = \mathbb{Q}$ and $Q_{\mathbb{C}} = \mathbb{Q}[i]$.

For the definition of a computable normed space it is helpful to have the notion of a computable vector space which we will define next.

Definition 21.5 (Computable vector space) *A represented space (X, δ) is called a* computable vector space *(over $\mathbb{F}$), if $(X, +, \cdot, 0)$ is a vector space over $\mathbb{F}$ such that the following conditions hold:*

1. $+ : X \times X \to X, (x, y) \mapsto x + y$ *is computable,*
2. $\cdot : \mathbb{F} \times X \to X, (a, x) \mapsto a \cdot x$ *is computable,*
3. $0 \in X$ *is a computable point.*

If (X, δ) is a computable vector space over $\mathbb{F}$, then $(\mathbb{F}, \delta_{\mathbb{F}})$, (X^n, δ^n) and $(X^{\mathbb{N}}, \delta^{\mathbb{N}})$ are computable vector spaces over $\mathbb{F}$. If, additionally, (X, δ), (Y, δ') are admissibly represented second countable T_0-spaces, then the function space $(\mathcal{C}(Y, X), [\delta' \to \delta])$ is a computable vector space over $\mathbb{F}$. Here we tacitly assume that the vector space operations on product, sequence and function spaces are defined componentwise. The proof for the function space is a straightforward application of evaluation and type conversion. The central definition for the present investigation will be the notion of a computable normed space.

Definition 21.6 (Computable normed space) *A tuple $(X, || \ ||, e)$ is called a* computable normed space *if*

1. $|| \ || : X \to \mathbb{R}$ *is a norm on X,*
2. $e : \mathbb{N} \to X$ *is a* fundamental sequence, *i.e. its linear span is dense in X,*
3. (X, d, α_e) *with* $d(x, y) := ||x - y||$ *and* $\alpha_e\langle k, \langle n_0, \ldots, n_k\rangle\rangle := \sum_{i=0}^{k} \alpha_{\mathbb{F}}(n_i)e_i$, *is a computable metric space with Cauchy representation δ_X,*
4. (X, δ_X) *is a computable vector space over $\mathbb{F}$.*

If in the situation of the definition the underlying space $(X, || \ ||)$ is even a Banach space, i.e. if (X, d) is a complete metric space, then $(X, || \ ||, e)$ is called a *computable Banach space*. If the norm and the fundamental sequence are clear from the context or locally irrelevant, we will say for short that X is a *computable normed space* or a *computable Banach space*. We will always assume that computable normed spaces are represented by their Cauchy representations, which are admissible with respect to the norm topology. If X is a computable normed space, then $|| \ || : X \to \mathbb{R}$ is a computable function. Of course, all computable normed spaces are separable. In the following proposition some computable Banach spaces are defined.

Proposition 21.7 (Computable Banach spaces) *Let $p \in \mathbb{R}$ be a computable real number with $1 \leq p < \infty$. The following spaces are computable Banach spaces over $\mathbb{F}$.*

1. $(\mathbb{F}^n, || \ ||_\infty, e)$ *with*
   - $||(x_1, x_2, ..., x_n)||_\infty := \max_{k=1,...,n} |x_k|$,
   - $e_i = e(i) = (e_{i1}, e_{i2}, ..., e_{in})$ *with* $e_{ik} := \begin{cases} 1 & \text{if } i = k \\ 0 & \text{else.} \end{cases}$
2. $(\ell_p, || \ ||_p, e)$ *with*
   - $\ell_p := \{x \in \mathbb{F}^{\mathbb{N}} : ||x||_p < \infty\}$,
   - $||(x_k)_{k\in\mathbb{N}}||_p := \sqrt[p]{\sum_{k=0}^{\infty} |x_k|^p}$,
   - $e_i = e(i) = (e_{ik})_{k\in\mathbb{N}}$ *with* $e_{ik} := \begin{cases} 1 & \text{if } i = k \\ 0 & \text{else.} \end{cases}$

We leave it to the reader to check that these spaces are actually computable Banach spaces. If not stated otherwise, then we will assume that $\mathbb{F}^n$ is endowed with the maximum norm $|| \ ||_\infty$. In the following definition we mention some typical examples of normed spaces which are non-separable in general (and hence not computable). Later, we will consider these as prototypes of general computable normed spaces.

Definition 21.8 (Further normed spaces) *Let* $(X, || \ ||), (Y, || \ ||')$ *be normed spaces.*

1. *Let* $\ell_\infty := \{x \in \mathbb{F}^{\mathbb{N}} : ||x||_\infty < \infty\}$ *be endowed with the supremum norm* $||(x_k)_{k\in\mathbb{N}}||_\infty := \sup_{k\in\mathbb{N}} |x_k|$.
2. *Let* $\mathcal{B}(\mathbb{N}, X) := \{x \in X^{\mathbb{N}} : ||x|| < \infty\}$ *be endowed with the supremum norm* $||(x_k)_{k\in\mathbb{N}}|| := \sup_{k\in\mathbb{N}} ||x_k||$.
3. *Let* $\mathcal{B}(X, Y) := \{T \in \mathcal{C}(X, Y) : T \text{ linear}\}$ *be endowed with the operator norm* $||T|| := \sup_{||x||=1} ||Tx||'$.

21.4 General Computable Normed Spaces

In this section we introduce a generalized notion of computability for non-separable normed spaces. The possibility to handle non-separable normed spaces is mainly limited by the following fact (cf. Lemma 8.1.1 in [1]).

Proposition 21.9 *If* (X, δ) *is a represented metric space with a* $([\delta, \delta], \delta_{\mathbb{R}})$*-continuous metric* $d : X \times X \to \mathbb{R}$*, then* $\delta :\subseteq \Sigma^\omega \to X$ *itself is continuous and this implies that* X *is separable.*

Here continuitiy of the metric can even be weakened to upper semicontinuity (that is, $\delta_{\mathbb{R}}$ can be replaced by the representation $\delta_{\mathbb{R}_>}$ which only requires us to enumerate all rational upper bounds). As a direct consequence one obtains the following corollary.

Corollary 21.10 *Each represented normed space* (X, δ) *such that*

1. *the norm* $|| \ || : X \to \mathbb{R}$ *is* $(\delta, \delta_{\mathbb{R}})$*-continuous and*
2. *the vector space subtraction* $- : X \times X \to X$ *is* $([\delta, \delta], \delta)$*-continuous,*

is separable.

The following result shows that the problem is not just a topological problem since, at least for certain spaces, it does not disappear if we are only interested in computable points.

Proposition 21.11 *There exists no representation δ of ℓ_∞ such that*

1. *the δ-computable points $x \in \ell_\infty$ are exactly the $\delta_{\mathbb{F}}^{\mathbb{N}}$-computable points with computable norm $||x||_\infty$,*
2. *the vector addition $+ : \ell_\infty \times \ell_\infty \to \ell_\infty$ maps pairs of δ-computable points to δ-computable points.*

Proof. Let us assume that δ is a representation of ℓ_∞ such that (1) and (2) hold. Let $x = (x_0, x_1, \ldots) \in \ell_\infty$ be a sequence which is $\delta_{\mathbb{F}}^{\mathbb{N}}$-computable but such that $||x||_\infty = \sup_{i\in\mathbb{N}} |x_i|$ is not computable; it is well-known that such sequences exist. Now let $r > ||x||_\infty$ be some rational number. Then $y := (r, x_0, x_1, \ldots) \in \ell_\infty$ and $z := (-r, 0, 0, \ldots) \in \ell_\infty$ and $||y||_\infty = ||z||_\infty = r$ and thus y, z are δ-computable by (1). Hence by (2) the sequence $y+z = (0, x_0, x_1, \ldots)$ is δ-computable but $||y+z||_\infty = ||x||_\infty$ is not computable in contradiction to (1). □

Regarding these negative results, we have to decide whether we want to keep computability of the norm or of the vector space operations. In order to remain open for both alternatives, we suggest the following definition for a general computable normed space.

Definition 21.12 (General computable normed space) *A tuple $(X, ||\ ||, \delta)$ is called a* general computable normed space, *if (X, δ) is a computable vector space, $||\ || : X \to \mathbb{R}$ is a norm on X and* Lim $:\subseteq X^{\mathbb{N}} \to X$ *is computable.*

We recall that $\mathrm{dom}(\mathrm{Lim}) = \{(x_n)_{n\in\mathbb{N}} : (x_n)_{n\in\mathbb{N}}$ converges and $(\forall i > j)\, ||x_i - x_j|| \leq 2^{-j}\}$, i.e. Lim is the limit operation restricted to rapidly converging Cauchy sequences. If in the situation of the definition $(X, ||\ ||)$ is even a Banach space, then we call $(X, ||\ ||, \delta)$ a *general computable Banach space*. Each general computable normed space gives rise to at least two canonical representations which yield certain computability properties of the norm.

Definition 21.13 *Let $(X, ||\ ||, \delta)$ be a general computable normed space. We define two representations $\delta^=, \delta^\geqslant$ of X by*

1. $\delta^=\langle p, q\rangle = x :\iff \delta(p) = x$ *and* $\delta_{\mathbb{R}}(q) = ||x||$,
2. $\delta^\geqslant\langle p, q\rangle = x :\iff \delta(p) = x$ *and* $\delta_{\mathbb{R}}(q) \geq ||x||$.

While $\delta^=$-names of points x are enriched by precise information on the norm $||x||$ (this includes all upper and lower bounds), $\delta^\geqslant$-names only include some upper bound on the norm $||x||$ in general. The reader should notice that this implies that the $\delta^\geqslant$-computable points coincide with the δ-computable ones (since for any norm of a point there is some computable upper bound), whereas the $\delta^=$-computable points only form a subset in general.

Example 21.14 The space $(\ell_\infty, || \ ||_\infty, \delta)$ with $\delta := \delta_{\mathbb{F}}^{\mathbb{N}}|^{\ell_\infty}$ is a general computable Banach space and $\delta^= < \delta^\geqslant < \delta$.

In Proposition 21.31 we will prove a more general fact. The next proposition shows that $\delta^=$ just contains sufficient information on the represented points to compute their norm.

Proposition 21.15 (Effective normability) *Let $(X, || \ ||, \delta)$ be a general computable normed space. Then $\delta^= \leq \delta^\geqslant \leq \delta$ and $\delta^=$ is maximal among all representations below δ which make the norm $|| \ || : X \to \mathbb{R}$ computable.*

Similarly, one could show that $\delta^\geqslant$ is maximal among all representations below δ which make the relation $||x|| < y$ r.e. We omit the straightforward proofs. The previous proposition relies on the fact that the norm is a unary function. A unary function can always be effectivized by a construction analogously to the definition of $\delta^=$, but no similar construction works for functions which depend on two or more inputs in general. This can be deduced from Proposition 21.9 using some non-separable metric. The following example is an example of a metric such that Lim is continuous with respect to any representation but the metric d is continuous with respect to none.

Example 21.16 Let $(\mathbb{R}, \delta)$ be a represented space and let $d : \mathbb{R} \times \mathbb{R} \to \mathbb{R}$ be the discrete metric on $\mathbb{R}$, i.e. $d(x, x) = 0$ and $x \neq y \Longrightarrow d(x, y) = 1$ for all $x, y \in \mathbb{R}$. Then the limit map $\text{Lim} :\subseteq \mathbb{R}^{\mathbb{N}} \to \mathbb{R}$ is simply the operation $\text{Lim}(x_n)_{n \in \mathbb{N}} = x_1$ and thus it is computable with respect to any representation of $\mathbb{R}$, in particular with respect to δ. But there is no representation δ' of $\mathbb{R}$ such that d becomes $([\delta', \delta'], \delta_{\mathbb{R}})$-computable by Proposition 21.9 since $(\mathbb{R}, d)$ is non-separable.

The next observation shows that each computable normed space is in particular a general computable normed space. Thus, the notion of a general computable normed space actually generalizes the ordinary notion.

Proposition 21.17 (Computable normed spaces) *If $(X, || \ ||, e)$ is a computable normed space with Cauchy representation δ, then $(X, || \ ||, \delta)$ is a general computable normed space and $\delta^= \equiv \delta^\geqslant \equiv \delta$.*

Proof. By definition of computable normed spaces it follows that the Cauchy representation δ yields a computable vector space (X, δ). It is known that the limit operation and the metric, induced by the norm, are computable with respect to the Cauchy representation δ (see Proposition 5.3 in [6]). The latter implies that the norm itself is computable since $0 \in X$ is computable. By Proposition 21.15 this implies $\delta \leq \delta^=$ and thus $\delta \equiv \delta^= \equiv \delta^\geqslant$. □

The next proposition shows that a general computable normed space keeps as much of the computability properties of a computable normed space as possible (in view of Corollary 21.10).

Proposition 21.18 *Let $(X, || \ ||, \delta)$ be a general computable normed space. Then*

1. *$|| \ || : X \to \mathbb{R}$ is $(\delta^=, \delta_{\mathbb{R}})$-computable,*

2. $\text{Lim} :\subseteq X^{\mathbb{N}} \to X$ *is* $(\delta^{=\mathbb{N}}, \delta^{=})$*- and* $(\delta^{\geqslant\mathbb{N}}, \delta^{\geqslant})$*-computable,*
3. $+ : X \times X \to X, (x, y) \mapsto x + y$ *is* $([\delta^{\geqslant}, \delta^{\geqslant}], \delta^{\geqslant})$*-computable,*
4. $\cdot : \mathbb{F} \times X \to X, (a, x) \mapsto a \cdot x$ *is* $([\delta_{\mathbb{F}}, \delta^{=}], \delta^{=})$*- and* $([\delta_{\mathbb{F}}, \delta^{\geqslant}], \delta^{\geqslant})$*-computable,*
5. $0 \in X$ *is a* $\delta^{=}$*- and a* $\delta^{\geqslant}$*-computable point.*

Proof. The norm is computable with respect to $\delta^{=}$ by definition. By assumption, the limit operation Lim is computable with respect to δ. We have to prove that it is computable with respect to $\delta^{=}$ and $\delta^{\geqslant}$ too. In case that $(x_n)_{n\in\mathbb{N}}$ is a convergent sequence with limit x and $||x_i - x_j|| \leq 2^{-j}$ for all $i > j$, we obtain $||x|| = ||\lim_{n\to\infty} x_n|| = \lim_{n\to\infty} ||x_n||$ since the norm $||\ ||$ is continuous and

$$\big|\ ||x_i|| - ||x_j||\ \big| \leq ||x_i - x_j|| \leq 2^{-j}$$

for all $i > j$. Thus, using evaluation and type conversion we can prove that the limit operation Lim is computable with respect to $\delta^{=}$, since it is computable with respect to δ and the limit operation on real numbers is computable with respect to $\delta_{\mathbb{R}}$. Obviously, Lim is also $(\delta^{\geqslant\mathbb{N}}, \delta^{\geqslant})$-computable since $||x|| \leq ||x_0|| + 1$. Similar to the case of the limit, computability of the scalar multiplication with respect to $\delta^{=}$ follows from $||ax|| = |a| \cdot ||x||$ because the scalar multiplication is computable with respect to δ and multiplication on the real numbers is computable with respect to $\delta_{\mathbb{R}}$. Analogously, computability of the scalar multiplication with respect to $\delta^{\geqslant}$ follows from $s \geq ||x|| \Longrightarrow |a|s \geq ||ax||$ and computability of addition from

$$s \geq ||x||, t \geq ||y|| \Longrightarrow s + t \geq ||x|| + ||y|| \geq ||x + y||.$$

Finally, the computable points with respect to $\delta^{=}$ are exactly the δ-computable points with computable norm. In particular, $0 \in X$ is $\delta^{=}$-computable because of $||0|| = 0$. □

Thus, on the one hand, $\delta^{=}$ is the optimal representation of X from the topological point of view, since norm and limit operation become computable (and the algebraic operations besides addition are computable too). On the other hand, $\delta^{\geqslant}$ is the optimal representation of X from the algebraic point of view, since all algebraic operations become computable and additional $\delta^{\geqslant}$ allows us to compute at least upper bounds on the norm of points. We obtain two elementary corollaries.

Corollary 21.19 *If* $(X, ||\ ||, \delta)$ *is a general computable normed space, then* $(X, ||\ ||, \delta^{\geqslant})$ *is a general computable normed space too.*

Thus, we can apply the construction schemes for the representations repeatedly. However, we obviously obtain $(\delta^{\geqslant})^{\geqslant} \equiv \delta^{\geqslant}$ and $(\delta^{\geqslant})^{=} \equiv \delta^{=}$. Corollary 21.10 yields the following result.

Corollary 21.20 *If* $(X, ||\ ||, \delta)$ *is a non-separable general computable normed space, then* $\delta^{=} \not\equiv_t \delta^{\geqslant}$*, the norm* $||\ ||$ *is not continuous with respect to* $\delta^{\geqslant}$ *and vector space addition is not continuous with respect to* $\delta^{=}$*.*

Although the norm is not computable with respect to $\delta^{\geqslant}$, it is frequently the case that the norm is at least lower semicomputable (with respect to δ and hence with respect to $\delta^{\geqslant}$ as well). Lower semicomputability means that all lower rational bounds of the norm can be computed. This is the case, for instance, for the norm $|| \ || : \ell_\infty \to \mathbb{R}$ and many other norms which are naturally defined as supremum. However, the following example shows that the norm of a general computable normed space is not necessarily lower semicomputable.

Example 21.21 Let $a > 1$ be a real number which is not left-computable (i.e. such that not all lower rational bounds can be listed computably). Then by $||x|| := a|x|$ a norm $|| \ ||$ on $\mathbb{R}$ is defined. Let $\text{Lim}_{|| \ ||}$ denote the limit operator with respect to this norm and $\text{Lim}_{| \ |}$ the limit operator with respect to the ordinary norm. If $\delta_{\mathbb{R}}$ denotes the usual Cauchy representation of the real numbers, then $(\mathbb{R}, \delta_{\mathbb{R}})$ is a computable vector space and $\text{Lim}_{| \ |}$ is computable with respect to $\delta_{\mathbb{R}}$. Since $a|x_i - x_j| = ||x_i - x_j|| \leq 2^{-j} \Longrightarrow |x_i - x_j| \leq 2^{-j}$ we obtain $\text{dom}(\text{Lim}_{|| \ ||}) \subseteq \text{dom}(\text{Lim}_{| \ |})$ and thus $\text{Lim}_{|| \ ||}$ is computable with respect to $\delta_{\mathbb{R}}$ as well. Altogether, $(\mathbb{R}, || \ ||, \delta_{\mathbb{R}})$ is a general computable normed space. However, $|| \ ||$ is not computable with respect to $\delta_{\mathbb{R}}$ and not even lower semicomputable (since a is not left-computable). Moreover, we obtain $\delta_{\mathbb{R}}^{=} \equiv_{\text{t}} \delta_{\mathbb{R}}^{\geqslant} \equiv \delta_{\mathbb{R}}$, but $\delta_{\mathbb{R}}^{\geqslant} \not\leq \delta_{\mathbb{R}}^{=}$.

The previous corollary shows that in non-separable normed spaces there is a 'computational gap' between the algebraic structure and the topological structure of the space. We have already seen in Proposition 21.17 that this gap does not exist in case of ordinary computable normed spaces. Next we will show that any effectively separable general computable normed space with $\delta^{=}$-computable addition is an ordinary computable normed space.

Definition 21.22 (Effectively separable) *A general computable normed space* $(X, || \ ||, \delta)$ *is called* effectively separable *with respect to* $e : \mathbb{N} \to X$, *if* e *is a* $\delta^{=}$*–computable sequence whose linear span is dense in* X.

An ordinary computable normed space X is endowed with the Cauchy representation δ_X. Under the aforementioned conditions a general normed space yields an equivalent computability structure.

Theorem 21.23 (Stability theorem) *If* $(X, || \ ||, \delta)$ *is a general computable normed space which is effectively separable with respect to* $e : \mathbb{N} \to X$ *and if the vector space addition is computable with respect to* $\delta^{=}$, *then* $(X, || \ ||, e)$ *is a computable normed space and* $\delta_X \equiv \delta^{=}$ *for the corresponding Cauchy representation* δ_X.

Proof. Let δ_X denote the Cauchy representation of $(X, || \ ||, e)$. We have to prove that (X, d, α_e) is a computable metric space, where $d(x, y) = ||x - y||$ is the metric induced by $|| \ ||$ and that (X, δ_X) is a computable vector space. Proposition 21.18 guarantees that the norm and scalar multiplication are computable with respect to $\delta^{=}$, by presumption the vector space addition and e are computable with respect to $\delta^{=}$. Altogether, this implies that $\alpha_e : \mathbb{N} \to X$ and d are computable with respect to $\delta^{=}$ as well and this in turn implies that $d \circ (\alpha_e \times \alpha_e)$ is computable. Thus, (X, d, α_e) is a computable metric space. Moreover, the limit operation is computable with respect to $\delta^{=}$ by Proposition 21.18

and together with the fact that α_e is computable with respect to $\delta^=$ one can directly conclude $\delta_X \leq \delta^=$. Since the metric d is $([\delta^=, \delta^=], \delta_{\mathbb{R}})$-computable, it follows that d is also $([\delta^=, \delta_X], \delta_{\mathbb{R}})$-computable and thus $\delta^= \leq \delta_X$ by Proposition 5.3 in [6]. Altogether, this implies $\delta_X \equiv \delta^=$ and this implies again by Proposition 21.18 and by presumption that (X, δ_X) is a computable vector space. Hence, $(X, || \ ||, e)$ is a computable normed space. □

This theorem is closely related to the stability lemma of Pour-El and Richards [4], which is a special case of more general stability results for topological structures [6, 7].

21.5 Topology on General Computable Spaces

In this section we briefly discuss the topologies induced by the representations $\delta^=$ and $\delta^\geqslant$ of a general computable normed space $(X, || \ ||, \delta)$. In particular, we are interested in the question of whether these representations are admissible. In case of a non-separable general computable normed space $(X, || \ ||, \delta)$ there does not exist any continuous representation by Proposition 21.9 and consequently $\delta^=$ and $\delta^\geqslant$ cannot be admissible with respect to the norm topology. For the following result we recall that for a given sequence $(X_i, \tau_i)_{i \in \mathbb{N}}$ of topological spaces (X_i, τ_i) with $X_i \subseteq X_{i+1}$ the *inductive limit topology* $\underrightarrow{\lim}\, \tau_k$ on $X = \bigcup_{k=0}^\infty X_k$ is defined by $\underrightarrow{\lim}\, \tau_k := \{U \subseteq X : (\forall k)\ U \cap X_k \in \tau_k\}$.

Theorem 21.24 (Admissibility) *Let $(X, || \ ||, \delta)$ be a general computable normed space and let δ be an admissible representation of X with respect to some topology τ. Then*

1. *$\delta^=$ is admissible with respect to the weakest topology $\tau^=$ such that the identity map $\mathrm{id} : (X, \tau^=) \hookrightarrow (X, \tau)$ and the norm $|| \ || : (X, \tau^=) \to \mathbb{R}$ become continuous,*
2. *$\delta^\geqslant$ is admissible with respect to the inductive limit topology $\tau^\geqslant = \underrightarrow{\lim}\, \tau_k$ of the subtopologies τ_k of τ on $X_k := \{x \in X : ||x|| \leq k\}$ for all $k \in \mathbb{N}$, provided that τ is a T_1–topology.*

Proof. We will use the definitions and closure properties of admissible representations provided by Schröder [2].

1. By assumption δ is an admissible representation of (X, τ). Hence, this space is a T_0-space which admits a countable pseudobase by Theorem 3.5 in [2] and thus the weakest topology $\tau^=$ induced by the mappings id and $|| \ ||$ on X is T_0 and admits a countable pseudobase by Theorem 4.1 in [2]. By the aforementioned Theorem 3.5 we can conclude that $(X, \tau^=)$ admits an admissible representation δ'. The construction of such a representation δ' in Section 4.2 of [2] yields a representation which is equivalent to $\delta^=$.
2. For the purposes of this proof we can, without loss of generality, replace $\delta^\geqslant$ by an equivalent representation defined by $\delta^\geqslant\langle p, n\rangle = x :\iff \delta(p) = x$ and $n \geq ||x||$. By assumption δ is an admissible representation of (X, τ) and hence $\delta_k := \delta|^{X_k}$ is an admissible representation of (X_k, τ_k). If (X, τ) is a T_1-space, then all the subspaces (X_k, τ_k) are T_1-spaces. By Theorem 4.6 in [2] it follows that the space

$(X, \tau^{\geqslant})$ admits an admissible representation δ''. Regarding the proof of Theorem 4.6 presented in [2] it turns out that δ'' is equivalent to $\delta^{\geqslant}$. □

In order to apply the previous theorem it is important to have some control on the sequentializations of the topologies $\tau^{=}$ and $\tau^{\geqslant}$. Recall that a topological space (X, τ) is called *sequential*, if all sequentially open sets are open. Moreover, a set U is called *sequentially open* if any sequence which converges to a point of U lies eventually in U. By $\mathrm{seq}(\tau)$ we denote the set of sequentially open subsets of X. It is known that $\mathrm{seq}(\tau)$ is a sequential topology on X with $\tau \subseteq \mathrm{seq}(\tau)$. Now we can deduce some facts on the considered topologies.

Lemma 21.25 *Let $(X, \|\ \|)$ be a normed space with limit map* $\mathrm{Lim} :\subseteq X^{\mathbb{N}} \to X$ *and with induced norm topology $\tau_{\|\ \|}$. Let δ be a representation of X which is admissible with respect to some topology τ. If* Lim *is $(\delta^{\mathbb{N}}, \delta)$-continuous, then $\tau_{\|\ \|} \supseteq \tau$ holds.*

Proof. First we note that $(\delta^{\mathbb{N}}, \delta)$-continuity of Lim implies that the map

$$L :\subseteq X^{\mathbb{N}} \times \mathbb{N}^{\mathbb{N}} \to X, ((x_n)_{n\in\mathbb{N}}, m) \mapsto \mathrm{Lim}(x_{m(n)})_{n\in\mathbb{N}}$$

with

$$\mathrm{dom}(L) := \{((x_n)_{n\in\mathbb{N}}, m) : (x_{m(n)})_{n\in\mathbb{N}} \in \mathrm{dom}(\mathrm{Lim}) \text{ and } m \text{ strictly monotone}\}$$

is also $([\delta^{\mathbb{N}}, \delta_{\mathbb{N}}^{\mathbb{N}}], \delta)$-continuous. Moreover, for any sequence $(x_n)_{n\in\mathbb{N}}$ in X which converges to some $x \in X$ with respect to $\tau_{\|\ \|}$ there is a modulus of convergence $m : \mathbb{N} \to \mathbb{N}$ such that $((x_n)_{n\in\mathbb{N}}, m) \in \mathrm{dom}(L)$ and we obtain $L((x_n)_{n\in\mathbb{N}}, m) = x$. Since δ is admissible with respect to τ, it follows by Section 4.3 in [2] that $\delta^{\mathbb{N}}$ is admissible with respect to the product topology $\tau^{\mathbb{N}}$ and by Theorem 2.4 in [2] we can conclude that L is sequentially continuous with respect to τ. In order to prove $\tau_{\|\ \|} \supseteq \tau$ it suffices to prove $\mathrm{seq}(\tau_{\|\ \|}) \supseteq \mathrm{seq}(\tau)$, i.e. that any sequence which converges with respect to $\tau_{\|\ \|}$ does also converge with respect to τ. This is because $\mathrm{seq}(\tau) \supseteq \tau$ and $\mathrm{seq}(\tau_{\|\ \|}) = \tau_{\|\ \|}$. The latter holds since $\tau_{\|\ \|}$ is metrizable and hence first countable and thus sequential [8]. Now consider some sequence $(x_n)_{n\in\mathbb{N}}$ which converges to some $x \in X$ with respect to the norm topology $\tau_{\|\ \|}$. Let $m : \mathbb{N} \to \mathbb{N}$ be some corresponding modulus such that $L((x_n)_{n\in\mathbb{N}}, m) = x$. Now define a double sequence for all $n, i \in \mathbb{N}$ by

$$x_{in} := \begin{cases} x_i & \text{if } i \leq n \\ x_n & \text{otherwise.} \end{cases}$$

Then obviously $((x_{in})_{n\in\mathbb{N}})_{i\in\mathbb{N}}$ converges to $(x_n)_{n\in\mathbb{N}}$ as $i \to \infty$ with respect to any product topology on $X^{\mathbb{N}}$ and especially with respect to $\tau^{\mathbb{N}}$. On the other hand, $(x_{in})_{n\in\mathbb{N}}$ converges to x_i for any $i \in \mathbb{N}$ with respect to $\tau_{\|\ \|}$ and m is still a modulus of convergence, i.e. $L((x_{in})_{n\in\mathbb{N}}, m) = x_i$ for any $i \in \mathbb{N}$. Since L is sequentially continuous with respect to $\tau^{\mathbb{N}}$ and τ, respectively, we can conclude that $(x_i)_{i\in\mathbb{N}}$ converges to x with respect to τ as $i \to \infty$. □

The analogous result for metric spaces (X, d) can be proved in the same way. Using the notation from Theorem 21.24 and the previous lemma we can conclude some facts on the relation of the considered topologies.

Theorem 21.26 *Let $(X, ||\ ||, \delta)$ be a general computable normed space and let δ be an admissible representation of X with respect to some topology τ. Then $\tau_{||\ ||} \supseteq \tau^= \supseteq \tau^{\geqslant} \supseteq \tau$.*

Proof. Let $U \in \tau$. Then we obtain $U \cap X_k \in \tau_k$ for all $k \in \mathbb{N}$ and hence $U \in \tau^{\geqslant}$. If $U \in \tau^{\geqslant}$, then for any $k \in \mathbb{N}$ there is some $V_k \in \tau$ such that $U \cap X_k = V_k \cap X_k$. This implies $U = \bigcup_{k=0}^{\infty}(V_k \cap X'_k)$ where $X'_k := \{x \in X : ||x|| < k\}$. Here '$\supseteq$' holds obviously and '$\subseteq$' holds since for any $x \in U$ there is some $k \in \mathbb{N}$ with $||x|| < k$. Since $V_k \in \tau \subseteq \tau^=$ and $X'_k \in \tau^=$, we can conclude that $U \in \tau^=$. Finally, $(X, ||\ ||, \delta)$ is a general computable normed space and hence Lim is $(\delta^{=\mathbb{N}}, \delta^=)$–computable by Proposition 21.18. Since $\delta^=$ is admissible with respect to $\tau^=$ by Theorem 21.24, we can conclude by the previous lemma $\tau_{||\ ||} \supseteq \tau^=$. □

As a consequence we also obtain $\tau_{||\ ||} \supseteq \text{seq}(\tau^=) \supseteq \text{seq}(\tau^{\geqslant}) \supseteq \text{seq}(\tau) \supseteq \tau$. In this sequentialized case, the second and the third inclusion can also directly be derived from $\delta^= \leq \delta^{\geqslant} \leq \delta$. If γ is an admissible representation of a topological space X and $(Y, ||\ ||, \delta)$ is a general computable normed space such that δ is admissible, then the previous proposition allows us to restrict the representations $[\gamma \to \delta^=]$ and $[\gamma \to \delta^{\geqslant}]$ to representations of the space of continuous functions $f : X \to (Y, ||\ ||)$. We will use this fact in the next section. By results of Schröder it follows that these function space representations are admissible again (see Section 4.4 in [2]).

21.6 Closure Properties of General Computable Spaces

In this section we will show that general computable normed spaces meet some helpful closure properties. Besides subspace and product space construction we will also prove that the dual space of an ordinary computable normed space is a general computable normed space. We start with the subspace construction.

Proposition 21.27 (Subspace) *If $(X, ||\ ||, \delta)$ is a general computable normed space with a linear subspace $Y \subseteq X$, then the subspace $(Y, ||\ ||_Y, \delta|^Y)$ is a general computable normed space too and the canonical injection $Y \hookrightarrow X$ is computable. Moreover, $\delta^=|^Y = \delta|^{Y=}$ and $\delta^{\geqslant}|^Y = \delta|^{Y\geqslant}$.*

Here $||\ ||_Y$ denotes the restriction of $||\ ||$ to Y in the source and $\delta|^Y$ denotes the restriction of δ to Y in the target. The proof follows from the fact that all subspace operations, including the limit, are defined as restrictions. We proceed with the product space construction.

Proposition 21.28 (Product space) *If $(X, ||\ ||_X, \delta_X)$ and $(Y, ||\ ||_Y, \delta_Y)$ are general computable normed spaces, then the product space $(X \times Y, ||\ ||, [\delta_X, \delta_Y])$ with the maximum norm, defined by $||(x, y)|| := \max\{||x||_X, ||y||_Y\}$, is a general computable normed space too. Moreover, $[\delta_X^=, \delta_Y^=] \leq [\delta_X, \delta_Y]^=$ and $[\delta_X^{\geqslant}, \delta_Y^{\geqslant}] \equiv [\delta_X, \delta_Y]^{\geqslant}$.*

Proof. By assumption, (X, δ_X) and (Y, δ_Y) are computable vector spaces with computable limit operations. Since the vector space operations are defined componentwise, it follows that $(X \times Y, [\delta_X, \delta_Y])$ is a computable vector space too. Now let $(x_n, y_n)_{n\in\mathbb{N}}$

be a rapidly converging sequence in $X \times Y$, i.e. $||(x_i, y_i) - (x_j, y_j)|| = \max\{||x_i - x_j||_X, ||y_i - y_j||_Y\} \leq 2^{-j}$ for all $i > j$. It follows $||x_i - x_j||_X \leq 2^{-j}$ and $||y_i - y_j||_Y \leq 2^{-j}$ for all $i > j$. Thus, the limit operation of $(X \times Y, ||\ ||)$ is also computable with respect to $[\delta_X, \delta_Y]$ since it suffices to compute the limit componentwise. We still have to prove the statements on reducibility. The proofs are straightforward and are based on the fact that $\max : \mathbb{R} \times \mathbb{R} \to \mathbb{R}$ is computable. □

It is easy to see that $[\delta_X^=, \delta_Y^=] \leq [\delta_X, \delta_Y]^=$ cannot be strengthened to an equivalence in general. The previous proposition holds analogously with the norm $||(x, y)|| := ||x||_X + ||y||_Y$ instead of the maximum norm.

Now we will discuss operator spaces. Even the dual space $X' = \mathcal{B}(X, \mathbb{F})$ of a separable normed space X is not necessarily separable. Thus it follows that computable normed spaces are not closed under dual space construction. However, the operator space $\mathcal{B}(X, Y)$ of computable normed spaces X, Y is at least a general computable normed space as the following result shows.

Proposition 21.29 (Operator space) *If $(X, ||\ ||)$ is a computable normed space with Cauchy representation δ_X and if $(Y, ||\ ||', \delta_Y)$ is a general computable normed space with some admissible representation δ_Y, then the space of linear bounded operators $(\mathcal{B}(X, Y), ||\ ||, \delta)$ with the operator norm $||T|| := \sup_{||x||=1} ||Tx||$ and representation $\delta := [\delta_X \to \delta_Y]|^{\mathcal{B}(X,Y)}$ is a general computable normed space. If $(Y, ||\ ||')$ is even a computable normed space with Cauchy representation δ_Y, then $\delta \equiv \delta^{\geqslant}$.*

Proof. We have to prove that the space $(\mathcal{B}(X, Y), \delta)$ is a computable vector space with a computable limit operation $\mathrm{Lim} :\subseteq \mathcal{B}(X, Y)^{\mathbb{N}} \to \mathcal{B}(X, Y)$. Since vector space addition $+ : Y \times Y \to Y, (x, y) \mapsto x + y$ and scalar multiplication $\cdot : \mathbb{F} \times Y \to Y, (a, y) \mapsto a \cdot y$ are computable with respect to δ_Y, it follows by evaluation and type conversion that the operator vector space addition $+ : \mathcal{B}(X, Y) \times \mathcal{B}(X, Y) \to \mathcal{B}(X, Y), (f, g) \mapsto f + g$ and scalar multiplication $\cdot : \mathbb{F} \times \mathcal{B}(X, Y) \to \mathcal{B}(X, Y), (a, f) \mapsto a \cdot f$ are computable with respect to $[\delta_X \to \delta_Y]$ too. Since $0 \in Y$ is a δ_Y-computable point, the zero function $z \in \mathcal{B}(X, Y)$ is a δ-computable point too. Now let us assume that $(T_n)_{n \in \mathbb{N}}$ is a sequence in $\mathcal{B}(X, Y)$ which converges to some $T \in \mathcal{B}(X, Y)$ such that $||T_i - T_j|| \leq 2^{-j}$ for all $i > j$. Then $||T_i x - T_j x|| \leq ||T_i - T_j|| \cdot ||x|| \leq 2^{-j}||x||$. Given $x \in X$ with respect to δ_X, we can effectively find some $k \in \mathbb{N}$ such that $||x|| \leq 2^k$ and thus $||T_{i+k}x - T_{j+k}x|| \leq 2^{-j-k}||x|| \leq 2^{-j}$ and $\lim_{i \to \infty} T_{i+k}x = Tx$. Since $\mathrm{Lim}_Y :\subseteq Y^{\mathbb{N}} \to Y$ is computable with respect to δ_Y, we can effectively evaluate the limit operation $\mathrm{Lim} :\subseteq \mathcal{B}(X, Y)^{\mathbb{N}} \to \mathcal{B}(X, Y)$. By type conversion it follows that Lim is computable with respect to $[\delta_X \to \delta_Y]$ too. If $(Y, ||\ ||')$ is even a computable normed space with Cauchy representation δ_Y, then $\delta \leq \delta^{\geqslant}$ follows from Theorem 4.1 in [9]. □

The reader should notice that we have tacitly applied Theorem 21.26 in order to guarantee that δ is actually a representation of $\mathcal{B}(X, Y)$. Virtually the same proof would apply to $\delta := [\delta_X \to \delta_Y^{\geqslant}]|^{\mathcal{B}(X,Y)}$ and in case of a general computable normed space $(X, ||\ ||, \delta)$ we could as well consider $\delta = [\delta_X^{\geqslant} \to \delta_Y^{\geqslant}]$. However, in this case it is not clear which space is represented by δ. Now, we only formulate a simple corollary of the previous result for dual spaces.

Corollary 21.30 (Dual space) *If* $(X, ||\ ||)$ *is a computable normed space with Cauchy representation* δ_X*, then the dual space* $(X', ||\ ||, \delta)$ *with* $X' = \mathcal{B}(X, \mathbb{F})$*, the operator norm* $||f|| := \sup_{||x||=1} |f(x)|$ *and* $\delta := [\delta_X \to \delta_{\mathbb{F}}]|^{X'}$ *is a general computable Banach space. Moreover,* $\delta \equiv \delta^{\geqslant}$.

We could extend this result to general computable normed spaces $(X, ||\ ||, \delta_X)$ with admissible representation δ_X in case that the corresponding topology $\tau^{\geqslant}$ is *consistent* (i.e. if $\tau^{\geqslant} \supseteq \tau_{\text{weak}}$ holds for the *weak topology* on X). As a last example of a closure property we discuss the space of bounded sequences. We omit the proof which is a simplified version of the previous proof.

Proposition 21.31 (Bounded sequences) *If* (Y, δ_Y) *is a general computable normed space, then the space of bounded sequences* $(\mathcal{B}(\mathbb{N}, Y), ||\ ||, \delta)$ *with the supremum norm* $||(y_n)_{n\in\mathbb{N}}|| := \sup_{n\in\mathbb{N}} ||y_n||$ *and* $\delta := \delta_Y^{\mathbb{N}}|^{\mathcal{B}(\mathbb{N},Y)}$ *is a general computable normed space.*

21.7 A Computable Version of Landau's Theorem

In this section we will study computable versions of Landau's theorem 21.1. Let $p, q > 1$ be real numbers such that $\frac{1}{p} + \frac{1}{q} = 1$ or let $p = 1$ and $q = \infty$. Landau's theorem states that under these assumptions ℓ_q is isometrically isomorphic to the dual space ℓ'_p. Given a sequence $a = (a_k)_{k\in\mathbb{N}} \in \ell_q$, let us denote by $\lambda_a \in \ell'_p$ the functional

$$\lambda_a : \ell_p \to \mathbb{F}, (x_k)_{k\in\mathbb{N}} \mapsto \sum_{k=0}^{\infty} a_k x_k.$$

This functional is well-defined, since $\sum_{k=0}^{\infty} a_k x_k$ converges for each $(x_k)_{k\in\mathbb{N}} \in \ell_p$ if and only if $a \in \ell_q$. This is a consequence of Hölder's inequality (at least for $p > 1$, see Section 46 in [10]). Moreover, it is known that $||\lambda_a|| = ||a||_q$ and the isometric isomorphism which exists by Landau's theorem is the map $\lambda : \ell_q \to \ell'_p, a \mapsto \lambda_a$. A natural computable version of Landau's theorem would state that λ is a *computable isomorphism* that is, λ as well as its partial inverse λ^{-1} are computable. However, the problem now is that we have different choices of representations for the dual space ℓ'_p and for the non-separable space ℓ_∞. Let us fix for the moment the ordinary Cauchy representation δ_{ℓ_p} of ℓ_p. Then the following simple example shows that it does not suffice to endow the dual space ℓ'_p with the representation $\delta'_p := [\delta_{\ell_p} \to \delta_{\mathbb{F}}]|^{\ell'_p}$ (we recall that $\delta'_p \equiv \delta'^{\geqslant}_p$) in order to obtain computability of λ^{-1}.

Example 21.32 Let $p, q > 1$ be computable real numbers such that $\frac{1}{p} + \frac{1}{q} = 1$. Let $a = (a_k)_{k\in\mathbb{N}}$ be a computable sequence of real numbers such that $||a||_q$ exists but is not computable. Then $\lambda_a : \ell_p \to \mathbb{R}$ is well-defined and it is a computable linear operator, since it is bounded by Hölder's inequality and $(\lambda_a e_k)_{k\in\mathbb{N}} = (a_k)_{k\in\mathbb{N}}$ is a computable sequence in $\mathbb{R}^{\mathbb{N}}$.

This example also shows that even simple computable functionals of Landau's type λ_a do not have a computable norm in general. Actually, one can use the idea of the previous example to prove that λ^{-1} is not continuous with respect to the compact open topology on ℓ'_p which coincides with (the sequentialization of) the weak* topology in this case. Therefore the following computable version of Landau's theorem is in a certain sense the best which one could expect.

Theorem 21.33 (Computable theorem of Landau) *Let $p, q > 1$ be computable real numbers such that $\frac{1}{p} + \frac{1}{q} = 1$ or $p = 1$ and $q = \infty$, and let $\delta_q := \delta_{\mathbb{F}}^{\mathbb{N}}|^{\ell_q}$ and $\delta'_p := [\delta_{\ell_p} \to \delta_{\mathbb{F}}]|^{\ell'_p}$.*

1. *$\lambda : (\ell_q, ||\ ||_q, \delta_q^{\geqslant}) \to (\ell'_p, ||\ ||, \delta_p'^{\geqslant})$ is a computable isometric isomorphism.*
2. *$\lambda : (\ell_q, ||\ ||_q, \delta_q^{=}) \to (\ell'_p, ||\ ||, \delta_p'^{=})$ is a computable isometric isomorphism.*

(Here the norm on the right hand side denotes the operator norm on ℓ'_p.)

Proof. 1. Given a sequence $a = (a_k)_{k\in\mathbb{N}} \in \mathbb{F}^{\mathbb{N}}$ and $s > 0$ such that $a \in \ell_q$ and $||\lambda_a|| = ||a||_q \leq s$, and given some $x = (x_k)_{k\in\mathbb{N}} \in \ell_p$ and some precision $m \in \mathbb{N}$ we can effectively find some $n \in \mathbb{N}$ and numbers $q_0, \ldots, q_n \in Q_{\mathbb{F}}$ such that $|| \sum_{i=0}^n q_i e_i - x||_p < \frac{1}{s}2^{-m}$. It follows

$$\left|\lambda_a\left(\sum_{i=0}^n q_i e_i\right) - \lambda_a(x)\right| \leq ||\lambda_a|| \cdot \left|\left|\sum_{i=0}^n q_i e_i - x\right|\right|_p < s \cdot \frac{1}{s} 2^{-m} = 2^{-m}.$$

By linearity of λ_a we obtain $\lambda_a(\sum_{i=0}^n q_i e_i) = \sum_{i=0}^n q_i \lambda_a(e_i) = \sum_{i=0}^n q_i a_i$ and thus we can evaluate λ_a effectively up to any given precision m. Using type conversion this proves that $\lambda : (\ell_q, ||\ ||_q, \delta_q^{\geqslant}) \to (\ell'_p, ||\ ||, \delta_p'^{\geqslant})$ is computable.

Now let us assume that $\lambda_a \in \ell'_p$ and $s > 0$ are given such that $||\lambda_a|| = ||a||_q \leq s$. By evaluation it follows that we can effectively compute $a = (a_k)_{k\in\mathbb{N}} = (\lambda_a(e_k))_{k\in\mathbb{N}}$. This shows that we can compute λ^{-1}.

2. This is a direct consequence of (1) and $||\lambda_a|| = ||a||_q$. □

If we accept our choice of $\delta_{\ell_q} \equiv \delta_q^{=}$ as standard representation of ℓ_q for real numbers $1 \leq q < \infty$ (and actually there is not much doubt that something could be wrong with this), then (2) leads naturally to $\delta_p'^{=}$ as standard representation of ℓ'_p for real numbers $1 < p < \infty$. Since $\delta_p'^{=} \not\equiv \delta_p'^{\geqslant}$, it is natural to choose $\delta_p'^{=}$ as representation of ℓ'_p also in case $p = 1$ and in this case (2) leads to $\delta_\infty^{=}$ as natural standard representation for ℓ_∞. In other words: if we want to keep the natural computable isometric isomorphism results of Theorem 21.33 (2), then there is no alternative to these choices of the representation of ℓ_∞ and of the dual space representations of ℓ'_p (up to computable equivalence).

21.8 Conclusion

In computable analysis the problem of handling non-separable spaces has been considered as open problem (see Problem C of Pour-El in [11]). Since sequential computability structures are tailor-made for the separable case, this problem can hardly be solved with such structures. In this chapter we have proposed one potential solution using the representation based approach.

The solution is based on the concept of a generalized computable normed space and we have seen that such spaces allow at least two natural concepts of computability. Which of these concepts is the appropriate one might depend on the application. Regarding the duality results provided by Landau's theorem as one such application, it turns out that we can identify the appropriate concept for this purpose. Other applications might lead to different conclusions.

Our concept of a generalized computable normed space is obviously related to the concept of a quasinormed space as used in constructive analysis [12]. A quasinorm on a linear space X is a family $(|| \ ||_i)_{i \in I}$ of norms such that $\{||x||_i : i \in I\}$ is bounded for every $x \in X$, and a point $x \in X$ is called normable if the aforementioned set admits a supremum. For the special case of classical norms which are defined as supremum over a family of norms, the concept of a quasi-normed space leads to a notion which corresponds to our representation $\delta^{\geqslant}$. This is because $\delta^{\geqslant}$ includes information on upper bounds of the classical norm, which just correspond to upper bounds of $\{||x||_i : i \in I\}$. The normable points are then those points with a computable norm in our terms. However, it is not quite clear how the theory of quasinormed spaces developed in [12] relates to our concepts. This would be an interesting topic for further study maybe from the point of view of realizability theory.

Another interesting line of open problems is related to the study of the underlying topological concepts. Our approach allows us to identify the classical topological counterparts of our computability structures (as expressed in Theorem 21.26). In this respect, a fruitful and important open question is related to the *weak topology* as considered in functional analysis (i.e. the weakest topology such that all linear bounded functionals become continuous). It would be desirable to identify the relations of our topologies $\tau^{=}$ and $\tau^{\geqslant}$ to the weak topology since this would allow us to construct representations of the dual space of general computable normed spaces in a natural way. However, first investigations in this direction indicate that the answers to such questions depend on the underlying set-theoretical axiomatic setting.

Finally, we mention that there are other interesting non-separable spaces which have not been studied in this chapter. For instance the standard example of a non-separable Hilbert space, the space of almost periodic functions, has not been investigated yet. This topic has been pointed out by the anonymous referee.

Acknowledgement

The author would like to thank the anonymous referee for several insightful and constructive remarks which helped to improve the chapter.

References

1. Weihrauch, K. (2000). *Computable Analysis*. Springer, Berlin.
2. Schröder, M. (2002). Extended admissibility. *Theoretical Computer Science*, **284**(2), 519–538.
3. Lacombe, D. (1957). Quelques procédés de définition en topologie récursive. In: *Constructivity in mathematics* (eds A. Heyting), pp. 129–158. Amsterdam, 1959. North-Holland. Colloquium at Amsterdam.

4. Pour-El, M. B. and Richards, J. I. (1989). *Computability in Analysis and Physics*. Perspectives in Mathematical Logic. Springer, Berlin.
5. Yasugi, M., Mori, T., and Tsujii, Y. (1999). Effective properties of sets and functions in metric spaces with computability structure. *Theoretical Computer Science*, **219**, 467–486.
6. Brattka, V. (2003). Computability over topological structures. In: *Computability and Models* (eds S. Barry Cooper and Sergey S. Goncharov), pp. 93–136. Kluwer Academic Publishers, New York.
7. Brattka, V. (1999). A stability theorem for recursive analysis. In: *Combinatorics, Computation & Logic*, Discrete Mathematics and Theoretical Computer Science (eds Cristian S. Calude and Michael J. Dinneen), pp. 144–158. Singapore, 1999. Springer. Proceedings of DMTCS'99 and CATS'99, Auckland, New Zealand.
8. Engelking, R. (1989). *General Topology*, volume 6 of *Sigma series in pure mathematics*. Heldermann, Berlin.
9. Brattka, V. (2002). Computing uniform bounds. In: *CCA 2002 Computability and Complexity in Analysis*, volume 66 of *Electronic Notes in Theoretical Computer Science* (eds V. Brattka, M. Schröder, and K. Weihrauch), Amsterdam, 2002. Elsevier. 5th International Workshop, CCA 2002, Málaga, Spain, July 12–13.
10. Heuser, H. (1986). *Funktionalanalysis*. B.G. Teubner, Stuttgart, 2 edition.
11. Pour-El, M. B. (1999). From axiomatics to intrinsic characterization: some open problems in computable analysis. *Theoretical Computer Science*, **219**, 319–329.
12. Bishop, E., and Bridges, D. S. (1985). *Constructive Analysis*, volume 279 of *Grundlehren der Mathematischen Wissenschaften*. Springer, Berlin.

INDEX